SNAKES AND REPTILES

OF THE WORLD

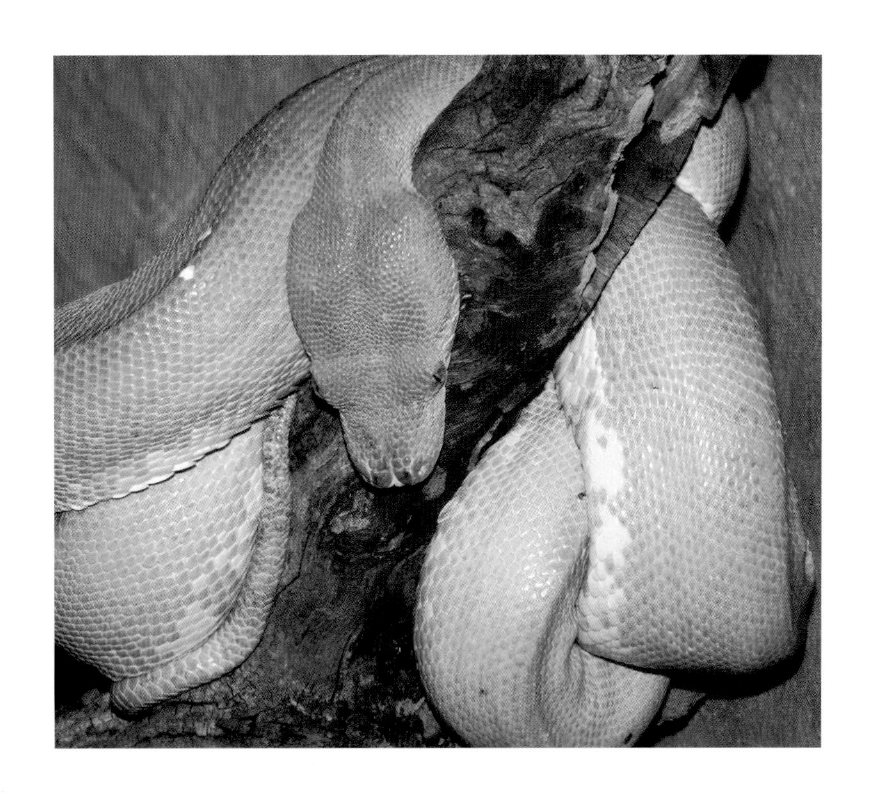

DAVID ALDERTON

VALERIE DAVIES

CHRIS MATTISON

Published 2007 by Grange Books
an imprint of Grange Books Ltd
The Grange
Kingsnorth Industrial Estate
Hoo, Near Rochester
Kent ME3 9ND
www.Grangebooks.co.uk

The Brown Reference Group plc
(incorporating Andromeda Oxford Limited)
8 Chapel Place
Rivington Street
London
EC2A 3DQ
www.brownreference.com

ISBN 978-1-84013-919-8

General Editor: Graham Bateman
Editorial Director: Lindsey Lowe
Copyeditors: Virginia Carter, Angela Davies
Art Editor and Designer: Steve McCurdy
Picture Researcher: Alison Floyd
Main Artists: Denys Ovenden, Philip Hood, Myke Taylor, Ken Oliver, Michael Woods, David M. Dennis, Steve McCurdy,
Maps: Tim Williams
Production: Alastair Gourlay, Maggie Copeland

Printed in Thailand

PHOTOGRAPHIC CREDITS
Ardea: Adrian Warren 94/95, 240/241; Adrian Watson 12/13; Alan Wearing 84/85; Andrey Zvoznikov 438/439; C. Clem Haagber 322t; D&E Parer-Cook 282/283, 284/285; Donald D. Burgess 413; Ferrero Labat 302/303; Francois Gohier 138/139; Jean-Paul Ferrero 57t, 318/319, 414/415; John Cancalosi 142/143, 150/151, 209b; John Daniels 212/213; Mary Clay 284/285; P. Morris 191b, 288b; R. J. C. Blewitt 18/19; Ron and Valerie Taylor 353b, 354/355; **Bruce Coleman Collection:** Luiz Claudio Marigo; **Chris Mattison:** 12t, 21, 25, 27b, 42t, 43b, 46/47, 48/49, 60/61, 62/63, 69b, 78/79, 83t, 86/87, 88/89, 88t, 96/97, 100/101, 102b, 113, 114, 115, 116/117, 123b, 124/125, 128/129, 132/133, 132t, 136/137, 140/141, 143t, 144/145, 152/153, 156/157, 157t, 162t, 166/167, 176/177, 186b, 192, 200/201, 218b, 250/251, 251t, 262/263, 268t, 268b, 272, 276/277, 283t, 295, 312/313, 340/341, 342/343, 357, 360/361, 362/363, 371l, 371r, 372/373, 374l, 374r, 376/377, 379, 382/383, 383t, 392/393, 401, 403, 414b, 426t, 428/429, 428t, 433r, 442/443; **Corbis:** Anthony Bannister/Gallo Images 85t, 331t; Bettmann 424r; David A. Northcott 58/59, 90/91, 131/132, 232/233, 434/435; Frank Blackburn 421r; George McCarthy 19t; Hulton Deutsch 122b; Jeff Rotman 53t, 98/99; Joe MacDonald 140/141, 154/155; Jonathan Blair 31b, Kevin Schafer 404/405; Martin Harvey 233t; Mary Ann McDonald 432/433; Michael & Patricia Fogden 56/57, 234b, 294, 338/339, 431t; Reuters 32/33, Tony Hamblin/FLPA 206/297; W. Perry Conway 432l; **Frank Lane Picture Agency:** B. Borrell 248/249; D. Zingel Eichorn 70/71; E&D Hosking 396/397; L. Chace 147b; L. Lee Rue 162/163, 292/293; Mark Newman 185t; Minden Pictures 182/183, 184/185, 267b, 280/281, 311t; Ron Austing 288/289; S&D&K Maslowski 146/147, Steve McCutchen 223, T. Whittaker 361b; Tony Hamblin 280/283, 380/381; Walter Rohdich 270/271; Wendy Dennis 24/25, 402; **Heather Angel/Natural Vision:** 230/231, 406/407; **NaturePicture Library:** Barry Mansell 37t, 364/365; Bernard Castelein 68/69; Brian Lightfoot 16/17; Bruce Davidson 362b; Claudio Vasquez 273; David Welling 34/35, 269; Doug Wechsler 296/297, 214/215; Elio Della Ferrera 188/189; Francois Savigny 54/55; George McCarthy 36/37; Georgette Dowma 108/109, 220/221, 221t; Ingo Arndt 160/161, 236/237; John Cancalosi 36t, 201t; Jurgen Freund 243t, 444/445, Mar MacDonald 92/93, 106/107, 322/323; Michael Pitts 242/243; Nick Garbutt 314tr, Nick Upton 394/395; Pete Oxford 14, 27t, 72/73, 104/141, 198/199, 254/255, 328/329, 338/339, 386/387, 446/447; Solvin Zankl 224/225, 257t; Stephen Bolwell 411b; Steven D. Miller 38/39, 39b, 77b, 390/391; Todd Pusser 274/275; Tom Vezo 186/187; Tony Heald 301t; Tony Phelps 20, 369b; **Natural History Photographic Agency:** ANT Photo Library 74/75, 348/349, 384b; Anthony Bannister 82/83, 234/235, 264/265, 440/441; Daniel Heuclin 47t, 126/127, 131t, 148/149, 170/171, 174/175, 194, 196/197, 202/203, 204/205, 210/211, 216/217, 244/245, 260/261, 290/291, 296t, 320/321, 330/331, 356/357, 388/389, 393t, 417r, 417r, 430/431, 436/437; Derek Karp 426/427; Gerard Lacz 217t; Image Quest 3-D 260b; James Carmichael, Jr., 28/29, 229, 258/259, 286/287, 358/359, Jany Sauvanet 168/169, 256/257, 443r; Karl Switak 180/181; Ken Griffiths 360b; Kevin Schafer 222; Laurie Campbell 268/269; Mark Bowler 366/367; Martin Harvey 346/347, 418/419; Martin Wendler 52/53; Ralph & Daphne Keller 66/67; Stephen Dalton 134/135, 246/247, 420/421; James Carmichael, Jr., 306/307, 318/319; **Photolibrary.com/OSF:** 30/31, 80; Ajay Desai 91t; Alan Root/SAL 298/299; Alastair Shay 334/335; Bert and Babs Wells 64/65; Breck P. Kent/AA 40/41; Brian Kenney 110/111, 164/165, 215t, 278/279, 333, 344/345, 422/ 423; Carol Farneti Foster 218/219, 326/327; Colin Milkins 424/425; David M. Dennis 310/311; Howard Hall 350/351, 355b; Professor Jack Dermid 308/309; Jacques Gowlett-Holmes 352/353; Marian Bacon 190/191, 314/315; Mark & Victoria Stone Deeble 304/305, 324t; Marl Pidgeon 119t; Michael Fogden 22/23, 23b, 44/45, 50/51, 76/77, 81, 175t, 252/253, 270b, 324/325, 336/337, 370/371; Mike Brown 7; Paul Franklin 376b, 378; Robin Bush 193, 408/409; Stan Osolinski 300/301; Stanley Breeden 121; Tom Ulrich 314tc; Tony Tilford 309t; Waina Cheng 363b, 398/399; Zig Leszczynski 102t, 231t; **Premaphotos Wildlife:** Ken Preston-Mafham 42/43; **Wilson, John D.:** 439b

Contents

Introduction

Snakes and Reptiles of the World combines exhaustive scholarship with appeal to the general reader to explain the diversity, behaviors, ecology, and anatomy of nearly every type of reptile imaginable, including alligators and crocodiles, lizards, snakes, turtles, and their diverse relatives.

The book is conveniently arranged into a series of alphabetical articles on key groups and subgroups, such as **SNAKES** and **Boas and Relatives** (denoted in boldface on the table of contents), as well as key species and closely related genera such as the king cobra and sea snakes. Articles on groups and subgroups present a general overview of the subject; articles on species and closely related genera provide fascinating detail on the species covered. Each of these species profiles begins with an illustrated fact file (see left) that profiles the group or species, including common name, scientific / family name, size, key features, habits, breeding, diet, distribution, conservation status, and range, which appears on a world map. Stunning full-color photography and artwork illustrate every page, making this book the very best introduction to the reptilian world.

Data panel presents basic statistics for each animal

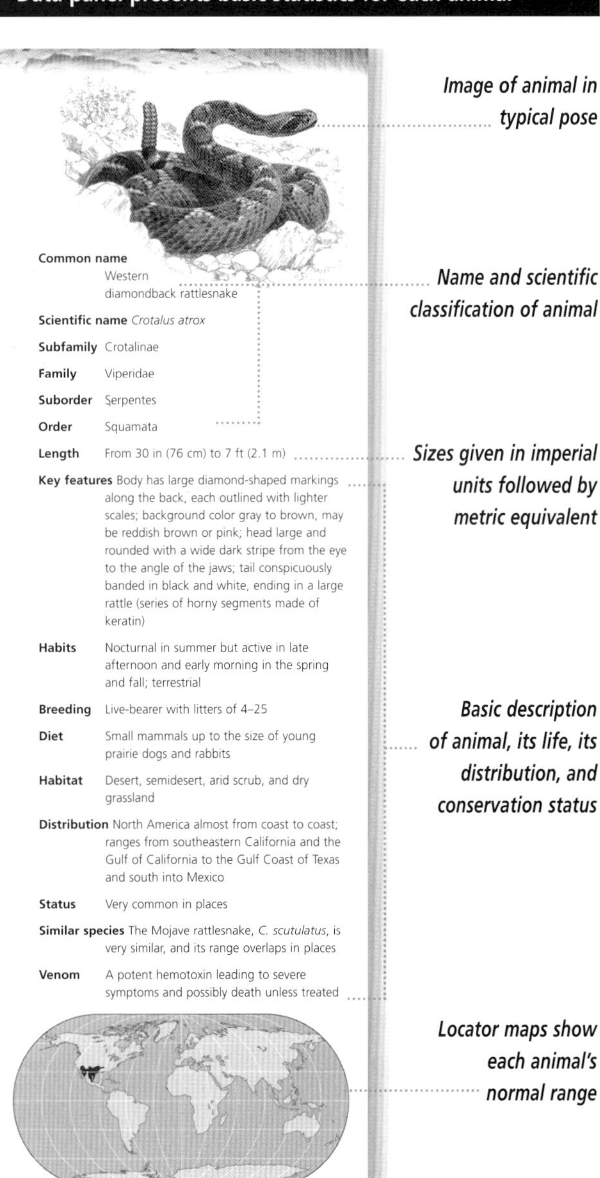

The Animal Kingdom

Kingdom Animalia, one of five living kingdoms, is divided into major groups called phyla. The phylum Chordata contains those animals that have a backbone—mammals, birds, reptiles, amphibians, and fish.

Snakes and Reptiles of the World explores one of the oldest lineages of land animals—the reptiles, which evolved from amphibians about 350 million years ago. Although reptiles are no longer dominant animals on earth (unlike the early reptiles typified by the dinosaurs), more than 8,000 species of reptiles still live on our planet. Most live in warmer or tropical regions of the world.

The kingdom Animalia is subdivided into phyla, classes, orders, families, genera, and species. Below is sthe classification for the western diamondback rattlesnake (right).

Rank	Scientific name	Common name
Kingdom	Animalia	Animals
Phylum	Chordata	Animals with a backbone
Class	Reptilia	Reptiles
Order	Squamata	Lizards, Snakes, Amphisbaenians
Suborder	Serpentes	Snakes
Family	Viperidae	Vipers and Pit Vipers
Genus	*Crotalus*	Rattlesnakes
Species	*Crotalus atrox*	Western diamondback rattlesnake

Classification of Reptiles

CLASS: REPTILIA—REPTILES

ORDER: Squamata—lizards, snakes, and amphisbaenians	
Suborder: Sauria	lizards
Family: Agamidae	agamas and dragon lizards
Family: Chamaeleonidae	chameleons and dwarf chameleons
Family: Iguanidae	iguanas, basilisks, collared lizards, and anoles
Family: Gekkonidae	"typical" geckos
Family: Diplodactylidae	southern geckos
Family: Pygopodidae	flap-footed lizards
Family: Eublepharidae	eyelid geckos
Family: Teiidae	tegus, whiptails, and racerunners
Family: Gymnophthalmidae	spectacled lizards
Family: Lacertidae	wall lizards
Family: Xantusiidae	night lizards
Family: Scincidae	skinks
Family: Gerrhosauridae	plated lizards
Family: Cordylidae	girdle-tailed lizards
Family: Dibamidae	blind lizards
Family: Xenosauridae	knob-scaled and crocodile lizards
Family: Anguidae	alligator and glass lizards
Family: Varanidae	monitor lizards
Family: Helodermatidae	beaded lizards
Family: Lanthonotidae	Borneo earless monitor
Suborder: Amphisbaenia	amphisbaenians (worm lizards)
Suborder: Serpentes	snakes
Family: Anomalepidae	dawn blind snakes
Family: Leptotyphlopidae	thread snakes
Family: Typhlopidae	blind snakes
Family: Anomochilidae	dwarf pipe snakes
Family: Uropeltidae	shield-tailed snakes
Family: Cylindrophiidae	pipe snakes
Family: Aniliidae	South American pipe snake
Family: Xenopeltidae	Asian sunbeam snakes
Family: Loxocemidae	American sunbeam snake
Family: Acrochordidae	file snakes
Family: Boidae	boas
Family: Bolyeriidae	Round Island boas
Family: Tropidophiidae	wood snakes
Family: Pythonidae	pythons
Family: Colubridae	colubrids
Family: Atractaspididae	African burrowing snakes
Family: Elapidae	cobras
Family: Viperidae	vipers and pit vipers
ORDER: Testudines	turtles, terrapins, and tortoises
ORDER: Crocodylia	crocodiles, alligators, and caimans
ORDER: Rhynchocephalia	tuataras

Naming Animals

To discuss animals, names are needed to identify the different kinds. Western diamondback rattlesnakes are one kind of snake, and sidewinders are another. All western diamondback rattlesnakes look alike, breed together, and produce young like themselves. This distinction corresponds closely to the zoologists' definition of a species.

Zoologists use an internationally recognized system for naming species, consisting of two-word scientific names, usually in Latin or Greek. The western diamondback rattlesnake is called *Crotalus atrox,* and the sidewinder *Crotalus cerastes. Crotalus* is the name of the genus (a group of very similar species); *atrox* or *cerastes* indicates the species in the genus.

The same scientific names are recognized the world over. However, a species may have been described and named at different times without the zoologists realizing it was one species. Classification allows us to make statements about larger groups of animals. For example, all rattlesnakes are vipers—along with other vipers they are placed in the family Viperidae. All vipers are placed with all other snakes in the suborder Serpentes; snakes are related to lizards, which are in the suborder Sauria, and so these two groups combine to form the order Squamata in the class Reptilia.

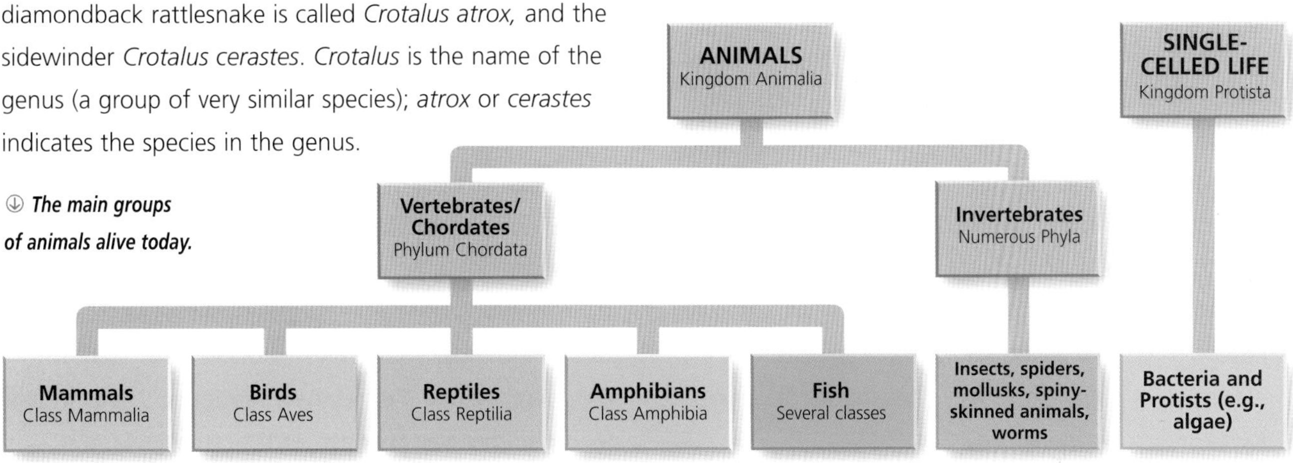

The main groups of animals alive today.

WHAT IS A REPTILE?

The reptiles form the class Reptilia. There are just over 8,000 species in total, and they are divided into four groups, or orders: the Testudines (turtles and tortoises), the Squamata (lizards, amphisbaenians, and snakes), the Crocodylia (crocodiles and alligators), and the Rhynchocephalia (tuataras). The numbers are unevenly divided among the orders, with the Squamata being the largest group in terms of numbers of species.
It is also the most widespread group with almost global distribution. It is divided further into three suborders: the Amphisbaenia (amphisbaenians, or worm lizards), the Sauria (lizards), and the Serpentes (snakes).

Although reptiles are less conspicuous than many other animal groups, they form a unit within the system of biological classification that puts them on a par with other major groups such as insects, birds, and mammals.

Like fish, amphibians, birds, and mammals, reptiles are vertebrates (animals with a backbone). However, they obtain their body heat from outside sources (they are ectotherms) rather than producing it metabolically from their food. This ability separates them from birds and mammals, which are endothermic. They are separated from the fish and the amphibians by their reproductive biology: Reptile embryos are surrounded by three special membranes: the amnion, chorion, and allantois. The evolution of the "amniotic egg," as it is called, was a significant step and one that led subsequently to the evolution of birds and mammals.

Reptiles lay shelled eggs or produce live young depending on species. That means they are not closely tied to water, unlike most amphibians (although perversely some reptiles, such as crocodilians and sea turtles, have become aquatic as adults and have to come back to the land to lay their eggs, the very opposite of amphibians). In contrast to amphibians, reptiles are covered in dry, horny scales that are relatively impermeable to water. These two factors (their scales and their amniotic eggs) allowed them to move away from watery habitats and colonize the interiors of continents, even though some still favor wet or aquatic habitats.

↑ ***Representatives of reptile groups: tuatara (*Sphenodon punctatus*, order Rhyncocephala) (1); female false gharial (*Tomistoma schlegelii*, order Crocodylia) (2); worm lizard (order Squamata, suborder Amphisbaenia) (3); Alabama red-bellied turtle (*Pseudemys alabamensis*, order Testudines) (4); green tree python (*Morelia viridis*, order Squamata, suborder Serpentes) (5); Madagascan day gecko (*Phelsuma laticauda*, order Squamata, suborder Sauria) (6).***

Temperature Regulation

Understanding how reptiles operate at different temperatures is the key to understanding their behavior, biology, and ecology. Each species has a "preferred body temperature" at which they are best able to move around to hunt for and digest their food, to produce eggs or sperm, and so on. The temperature varies according to species but is often about 85 to 100°F (30–37°C). They may still be active at lower temperatures but they slow down, and at some point conditions will become too cold for them to move at all. This is known as the "critical lower temperature." If they have not found shelter at this point, they become stranded and are vulnerable to

Who's Who among the Reptiles?

Class Reptilia Over 8,000 species in 4 orders:

Order Squamata 3 suborders:
- **Suborder Sauria**—over 4,700 species of lizards in about 20 families
- **Suborder Amphisbaenia**—168 species of amphisbaenians and worm lizards in 3–4 families
- **Suborder Serpentes**—2,900 species of snakes in 18 families

Order Testudines—238 species of turtles and tortoises in 12 families

Order Crocodylia—22 species of crocodiles and alligators in 1 family and 3 subfamilies

Order Rhynchocephalia—2 species of tuataras

predation. If the temperature continues to fall, they are at risk of freezing (they reach their "lethal minimum temperature"). The same thing happens with rising temperatures. Critical maximum temperatures are often quite close to preferred body temperatures, so even a small rise can spell trouble. The reptile must find a place away from the heat (in the shade, under a rock, or in a burrow) to keep its body from becoming overheated. Reptiles trapped in the sun (if they fall into a trench, for instance) succumb and die in minutes on hot days.

Reptiles living in different climates clearly need to use different strategies to maintain a suitable temperature. In the tropics they may need to do little by way of thermoregulation because the ambient temperature may be close to their optimum for much of the time. In cooler climates, such as North America, South Africa, and Europe, they can raise their body temperature during the warmest part of the day in spring and summer, perhaps by basking, but nighttime temperatures may be too cool for them on all but the warmest nights.

Species living in these climates tend to be diurnal, but depending on their locality, they may become nocturnal during midsummer. Of course, different species have different preferred body temperatures, so even in the same locality some may be diurnal and some nocturnal. In winter none of them can reach their preferred temperature at any time of the day, and they need to retreat to a safe place and remain there in hibernation until the following spring.

At the other extreme, reptiles living in very hot places are often active at night and seek shelter during the day. Another advantage is that they can reduce the risk of predation by daytime hunters such as birds of prey (but not by nocturnal predators, of course). Where conditions are too harsh, they may retreat underground for days or even weeks at a time to "sit out" the worst excesses of the climate before returning to the surface (this is known as estivation). In practice, most reptiles that estivate usually do so in order to avoid extreme dryness rather

Many reptiles need to bask in the sun to reach their body's preferred temperature. In central Oman a spiny-tailed lizard, **Uromastyx aegyptia microlepis,** *stretches out on a rock to warm up.*

Right: A chart demonstrating the main lines of reptilian evolution. The four subclasses of reptiles (Euryapsida, Anapsida, Diapsida, and Synapsida) are distinguished by the arched recesses, or apses, in the skull behind the eye sockets.

Below right: Anapsida (1) have no apses, and today they are represented by the turtles and tortoises. Synapsida (2) have one apse, and this line led to the evolution of mammals. Diapsids (3), with two apses, included the now extinct dinosaurs and are represented today by all other reptile groups apart from turtles and tortoises. The Euryapsids (4) had one apse high on the skull and are represented by the now extinct marine reptiles of the Mesozoic.

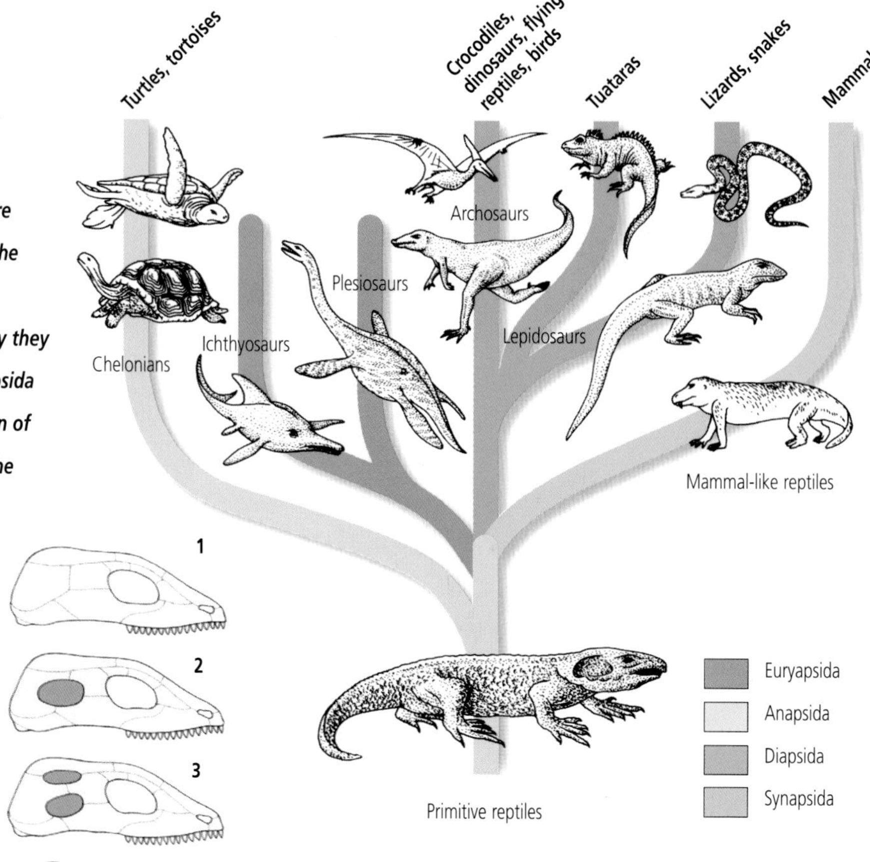

than heat, and they tend to be species that normally rely on water, for example, crocodilians and freshwater turtles.

Maintaining the correct body temperature by thermoregulation takes several forms in reptiles. Species that live in open environments shuttle backward and forward between areas of sun and areas of shade. Some of them are dark in color so that they absorb heat more quickly. They may also flatten or orient their body in a particular way to help absorb heat more quickly: American fence lizards, *Sceloporus* species, side-blotched lizards in the genus *Uta*, and European vipers, *Vipera* species, are good examples.

Forest dwellers have less opportunity to bask, but they can move into small open areas where the sun gets through or move up into the canopy. Aquatic species, such as crocodilians and turtles, have limited opportunities to control their body temperatures and must often "make do" with whatever temperature the water happens to be; most of them come from warmer climates where typical temperatures are suitable. Crocodiles, alligators, and freshwater turtles may "haul out" for long periods to bask on the riverbank or on logs or rocks that stick out. Sea turtles do not usually bask, except perhaps at the water's surface, and so they are restricted to the tropics. Burrowing species, such as the worm lizards, may be able to move up and down through the levels of their tunnels; but by and large they do not actively thermoregulate, and they are sometimes known as "thermal conformers."

Origins of Reptiles

Reptiles evolved from four-legged amphibians about 350 million years ago. However, species that would pinpoint their exact origins have not been positively identified from fossil records. They are known to have laid amniotic eggs, which was a significant development. By 310 million years ago the early land-dwelling animals that laid amniotic eggs split into two branches, one that would lead to the mammals and the other to reptiles and birds. The implications of this division are, perhaps surprisingly, that birds are more closely related to reptiles than to mammals—some scientists even maintain that they are reptiles.

1

A scene from an early Jurassic landscape shows the diversity of reptilian life. A pterosaur, **Rhamphorhyncus** *(1); a stegosaur,* **Kentrosaurus** *(2); theropod dinosaurs,* **Elaphrosaurus** *(3) and* **Ceratosaurus** *(4); sauropod dinosaurs,* **Dicraeosaurus** *(5), and* **Brachiosaurus** *(6).*

By the end of the Triassic Period (about 208 million years ago) the oldest lineages of reptiles that we know today had appeared. They were the early chelonians, or shelled reptiles (turtles and tortoises), the crocodilians, and the rhynchocephalians (the ancestors of the tuataras). The lizards, worm lizards, and snakes came later, first appearing during the Jurassic Period about 208 to 144 million years ago. The worm lizards are thought to be the most recently evolved of the major groups. Many other branches of the reptile lineage led to evolutionary dead ends but only after they had been highly successful for very long periods of time before eventually dying out.

Form and Function

Compared with birds or mammals, living reptiles form a diverse group. There are species with and without shells. Some have four limbs, some have two, and some have none. All are covered in scales, but the scales can be massive, knobby, and stonelike or tiny, granular, and silky to the touch. Compare a snapping turtle or an alligator with an anole lizard or a gecko to get an idea of the wide range of forms and sizes in the order. The sizes, shapes, and colors of reptiles are not there to help us tell one species from another: They have been finetuned through the evolutionary process to help each species adapt to its particular place in the scheme of things. Even small groups of closely related species contain very diverse species occupying different ecological niches.

We know that stout snakes with short tails are likely to be slow-moving, burrowing reptiles, and that long, thin ones with long tails are likely to be fast-moving, terrestrial types (unless they have prehensile tails, in which case they will be climbers). Flattened turtles with streamlined shells are aquatic, while species with domed shells and elephantlike feet are terrestrial, and so on.

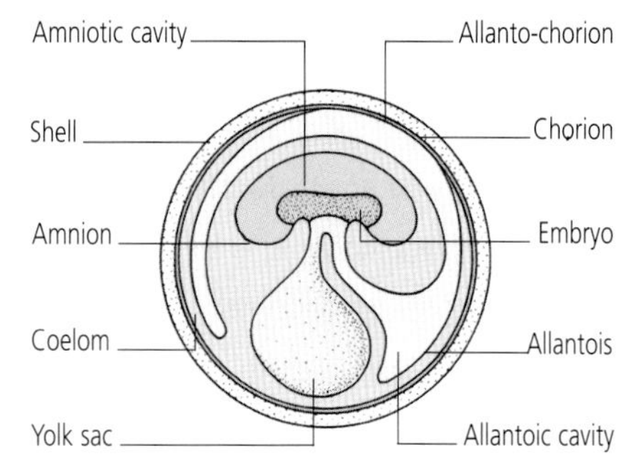

Developing egg, showing layers of membrane between shell and embryo. The partly fused chorion and allantois on the inner surface of the shell are supplied with blood vessels, enabling the embryo to breathe through pores in the shell. The allantois also acts as a repository for the embryo's waste products. The amnion is a fluid-filled sac around the embryo that keeps it from drying out. The yolk sac contains the embryonic food supply, rich in protein and fats. Eggs of this type, such as those of birds, are called cleidoic ("closed-box") eggs, since apart from respiration and some absorption of water from the environment, they are self-sufficient. Water absorption by the eggs of many reptiles, especially the softer-shelled types, is higher than by birds' eggs.

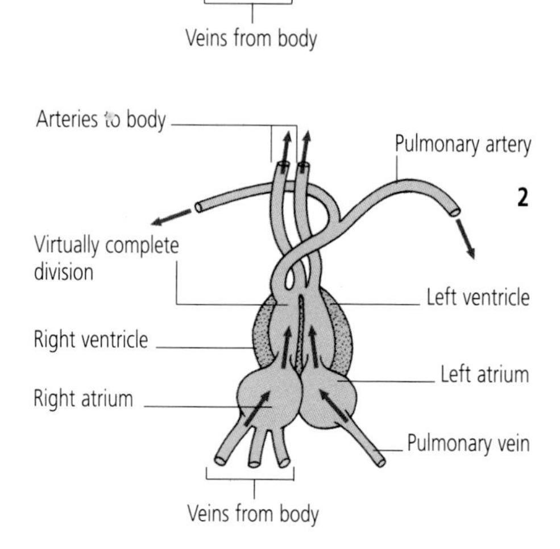

Reptilian hearts. In most reptiles the chambers of the ventricles are incompletely separated (1). In crocodilians complete separation exists, although there is a small connection, the foramen of Panizza, between the outlet vessels (2). Even in the unseparated ventricle a system of valves and blood pressure differences ensures that there is little mixing of arterial and venous blood under normal conditions. In all reptiles, however, the potential exists to shunt the blood from one side of the heart to the other. This helps them adapt, especially aquatic animals, since blood can be recycled when breathing is interrupted.

Within several families of lizards there has been a tendency to lose limbs, which is also related to lifestyle. The skinks are a particularly good example of "adaptive radiation"—the process by which related species evolve in different ways to suit different conditions. Skinks range from tiny, legless, burrowing forms through elongated, grass-dwelling species with tiny reduced limbs to chunky, heavy-bodied, terrestrial forms.

Coloration is usually related to defense. Cryptic, or disguised, species are colored to match their surroundings, and other species have disruptive geometric markings that help break up their outline. Some species are so well camouflaged that they are nearly impossible to make out even when you know where they are. Other reptiles, however, are brightly colored to warn that they are venomous. On the other hand, there are species that are harmless but also brightly colored to fool predators into thinking that they are venomous (known as Batesian mimicry). Some camouflaged species have brightly colored patches or extensions to their body that they can flash when they want to display to other members of their own species or to deter potential predators.

Who Lives Where and Why?

Because they are dependent on temperature, reptiles are most at home in tropical and subtropical regions, with the greatest numbers of species and individuals occurring in the tropical rain forests of Central and South America, West and Central Africa, and Southeast Asia. Farther away from the tropics toward the poles the numbers fall dramatically. Species that do not have the ability to thermoregulate, such as burrowing and aquatic reptiles, are even more restricted to warm places.

Superimposed on this pattern are the historical events that have affected the way in which reptiles have been able to spread across the globe. At about the time they were diversifying most rapidly, the landmasses and "supercontinents" were changing shape through continental drift. Areas that had been connected were breaking apart, while in other places landmasses collided. Evolving reptiles were "passengers" on these landmasses; and lacking the ability to cover large tracts of water (with the obvious exception of the sea turtles), they became isolated in some places but were presented with opportunities to expand in others.

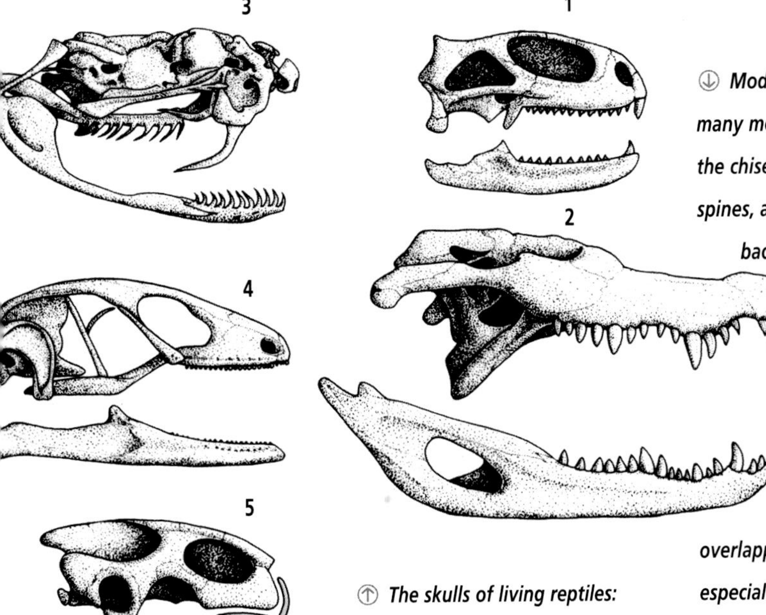

Ⓣ *The skulls of living reptiles: tuatara (1), crocodile (2), snake (3), lizard (4), and turtle (5).*

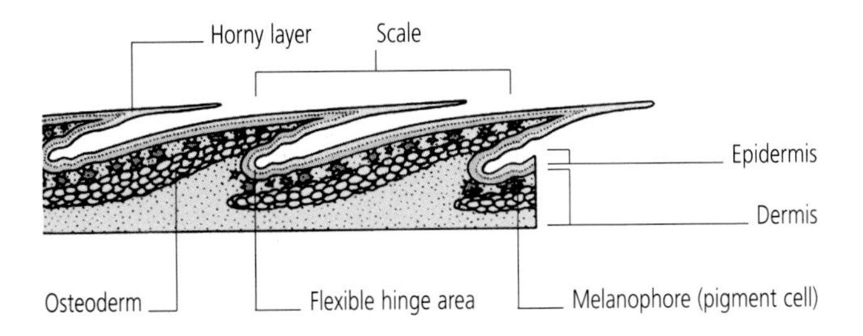

Ⓣ ***Cross-sectional diagram of the skin of a slowworm. All anguimorphs such as this are heavily armored, having mostly nonoverlapping scales with underlying osteoderms.***

Ⓓ *Modifications of the skin. The skin, particularly the epidermis, shows many modifications in reptiles. It can be raised up into tubercles, as in the chisel-toothed lizard,* Ceratophora stoddari *(1), or into defensive spines, as on the tails of certain lizards. It can form crests on the neck, back, or tail, often better developed in the male and perhaps helping in sexual recognition, as in* Lyriocephalus scutatus *(2). The rattlesnake's rattle (3), made up of interlocking horny segments, is a unique epidermal structure; a new segment is formed at each molt, but the end segment tends to break off when the rattle gets very long. In most snakes the underbody scales are enlarged to form a series of wide, overlapping plates that assist in locomotion, especially in forms such as boas that can crawl stretched out almost straight. The modified scales, or lamellae, on the toe pads of geckos (4) have fine bristles (setae) that allow them to climb smooth surfaces.*

"Older" lineages (those that appeared early on) were able to spread onto most landmasses and travel with them. Some of them subsequently thrived and became widespread (the geckos, for example), while the success of others diminished as more adaptable groups ousted them. For the "newer" families, however, some options were closed because they were already isolated by the time they appeared. That is why there are no vipers in Australasia and no monitor lizards in the Americas.

Isolated island groups are often very poor in reptile species, but their patterns of colonization and speciation (species formation) are especially interesting to biologists because they can add to our understanding of the processes of evolution and natural selection. The most obvious example is the Galápagos archipelago. Reptiles first spread to places like this accidentally (and perhaps only on one or two occasions), possibly by "rafting" on mats of floating vegetation.

Reproduction

Unlike the amphibians, reptiles have internal fertilization. The male introduces the sperm directly into the female's reproductive tract through the cloaca, which is the opening for the digestive and reproductive systems. In all reptiles except the tuataras males have copulatory organs. In lizards, snakes, and worm lizards they are paired and are called the hemipenes. There is usually competition among males for access to females, which can take various forms. In species that use visual clues for communication, such as some lizards, males often display crests, frills, or brightly colored parts of their anatomy to

A clutch of eggs laid by a female milksnake hatches out. Most snakes lay eggs, but other reptiles, including some snakes, are live-bearers. Reproductive patterns are often determined by lifestyle and habitat.

attract mates and to advertise their ownership of territory. Color change can also be involved, most famously in chameleons. The courtship process in many species, especially secretive ones such as skinks and worm lizards, is poorly known, but chemical communication almost certainly plays an important part.

Reproductive cycles vary greatly according to species and where they live. Some tropical species breed all year around, while some from colder climates breed only once every two or three years. Tuataras breed only every five or more years. Most temperate species breed in the spring and summer, but again there is some variation.

Reptiles may lay eggs or give birth to live young. This is an evolutionary "decision" with important tradeoffs. Laying eggs frees the female to continue feeding and may enable her to produce a second clutch quickly; on the other hand, the eggs are vulnerable to predation and are at the mercy of the elements. Giving birth to live young enables the female to care for her developing embryos more effectively because she is carrying them around with her, but it is an added burden for several months. The "choice" she makes (in evolutionary terms) will depend on factors such as climate and lifestyle.

Superimposed on this, however, is an ancestral element: Reptiles in some families seem "locked into" a particular reproductive mode (for example, all pythons lay eggs, but nearly all boas give birth to live young). The crocodilians, turtles, and tuataras do not seem to have evolved the facility to give birth—they lay eggs, which is the "ancestral mode" for all reptiles. Among lizards and snakes most lay eggs, but a significant proportion are live-bearers. Many aquatic snakes, including the sea snakes, give birth because finding a suitable place to lay their eggs presents a problem (although some species, notably the sea kraits, come ashore to lay eggs). Worm lizards are all egg layers as far as is known, but the natural history of many of these obscure reptiles remains a mystery.

Food and Feeding

Among them reptiles eat just about anything organic. There are divisions along taxonomic lines, however. All snakes are carnivorous, for instance, although their prey can vary from ants to antelopes. Crocodilians are also carnivorous—their prey ranges from insects to large mammals such as zebras and wildebeests. Worm lizards are probably all carnivorous too and feed largely on burrowing insects such as ants and termites, but the larger species also take small vertebrates, including lizards. Lizards feed on a wide variety of items—many eat insects, but many others are herbivores. Large monitors are ferocious predators of vertebrates such as other lizards, birds, and mammals. Marine and freshwater turtles eat animal and plant material depending on species, and some eat both. Land turtles tend to be herbivores but are not averse to eating animals when they can catch them, which is not very often. A number of reptiles, perhaps more than we realize, eat carrion as a sideline.

Methods of finding and overcoming food are equally diverse. Finding and catching plant material is not very hard, although plants are well known for producing toxins to deter grazers and browsers. Many reptiles are amazingly oblivious to spines and bitter substances, and can eat plant species that other herbivores reject.

Catching animal prey takes a number of different forms. Many are "sit-and-wait" predators, setting themselves up in a likely place and waiting for prey to blunder past. American horned lizards, *Phrynosoma* species, and Australian thorny devils, *Moloch horridus*, position themselves next to ant trails and simply mop up the ants as they walk past. Overpowering larger prey—especially if it can fight back—calls for more cunning and specialized equipment. The ultimate weapon is the evolution of venom in some snakes, which enables them to dispense death in the blink of an eye even to animals many times their own size.

Classification of Species

Compared with other groups of zoology, the naming and reclassification of reptiles seems to be always in a state of change. In 2003, for example, there were 59 new species described, 3 subspecies were elevated to full species, and 18 species were suppressed (because it turned out that they had been named twice). In 2002 there were 60 new species, in 2001 80 new species were named, and 72 new species were described in 2000. In just four years, then, 271 completely new species were added to the reptiles. In the same period there were few, if any, new birds or mammals. What is more, scientists are constantly reclassifying and renaming existing species to try to represent more accurately the relationships between them. This can make life difficult for those studying or writing about reptiles.

Many new reptile species are discovered in places that have hardly been explored from a herpetological point of view, such as Madagascar. Other species belong to groups that are hard to find or difficult to work with (or both), such as the blind and thread snakes, Leptotyphlopidae and Typhlopidae. Others are still turning up in parts of the world where herpetologists have been working for years and belong to conspicuous groups of reptiles. The reptile lists of South Africa and Australia, for example, have grown significantly in the last 10 years or so—by 20 percent in South Africa, an average of one new species every 44 days!

The Galápagos Islands are home to some of the most unusual reptiles, including the marine iguanas, **Amblyrhynchus cristatus.** ***They are the only lizards that enter the sea and feed on seaweed.***

Declining Species

Until 200 years ago the Galápagos Islands were home to hundreds of thousands of giant tortoises. During the 19th century visiting whaling ships began to collect the tortoises to stock their holds with fresh meat. They left behind a number of destructive, introduced mammals—rats, cats, pigs, and goats—that preyed on the tortoises' eggs and young or competed with them for food. By the mid-20th century three of the original 14 subspecies of giant tortoise were extinct. Only four subspecies are considered to be safe from extinction. Six out of a total of seven marine turtle species are classed as Endangered or Critically Endangered (IUCN) as are seven of just 22 surviving species of crocodilians.

According to the IUCN 21 species of reptiles have become extinct in recent times. Sixteen of them lived on islands. Island species are especially vulnerable because their environment is easily affected by human impacts, especially the introduction of predatory animals. On Round Island in the Indian Ocean every native reptile species is extinct or on the brink of extinction, while Mauritius has lost eight species.

Not all the news is bad, however. The surviving Galápagos tortoises are being bred successfully in captivity, goats and rats have been eliminated on some islands, and the vegetation is beginning to recover. The Jamaican iguana, *Cyclura collei*, was believed extinct since the 1940s but turned up in small numbers in 1990 on a

Reptiles as Pets

Reptiles have become popular pets. Species available range from small geckos to huge pythons and the common boa (*Boa constrictor*). Many are now being selectively bred to give the enthusiast a wide selection of color and pattern forms. Their care varies greatly according to the species, so always seek the advice of the vendor, and consult a specialist book for the relevant information.

Captive-bred animals should be obtained wherever possible. There are a number of reasons for this. First, they will be better adapted to captivity than wild ones and will therefore calm down sooner and be more inclined to accept an unnatural diet. Second, they are likely to be free from parasites and infections that often plague specimens captured from the wild. Third, the fact that they were produced in captivity means that they are an adaptable species. Finally, many wild reptile populations are under threat, and to encourage trade in them is irresponsible. Many species are protected internationally, nationally, or locally, and you may be breaking the law by keeping them. Similarly, collecting species from national or state parks is not allowed.

Accommodation

Accommodation can range from plastic containers for the smallest species (or for rearing juveniles of some of the larger species) to huge, room-sized cages that will

Green iguanas lie trussed up ready for sale at a market in Guyana. These animals are destined for the cooking pot; their flesh is often used in stews and curries in that part of the world.

be necessary to house the large constricting snakes or large, active lizards such as iguanas and monitors. As a rule, however, beginners are advised to avoid any large, active species. Venomous snakes and lizards do not make good pets either, for obvious reasons.

Environment

Background reading about the natural history of your chosen species will provide clues to its requirements. Heating of some sort will probably be necessary depending on the species you keep and where you live. Diurnal lizards and snakes prefer an overhead light source such as a heat lamp or spotlight because they are used to basking in the sun. Others fare better if a gentle heat is applied under their cage by means of a heat mat or heat strip. The best plan is to arrange the heating at one end of the cage only: That will create a thermal gradient, and the reptile will be able to move from one part of the cage to another to take advantage of different temperatures.

In addition, many lizards and turtles require a source of ultraviolet light because it enables them to synthesize vitamin D, which they need in order to absorb calcium into their skeleton. In the wild they would obtain vitamin D from sunlight, but in captivity special lights, together with dietary supplements, are often necessary to provide the correct nutritional balance. Lighting is not normally required for most of the more popular snakes such as corn snakes because in nature they shun the light and are most active in the evening and at night. Garter snakes, however, do like to bask.

Some species are very sensitive to humidity, and it is important to make sure that they are neither too damp nor too dry. Some species are particularly susceptible to shedding problems if they are kept too dry. If their skin becomes dry or comes away in many pieces, that is a sure sign they are being kept in conditions that are too dry. Clearly, freshwater turtles require an area of water, and some species can be kept in totally aquatic accommodation such as an aquarium; but most will need an area where they can crawl out to bask under a light source.

Feeding

There are almost as many types of reptile food as there are reptiles. However, it is best to choose a species whose diet is easily catered to in captivity. Insectivorous lizards, snakes, and turtles will usually eat crickets, which can be bought from pet stores, or you may be able to collect enough insects, at least during the summer.

Earthworms are another good source of food for species that will eat them. Many snakes require vertebrate food, of which the most convenient is rodents, which can be bought frozen and then thawed out as required. Again, captive-bred individuals are more likely to accept food that has been stored in this way; wild snakes often insist on having, at best, freshly killed prey, which is not always convenient (or legal).

---

remote hillside. Eggs have been collected, a captive-breeding program is underway, and young iguanas will be released into the wild once they are no longer vulnerable.

Cause for Concern

Despite these measures hundreds of reptile species may disappear over the next century. Habitat destruction through agricultural development, urbanization, mineral extraction, erosion, and pollution, is the most important cause. On top of this thousands of reptiles are killed by traffic on the roads every day, and several populations have been lost through the flooding of valleys for hydro-electrical projects. Reptiles are also hunted for food, their eggs, or the pet trade. Sea turtles enjoy total protection throughout the world but poachers still take adults and eggs in many of the poorer parts of the world, and wild crocodilians are still hunted illegally for their skins.

Not only rare species are affected. Some species that were widespread a few decades ago are becoming scarce. Many people will grow up without ever seeing a wild lizard, snake, or turtle. The challenge for the future will be to find ways to reconcile the human race's need to expand and feed itself with the preservation of the wild places needed by reptiles and other animals.

Common name Adder (northern viper)

Scientific name *Vipera berus*

Subfamily Viperinae

Family Viperidae

Suborder Serpentes

Order Squamata

Length From 24 in (61 cm) to 30 in (76 cm), exceptionally to 36 in (91 cm)

Key features Body stocky; head flat and covered with several large scales; dorsal scales are keeled, as in most vipers; basic pattern consists of a dark zigzag stripe on a lighter background, which may be gray, light brown, or reddish-brown; males' markings contrast more than those of females; melanistic (all-black) individuals are known; females are larger on average than males

Habits Diurnal; terrestrial, occasionally climbing into low bushes; swims well and may take to the water voluntarily

Breeding Live-bearer with litters of 3–18 young; gestation period about 112 days

Diet Small rodents, lizards, frogs, and invertebrates

Habitat Varied but typically dry meadows or dry sunny banks

Distribution Europe and Asia from the British Isles to the Pacific coast of Russia, north into Scandinavia, and south to northern France and Italy

Status Very common in places

Similar species Several other small vipers are similar

Venom Quite potent, but bites are very rarely fatal

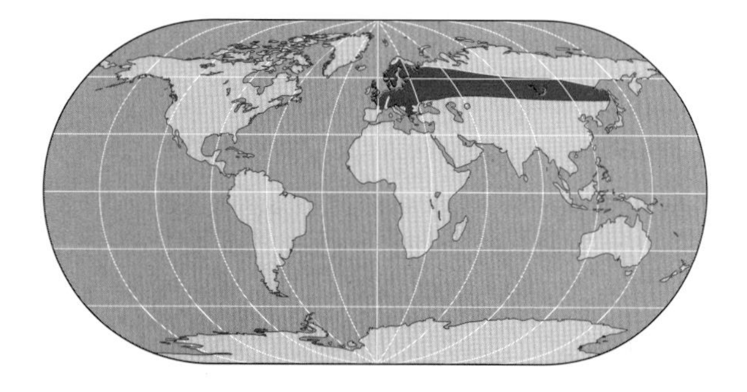

Adder

Vipera berus

The adder has the largest range of any land-dwelling snake, occurring in a broad swathe across Europe and Central Asia. It is also the only snake to occur inside the Arctic Circle.

THE ADDER IS AMONG the world's most successful snakes. Its range is only exceeded in size by that of the sea snake, *Pelamis platurus*, which can be found in open seas over a slightly larger area in total.

The adder occurs north from Britain (but not Ireland) and Scandinavia, through much of Central Europe, and across northern Asia as far east as the Pacific Ocean. Its range also includes the island of Sakhalin north of Japan, where it takes the form of a subspecies, *Vipera berus sachalinensis*. Some herpetologists, however, consider this to be a separate but very closely related species, *Vipera sachalinensis*.

Similarly, the population along the north coast of Spain has been given species status as *V. seoanei*, or Seoane's viper. Other than those forms there is little variation within *V. berus*, and only one subspecies, *V. b. bosniensis* from the Balkan region, is known.

Habitat Preferences

In the southern parts of its range the adder lives mainly in mountain ranges up to 8,500 feet or more (2,600 m). Elsewhere it has an almost continuous distribution, living in a variety of habitats, including moors and heaths, meadows, and woodland clearings. It also likes sunny south-facing banks, even when surrounded by damp meadows and marshes. It can be common along railway cuttings, which are well-drained suntraps and suffer little disturbance from people. A typical adder habitat would be a dry meadow with plenty of open areas for basking and sprinkled with cover in the form of bramble, thorn, or gorse bushes.

Saint Patrick and the Snakes

There is much folklore surrounding Saint Patrick, the patron saint of Ireland, although not much of it is substantiated. As well as the belief that he raised people from the dead, he is also said to have given a sermon from a hilltop that drove all the snakes from Ireland. It is true that there are no snakes in Ireland, but it wasn't due to the work of Saint Patrick.

As the ice sheets receded at the end of the last ice age, about 12,000 years ago, birds and animals moved north to reinhabit the ice-free countryside. But traveling at its own pace, the adder was not quick enough to reach what is now Ireland before rising sea levels flooded the land and created the uncrossable Irish Sea.

Adders frequently bask at the base of bushes under which they can retreat if necessary. When the adders are coiled up and tucked under the lowermost branches, shadows falling across their body enhances the disruptive effect of their markings and make them hard to see. They are also difficult to make out if they are among dead bracken, which they often are early in the year, or in the shade of fresh bracken, which often grows up over their favored places later in the summer.

Adaptations to Cold

The adders occur farther north than any other snake and have been recorded at 69°N in Scandinavia, well inside the Arctic Circle. In the northern parts of their range, including Scotland, they are the only snakes to be found, and in many other places they are the most common species. The average population density in good habitat is reckoned to be about 12 adders per acre (30/ha), and under exceptionally favorable conditions numbers can reach 100 per acre (250/ha). Populations fluctuate with the availability of prey.

The fact that an ectothermic animal can be successful in cold northerly regions is the result of a combination of heat-absorbing adaptations, including its small size (which allows it to heat up more quickly), dark coloration, and live-bearing breeding habits. In parts of its range the adder may be forced to hibernate for up to eight months of the year. It often emerges in early spring when patches of snow still lie on the ground.

Two adders bask on a rock in Scotland. Adders have a dark zigzag running along the back and a dark "v" shape on the back of the head.

Variations

Leaving aside the question of subspecies and possible subspecies, the adders show little variation across their range, although the background color and the contrast between the zigzag stripe and its background may vary slightly. There is a rare form in which the zigzag is replaced by a continuous vertebral line or is sometimes broken up into separate blotches.

There is also a totally black (melanistic) form, more common in some places than others. Black adders have a velvety texture and sometimes crop up in ones and twos in colonies with normal-colored snakes. More commonly, however, they make up a substantial proportion of certain populations toward the north of the range, particularly on some of the small islands in the Baltic Sea, where they are common. It seems that the dark coloration allows them to warm up more quickly, and that more than makes up for their poor camouflage.

Hibernation

Adders typically hibernate for about five months, with males emerging slightly earlier than females. Banks of earth, especially if they are south facing, are popular places in which to hibernate, provided there is access to frost-free depths where the snakes can remain dry.

Origins of the Name

The name "adder" that is applied to this species and to several of the African vipers comes from the Anglo-Saxon word *naedre*. Over the years this word became *nedder*, or a *nadder* in Middle English. It meant "a creeping thing." Eventually the "n" moved, and "a nadder" became "an adder." Strangely, the word "newt" arose through the opposite process, when the "n" moved in the other direction: "An ewt" became "a newt."

In England the word "snake" was reserved for the grass snake, *Natrix natrix*, in order to distinguish it from the adder.

Ⓣ ***The "dance of the adders" is, in fact, a courtship tussle in which a larger male attempts to frighten off a smaller one. The snakes writhe around each other and raise the front part of their body, making an impressive display.***

Ⓛ ***A pair of adders indulge in prenuptial behavior. In April or May the male follows a female around until she allows him to mate with her.***

For the first two or three weeks after they emerge, males stay near the hibernation site to bask, often in heaps containing several individuals. Females emerge later and move through the areas occupied by the males on their way to find feeding areas. Males shed their skin around this time and follow the pheromone trails left by the females. After catching up with them, courtship begins.

Frenzied "Dance"

Courtship consists of the male crawling along the female's back, rubbing her with his chin, and attempting to push his tail underneath her so that copulation can occur. If there is more than one male present, they will engage in combat. This combat is the well-documented "dance of the adders," which occurs in many other vipers and snakes of other families. The pair of fighting males rear up with the front part of their bodies intertwined and try to force each other to the ground. In some seasons such bouts occur repeatedly, while at other times they occur rarely. It probably depends on the weather and the sequence in which individual snakes emerge from hibernation. If the weather turns warm quickly, all the males emerge within a few days. Otherwise they emerge in ones and twos, and disperse immediately.

During intense bouts of courtship adders can become quite frenzied in combat and are oblivious to their surroundings. An observer can even sit and watch within a few yards of them. The winner is nearly always the larger of the two snakes. After chasing off his rival, he returns to the female and continues courtship.

Despite the fact that the males remain near the females after they have mated, it is not unusual for individual females to mate with more than one male: In a recent study 17 percent of adder litters contained young that had more than one father. Litters with multiple fathers apparently contain fewer dead or weak young. This could be due to sperm competition—the most vigorous sperm being successful in fertilizing the ova and therefore producing the most vigorous offspring.

Pregnant females spend much of their time basking: Since they are duller in color than the males, they can risk being exposed for longer periods. The developing embryos are partly nourished by nutrients from their mother's bloodstream. Females give birth in late summer after a gestation period of about four months, by which time they have often migrated back to the hibernation site. They feed heavily for the remaining few weeks of warm weather but rarely put on enough weight to see them through the winter and enable them to breed again the following year. For that reason adders tend to be biennial or even triannual breeders.

Once the mating season is over, male adders are free to move off in search of productive hunting areas. They may be up to 1.3 miles (2 km) from the hibernation site and often consist of lower-lying fields and meadows where prey is abundant. Their summer regime consists of basking and feeding.

Adders usually bask in a coiled position, loosely coiled at first, with their bodies flattened to absorb as much heat as possible. As they warm up, they become more tightly coiled. On hot days they move under cover during the middle of the day, emerging again in the late afternoon and evening, when they often lie on rocks or logs that have been warmed by the sun. By doing this, they can absorb heat through their underside. In cooler weather, such as in early spring, their behavior pattern is the opposite: They emerge late in the morning, bask throughout the middle of the day, and retreat back under cover early in the evening.

Hunters and Foragers

Male and nonbreeding female adders combine "sit-and-wait" hunting with active foraging. Prey items taken by adders include mice, voles, shrews, lizards, and occasional nestling birds. While they are basking, they may ambush prey that comes by. The coiled basking position is an ideal one from which to strike, which is probably why they adopt it. If they notice prey within a few feet while they are basking, they may move forward slowly and stalk it.

Once they are warmed up, however, they often go actively in search of food, poking their heads into burrows, crevices, and the nests of ground-nesting birds. Occasionally they climb into low bushes in search of nestling birds. Large prey, such as shrews and small rodents, is struck and released, and the adder follows the dying animal's scent trail.

Sometimes the snake that delivered the original bite arrives on the scene after another adder has found the corpse, and a struggle similar to the courtship combat may take place. Again, the larger snake usually wins, and in this case it may well be a female.

The estimated annual food consumption of an individual adder is nine adult voles or the equivalent. Juvenile adders take nestling mice as well as small lizards, indicating that they too hunt actively.

Life Span

Newborn adders do not feed in the year of their birth. They live off the yolk (stored in their

↑ A female adder gives birth to live young in the United Kingdom. When they are born, young adders are about the size and shape of an earthworm, but a perfect miniature of the adult snake.

Viper Relatives

Apart from the Sakhalin Island viper, *Vipera sachalinensis*, and Seoane's viper, *V. seoanei*, there are a number of other small vipers from the region that are clearly close relatives of the adder.

Orsini's viper, or the meadow viper, *Vipera ursinii*, has several separate populations in Central and southeastern Europe. This small species is rarely more than 24 inches (61 cm) long and is usually less than 19 inches (48 cm). Apart from its size, it differs from the adder only in minor details of its scales. It lives at about 3,000 feet (1,000 m) in the south of its range, but farther north it lives at lower altitude, its main requirement appearing to be open grassland. It is very docile and hardly ever attempts to bite. Some populations eat a high proportion of grasshoppers, while others eat lizards.

The asp viper, *V. aspis*, lives in Central Europe and Italy, and is a more dangerous species than the adder. Geographically its venom varies in composition, and in parts of southwestern France it is largely neurotoxic. *Vipera aspis* may also be totally black in color.

Many other small vipers occur in the Caucasus region in Turkey and the Middle East. Some have very limited ranges, and others are only recently described. *Vipera latifii*, Latifi's viper, occurs in four distinct color forms in the Lar Valley of Iran, which is now flooded. Current assessments of its present and future prospects for survival are unknown, but they are probably poor.

stomachs) with which they were born, growing nearly 0.5 inches (1 cm) in the process. After this they grow about 3 to 5 inches (8–13 cm) each year, and they mature at three to four years of age. Females usually take a year longer than males to mature. About 10 percent of the young reach sexual maturity, although this figure is higher in some habitats. After maturity their lifespan is not very long. On average, male adders live for about three breeding seasons, and females only manage two. Females are at a disadvantage because they are obliged to bask when they are pregnant and therefore to expose themselves to danger for longer periods.

Predators of the adder include birds of the crow family, birds of prey, hedgehogs, foxes, and, of course, humans.

Orsini's viper,* Vipera ursinii, *is a relative of the adder. Some European populations of this small viper are in danger of becoming extinct.

African Burrowing Snakes

The snakes currently included in the family Atractaspididae have been a thorn in the side of taxonomists for many years. At various times some or all of them have been included in the Viperidae, the Elapidae, and the Colubridae, a situation that arose because earlier classification systems were based largely on fangs and teeth. Now they have a family of their own, but their relationships with other snakes are still confused. There are 60 to 64 species, all occurring in Africa, although one or two species also enter the Near East. Despite their affinities to each other, there is no single characteristic that links them, and it is possible that some are derived from different ancestors (in which case they do not really belong in the same family and will have to be removed at some point in the future).

Common name African burrowing snakes **Family** Atractaspididae

Family Atractaspididae 2 subfamilies, 8 genera, 60–64 species

Some authorities include in this family several other genera, such as *Brachyophis* and *Hypoptophis*, both containing a single species. Unfortunately, they are obscure species that live in parts of Africa that are rarely visited and have not been collected for many years, so their classification needs to be clarified.

Subfamily Atractaspininae 1 genus:
Genus *Atractaspis*—18 species of stiletto snakes, including Bibron's stiletto snake, *Atractaspis bibroni*

Subfamily Aparallactinae 7 genera:
Genus *Amblyodipsas*—9 species of purple-glossed snakes, including Jackson's centipede-eater, *Aparallactus jacksoni*
Genus *Aparallactus*—11 species of centipede-eating snakes, including the Cape centipede-eater, *Aparallactus capensis*
Genus *Chilorhinophis*—3 species of black-and-yellow burrowing snakes
Genus *Macrelaps*—1 species, the Natal black snake, *M. microlepidotus*
Genus *Micrelaps*—3 or 4 species without a collective common name
Genus *Polemon*—13 species of snake-eating snakes
Genus *Xenocalamus*—5 species of quill-snouted snakes

The bicolored quill-snouted snake,* Xenocalamus bicolor *from Namibia, is one of Africa's burrowing snakes. The unusual pointed snout, shaped like an old-fashioned quill pen, gives the snake its common name.

All eight genera are venomous, but their venom-delivery apparatus varies. The harlequin snakes, *Homoroselaps*, have fixed venom fangs in the front of the mouth, as in the cobras, while the stiletto snakes, *Atractaspis*, have large folding fangs on the front of the upper jaw, but no other teeth. *Polemon*, *Amblyodipsas*, *Chilorhinophis*, *Macrelaps*, and *Xenocalamus* all have short upper jawbones that bear three to five normal teeth and a pair of grooved teeth under the eyes. Most *Aparallactus* species, which specialize in eating centipedes, also have enlarged fangs below the eyes, but the fangs of some species lack grooves. Fangs are lacking altogether in *A. modestus*, so this species can be regarded as nonvenomous.

All the atractaspids are burrowers. Some live in permanent underground tunnel systems, either their own or those made by their prey. Others push their way through loose soil, leaf litter, or sand. They feed mainly on other subterranean animals, especially other burrowing snakes, legless lizards such as skinks and worm lizards, small mammals, and burrowing frogs. Several turn up regularly in termite nests, attracted by the constant conditions and a supply of food in the form of other small animals that are also attracted there for the same reason.

In the stiletto snakes the arrangement of teeth is specially adapted for feeding in narrow spaces. Details of mating and courtship are nonexistent; but as far as anyone knows, stiletto snakes all lay eggs except Jackson's centipede-eater, *Aparallactus jacksoni*, and some populations of *Amblyodipsas concolor,* the Natal purple-glossed snake. This species is reported to have laid eggs in some regions and given birth to live young in others.

Atractaspis duerdeni, ***Duerden's stiletto snake (also known as Duerden's burrowing asp), in savanna in southern Africa. This snake can inflict painful but not fatal bites, since its venom is weak.***

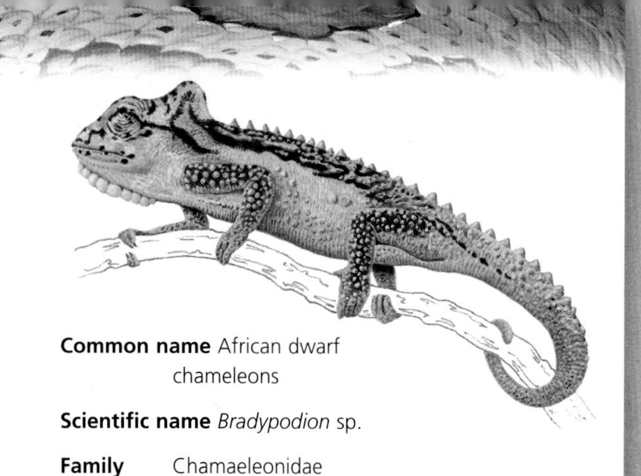

Common name African dwarf chameleons

Scientific name *Bradypodion* sp.

Family Chamaeleonidae

Suborder Sauria

Order Squamata

Number of species 21

Size From 4.5 in (11 cm) to 8 in (20 cm)

Key features Crests of small tubercles present along the back and throat; larger tubercles on body, legs, and tail; small to medium casque; faint dorsal and gular crests; females, juveniles, and males are usually mottled greens and browns in color outside the breeding season; displaying males can be spectacular; *Bradypodion* sp. differ from other chameleons in that they have single-lobed lungs

Habits Arboreal, climbing to the top of vegetation to bask; nights spent in denser vegetation for safety and warmth

Breeding Live-bearers; females produce 3 litters a year with about 12–15 babies in each litter; gestation lasts 3–5 months depending on species; species living at higher altitudes have the longer gestation period

Diet A wide variety of crawling and flying insects

Habitat From montane forest to coastal fynbos that offer grass, heathers, and low bushes; some species inhabit parks and gardens

Distribution South Africa

Status Setaro's dwarf chameleon, *Bradypodion setaroi*—Endangered (IUCN); Smith's dwarf chameleon, *B. taeniabronchum*—Critical (IUCN); others common locally

Similar species None

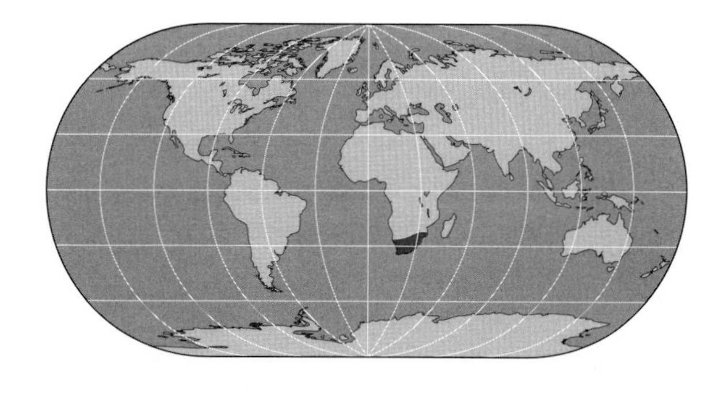

African Dwarf Chameleons

Bradypodion sp.

Bradypodion *species are unusual among chameleons in that they give birth to live young rather than lay eggs. Identification of species is difficult, but each is confined to and named after the specific locality in which it occurs.*

For several years there has been confusion over the exact number of species and subspecies within the genus *Bradypodion,* but it is now generally accepted that there are 21 species. They are sometimes divided roughly into two groups. First, there are the larger species, which are found predominantly in montane forests in Natal, Cape, and Transvaal. The second group, consisting of the smaller species, inhabits fynbos vegetation and coastal scrub areas. Old texts dealing with chameleons may use the name *Chamaeleo pumilus* to include all species or the name *Microsaura* when referring to the genus.

Temperature Control

Many regions inhabited by *Bradypodion* species experience relatively cold winters. During cold spells the chameleons bask on sunny days, but their activity levels fall. In evergreen areas they retreat into deep foliage, and in other areas they descend to the floor to find warmth and protection among leaf litter. All *Bradypodion* species flatten their bodies and turn dark to absorb heat. To avoid overheating, they retreat into shrubs and trees. They also adopt pale coloration to reflect heat and open their mouth to gape and lose heat. At night their heart slows down, and they turn a grayish color. This helps conserve heat during the colder nights.

Bradypodion damaranum, the Knysna dwarf chameleon, climbs high into the canopy to sleep in the center of ferns, its coiled tail resembling a newly emerging fern frond. Rain does not seem to affect this small chameleon. It takes advantage of it by lapping water from

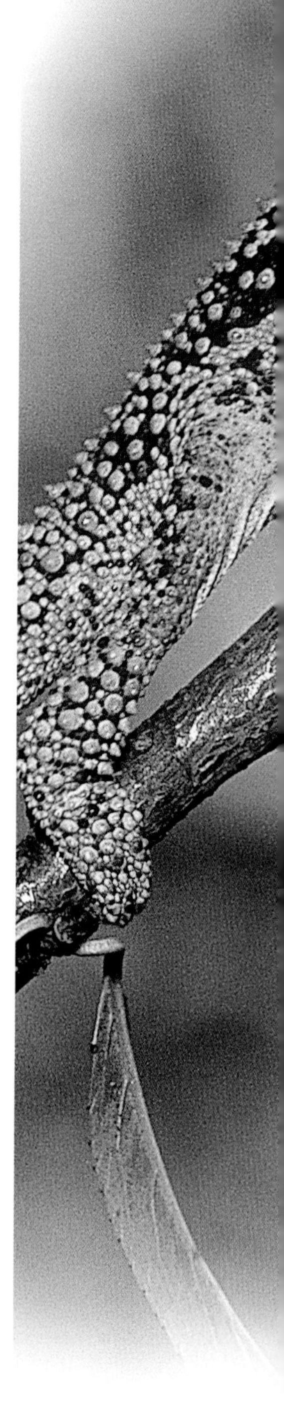

leaves. Like most chameleons, *Bradypodion* prefer to drink in the morning, since a moist tongue works better for catching prey.

Bradypodion means "slow foot," a good description, since all species in the genus have a hesitant stop-go gait that is particularly effective when stalking insects. Their jerky walk matches the movement of the foliage. Their diet changes according to season. For example, Smith's dwarf chameleon, *B. taeniabronchum*, moves into protea bushes when they begin to flower and feeds on flying insects attracted by the pollen. When the flowering season is over, it seeks another type of plant.

↑ The Natal midlands dwarf chameleon, **Bradypodion thamnobates*****, comes from the South African province of Kwazulu-Natal. Its preferred habitat is in bushes and scrubs along roads, fences, and gardens.***

Keeping a Low Profile

Predators of *Bradypodion* chameleons include the boomslang, *Dispholidus typus*, the spotted bush snake, *Philothamnus semivariegatus*, larger lizards, and the long-tailed shrike, *Corvinella melanoleuca*, which impales the chameleons on a spike or large thorn before eating them. Since some species in the genus have adapted to urban parks and gardens, dogs and cats are also potential hazards. Some spiders prey on baby chameleons. The chameleons respond to threats by retreating slowly into foliage or "squirreling"—by swiveling around, they put the branch between themselves and the danger. If little foliage cover is available, they drop into the leaf litter. Occasionally a specimen will make an aggressive show by expanding its gular pouch and bobbing its head, which is not very effective against its enemies.

Breeding

Bradypodion breed for eight to nine months of the year. When defending territory or courting a female, the male's color intensifies, and he nods his head up and down several times in quick succession. A receptive female often allows a male to mate several times a day for a few days. This ensures that more live young and fewer unfertilized ova are produced. Gravid females are intolerant of other chameleons, including members of their own sex.

Bradypodion bear live young. The eggs continue developing in the female's body, which is important for creatures living in areas with temperatures that fluctuate. Low temperatures at night would kill off eggs incubating in soil; but inside the female they stay warm, and development is not slowed down. Females give birth 90 to 150 days after mating. As they are born, the babies break out of the membrane and are deposited on a branch.

Southern stripe-bellied grass snake ***(Psammophis subtaeniatus subtaeniatus)***

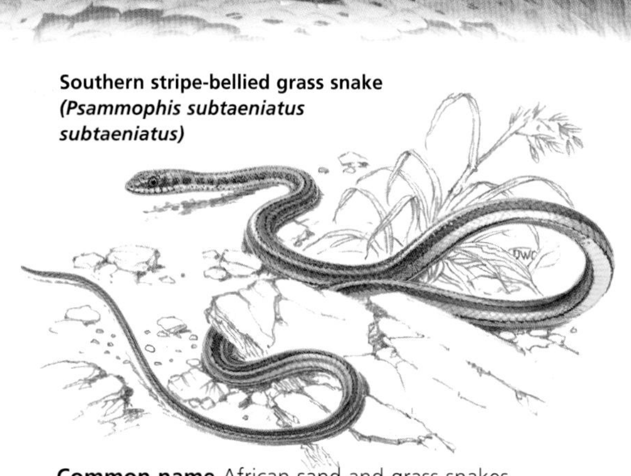

Common name African sand and grass snakes

Scientific name *Psammophis* sp.

Subfamily Psammophinae

Family Colubridae

Suborder Serpentes

Order Squamata

Length 31 in (80 cm) to 5.9 ft (1.8 m)

Key features Long, slender snakes with long tails and narrow, streamlined heads; eyes large with round pupils; smooth scales are mostly gray, brown, or olive in color, often with longitudinal stripes in lighter or darker shades; some species have intricate markings on their head and neck

Habits Fast-moving, alert snakes that hunt during the day; appear to rely heavily on sight and sometimes raise their heads while moving

Breeding Egg layers with clutches of up to 30 eggs; largest species lay the largest clutches; eggs hatch after 45–70 days

Diet Lizards, birds, small mammals, and frogs

Habitat Open country such as savanna, dry scrub, and montane grassland

Distribution Africa from the Mediterranean coast to the Cape, extending into the Middle East and as far as Pakistan in the case of *Psammophis schokari*

Status Varies, but most species are locally abundant

Similar species Related species in the region include the skaapstekers, *Psammophylax*

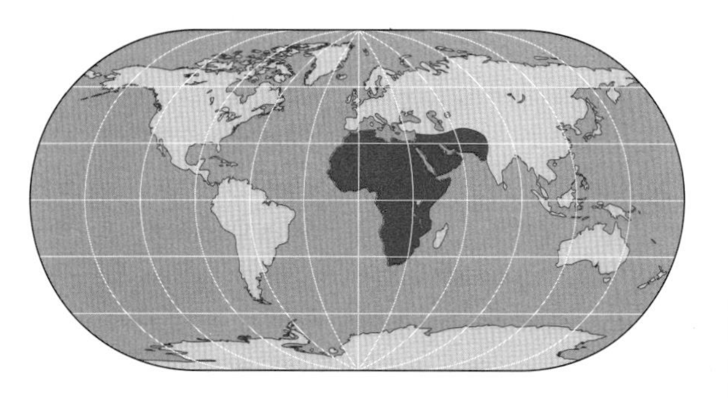

African Sand and Grass Snakes *Psammophis* sp.

Almost unbelievably fast, African sand and grass snakes hardly seem to touch the ground when they are at full speed, and it can be difficult to follow their progress through scrub and tussocks of grass.

SPEED IS THE GRASS AND sand snakes' main weapon, both in defense and attack. Their slender bodies and long tails help them propel themselves across the ground, changing direction rapidly as necessary, and sustaining high speeds over considerable distances. Sand snakes are among the few snakes that are active during the heat of the day and are frequently seen streaking across quiet roads and tracks around midday. As they move through sparse vegetation in search of food, they often pause to raise their heads to get a better view of their surroundings, and they sometimes move forward in this position. If they are surprised in the open, they usually turn quickly and shoot off at great speed—catching a sand snake as it travels over broken ground is extremely difficult.

Chasing Prey

African sand and grass snakes hunt mainly lizards, especially skinks and wall lizards. They find them by sight, then chase them down, or if necessary, follow them into cracks and crevices. Some species also climb into shrubs to take nestlings birds, while larger species can tackle rodents and shrews. The few species from damper habitats feed on frogs. Having caught up with their prey, they subdue it by gripping it firmly in their jaws and making wounds with their enlarged and grooved rear fangs, which are situated just behind their eyes. Their venom is potent and fast acting, although rarely dangerous to humans—only the larger species such as *P. sibilans, P. mossambicus,* and *P. phillipsi* have bites that occasionally produce symptoms such as swelling, bleeding, and pain.

→ A Namib sand snake,* Psammophis leightoni, *in Namaqualand, South Africa.

An olive grass snake, **Psammophis phillipsii,** *feeds on a skink in the Okavango Delta, Botswana. Skinks form a large part of the snakes' prey.*

There are 23 species altogether, mostly in Africa from the Cape to the Mediterranean coast, and about five species also range into the Middle East. *Psammophis schokari* occurs as far east as Pakistan. Sand and grass snakes are closely related to the European Montpellier snake, *Malpolon monspessulanus,* and like that species, at least some species (possibly all of them) use scale polishing to increase their protection from drying out. Another closely related genus is *Psammophylax*, the skaapstekers, of which there are three species, all superficially similar but not as slender. These species are also rear fanged; and although their venom is often considered to be among the most potent of all snakes, that has never been proven. In any case, they are generally inoffensive and reluctant to bite.

Broken Tails

Psammophis species often bite vigorously if they are captured, although their temperament varies from species to species. They are also among the few snakes that are able to break off their tails as a means of defense. If they are held by the tail, they twist and spin frantically until it snaps off. Unlike lizards, they cannot regrow their tail, although it may heal by forming a small cone-shaped scale at the tip.

The proportion of snakes with broken tails can be a good way of indicating how heavily they are preyed on: In some species there are very few individuals with broken tails, whereas in others over half of individuals have broken tails. Important predators include the African file snakes, *Mehelya*, which specialize in eating snakes, and a host of predatory birds, as well as many small carnivorous mammals. Today cars probably kill large numbers of sand snakes.

Alligator Snapping Turtle

Macroclemys temminckii

The alligator snapping turtle ranks among the largest of all freshwater turtles, as well as being the biggest found in the United States. However, giant specimens are very rarely encountered these days.

Common name Alligator snapping turtle

Scientific name *Macroclemys temminckii*

Family Chelydridae

Suborder Cryptodira

Order Testudines

Size Carapace about 26 in (66 cm) in length

Weight 219 lb (99.5 kg)

Key features Head large; jaws prominent and hooked; tail long; carapace varies in color from brown to gray depending on the individual and has 3 distinctive keels arranged in ridges, resembling those on the back of an alligator; feet on all four limbs end in sharp claws; lure present in mouth to attract prey

Habits Sedentary predator usually found in deep stretches of water; lures prey within reach especially during the daytime; may become more active as a hunter at night; strictly aquatic, but females leave the water to lay their eggs; relatively weak swimmer

Breeding Occurs in spring and early summer; clutches contain up to 50 eggs that hatch after about 100 days

Diet Eats anything it can catch, including birds, small mammals, other turtles, fish, and mussels where available; also eats fruit and nuts

Habitat Relatively sluggish stretches of water

Distribution North America from Kansas, Illinois, and Indiana to the Gulf of Mexico, including Florida and eastern Texas

Status Declining; now rare in many parts of its range; protected locally in parts of United States; Vulnerable (IUCN)

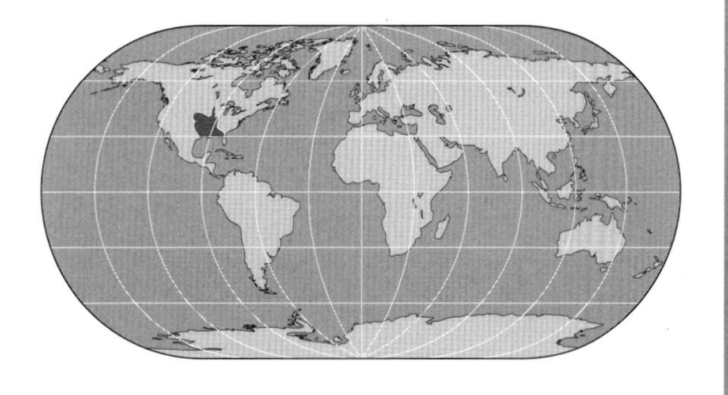

ALLIGATOR SNAPPING TURTLES get their common name from the keeled ridges on their carapace, which resembles the back of an alligator. They used to be heavily hunted to make turtle soup, which was a popular dish in the southern states. During a three-year period in the mid-1980s more than 37,736 pounds (17,117 kg) of their meat was bought by a single dealer in Louisiana. Even today hunting is a potential problem in some areas, and habitat change has generally had a harmful effect on the populations of alligator snapping turtles.

An exception has occurred in Florida, however, as a result of drainage of the Apalachicola River. The silt that was dredged out of the river was deposited on the floodplain. Clutches laid there by alligator snapping turtles were subsequently exposed to raised temperatures during the incubation period, giving rise to a higher percentage of female offspring among the hatchlings.

This occurred as a result of temperature-dependent sex determination (TDSD), in which ambient temperature during incubation plays an important part in determining the sex of the developing young. The extra females should help increase the reproductive potential of the species in this area, since a greater number of females in the population will mean that more eggs are laid.

Luring Prey

The alligator snapping turtle's bulky body means that it is not well suited to being an active predator. As a result, the species has developed a very distinctive method of obtaining prey.

The alligator snapping turtle is unusual in being able to pump blood to its tongue, creating a lure to entice fish into its mouth. It feeds mainly on fish but will also capture and eat small turtles.

These turtles are well camouflaged in their surroundings thanks in part to their dark coloration, and they often feed during the day. An individual will rest in a characteristic hunting pose on the muddy bottom with its mouth open. A projection on its tongue turns pink as it fills with blood and acts as a lure to entice prey into its jaws. The turtle can even move this structure to make it look like a wriggling worm. As soon as the prey enters its mouth, the turtle snaps shut its jaws. At other times, however, the lure is relatively inconspicuous and whitish in color, drained of blood and lying on the floor of the mouth.

This method of feeding is especially common in juvenile alligator snapping turtles. As they grow older, their feeding preferences and tastes change. It has been suggested that the main type of food in the diet of these turtles was once freshwater mussels; but thanks to the effects of water pollution and overexploitation of stocks the mussels have become rare, and the turtles have been forced to switch to other food. In areas where the mussels are available, however, they feature significantly in the turtles' diet.

Their powerful jaws also enable them to feed on smaller turtles occurring in their habitat, such as the common musk turtle, *Sternotherus odoratus*, and even their smaller relative, the common snapping turtle, *Chelydra serpentina*. There is virtually nothing that large alligator snapping turtles will not prey on—they eat all types of creatures, including birds such as wood ducks, *Aix sponsa*, and even mammals such as raccoons, which they seize in their massive jaws. The turtle drags them under water and drowns them before eating them.

The alligator snapping turtle is not exclusively predatory by nature, however. It has a keen sense of smell, which makes it an effective scavenger. One turtle was even trained to find human corpses in the waterways of Indiana. It was released on a wire leash into the water close to where a person had disappeared, and it was followed by observers in a boat as it picked up the corpse's scent. It also feeds opportunistically on vegetable matter such as persimmons and acorns, gathering these seasonal foods as they fall

↑ ***The three dorsal ridges can be seen clearly on the carapace of this alligator snapping turtle. The irregular outline together with the brownish body color give it a degree of camouflage on the riverbed.***

from trees and bushes overhanging the water—acorns in particular form a significant part of the turtle's diet in some places.

Breeding Behavior

The mating period of alligator snapping turtles begins in February and usually lasts until April. Where several males congregate hoping to mate with a single female, they often behave aggressively toward their potential rivals in an attempt to drive them away. A male that wants to mate with a female first sniffs her body carefully starting in the vicinity of her head. He then moves down the side of her body to the cloaca before mounting her under water. He grips her with his claws, anchoring on slightly to one side of her body. This enables him to direct his tail beneath the female's so that he can introduce his sperm into her body.

As in many other chelonians, the gap between the base of the tail and the opening in the anogenital region is longer in males, which aids mating. Copulation itself can last anywhere from five to 25 minutes, with the male

releasing a steady stream of air bubbles out of his nose during this period.

When she is ready to lay, the female alligator snapping turtle hauls herself onto land and digs a nesting chamber with her hind feet. This activity often takes place during the day. The nest is enlarged at the base to accommodate the eggs. It may extend over 12 inches (30 cm) down into the ground.

The female starts by digging a pit into which she can lower much of her body. She then creates a smaller hole beneath, into which she can deposit her eggs. It seems that the number laid depends to a significant extent on the size of the female, with larger individuals laying comparatively bigger clutches of up to 50 eggs. The eggs themselves are hard shelled and relatively spherical in shape.

The nests of alligator snapping turtles can sometimes be raided by predators like raccoons. It typically takes about 100 days or so for the eggs to hatch. The carapace of the hatchlings measures about 1.8 inches (4.6 cm) in length at this stage.

Mossbacks

Although alligator snapping turtles naturally inhabit rivers in the Mississippi drainage area of southern parts of the United States, they are relatively weak swimmers. They prefer to move by walking on the riverbed. They do not come onto land to bask, yet the heavy growth of green algae present on the carapace of many larger individuals suggests that they regularly spend time in shallow areas of water. Relatively intense sunlight falling on their backs is responsible for triggering the development of the plant growth, and it may even spread farther along the upper surfaces of the head and tail in some cases. As a result, the turtles are often referred to as "mossbacks" by people in the Deep South.

A dense covering of algae (not, in fact, moss as suggested by its nickname) often coats the shells of alligator snapping turtles that frequent shallow waters.

Giants of the Past

Although there are only two surviving species in the family Chelydridae (the other being the snapping turtle, *Chelydra serpentina* from Canada to Ecuador), snapping turtles used to be much more widely distributed, including in Europe. Another species of alligator snapping turtle, *Macroclemys schmidti*, lived in North America about 26 million years ago near present-day South Dakota. Records suggest that the largest alligator snapping turtles were found in northern parts of the species' range, possibly migrating there from farther south. It is also likely that much larger specimens than those officially known to zoologists existed.

One of the most celebrated "giants" was the so-called "Beast of Busco," or Oscar, as it became known. It was originally reported by a farmer in the summer of 1948. He spotted the monstrous turtle in Fulk's Lake, a stretch of water covering some 7 acres (2.8 ha) near the town of Churubusco in Indiana. It was seen again in March 1949, and some townsfolk made an attempt to corral the turtle in a small area of the lake. They constructed a stockade using stakes and managed to keep the giant reptile confined in 20 feet (6 m) of water. Unfortunately, it managed to break out of the enclosure. Those who observed the turtle said that it was about the same size as a dinner table, with a heavy covering of algal growth on its back. Its weight was estimated as being about 500 pounds (227 kg). A film of the event was taken at the time but has subsequently been lost.

More than 200 witnesses saw the turtle try to seize a duck that was being used as a lure to catch it. It was then decided to drain the lake to expose the "Beast of Busco," but the attempt nearly ended in tragedy when two people became trapped in the treacherous mud that coats the bottom of the deep lake and almost drowned. After that the turtle was left alone, and nothing more appears to have been written about it. However, the story has been immortalized in a unique annual turtle festival held in the town that takes place during June. It lasts for four days and includes a carnival parade as well as turtle racing, and now even the town's official logo features a turtle!

Other myths surround these turtles, not least that their jaws are reputed to be strong enough to break a broom handle with a single bite. Tests have shown that even a large alligator snapping turtle weighing 40 pounds (18 kg) would have difficulty snapping a pencil in this way, although the shearing effects of the jaws are such that they can bite chunks out of boats when lifted aboard. Big specimens are very dangerous to handle not just because of their strong jaws but because of their powerful flippers, which have sharp claws.

Once they are in the water, the young turtles may occasionally fall victim to larger individuals of their own species, but they are more likely to be caught and eaten by alligators. While these turtles will prey on small gars, *Lepisosteus* species, larger examples of these fish, which can reach 10 feet (3 m) in length, regularly hunt small alligator snapping turtles in return.

Forced to Move

There is some evidence that alligator snapping turtles are territorial. Established individuals may actively resent the incursion of smaller turtles into their territory. This may be related to the fact that the turtles can be forced to shift regularly from one locality to another in order to guarantee a food supply. In some areas at least it appears that the lure in their mouth often becomes less effective at attracting prey over the course of several years, and fish tend to avoid it. This may be particularly significant in view of the fact that these turtles are potentially very long-lived. There are reliable records of individual alligator snapping turtles in zoological collections living for over 60 years, and it is thought that their life expectancy could be much longer, possibly more than 100 years.

↑ ***Twenty-one years after it was stolen from a reptile park, this 110-pound (50-kg) alligator snapping turtle was found in sewers in Sydney, Australia. It has since been returned to the park.***

Common name American alligator

Scientific name *Alligator mississippiensis*

Subfamily Alligatorinae

Family Crocodylidae

Order Crocodylia

Size Large specimens measure up to 13 ft (4 m); reports of individuals up to 20 ft (6 m) long are unsubstantiated

Weight Can exceed 550 lb (249 kg)

Key features Body almost black; snout relatively long, wide, and rounded; front feet have 5 toes on each; hind feet have 4; when the mouth is closed, only upper teeth visible (which distinguishes alligators from crocodiles)

Habits Active during the summer; may hibernate during the winter, especially in northern areas; semiaquatic, emerging to bask on land; can move quite fast on land and will search for new habitat when pools dry up

Breeding Female lays clutches of 30–70 eggs; hatchlings emerge after about 2 months

Diet Carnivorous; feeds on prey ranging from crustaceans to much larger aquatic life, including fish, turtles, and wading birds, as well as mammals

Habitat Rivers, marshland, and swamps; sometimes in brackish water; rarely seen at sea

Distribution Southeastern United States from Texas to Florida and north through the Carolinas

Status Delisted from being an endangered species in 1985, having been the subject of a successful recovery program; listed on CITES Appendix II

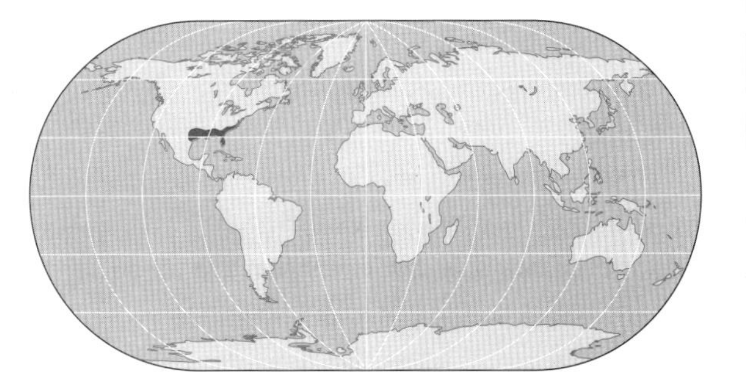

American Alligator

Alligator mississippiensis

Once on the verge of extinction, the American alligator has made a real comeback. Trade in its skin and meat is now strictly controlled, and the alligator has been reinstated as a vital part of the entire ecosystem.

THE AMERICAN ALLIGATOR USED TO range over a much wider area. About half a million years ago it reached as far north as the present-day state of Maryland. Then climatic changes occurred, and its range started to contract. However, when European settlers reached the southeastern area of the country, they found that these reptiles were still very common. Habitat modifications and hunting pressures subsequently reduced their range even more, and American alligators disappeared from the southeastern parts of Virginia and Oklahoma. Today their range includes Mississippi, Arkansas, eastern Texas, the Carolinas, and Alabama, although the species' main strongholds are southern Georgia, Louisiana, and Florida.

The relatively large size of these reptiles has given them a critical role in maintaining the entire ecosystem in which they occur because they dig so-called "gator holes" using their tail and snout. These provide temporary reservoirs of water and therefore maintain suitable aquatic habitats for various other animal and plant life. Vegetation around "gator holes" always tends to be lush thanks to the silt that the alligators deposit on the banks. The movements of alligators on regular paths can also create additional channels that enable water to run into marshlands more easily during periods of heavy rain.

Disguised as Logs

These reptiles often spend long periods floating motionless on the surface of the water, where they resemble partially submerged logs. They lie

with their nostrils above the water's surface so that they can breathe easily. This behavior allows them to spot and ambush prey and also helps them maintain their body temperature, since they can warm themselves up from the sun without leaving the water. This is achieved by means of the osteoderms, or bone swellings, along the back that are linked with blood vessels. Heat is absorbed into the body there, helped by the alligators' dark coloration, and then circulated through the bloodstream.

A similar method is used by many other crocodilians, but it is especially important in this species because of the relatively temperate areas in which the alligators are found. During the winter they become sluggish. They retreat to the bottom of the waterway or burrow into a riverbank below the waterline and only emerge when the weather is warm. At this time their heart rate can reduce to just one beat per minute. The heart has a complex four-chambered structure more like that of a mammal than a reptile. American alligators can also survive being trapped in ice, provided that their nostrils are not submerged. It has even

Almost black in color, the American alligator has prominent eyes and nostrils and coarse scales all over its body. Its upper teeth are visible along the top jaw.

It Just Takes Two

The only other species of alligator is the smaller Chinese alligator, *A. sinensis*, which reaches a maximum size of approximately 7 feet (2.13 m). It has a very limited area of distribution today in China's Yangtze Valley and is highly endangered, with an overall population of only about 300 individuals. Captive-breeding programs in China and overseas, particularly in the United States, are underway with the aim of creating a more viable population. In terms of its habits this species appears to have a lifestyle similar to that of the American alligator, hibernating in burrows over the winter. The young mature slightly earlier, however, typically at about four years old.

been known for them to recover after over eight hours frozen beneath the water's surface without breathing thanks to their low oxygen requirement under these conditions. Alligators lose their appetite dramatically in the winter, and they are likely to stop eating altogether simply because their slow metabolism does not allow them to digest food at this stage.

Their relatively wide snout allows them to tackle a variety of prey, and the mouth itself contains about 80 teeth. They are constantly replaced throughout the alligator's life as they become worn or even broken, but the rate of growth slows markedly in old age. As a result, older alligators may have difficulty catching prey to the point of facing starvation. Older individuals are more likely to resort to attacking people for this reason, since they often represent a relatively easy target.

Encounters with Humans

With the ability to swim and run over very short distances at speeds of up to 30 mph (48 kph)—significantly faster than a human—American alligators will take a wide variety of prey. Generally they do not pose a major threat to humans. But as they have increased in numbers again over recent years and development has encroached farther into the swamps of Florida, for example, greater conflicts have arisen. They often take the form of an alligator emerging onto the green of a golf course or roaming into a backyard area rather than actual attacks. Unfortunately, chain-link fencing is not an effective barrier, since these alligators can climb fences up to 6 feet (1.8 m) tall without a problem. Those that threaten or harm the public are caught under a nuisance alligator program and may be moved elsewhere.

When attacks on people occur, they are often the result of the reptile being threatened or caught unawares. Feeding alligators is

Mutant Alligators

Two very rare color mutations of the American alligator have been documented. There is a pure albino form, characterized by its reddish eyes and white body. There is also a separate leucistic variant, in which the alligators have an attractive pale yellow body color. They can be further distinguished from the albino form by their blue eyes.

There are an estimated 70 albinos, and many of them are exhibited in zoological collections or breeding farms that are open to the public. This is because their coloration makes them so conspicuous that they would be extremely vulnerable to predators in the wild. Leucistic alligators are even rarer, known from a clutch of just 17 individuals that were discovered in Savoy, Louisiana, in 1987. A single female was then found at a site 100 miles (160 km) away in 1994. Both these mutant forms are also vulnerable to skin cancer because of the lack of protective melanin pigment in their bodies, and so they need to be kept in shaded surroundings.

Albino American alligators are quite rare. Most of them, including this individual from Los Angeles, are kept in captivity because they would be unable to survive in the wild.

American alligators will eat anything they can catch. In the Florida Everglades a raccoon is this alligator's next meal.

especially dangerous, since they soon come to associate people with being a source of food. Children are more vulnerable to attack than adults because of their smaller size, but dogs are especially at risk. Alligators appear to have a particular dislike for them, possibly because they regard their barking as a threat.

American alligators communicate with each other by letting out a roar that can be heard over 1 mile (1.6 km) away. They also make a noise by slapping down their jaw on the water's surface. In addition, they keep in touch with each other by means of vibrations transmitted through the water using their throat and stomach. These sounds are made more frequently in the spring—males call to attract females in their territories, which may extend over an area of up to 10 square miles (26 sq km). They also track each other by means of special scent glands located in the cloaca and on each side of the jaw.

Dry Nesting Sites

The mating period is influenced by locality but typically lasts from March to May, with egg laying occurring a month later. The female will seek out a spot that is unlikely to flood but that is nevertheless located close to the water and often partially concealed among trees and other vegetation. The eggs will not survive in flooded ground and will be ruined if they are immersed for more than 12 hours. The female constructs a nest for her eggs by piling plant matter up to a height of 36 inches (90 cm). As the vegetation rots, it emits heat and warms the eggs, which measure about 3 inches (7.5 cm).

The incubation temperature is critical in determining the gender of the hatchlings. At temperatures below 85°F (29.5°C) the majority of hatchlings will be female, whereas above that figure males will predominate. It will be about two months before the young

↑ Although as adults they are among the largest reptiles and can grow up to 13 feet (4 m) long, American alligators are only about 9 inches (23 cm) in length when they hatch.

alligators emerge from the nest. Their mother hears them uttering their distinctive "yipping" calls and helps them out. She carries them to the water in her mouth, with her tongue serving as a pouch.

The young alligators measure about 9 inches (23 cm) long when they hatch and are much more brightly colored than the adults, with a black-and-yellow banded patterning on their body. They stay together as a group (known as a pod) in close proximity to the female until they are two years old. During this time the mother will try to protect them. A number will be lost, however, sometimes even to large males. Other potential predators include wading birds, gars, and other large fish. By the time they are six years old, the young alligators are likely to have reached about 6.8 feet (2.1 m) in length, after which their growth rate slows significantly.

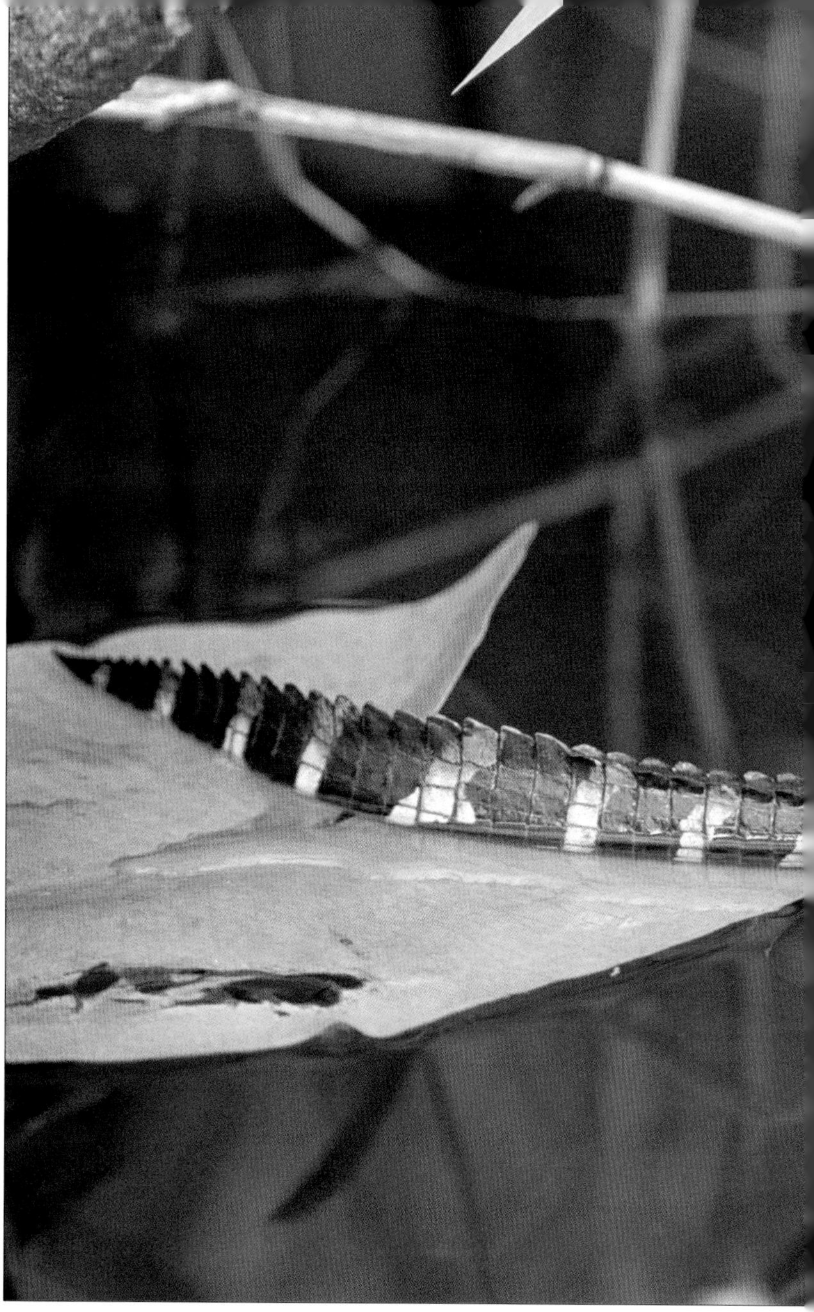

Alligator Recovery Programs

In the first half of the 20th century American alligators were killed in large numbers. Some estimates suggest that more than 10 million of these reptiles were hunted and killed for their skins between 1870 and 1970. Since that time, however, their numbers have increased dramatically thanks to effective conservation measures based partly on an acknowledgement of the commercial value of the alligators.

There are now over 150 alligator farms in various states, including Louisiana, Florida, and Texas. In the early days especially they helped restore wild populations. Farmers were permitted to remove a percentage of eggs from the nests of wild alligators, which they could hatch artificially, but a significant percentage of the resulting offspring had to be returned to the wild to repopulate areas where the species had disappeared or become very scarce.

An incidental but important benefit of these recovery programs has been our increased knowledge of the biology of the alligators. In turn this has helped develop effective management plans for wild populations. In Florida, for example, it has been shown that the alligators' reproductive potential is such that eggs could be taken from half of all nests with no adverse effects on the overall population.

Because of better habitat management larger areas are available to alligators and the other creatures living alongside them. There are some new concerns, however, notably about the rising level of mercury in certain alligator populations as a result of industrial pollution. Since the alligators are at the top of the food chain, this contaminant accumulates in their bodies from their prey. Its long-term effects are as yet unclear because, once they have survived the vulnerable stage as hatchlings, alligators can live for at least 50 years and possibly closer to a century in some cases.

This eight-week-old hatchling in the Florida Everglades is vulnerable to predation by larger aquatic animals. Juveniles usually stay in small groups close to their mother for the first two years.

Skin Tagging

Careful monitoring by means of tagging ensures that skins of illegally killed alligators cannot be traded. The success of this program has been shown by the fact that the alligator population in Louisiana has grown today to just below that of a century ago in spite of the massive development that has occurred during this period.

Although the skins are the most valuable items and are exported worldwide (especially to markets in Europe, such as Italy, as well as to Japan), alligator meat has also acquired something of a gourmet reputation and can be found on the menus of fashionable restaurants in many cities. Even the teeth of these reptiles are in demand and are made into jewelry or simply sold as curios.

American alligators often live in close proximity to humans and are an important attraction on the itinerary of many tourists visiting the southeastern states.

Sonoran coral snake
(Micruroides euryxanthus)

Common name Coral snakes

Scientific names *Micrurus* sp. and *Micruroides euryxanthus*

Subfamily Elapinae

Family Elapidae

Suborder Serpentes

Order Squamata

Number of species 54

Length From 10.5 in (27 cm) to 5 ft (1.5 m) depending on species

Key features Cylindrical snakes; usually slender, but a few species are stocky; head small and a little wider than the neck; tail short; eyes small and black; scales smooth and shiny; typically brightly colored in rings of red, black, and yellow (or white), but there is some variation

Habits Mostly burrowing and secretive, although they can be seen on the surface at least occasionally; 1 species is semiaquatic

Breeding Egg layers, but details of breeding behavior are lacking for most species

Diet Snakes, burrowing lizards, and worm lizards; *Micrurus surinamensis* eats eels

Habitat Varied; from deserts to rain forests

Distribution North America (2 species) through Central America (many species) and into South America as far south as central Argentina

Status Common in places, but some species are known from only a few specimens

Similar species Many, all of which may be mimics

Venom Dangerously neurotoxic; bites may be life threatening unless treated

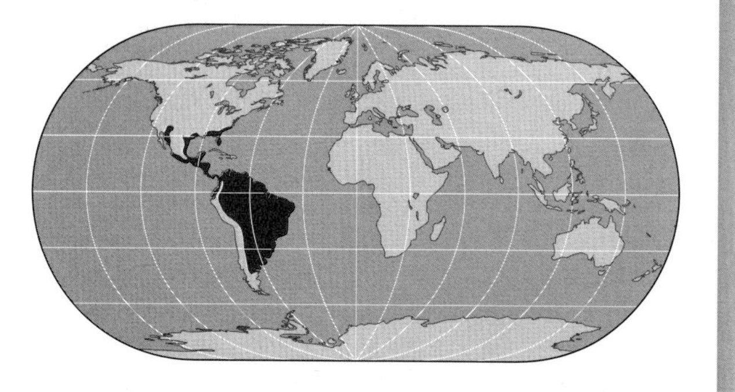

American Coral Snakes

Micrurus and *Micruroides*

"Red and yellow kill a fellow, red and black venom lack." The rhyme reminds us how coral snakes in North America can be distinguished from nonvenomous snakes by the colored rings around their bodies—if red touches yellow, the snake is usually venomous.

Coral snakes have much the same lifestyles as many other secretive species of snakes. They are most often associated with rain forests, but some species live in deserts, scrub, or dry deciduous and thorn forests.

They may be active by day or by night; and in places where there are distinct seasonal differences (such as Arizona and parts of northern Mexico), their activity pattern may change. It may go from day-active in spring and fall to nocturnal in summer.

Forest Foragers

Coral snakes forage through leaf litter, looking for smaller snakes to eat, although one has been seen several feet up in a tree eating a ratsnake. Many Central and South American species eat burrowing reptiles, especially worm lizards, which they follow through their tunnels.

Coral snakes typically bite their prey and then hang on until the venom begins to take effect. They then manipulate the prey in their mouth to swallow it headfirst, using the direction of their prey's scales to find the head.

One coral snake species, *Micrurus surinamensis*, is semiaquatic and eats eels. It is one of the largest coral snakes, reported as reaching up to 6 feet (1.8 m) in length and regularly growing to 5 feet (1.5 m).

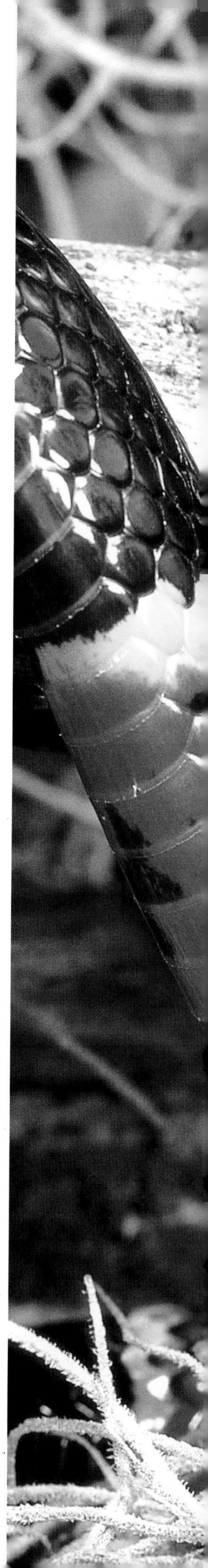

An eastern coral snake,* Micrurus fulvius fulvius, *in Florida. The color sequence reveals it as a true, venomous coral snake (because the yellow and red bands are touching each other).

Other large species that also grow to 5 feet (1.5 m) include the regal coral snake, *M. ancoralis* from Colombia and Ecuador, and the Venezuelan coral snake, *M. isozonus* also from Colombia. The smallest species may be the Colombian coral snake, *M. spurrelli,* at 10.5 inches (27 cm); but since the species is only known from three specimens, it is possible that it grows larger. A number of other species are rare and known from just a handful of species with a limited distribution. However, Spix's coral snake, *Micrurus spixi*, has a huge range that covers the whole of the Amazon Basin. Others are also widespread.

Little is known about courtship in coral snakes, but all species lay eggs. One relatively well-studied species, *Micrurus fulvius* from southeastern North America, has clutches of between three and eight eggs. The hatchlings are colored in the same way as the adults.

Venom

Although they do not attack humans very often, coral snakes have powerful venom that affects the nervous system. Often there are no

Mimicry in Animals

Mimicry in animals was first described by the English naturalist and explorer Henry Bates in a paper in 1861, although his subject was Amazonian butterflies. He noticed that edible species often had colors and markings almost identical to distasteful ones. He theorized that the predators avoided the edible butterflies after bad experiences with the nasty ones. Since Bates's time many other examples of mimicry have been described, including a species of African swallowtail butterfly, *Papilio dardanus*, in which the females mimic any one of five different distasteful species, beetles that mimic wasps, lizards that mimic beetles, and caterpillars that mimic snakes.

Nelson's milksnake, **Lampropeltis triangulum nelsoni** *from Mexico, is a nonvenomous mimic of the coral snakes.*

immediate effects in humans, but symptoms begin after two to six hours, and sometimes up to 48 hours, later. The first symptom is double vision followed by general muscle weakness. In the most serious cases the symptoms can lead to respiratory arrest and death. Bites containing venom may prove fatal in about 20 percent of all cases.

Color Mimicry

Many snakes (and other reptiles and amphibians for that matter) use color to increase their survival prospects. There is a theory that some harmless snakes gain protection from predators by mimicking the dangerous coral snake species. (They are sometimes called "false coral snakes.") Predators that have had a bad experience with a venomous coral snake are reluctant to attack any snake that resembles it. Ever since the theory was first put forward, there have been a number of objections to it; but what is not in doubt is that coral snakes have vivid colors and markings to protect themselves in some way.

Of the 50 or so species most have bands arranged into groups of three, or triads, with the sequence red-yellow-black-yellow-red. There are other arrangements, but they are in the minority. A few species, for example, have only two colors, such as yellow and red or black with white rings.

The bright colors probably work to the snake's advantage in several ways. First, despite their bright colors, coral snakes are not always easy to see in the broken light of a rain forest.

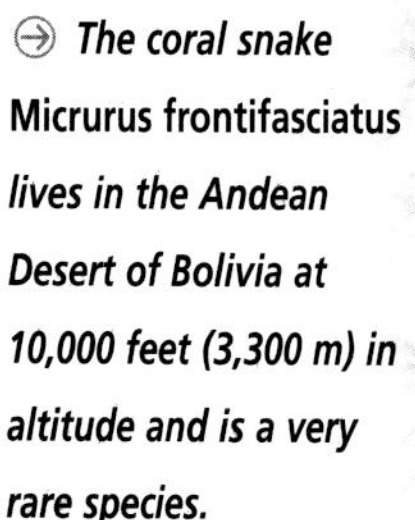

The coral snake **Micrurus frontifasciatus** *lives in the Andean Desert of Bolivia at 10,000 feet (3,300 m) in altitude and is a very rare species.*

Innate Aversion

In a series of experiments in the 1970s a researcher named Susan Smith showed how birds bred in captivity avoided wooden rods painted with black, yellow, and red rings. However, the birds were indifferent to other rods painted in different color schemes. She called this "innate aversion." In other words, the birds (which had never experienced the real thing in nature) had an inbuilt sense that objects with this coloration were best avoided. Humans are the same: We use similar color patterns (often red) on signs that warn of danger. Having an instantly recognizable signal is important because predators often need to make immediate decisions whether to attack or retreat when they uncover potential prey.

The "Other" Coral Snakes

Of the 54 species of American coral snakes 53 are in the genus *Micrurus,* and the remaining one is the only species in the genus *Micruroides.* It is the Sonoran coral snake, *M. euryxanthus*, which lives in the southwestern United States (mostly in Arizona but also in a small part of New Mexico) and in northwestern Mexico, mainly in the state of Sonora. It is a small species, growing to about 18 inches (46 cm) in length. Its black, red, and yellow or white bands are of roughly equal width.

The Sonoran coral snake has several mimics, including the shovel-nosed snake, *Chilomeniscus palarostris*, and the long-nosed snake, *Rhinocheilus lecontei*. It lives in dry areas, often semidesert habitats, and is active mainly at dusk. It eats small snakes, including quite a high proportion of *Leptotyphlops* species thread snakes. The venom of *M. euryxanthus* is not very potent and usually affects only the area around the bite. There have been no recorded fatalities.

It is important not to confuse the American coral snakes with other coral snakes from around the world, which include the Malaysian species in the genera *Maticora* and *Calliophis*, the South African *Aspidelaps lubricus*, and the Australian *Simoselaps australis*, all of which are also called coral snakes.

↑ *A Sonoran coral snake,* Micruroides euryxanthus *from Arizona. The Sonoran species is the only member of its genus.*

Next, there is the startle element: A predator picking through the leaf litter will be surprised when it uncovers a brightly colored coral snake and may hesitate before attacking it, giving the snake time to escape. Third, when a coral snake moves quickly through leaves or vegetation, the colored bands flicker as they pass before the eyes, creating a kind of optical illusion. An observer can still be trying to make out the shape of the snake and figure out which direction it is going when it is suddenly gone.

If an attacker persists, coral snakes all have similar display behaviors. They raise their tail, coiling and uncoiling it repeatedly to divert attention away from the head, which they often hide under their body (the Latinized name *Micrurus* means flickering or flashing tail). In the final stages of the display they writhe around erratically, thrashing their head from side to side, and snapping at anything they touch. The bright alternating colors enhance this display and may intimidate predators.

Once the snake has convinced a predator that it is dangerous, the snake needs to make sure that the predator recognizes it again in the future, hence the distinctive markings.

Types of Mimicry

Taking this one stage farther, if there are several species of coral snakes in an area and they are all similar in appearance, they will all benefit from each other's signals. This is known as Müllerian mimicry after the German naturalist Fritz Müller, who first described it in 1879.

False Pretenses

But if harmless snakes also look like the venomous ones, they can also benefit under false pretenses. This is known as Batesian mimicry. Many of the coral snake mimics, such as the milksnake, *Lampropeltis triangulum*, also thrash around if they are disturbed, adding to the similarity. The sequence of colored rings on some of the false coral snakes is different from that of real coral snakes—but predators would be unable to analyze such detailed information. (In Batesian mimicry the mimicking snake is known as the mimic, and the snake it is mimicking is called the model.)

There are some objections to the mimicry theory. The first is that coral snakes are so venomous that predators would be killed in their first encounter, so the lesson would be wasted. In the past some scientists argued that the real models were not coral snakes at all but mildly venomous rear-fanged species—such as *Erythrolamprus* from Central and South America. In fact, the explanation is probably simpler than this.

It seems that predators have an "innate aversion" to coral snakes and all other brightly colored animals. In other words, they avoid them whether they have seen them before or not. Perhaps they learned their lesson from brightly colored caterpillars that are distasteful but not deadly, or perhaps they are genetically programed to avoid brightly colored animals. That may be because in previous generations the individuals that did not avoid them were killed and were therefore unable to pass on their genes. The theory of innate aversion has been established scientifically in a number of different experiments.

Other scientists have objected because some of the brightly colored mimics, such as the California mountain king snake, *Lampropeltis zonata*, live in places where there are no coral snakes. Even so, they may have lived side by side in the past. In any case, their predators will include migratory birds that move from place to place and therefore may have seen coral snakes elsewhere.

The final objection is that coral snakes are nocturnal, and predators would never see their bright colors anyway. That is simply wrong. Coral snakes are just as likely to be on the surface during the day as at night. They are especially active in the evening, on overcast days, or during rainstorms.

If brightly colored harmless snakes existed by chance (rather than because they were coral snake mimics), they would be expected to occur randomly in parts of the world far from true coral snakes, such as Europe, Asia, and Africa—but they do not. False coral snakes have been identified in 35 genera of harmless or mildly venomous snakes, all confined (broadly speaking) to the southwestern parts of the United Sates, Mexico, and Central and South America, where coral snakes also live.

Changing Colors

An even more convincing argument points out that if a number of different coral snakes live within the range of a single harmless mimic, the mimic's markings will alter from place to place so that it looks almost exactly like the species with which it is found. The opposite is also true: A single coral snake species that varies from place to place may have different mimics that resemble its various forms. Three false coral snakes belonging to the genus *Pliocercus*, for example, all resemble the widespread *Micrurus diastema* from Central America, but each *Pliocercus* species resembles the particular color form of *M. diastema* with which it is found.

To sum up, coral snakes avoid predation by a variety of different methods. They may startle predators, intimidate them, create optical illusions that make it hard to be handled, or they may play on the innate aversion that most animals have for brightly colored animals. In fact, they probably do all these things.

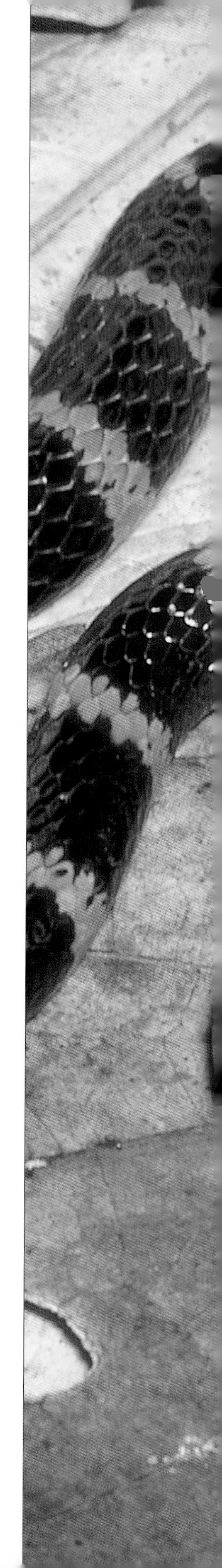

In a Costa Rican rain forest* Micrurus mipartitus *raises its tail in a defensive posture. It is hoping that the predator will attack its tail end rather than its head, which is hidden under one of its coils.

Black ratsnake
(Elaphe obsoleta obsoleta)

Common name American ratsnake (black, Everglades, gray, Texas, or yellow ratsnake)

Scientific name *Elaphe obsoleta*

Subfamily Colubrinae

Family Colubridae

Suborder Serpentes

Order Squamata

Length From 39 in (1 m) to 8.1 ft (2.5 m)

Key features Long, muscular body; color varies depending on subspecies; juveniles marked differently than adults except in the gray ratsnake; scales weakly keeled on the back but smooth on the sides; distinct ridge along either side of the belly, which helps them climb

Habits Terrestrial and arboreal; black ratsnake swims well

Breeding Egg layer with clutches of 10–20 eggs, exceptionally up to 40 or more; eggs hatch after about 70 days

Diet Small mammals and birds

Habitat Varied but usually in places with trees; forest edges, thinly wooded hillsides, and hammocks

Distribution Throughout the eastern half of the United States

Status Common

Similar species Black form, *E. obsoleta obsoleta*, may be mistaken for the black racer, *Coluber constrictor*, but that species has smooth scales and is fast moving

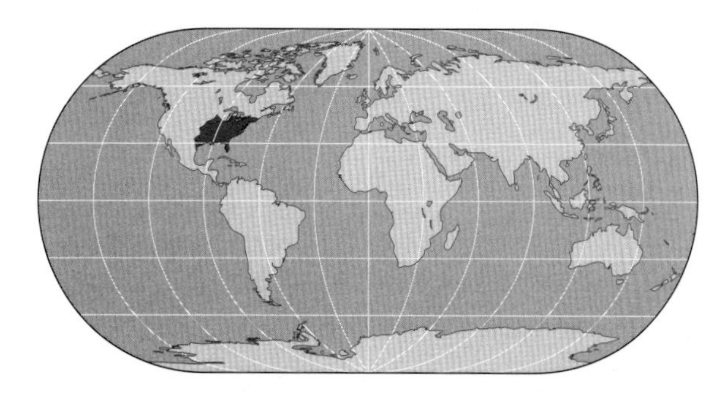

American Ratsnake

Elaphe obsoleta

The American ratsnake is a powerful, impressive snake that manages to live in almost any situation owing to its great adaptability. In its various forms it is found throughout the eastern half of North America.

THERE ARE FIVE FORMS OF AMERICAN ratsnake. In the north the black form, *Elaphe obsoleta obsoleta*, lives on rocky, timbered hillsides and forest edges. Farther south the yellow and Everglades ratsnakes (*E. o. quadrivittata* and *E. o. rossalleni*) live in coastal and inland swamps of the Carolinas and Florida, where they forage on drier ground among the cypress trees and live oaks. The other two forms, the gray and Texas ratsnakes (*E. o. spiloide and E. o. lindheimeri*), live in a range of intermediate habitats. All ratsnakes seem partial to abandoned buildings, which are the easiest places to find them.

Ratsnakes usually stop moving when disturbed, in the hope that they will escape notice. They sometimes draw their body up into a series of bends, perhaps to form a less regular shape and blend into the background better or perhaps to be ready for "flight or fight."

Elaphe obsoleta rossalleni, ***the Everglades ratsnake, with jaws agape in a display of aggression. The true, nearly patternless Everglades ratsnakes have a very small distribution. They intergrade with the yellow ratsnake where the ranges of the two subspecies meet.***

Pretend Rattlesnakes

If they think they have been discovered, they often raise the front third of the body and draw back their head, forming an "s"-shaped curve with the neck. At the same time, they open the mouth slightly and make a loud, prolonged hiss and vibrate their tail rapidly. If they happen to be resting among dead leaves, they produce a sound not unlike that of a rattlesnake, designed to intimidate their enemies. If all else fails, they lunge out, often so forcefully that they almost leave the ground. However, there seems to be a difference in temperament between the subspecies, with the Texas form having the reputation for being the most aggressive, and the gray ratsnake usually being relatively calm.

Ⓘ ***Yellow ratsnakes,* Elaphe obsoleta quadrivittata, *occur in the coastal regions of the Carolinas, central Georgia, and Florida. They are often found in citrus groves, pasture lands, and abandoned buildings.***

All American ratsnakes are great climbers. They use ridges at the edges of their ventral scales to hook onto irregularities in tree bark and wind slowly upward by concertina locomotion. They feed on arboreal mammals such as squirrels and nestling birds. They are efficient nest raiders and are frequently mobbed by the parents of the nestlings and other nearby birds. They often rest in tree hollows and return to the same place time and again.

Adult black ratsnakes patrol a well-defined home range. They use different parts of it at different times of the year according to whether they are actively feeding, looking for a mate, or hibernating. In places they switch from hunting for arboreal birds' nests to those of bank swallows, *Riparia riparia*. The birds nest in colonies that provide the snakes with the opportunity to eat well for very little effort. Apart from climbing, they also swim well and can stay underwater for up to an hour.

American ratsnakes are active mainly in the day, with peaks of activity in the early morning and late afternoon. In midsummer they may become more active later in the evening to avoid the heat of the day. In Canada they may hibernate for up to seven months, secreting themselves in hollow trees, caves, or old buildings. Several individuals may use the same hibernaculum, and it is not unusual to find them sharing with other species of snakes. Farther south the period of hibernation gradually reduces: In Florida, for example, they can be active all year long, sheltering for a few days at a time if the weather turns cold.

Mating Rivalry

In populations that hibernate, breeding activity typically begins immediately after the snakes emerge (in April or May), once they have shed their skin. Males are very competitive and engage in vigorous combat. They raise the front half of their bodies, intertwining them, and trying to force their opponent to the ground. Mating takes place after these bouts, and the female lays her eggs about

40 days later. Rotting tree stumps or piles of dead leaves are favorite places. A typical clutch contains 10 to 20 eggs, although much larger clutches have been recorded—the record was 44. As in all snakes, larger females lay larger clutches. The pure white, slightly oval eggs are tacky when laid so that they stick together and form a large single cluster. Sometimes several females lay their clutches in the same place—a communal nest of 76 eggs was once recorded.

The eggs hatch about 70 days after they have been laid, and the young snakes measure about 10 to 12 inches (25 to 30 cm) in length. Their first food is usually nestling mice, but they may also take small lizards and frogs. They reach maturity in about two to four years depending on where they live. Snakes in the northern part of the range that hibernate for up to half of each year grow more slowly than those from the south, which feed more or less throughout the year. Southern snakes also have an extended breeding season and may even lay more than one clutch of eggs in a single year.

Changing Colors

American ratsnakes undergo a dramatic color change (known as ontogenetic color change). Hatchlings are pale gray with darker blotches along their back. As they grow, the background color changes gradually, and the blotches fade (except in the gray ratsnake, which does not have a color change). In the black ratsnake the overall color darkens until it is nearly uniform black by the time it is about two years of age.

The Texas ratsnake goes through a similar change, but its blotches never disappear completely. In the yellow and Everglades ratsnakes the background color slowly changes from light gray to pale yellow or orange respectively as the blotches fade. Traces of four longitudinal stripes gradually become apparent. By the time they are adult, there is no sign of the blotches, but the stripes are well defined.

A young Texas ratsnake, **Elaphe obsoleta lindheimeri.** *The opaque cast on its eye indicates that it is ready to shed its skin.*

American Ratsnake Subspecies

There are five subspecies of *Elaphe obsoleta*. From north to south of their range the first is *E. obsoleta obsoleta,* the black ratsnake. Adults are totally black or black with traces of lighter markings. Next is *E. o. spiloide,* the gray ratsnake or oak snake. It is light- to medium-gray with dark brown or gray blotches along its back. It is the only form of *Elaphe obsoleta* in which adults and juveniles look the same.

The third subspecies is *E. o. lindheimeri*, the Texas ratsnake. It is a blotched form, with an indistinct pattern. It often has reddish areas between its scales. *E. o. quadrivittata,* the yellow ratsnake or chicken snake, is the fourth subspecies. It is yellow or pale brown with four well-defined dark lines running from its neck to its tail. Finally, there is *E. o. rossalleni,* the Everglades ratsnake. It is similar to the yellow ratsnake, but with an orange background color. This form was named after Ross Allen, a snake expert from Florida.

Other American Ratsnakes

There are several other separate species of American ratsnake, including Baird's ratsnake, *Elaphe bairdi* from southern Texas and adjacent parts of Mexico. It is gray, and each of its dorsal scales has an orange base. It sometimes has four dusky lines down its back. Juveniles are gray with darker blotches. *Elaphe vulpina,* the fox snake, is from the north-central United States and adjacent parts of southern Canada. It is a thickset species with rich brown blotches on a yellowish to light-brown background.

Elaphe guttata, the corn snake or red ratsnake from eastern North America, is a distinctive snake with red or orange blotches on a background that may be yellow or gray. Its underside is checkered black and white. The Central American ratsnake, *Elaphe flavirufa*, has large reddish-brown blotches on a dirty yellow or gray background and occurs from southeastern Mexico, including the Yucatán Peninsula, to Honduras.

Three other ratsnakes, formerly included in *Elaphe* but now in different genera, are the trans-pecos ratsnake, *Bogertophis subocularis*, the Baja ratsnake, *B. rosaliae*, and the green ratsnake, *Senticolis triaspis*.

Dipsas indica

Common name American snail-eaters

Scientific name *Dipsas* sp.

Subfamily Dipsadinae

Family Colubridae

Suborder Serpentes

Order Squamata

Length From about 18 in (45 cm) to 30 in (75 cm)

Key features Very slender with short, wide head and narrow neck; snout blunt; eyes large and protruding with vertical pupils; body compressed from side to side, producing a ridge along the backbone; scales are large and show a range of colors and markings depending on species

Habits Arboreal; nocturnal

Breeding Egg layers with small clutches of 2–4 eggs, but details of where they lay them in the wild are not known

Diet Slugs and snails

Habitat Rain forests, living in the forest canopy; rarely seen on or near the ground

Distribution From western Mexico and the Yucatán Peninsula down through Central America into northern South America as far as extreme northern Argentina

Status Probably very common but hard to find and therefore underrecorded

Similar species Some of the 31 species in the genus are superficially similar; other thin, arboreal snakes in the region

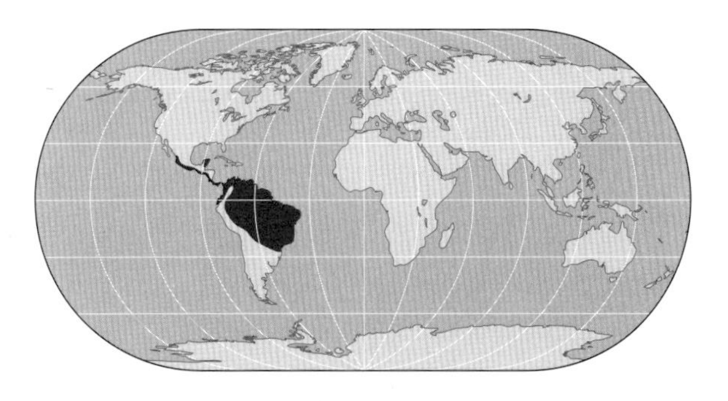

American Snail-Eaters

Dipsas sp.

Living secretive lives in the rain-forest canopy and being active only at night mean that the snail-eating snakes of the American tropics rarely come into contact with humans.

SNAIL-EATERS ARE HARD TO FIND. Following their behavior is even harder, which is why most of our knowledge comes from museum specimens and observation of captive specimens. Their most notable feature is their ability to eat snails. To do this, they have evolved a jaw arrangement that is completely different from that of other snakes (apart from another group of snail-eaters that live in Southeast Asia).

In *Dipsas* species the upper teeth point inward toward the midline of the mouth, not backward as in most other snakes, and the teeth in the lower jaws are long and needle-like. They have an extra joint in the lower jaw so that the front part can move independently of the more rigid rear part. This allows the snake to push each side of its lower jaw forward alternately.

A banded snail-eater, Dipsas bicolor, on a leaf in rain forest in Costa Rica. Snail-eaters rarely come down to the ground, since the rain-forest canopy is an ideal place for finding snails and slugs. For the same reason they are also more active during or after rain.

Expert Technique

The snake locates a snail by following its slime trail, using its tongue to pick up a scent. It bites into the snail close to its shell. The snail reacts by withdrawing its body into the shell, dragging the snake's lower jaw with it. The upper jaw slides over the outside of the shell, preventing the snail from withdrawing completely. As this happens, the bones holding the snake's upper rows of teeth bend inward so that the teeth are not damaged. The snake then begins the extraction process by bracing its upper jaw on the snail's shell. Using specialized abductor muscles, it pulls on each lower jawbone alternately. Withdrawing the snail from its shell is the hardest part of the procedure; but once it is out, the snake still has to cope with holding

and swallowing a wriggling, slippery animal. Its jaw modifications help with this.

Other behavioral aspects of *Dipsas* are not as well known. A female *Dipsas articulata* laid clutches of two and three eggs, one of which hatched in 85 days. This species, together with others that live in tropical regions where the seasons are not well defined, may well breed throughout the year when conditions are suitable. Nothing is known about the egg-laying sites, but it seems unlikely that the snakes come down to the ground. Tropical trees are festooned with epiphytic plants such as orchids, bromeliads, and ferns that accumulate forest debris and moss around them. These aerial compost heaps probably provide the right conditions for the developing eggs.

When captured, snail-eaters do not attempt to bite, but they obtain protection from some predators by resembling venomous snakes. Some species are brightly banded like coral snakes, and others are very good likenesses of arboreal pit vipers that live in the same parts of the world.

Other American Snail-Eaters

The subfamily to which the *Dipsas* species belong, the Dipsadinae, includes many snakes that do not eat mollusks but have more generalized feeding habits. Many of them eat lizards, for instance. The most closely related genus is *Sibynomorphus*, which shares many of the jaw modifications with *Dipsas* and probably feeds in the same way. Two other genera, *Sibon* and *Tropidodipsas*, also eat slugs (and possibly snails) but have to make do without the specialized toolkit of the other species.

Common name Anaconda (green anaconda)

Scientific name *Eunectes murinus*

Family Boidae

Suborder Serpentes

Order Squamata

Length Up to 33 ft (10 m) or more, but typically from 16.5 ft (5 m) to 19.7 ft (6 m)

Key features Body massive; head narrow; neck short and thick; head and body greenish in color; body has dark oval spots along either side of the dorsal midline; a bold black line runs diagonally backward from each eye to the angle of the jaw; eyes small and positioned toward the top of the head; nostrils also positioned on top of the head; heat pits absent; females more than twice the size of males

Habits Semiaquatic

Breeding Bears live young with litters of up to 80, although litters of 10–20 are more usual

Diet Mammals, birds, and other reptiles

Habitat Swamps, slow-moving rivers, shallow lakes, lagoons, and flooded grasslands

Distribution South America, mostly within the Amazon and Orinoco Basins, but extending into coastal southeastern Brazil

Status Common

Similar species Yellow anaconda, *E. notaeus*, is smaller with overall yellow coloration; De Schauensee's anaconda, *E. deschauenseei*, is similar to the yellow anaconda but not always recognized as separate species; 2 subspecies, *E. m. murinus* and *E. m. gigas*, differ mainly in head coloration

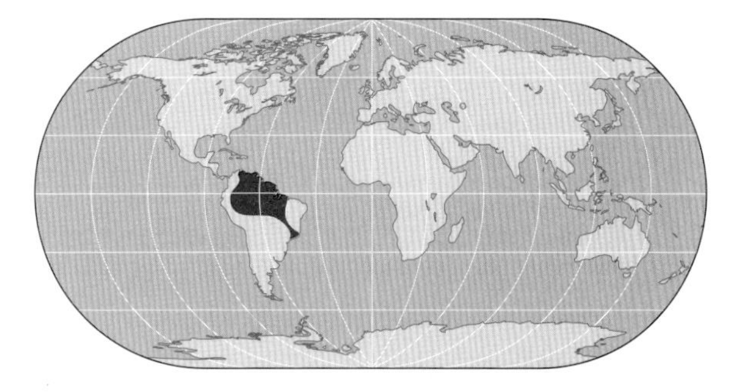

Anaconda *Eunectes murinus*

The anaconda is the world's largest snake. However, only the females reach truly enormous sizes, sometimes feeding on the much smaller males.

THE ANACONDA, OR GREEN ANACONDA as it is sometimes known, is a South American snake of almost legendary status. This gigantic serpent frequently reaches over 19 feet (6 m) in length, with a girth of 39 inches (100 cm) or more and a weight of over 440 pounds (200 kg).

Large adult anacondas are so huge that their bodies need the support of water, so they live most of their lives in swamps, shallow lakes, and rivers. They can also occur in the drier regions known as *llanos*, however, because that grassland habitat is seasonally flooded and remains under water for six to eight months each year.

Although there is little doubt that the anaconda is the world's largest snake in terms of weight, there is some disagreement over its length and whether it is longer than the reticulated python. The longest specimen ever reported was 62 feet (18.9 m) in length and was shot by the British officer Colonel Percy Fawcett of the Royal Artillery in Brazil in 1907.

Tall Tales

There is a theory that the reports of very large anacondas measuring 49 feet (15 m) or more are based on encounters with a rare undescribed species, sometimes called the *Sucuriji gigante*—the name sucuri being used locally for the green anaconda. Trails through the forest up to 6 feet (2 m) wide are sometimes attributed to this mythical "species," giving some support to local Indians and rubber tappers who claim to have seen it. However, herpetologists do not take these tales seriously.

One of the problems of studying any large snake is that of preservation. It is hard to imagine being able to get enough preservative into such a bulky animal to keep it from decomposing in the hot and humid conditions of the Amazon rain forest. Then it has to be transported a long way to a suitable site for studies to be carried out. Museum specimens of large anacondas are rare, and most are of moderate size.

Anacondas of any size are formidable predators. A common strategy is for the snake to lie in wait for prey concealed by the abundant vegetation but ready to explode into action should any animal stray within range. Once captured, the prey is enveloped in several coils of the anaconda's muscular body. It is swiftly asphyxiated or drowned. Prey normally consists of mammals such as white-tailed deer and rodents, including the capybara. When the opportunity arises, however, anacondas will tackle fully grown caiman and even the large freshwater river turtles of the Amazon Basin. The name given to the anaconda by early Spanish travellers was *matatoro*, or "bull killer."

Ⓤ Matses Indian hunters in Peru hold up the body of an 18-foot (5-m) anaconda. Much fear surrounds these snakes, and they are often killed on sight.

Ⓓ An anaconda constricts its prey of a caiman crocodile—itself a powerful aquatic predator.

Human Victims?

The possibility of anacondas eating humans cannot be discounted. However, the opportunities are limited, owing to the relatively sparse human populations in areas inhabited by the anaconda and to the fact that humans rarely wander around in swamps. There is a theory, however, that anacondas that come into frequent contact with humans may eventually see them as potential prey. Anacondas are frequently found around villages and plantations, where they are attracted by the relatively easy prey of domestic animals such as pigs, dogs, and chickens. After a large meal anacondas may go for several months before feeding again, and captive anacondas have fasted for up to three years. Pregnant females do not feed during the six months or more of the gestation period.

Formidable Females

All the largest specimens are females. Male anacondas are much smaller and may be less than half the length of a female of equivalent age and only a fraction of the weight.

During mating, which usually takes place in shallow water in the dry season, several males may be attracted to the same female. A mating "ball" may form, with each male trying to gain access to the female. Up to 11 males have been found crawling over one female, and they may remain near the female for up to four weeks. However, males do not fight directly with one another as in some other snakes.

After mating, the females stop feeding altogether, but there are plenty of reports of females eating the smaller males. That could happen because, having mated, the females need nourishment to see them through the

A breeding ball of anacondas in Venezuela. Up to 11 smaller-sized males at one time may be found crawling over one female, attempting to mate with her.

A Meal of a Male

Female anacondas have a tendency to eat smaller males, but this example of cannibalism is highly unusual among snakes, and perhaps even unique. Many snakes eat other snakes. Small snakes may stand more chance of being eaten by another snake than by any other type of predator, but this is not cannibalism any more than when one mammal eats another. Cannibalism, when one animal eats another of the same species, is rare in nature and only occurs under unusual circumstances. For example, male lions kill and eat cubs when they join a new pride. That is to ensure the females will come back into breeding condition quickly, and the new males will be able to mate with them and produce offspring of their own right away. Clearly, this does not apply to anacondas.

Another well-documented example is that of the praying mantis, *Mantis religiosa*, where the female usually eats the male that has mated with her. This is thought to provide her with a convenient protein boost just when she needs it most to produce eggs. Although nobody has claimed that anacondas adopt a similar strategy, a newly mated female may "decide" to make a meal of the male to help her through her pregnancy.

gestation period. Another theory is that females that are not in breeding condition are nevertheless attracted to the mating balls because they know there will be an abundance of males. Breeding females that put pheromones out into the atmosphere to attract males may inadvertently act as "bait" for other females that are in need of a meal.

The gestation period can last from 182 to 280 days. The embryos develop faster in warm conditions, and pregnant snakes often bask in sunlight to raise their body temperature and therefore that of their developing young. When they are born, the young anacondas measure anything from about 24 to 39 inches (60 to 100 cm), and litter size varies from four to 82. The variation is partly accounted for by the differing sizes of the female— larger females tend to have larger litters—but may also be due to varying conditions such as food supply. Captive anacondas, for example, live pampered lives with plenty of high-quality food and freedom from parasites and disease; as a result, they have larger litters. Figures from snakes that have mated in the wild come mainly from museum specimens that have been dissected, but they tend to be mainly smaller individuals.

Young anacondas are capable of eating quite large prey animals right from the start, but they have a more varied diet than their parents and will eat fish and frogs as well as birds, mammals, and reptiles. They grow rapidly, and males probably reach reproductive size at two to three years. Females, because they grow much larger, probably take a year or two more.

Related Species

There are two other species in the genus *Eunectes*. The yellow anaconda, *E. notaeus*, is from the southwestern parts of the Amazon Basin (southwestern Brazil, Paraguay, Bolivia, and northern Argentina). It is smaller than the green anaconda—10 feet (3 m) at most—and its background color is yellow. The second species is De Schauensee's anaconda, *E. schauenseei*, from the island of Marajó, at the mouth of the Amazon, and the Guianas.

Asian Brown Tree Snake

Boiga irregularis

The brown tree snake is one of the few snakes to have benefited from human activities by stowing away in cargo ships. It arrived on the island of Guam, where it has been responsible for the extinction of several species of native birds.

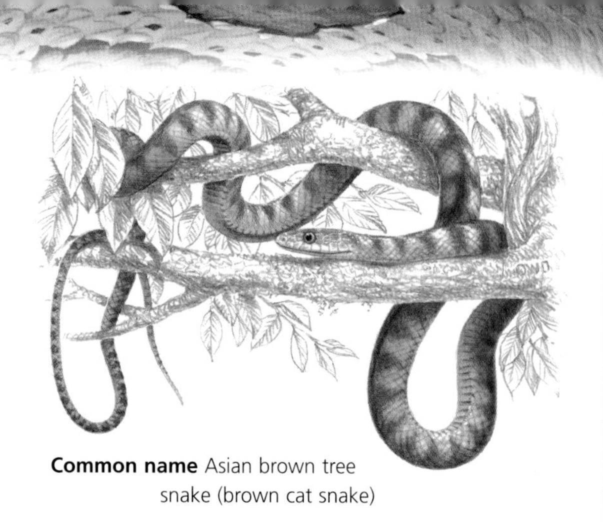

Common name Asian brown tree snake (brown cat snake)

Scientific name *Boiga irregularis*

Subfamily Colubrinae

Family Colubridae

Suborder Serpentes

Order Squamata

Length From 6.5 ft (2 m) to 7.5 ft (2.3 m)

Key features Very long, slender snake; body compressed from side to side; usually brown but occasionally yellowish- or reddish-brown; may be traces of bars on its sides; eyes large with vertical pupils; scales down the middle of its back are larger than the other dorsal scales

Habits Arboreal, climbs in trees, shrubs, buildings, and other structures

Breeding Egg layer with clutches of about 6 eggs; hatching time unknown

Diet Anything, including birds, eggs, lizards, and small mammals

Habitat Lowland forests, plantations, and gardens

Distribution Southeast Asia, including parts of eastern Indonesia, New Guinea, the Solomon Islands, and northern Australia; introduced to the Pacific island of Guam

Status Abundant, especially on Guam, where it has no predators

Similar species There are a number of similar *Boiga* species, but *B. irregularis* is the most widespread and common

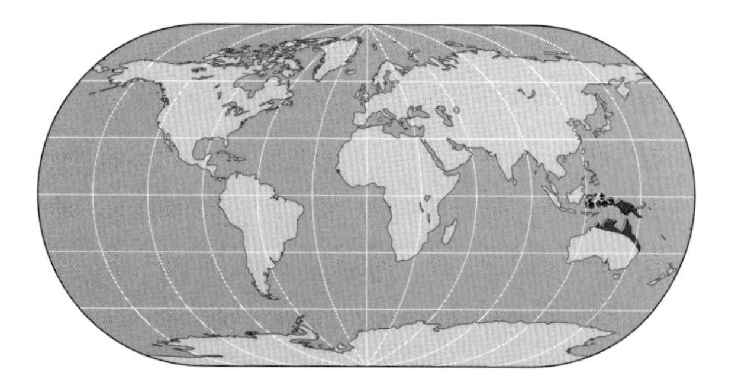

IN THE LATE 1940S the Asian brown tree snake was accidentally introduced to the small island of Guam by stowing away in cargo delivered to the United States military base on the island. It moved rapidly inland; and since it had no natural predators on the island, it thrived. By 1982 it had spread to every part except a few areas of treeless savanna. In some forested areas population densities had reached 13,000 per square mile (5,000 per sq. km) by 1978.

Driven to Extinction

By the 1960s people noticed that populations of several forest birds on the island were declining, and by 1987 all 10 of the forest species were in trouble. Some had not been seen for several years and were presumed extinct, while the few remaining species had retreated to a small part of the island farthest away from where the snake was introduced.

Several of the extinct species had been unique to Guam: the Guam flycatcher, *Myiagra freycineti*, the Guam rail, *Rallus owstoni*, the rufous fantail, *Rhipidura rufifrons*, and the Guam subspecies of the bridled white-eye, *Zosterops conspillatus*. Another species, the Micronesian kingfisher, *Halcyon cinnamomina*, only survives today in zoos. Seabirds nesting on Guam's rocky coastlines were not immune, and local populations were wiped out or reduced dramatically. Apart from birds, the snakes were eating small lizards, including skinks and geckos, and the only three species of native mammals on the island. One of them, the Little

In Australia the spread of brown tree snakes such as this one is controlled by natural predators, but on Guam population levels have spiraled out of control.

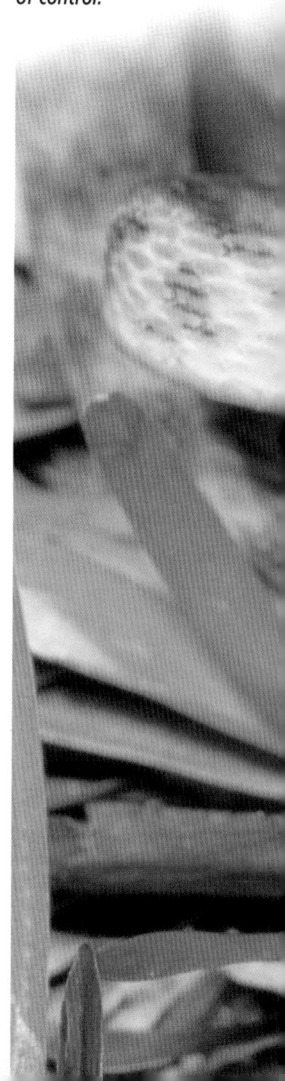

Marianas fruit bat, *Pteropus tokudae* —which was endemic to Guam—is now extinct.

The snakes also caused economic havoc, to say nothing of alarm and inconvenience. By climbing into overhead cables, they cause, on average, one power cut every four days. The resulting damage to frozen foods and computers is estimated at $1 million to $4 million per year. The brown tree snakes were also eating domestic chickens. More worryingly, they were regularly attacking sleeping babies, some of whom suffered serious bites to their heads. Unsurprisingly, the snakes' presence was also having an impact on tourism.

Experiments in 1987 showed conclusively that the introduced snakes were responsible for these problems. Measures initiated to control snake populations (mainly trapping) have only been partially successful. Cargo leaving the island is now inspected to stop the snakes from being transported to other small islands. The rare bird species that remain are maintained and bred in zoos in the hope that they can be reintroduced at some time in the future if the snakes can be eliminated.

The Asian brown tree snake is an excellent climber that lives in trees and shrubs. Using its keen sense of smell and sensitive night vision, it forages by night for lizards, birds, and mammals.

Equipped for Success

The brown tree snake is well equipped for life as a nocturnal, arboreal predator: Its long, slender body is compressed from side to side and can be held rigid when stretched out to span the gap between two branches. Its tail is long and prehensile, and its eyes are large and bulbous, giving it good nocturnal vision. It is only mildly venomous to humans, but its venom is effective against small animals and is delivered by enlarged fangs situated toward the back of the mouth.

Its method of hunting is to advance slowly along a branch on which a bird, lizard, or small mammal is sleeping and then to strike suddenly, grasping the prey

firmly and chewing on it to force as much venom into it as possible. (This can amount to 55 percent of its total venom supply, after which it needs to produce more venom before hunting again.) At the same time, it throws several coils of its body around the prey to constrict it.

As far as humans are concerned, the brown tree snake is an unpleasant species to tangle with. The long fangs of large adults cause deep lacerations; and because they tend to chew once they have a grip, they can be difficult to dislodge. Simply pulling the snake off will add to the damage and may result in teeth being embedded at the site. The best method is to unhook the teeth as patiently as possible, a process not made any easier by the fact that the snake will probably have coiled around your arm in the meantime. The venom

A Blanding's tree snake,* Boiga blandingi *from Central Africa, opens its mouth in a threatening gesture to reveal its sharp rear fangs.

The Mangrove Snake

The genus *Boiga* is a large one consisting of 30 species. They are all fairly slender climbing snakes with large eyes and vertical pupils, indicating that they are nocturnal hunters. All the species are rear-fanged, but most are not normally considered dangerous to humans. The Asian mangrove snake, *Boiga dendrophila*, is one of the brown tree snake's closest relatives. It is a wide-ranging species found on many Indonesian islands (including Borneo, Java, Sulawesi, and Sumatra) as well as India, peninsular Malaysia, Singapore, Thailand, Vietnam, and some Philippine islands.

In contrast to the brown tree snake's dull-brown coloration, the mangrove snake is black with bright-yellow rings around its body. Its shiny scales and sharply contrasting colors give it an almost enamel-like appearance, but its pattern makes it surprisingly difficult to see when resting in dappled light among the foliage of forest trees. Its upper and lower lips are also bright yellow. If it feels threatened, it flattens its head, exposing as much of these areas as possible—a strategy that is also found in several other harmless rear-fanged snakes. It also takes on a threatening posture, with its head drawn back and its neck bent to form an "s"-shaped coil, ready to strike. If the display does not intimidate its tormentor, it does not hesitate to strike and, like the brown tree snake, can give a painful bite.

fangs are easily brought into play because they are large and positioned just under the eyes. The effects of the venom are not serious in adult humans but can cause swelling and discoloration around the bite area in children. In the worst cases victims have breathing difficulties, and their heart rate slows down.

Although rain forests are their preferred habitat, brown tree snakes are very adaptable. They also live in mangroves and even in dry, sparsely forested regions in Australia. They are commonly seen around farms and buildings, which increases the chances of contact with humans. During the day they rest motionless in the forest canopy or in the timbers of buildings, in rock crevices, and even under stones.

Where Next?

The brown tree snake lives in perfect ecological balance with its prey species in parts of the world where it occurs naturally, so why has it caused such havoc on Guam? The first reason is that there are no native snakes on Guam and therefore no snake-eating predators. Second, the forests of Guam have a different "character" from those of its native habitat. In particular, the canopy layer is lower, and birds cannot find roosting places out of reach of the snake. Because there are no other snakes, the birds living on the island have not evolved antipredator strategies against them, whereas birds in other parts of the world often build nests, for instance, that are more or less "snake proof." Third, small lizards, which are common on Guam, provide a food source for young tree snakes and for adults if birds become scarce: If the bird population starts to recover, the snakes simply switch their feeding habits again.

The greatest worry is that brown tree snakes will spread to other islands and have a similar effect. They have already made their way on aircraft to Hawaii on six occasions. Fortunately, inspectors discovered the snakes before they could move out into the wild. Brown tree snakes have also been sighted on eight other Pacific islands but do not appear to have become established—yet.

Asian Slug Snakes

Pareas sp.

The Asian slug snakes closely parallel the South American snail-eaters; but despite their superficial similarity, they are now placed in a separate subfamily—the Pareatinae.

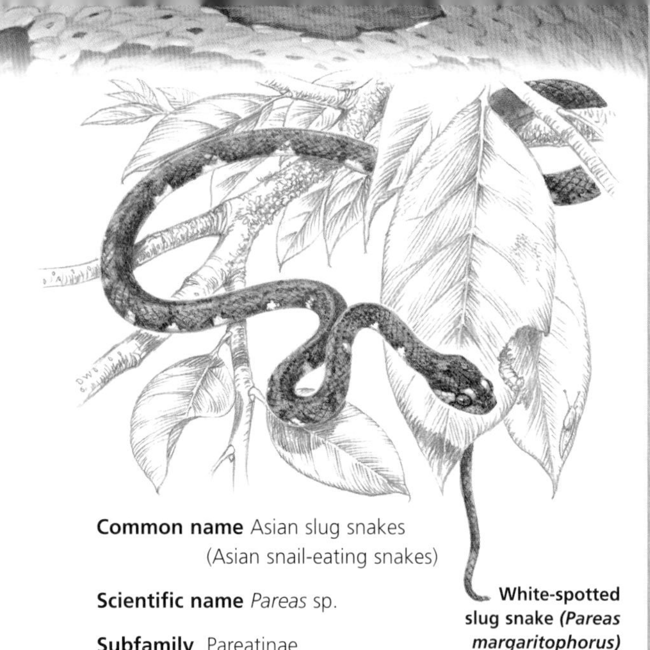

White-spotted slug snake *(Pareas margaritophorus)*

Common name Asian slug snakes (Asian snail-eating snakes)

Scientific name *Pareas* sp.

Subfamily Pareatinae

Family Colubridae

Suborder Serpentes

Order Squamata

Length 12 in (30 cm) to 30 in (76 cm)

Key features Long and slender or more robust depending on habits; most have a broad head, narrow neck, and large eyes; range of colors and markings, but most are some shade of brown or gray with small, indistinct dorsal spots and bars

Habits Strictly nocturnal; some species climb, and others live on the ground

Breeding Females lay small clutches of 2–4 eggs

Diet Slugs and snails; possibly other soft-bodied invertebrates

Habitat Varied, including rain forests, montane forests, plantations, and gardens

Distribution Southeast Asia

Status Some are common, but others are rare and poorly known

Similar species Many other small Asian snakes can look similar, but the scale arrangement on their chin makes them readily identifiable

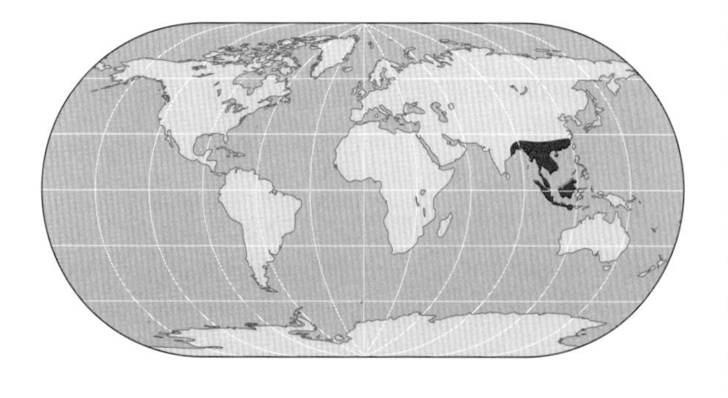

THE SUBFAMILY CONTAINS 12 or 13 species in the genus *Pareas* and a genus, *Aplopeltura*, with just a single species (the blunt-headed tree snake, *A. boa*). Some authorities place three of the *Pareas* species in a third genus, *Internatus*, but that is open to debate. Taken as a whole, *Pareas* species are not as similar to each other as the South American snail-eaters, *Dipsas*, are to each other because they have diverged.

Taking to the Trees

One group of *Pareas* species has taken to the trees and evolved the characteristics of arboreal snakes—long, thin, elongated bodies that are compressed from side to side, allowing them to reach out and span the spaces between trees. The other group lives on the ground and has more normal proportions. Tree snails are quite common in Southeast Asia, but other mollusks live on the ground: *Pareas* species exploit both resources. They find slugs and snails by following their slime trails, and they extract snails from their shells in exactly the same way as their South American counterparts. There are also suggestions that at least some species will eat other prey, such as earthworms, and *Aplopeltura boa* may even eat lizards.

Arboreal species such as the montane slug snake, *Pareas vertebralis*, and Hampton's slug snake, *P. hamptoni*, live in montane forests up to 5,000 feet (1,500 m) or more in altitude, where the rainfall is high throughout the year, and the air temperature can be low—ideal conditions for slugs, the snakes' main prey. Both species have slender, compressed bodies, and their backbone forms a sharp ridge along the

back, where there is a row of especially large dorsal scales. The montane slug snake's eyes are huge and blood-red in color, giving it good night vision. It is a slow-moving snake that crawls along thin branches and vines, relying on the cover of darkness to escape detection from predators. The lowland forms, of which the most common is the white-spotted slug snake, *P. margaritophorus*, are not so obviously specialized but live in moist fields and lowland forests, including plantations and parks.

What all *Pareas* species have in common is an unusual arrangement of scales on their chin. In most snakes the scales on the chin are divided by a gap called the mental groove. *Pareas* (and *Aplopeltura*) do not have a mental groove. Instead, their chin scales are rectangular rather than elongated, and there is an irregular furrow between them down the midline. This arrangement is the result of modifications to the skull bones that allow the snakes to extract snails from their shells. It is also present in *Dipsas* and its relatives (and, for a different reason, in egg-eating snakes, *Dasypeltis*).

Pareas species slug snakes and the blunt-headed tree snake are typically seen at night (especially during light rain when they are hunting) or during the day beneath logs and piles of vegetation. The white-spotted slug snake often hides under the discarded palm fronds in oil palm plantations.

None of the snakes attempt to bite if held, and they appear to have no obvious defensive strategies. Little is known of their breeding habits, but it is likely that they breed at any time of the year if conditions are right and they have fed well. Females lay small clutches of eggs, but details of their development and the size and behavior of the young are lacking.

Pareas vertebralis, *the montane slug snake in Bukit Larut, Malaysia. Against the forest foliage these snakes are very hard to spot.*

Common Slug-Eater

The common slug-eater, *Duberria lutrix* from southern and Central Africa, is often referred to as the gardener's friend because it eats large numbers of slugs and snails. Unfortunately, the gardener is not the snake's friend, because it often dies from eating slugs and snails poisoned with bait.

It is a stocky snake measuring about 17 inches (43 cm) long, with a short head and no distinct neck. It is usually brick-red in color along its back, with paler gray or light-brown flanks. A darker broken line occurs at the point where the flanks meet the ventral scales, which are creamy white in color. It has one of the smallest number of ventral scales seen in snakes—usually between 118 and 132, but sometimes fewer than 100.

This secretive snake hides beneath logs or stones among grasslands or scrub and emerges to feed only at night. It follows the trail of slime made by slugs and snails. It eats snails by grasping the soft parts and pulling them out of their shell—not always successfully: Dead snakes have been found with their heads jammed inside a snail's shell, because a more powerful snail managed to withdraw its body (along with the snake) into its shell, where the snake suffocated. The large garden snails, *Helix aspersa*, are the usual culprits.

The slug-eater is not dangerous (except to slugs and snails) and has many enemies, such as birds of prey, crows, ravens, other snakes, and small mammals. Its only defense is to roll itself into a tight spiral like a spring and produce a foul-smelling fluid from its cloaca. The coil is similar in shape to old-fashioned tobacco rolls—the Afrikaans name for the snake, *tabakrolletjie*, refers to this.

Slug-eaters mate in the spring and give birth to live young in late summer. Each youngster is relatively large compared to the mother, and their numbers are usually limited to between six and 12, although very large females may produce up to 22. Females are slightly longer than the males, and in the late stages of pregnancy they become hugely distended. The total weight of the litter can exceed that of the female, representing a huge reproductive effort. Each young snake measures between 3 and 4 inches (7–10 cm).

There is only one other member of the genus *Duberria*—the variegated slug-eater, *Duberria variegata*. It is slightly smaller than the common slug-eater and has a more pointed snout and three rows of small blotches down its back. Its habits are similar, although nobody has seen it make the defensive coil. It has a more restricted range in southeastern South Africa.

The common slug-eater preys on slugs and small, thin-shelled snails. After swallowing its prey, it rubs its head against blades of grass, twigs, or stones to remove any excess slime.

***Pareas vertebralis*, the montane slug snake, coiled around a tree in Malaysia. This species is only active at night during or immediately after rain, when temperatures can be cool.**

Australian Tiger Snakes

Notechis scutatus and *Notechis ater*

Despite their relatively small geographical range, the mainland tiger snakes are responsible for most of the snakebite accidents in Australia, but they amount to only a handful of fatalities.

Common name Australian tiger snake, common tiger snake, mainland tiger snake (*N. scutatus*), black tiger snake (*N. ater*)

Scientific name *Notechis scutatus* and *Notechis ater*

Subfamily Hydrophiinae

Family Elapidae

Suborder Serpentes

Order Squamata

Length From 4 ft (1.2 m) to 7 ft (2.1 m)

Key features Body stout with large satiny scales; head large with a blunt snout; eyes small; may flatten its neck when annoyed but does not rear up and spread a hood as in the Asian and African cobras; body dull black or dark brown with a yellow underside; distinctive bands of yellow extend from beneath onto the flanks, becoming narrower toward the midline; the black tiger snake is almost always completely black

Habits Diurnal; terrestrial

Breeding Live-bearers with large litters usually of 10–40

Diet Frogs, lizards, birds, and small mammals

Habitat Varied, including rain forest, grassland, swamps, and around farms

Distribution Southeastern and southwestern Australia (*N. scutatus*), Tasmania, and a small part of South Australia, including Flinders Range, Eyre Peninsula, Yorke Peninsula, and Kangaroo Island (*N. ater*)

Status Common

Similar species None in the region

Venom Very potent

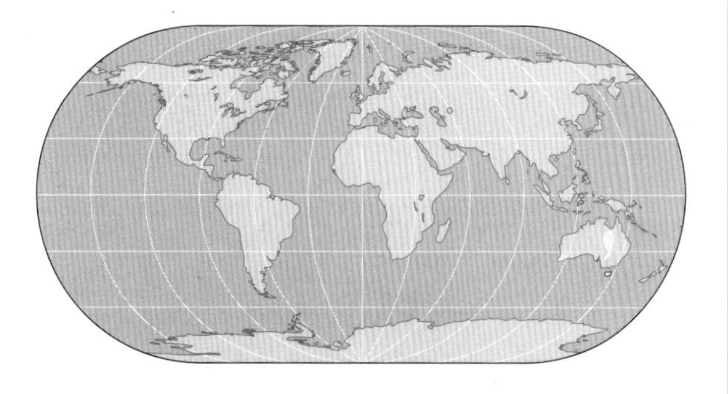

AUSTRALIA IS UNIQUE IN having more venomous than nonvenomous snakes. About 120 species are technically venomous, of which 20 to 25 are considered highly dangerous to humans. Even so, snakebite casualties are rare due to the awareness of the hazard and therefore the wearing of protective clothing, the availability of antivenom, and good medical facilities.

In parts of the temperate southern region the tiger snake is the only venomous species. The tiger snake is the fourth most venomous snake in Australia, but the strength of its venom varies between populations, as do its effects. Tiger snakes can be bold, prepared to stand their ground, and ferocious when annoyed. They flatten their neck slightly and raise their head off the ground in a pale imitation of the display of Asian and African cobras, but they are willing to bite.

Similar Habits

Notechis ater and *N. scutatus* used to be classified together. Some scientists still think they are simply regional color forms. Much information was gathered before they were classified separately and applies to both species. The mainland form, *N. scutatus*, lives in the cool, moist habitats along river courses, but it is also found in some inland regions. The black tiger snake, *N. ater*, lives mostly on Tasmania and small offshore islands where the habitat tends to be somewhat drier.

Both species bask, especially on overcast days, and are diurnal for most of the year. During the hottest months they become active at dusk or by night. The evolution of the black

In western Australia some forms of the black tiger snake,* Notechis ater*, are in fact banded, as in this individual from the upper Swan River.

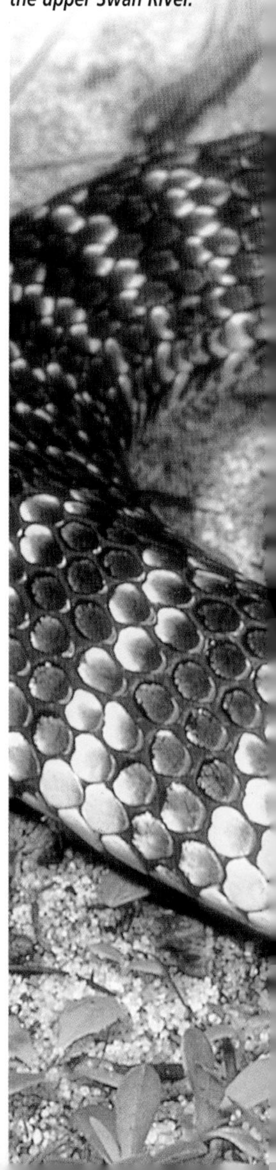

species is a good example of adaptation. In order to absorb heat more efficiently, snakes from very high or low latitudes tend to be darker in color than those from tropical or subtropical regions. As expected, the black tiger snake is found farther south (farther from the tropics) than the mainland form.

Tiger snakes are not great travelers and tend to remain in the same place except to find food or a mate. Pregnant females do not feed and may use the same hiding place for weeks on end, only emerging during the day to bask. After they give birth, they usually move away in order to find food. Males are more mobile and have home ranges of about 2 acres (0.75 hectares). They are sexually active throughout much of the year, with courtship beginning in fall and continuing through the

Family Tree

The elapids of Australasia seem to have come from a different branch of the family than those of the rest of the world. That is why scientists place them in the subfamily Hydrophiinae along with the sea snakes and sea kraits.

However, once they were isolated from the other branch of the family (the subfamily Elapinae) due to continental drift, they underwent another divergence. One group of species evolved a live-bearing (viviparous) mode of reproduction, while the others continued to lay eggs, as in the Asian and most of the African species. (Live bearing evolved independently in a small number of species from the African branch of the family, such as the rinkhals, *Hemachatus haemachatus*.)

The live-bearing habit allowed Australian elapids to move into cooler parts of the continent because live-bearing snakes can help the development of their young by basking in the sun and speeding up their development. The most significant implication of becoming live bearers, however, was that elapids in the Australasian region could exploit the marine environment.

winter for as long as the weather is warm enough. At the height of the breeding season, in spring, males fight each other, intertwining their bodies and trying to force their opponent's head to the ground. Males are larger than females, a common characteristic of species in which combat is part of breeding behavior.

Females bask frequently to raise their body temperature and speed up the development of their young. They give birth to large litters, typically 10 to 20, but litters of 50 are common. The largest recorded litter size was 109. For reasons that have not been established, litters contain, on average, more males than females in the ratio of 3:2. The young grow quickly and can mature in under two years.

Adaptable Feeders

Mainland tiger snakes feed mostly on frogs, hence their preference for moist places. Their method of eating frogs is straightforward: They simply grasp the frog and swallow it without waiting for their venom to take effect. They also eat rodents and small marsupials, but the way in which they capture the larger animals is different. The snakes strike, then release their prey, tracking and consuming it later.

When the global climate warmed up and part of the polar ice caps melted, sea levels rose. Areas of what was then southern Australia were flooded. A trail of large and small islands was left at the southeastern tip. Tasmania was the largest of them, but there were many smaller ones, each with its resident population of tiger snakes. The snakes had to adapt to changing conditions if they were to survive. For example, there were few frogs on these islands, and the snakes had to subsist on whatever food was most readily available.

Over time this affected their relative sizes. On islands where large prey such as bandicoots, rats, and large seabirds occur, the tiger snakes are large; but on islands where only small lizards live, they are small. On islands with medium-sized prey, such as small seabirds and mice, the tiger snakes are also medium sized. This could be a simple matter of the snakes growing as large as their food supply allows. Or they may have been genetically programmed over generations to reach different maximum sizes in order to live on the available food supply. Size variation can be extreme, with snakes on one island weighing five times more than those on an island just a few miles away.

The Chappell Island Tiger Snakes

Chappell Island in the Bass Strait between the Australian mainland and Tasmania is the breeding ground for short-tailed shearwaters, *Puffinus tenuirostris*, known locally as mutton birds. They are large seabirds that nest in burrows. The tiger snakes on the island prey on their chicks. The food from the birds is highly nourishing but seasonal, and the snakes must gorge themselves for about six weeks while the chicks are young to store enough energy to see them through the rest of the year. After that the chicks grow too big for them. Fortunately, there are also small skinks on the island for the snakes to fall back on. (More importantly, the skinks provide food for young tiger snakes until they grow large enough to swallow the chicks.) However, juvenile tiger snakes are heavily outnumbered by adults, so it seems that very few each year find enough skinks to reach the necessary size threshold.

The Chappell Island tiger snakes are the largest tiger snakes, with a maximum weight of 3.8 pounds (1.7 kg). They have to be large, since a small race could not tackle the young mutton birds, nor could they store enough fat to see them through many months of near famine. Furthermore, Chappell Island tiger snakes occur in very high densities, because when food is available, there is plenty to go around. They also make use of the mutton bird burrows for sheltering on cool nights and for hibernating. Several snakes may share the same burrow.

The large head, blunt snout, and small eyes that are typical features of the tiger snake can be seen here in* Notechis scutatus *from southeastern Australia.

Bengal or Indian Monitor *Varanus bengalensis*

For many years the meat and skin of Bengal monitors have been highly prized in areas of India and Asia, a situation that has led to a decline in numbers over parts of their range.

Common name Bengal monitor (Indian monitor)

Scientific name *Varanus bengalensis*

Family Varanidae

Suborder Sauria

Order Squamata

Size 6 ft (1.8 m)

Key features Head relatively small and pointed; neck thick and muscular; limbs short; tail long and slightly compressed from side to side; dorsal pattern varies, but basic coloration is black, gray, or brown with lighter markings

Habits In wet habitats rests on submerged vegetation; in other places climbs trees or forages on the ground; takes refuge in tree hollows, burrows, or termite mounds during inactive periods

Breeding Mating, egg laying, and hatching vary from region to region; on average female lays 1–3 clutches of a maximum of 30 eggs that hatch after 170–250 days

Diet Insects, beetles, snails, small amphibians, small mammals, and lizards

Habitat Variable from rain forest to swamps to more arid, rocky regions as well as cultivated areas

Distribution Eastern Iran, Afghanistan, Pakistan, India, Nepal, Sri Lanka, Bangladesh, Mayanmar, Malaysia, Sumatra, Java, and the Sunda Islands

Status Endangered (IUCN) in some parts of its range; listed in CITES Appendix I

Similar species None

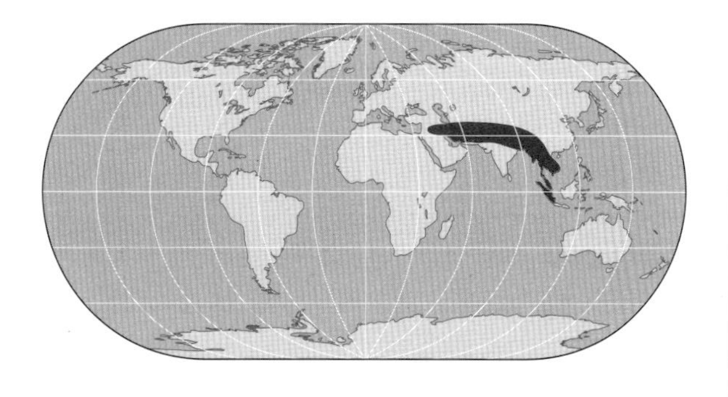

DESPITE ITS COMMON NAME, the Bengal (or Indian) monitor has one of the widest distributions of all monitors. It is found in river valleys in eastern Iran, Afghanistan, and western Pakistan. In the rest of Pakistan it is widespread in various habitats but especially abundant in agricultural areas. It is also found in northern India, Nepal, Sri Lanka, Vietnam, peninsular Malaysia, Java, Sumatra, and the Sunda Islands, where it occurs in habitats as diverse as desert fringes, pristine rain forest, dry open forests, and farmland.

There are two forms of Bengal monitor. *Varanus bengalensis bengalensis*, the nominate form, has larger scales than those of the other subspecies arranged in irregular rows above its eyes. Adults are black, dark gray, or tan to brown with cream or yellowish spots on the back, which vary among individuals. The second form is *V. b. nebulosus*. The scales above its eyes are of differing sizes. Its dorsal pattern is lighter and less distinct, resembling speckling on a light gray to yellowish background. The name *nebulosus,* meaning cloudy, refers to this effect. If there are spots of color, they are tiny, single-scale size. The amount of patterning on an individual seems to be determined by the level of rainfall in its habitat. The plainest individuals are found in desert areas (a feature shared with other monitors in which the lack of any distinct pattern provides better camouflage).

Versatile Lizard

Although mainly terrestrial, Bengal monitors are adept at climbing trees and have been seen dropping to the ground from heights of

between 32 and 50 feet (10–15 m) without sustaining any injury. They are also good at running and swimming, and can remain submerged for about an hour. In the more temperate parts of their range activity reduces or halts during the cooler months, although *V. b. nebulosus*, the more tropical form, is active for most of the year.

Bengal monitors shelter either in burrows that they dig using their sturdy front legs and claws or in crevices in rocks and buildings. *Varanus bengalensis nebulosus* prefers to seek shelter in tree hollows. Both subspecies also use abandoned termite mounds as hides. Males of the species eat more food than females and are consequently more active. They also grow faster.

The mainly terrestrial Bengal monitor, Varanus bengalensis bengalensis, *will climb into trees in search of food or refuge.*

Compatible Neighbors

Bengal monitors are active in warm habitats. They spend the nights in refuges such as burrows and tree hollows, where their body temperature falls below that of the ambient temperature. Therefore, after emerging from their retreats the following morning, they have to bask for four or more hours to raise their active body temperature to 93°F (34°C). As a result, Bengal monitors tend to be more active and hunt for food in the afternoon, when daytime temperatures are at their highest.

The water monitors, *Varanus salvator*, inhabit the same areas. They spend the night in the water and are able to keep their body temperature higher than the ambient temperature. Since their active body temperature is lower than that of the Bengal monitors—86°F (30°C)—they take less time to warm up and are able to forage for food in the morning. In the afternoon, when the daytime temperature is at its highest, they take refuge in the water to cool off, emerging in the late evening for a final forage. The heat-regulation patterns of these two monitors allow them to coexist in the same area without competing directly with one another.

Water monitors, such as Varanus salvator kabaragoya *from Sri Lanka, live alongside Bengal monitors in many parts of their range. These large monitors can grow up to 8 feet (2.4 m) long.*

Muscular Body

Bengal monitors vary in size according to region and habitat. The total length of specimens in Bangladesh is about 39 inches (100 cm); in Sri Lanka it is 4.6 feet (1.4 m); in Malaysia it is 5.3 feet (1.6 m); and other areas have records of specimens reaching 6 feet (1.8 m) long.

Bengals are characterized by a relatively small, pointed head, a thick, muscular neck, strong limbs, and sharp claws for digging extensive, deep burrows. They can also wedge themselves in the burrows by inflating their body and fixing their claws to the wall. When swimming, they hold their limbs close to their body and propel themselves through the water using undulating body movements and the laterally compressed tail. Their muscular limbs and claws are useful for climbing trees or houses. The tail can be used as a weapon or as a counterbalance when the monitor stands on its hind legs to peer over tall grass.

Foraging for Prey

Although both subspecies feed on the ground, *V. b. nebulosus* will climb trees and is capable of capturing roosting bats. Despite its size and formidable jaws, the creature seems to prefer smaller prey items such as beetles, grubs, grasshoppers, crickets, scorpions, and snails. It licks up ants with its long tongue. Vertebrate prey such as frogs, lizards, small snakes (including cobras), and small mammals appear to be a second choice.

Bengal monitors spend long periods of time rooting in leaf litter for small prey items and have been seen foraging in human garbage. They crush small prey with their powerful jaws, raising their head and throwing the food back into their throat. When eating carrion, they lap up the contents of the digestive tract of herbivorous animals. This is thought to provide additional vitamins and minerals.

Reproduction

Sexual maturity is determined by habitat, food availability, and more importantly, by size. Individuals can take from three to five years to mature (the longer period in less favorable habitats). Females tend to have shorter tails than males; while males have patches of scales arranged as flaps around the anus.

Breeding times vary across the species's range: In northern parts the main reproductive period is during the wet season; in Sri Lanka

Human Enemies

Although young and subadult Bengal monitors may be attacked by other monitor species or by large predatory birds, adult Bengals have few enemies. The greatest danger comes from humans. The monitors are attracted into villages by the presence of poultry and rodents, which make easy pickings. As a result, they are killed by local people protecting their property.

Bengal monitors have been collected on a large scale for their meat and skins. Specimens have been dragged from burrows; their flesh when cooked is supposed to be easy to digest and is used to cure various illnesses. In India, Sri Lanka, Thailand, and Vietnam water monitors, *Varanus salvator*, inhabiting the same range are thought to be inedible, but the Bengal is consumed in large numbers and often forms a major part of the diet of local people. Dark-skinned specimens are particularly sought after. As a result of man's predation, the Bengal monitor can no longer be found in some parts of Sri Lanka, India, and Bangladesh. It is listed in CITES Appendix I, and commercial trade was banned in 1975. However, many countries ignored the ban, and a number of years later Japan was found to be importing hundreds of thousands of skins from Pakistan, Bangladesh, Thailand, and Malaysia.

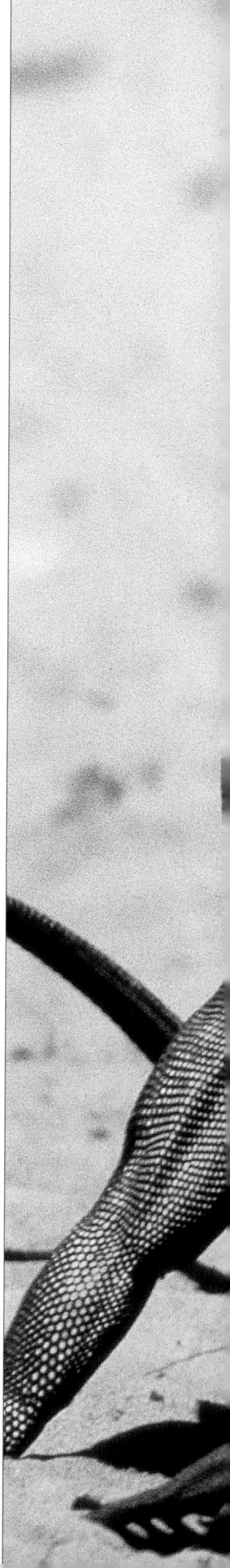

eggs are laid between January and April; in Thailand, where there is little temperature variation, eggs are produced throughout the year, with some females laying up to three clutches a year. Studies using radio-tracking equipment have revealed that some female Bengal monitors return to the same place each year to lay their eggs.

Depending on the habitat and terrain, eggs may be laid in burrows, termite mounds, tree hollows, or rotten logs. Although the maximum clutch size is 30, *V. b. nebulosus* can average 70 eggs in a year by laying two to three clutches. The incubation period varies between five and nine months.

After hatching, the young monitors tend to hide, often seeking safety in trees and feeding on insects found in the foliage. Juveniles exhibit brighter coloring than their parents, being brownish orange to light brown with bands of black and yellow on the tail and body. It is interesting to note that a clutch of young Bengal monitors remains together as a group for several months before dispersing—there is probably greater safety in numbers.

Sparring Males

Frequent wrestling contests take place as younger males try to secure territories and females. When a rival is spotted, a male rears up on its hind legs, arches its body toward the intruder, lunges, and lashes its tail from side to side. The pair wrestle using their forelimbs.

The scent of a receptive female is present in her feces and is picked up by males in the area. Mating is vicious, the male biting and holding onto the female's neck in order to immobilize her. This means that the most successful males are usually the strongest. In the wild studies have shown that pair bonding plays an important role in the reproduction of Bengal monitors, and there have been many reports of pairs of monitors found together.

Looking for all the world as though they were dancing, a pair of Bengal monitors wrestle. The strongest male will win the right to mate with a receptive female.

Common name Black caiman

Scientific name *Melanosuchus niger*

Subfamily Alligatorinae

Family Crocodylidae

Order Crocodylia

Size Up 20 ft (6.1 m), making it the largest of all South American crocodilians

Weight Approximately 500 lb (227 kg)

Key features Body black with dots that are especially evident in young hatchlings; head gray in youngsters, becoming reddish-brown as they age; snout relatively wide at the base, rapidly narrowing along its length; an obvious bony ridge is evident above the eyes, continuing down the snout; protective body casing on the neck and back is the thickest of all crocodilians

Habits Nocturnal hunter; often encountered in flooded areas of forest during the wet season

Breeding Female produces clutch of 50–60 eggs that are deposited in a nest mound; eggs hatch after 6 weeks

Diet Young feed on aquatic invertebrates and small fish; larger individuals eat bigger prey, including some mammals

Habitat Shallow areas of water in rain-forest areas

Distribution Confined to the Amazon drainage basin; appears to be conspicuously absent from Surinam

Status Has declined in many parts of its range as the result of heavy hunting for the leather trade during the second half of the 20th century; IUCN Lower Risk; CITES Appendices I and II

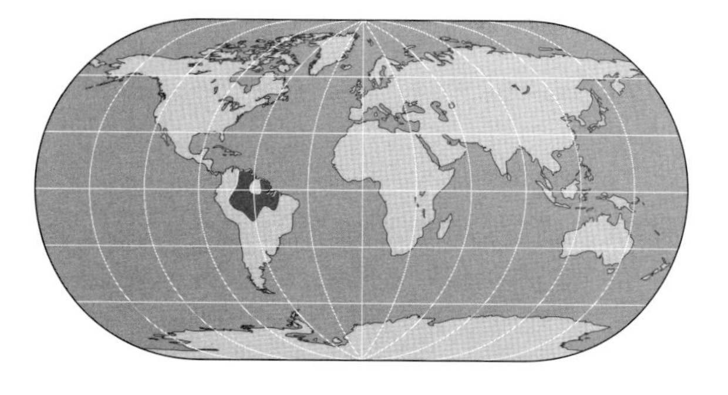

Black Caiman

Melanosuchus niger

In spite of its name, the black caiman is one of the more colorful crocodilians. The white spots on its black body are particularly bright in young hatchlings.

THE HABITS OF THE BLACK CAIMAN appear to be closest to those of the land crocodiles of the past. It is an agile hunter found in shallow stretches of water. It finds its prey by a combination of sight and sound—it has very acute hearing. As they grow larger, the caimans feed on mammals. They sometimes take domestic livestock such as pigs and dogs, but they rarely attack humans. In some areas they have been known to prey regularly on cattle.

The black caiman's relatively large size has made the species an attractive target for hunters, and a huge trade in skins developed out of Colombia from the 1940s onward. About 60,000 hides were exported annually from this country alone right up until the 1970s, and a survey at the end of that decade showed that the species had been virtually wiped out in Colombia. Overall, the total population of black caimans throughout South America is now believed to be just 1 percent of the numbers that existed there a century ago.

Unfortunately, even when given protection, it is difficult for the caiman to recolonize former habitats. This is because it faces competition from the smaller, more adaptable common caiman, *Caiman crocodilus*, which has become more widespread. Common caimans adjust well to habitat changes, they are less conspicuous by nature, and they breed more rapidly. The last remaining stronghold of the black caiman appears to be French Guiana, particularly around Kaw, where the largest surviving examples are most likely to be found today.

Breeding and Lifestyle

The nesting season of black caimans usually begins during the dry season throughout their range. The female constructs a mass of

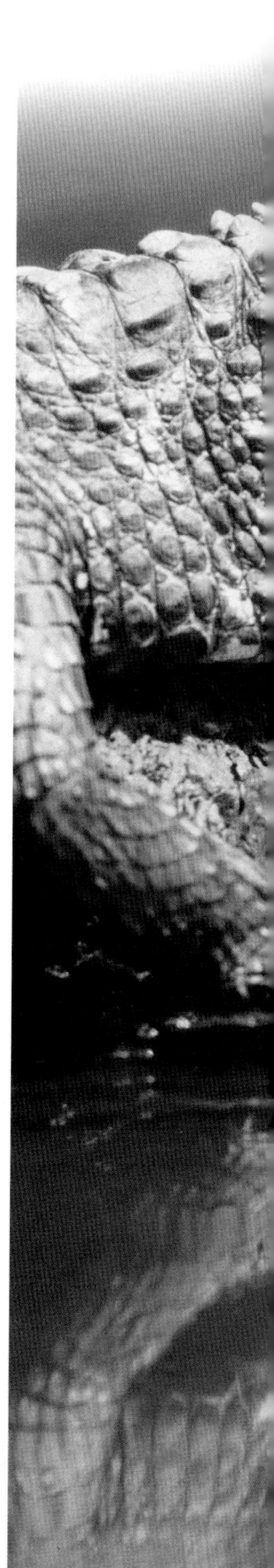

vegetation that can measure up to 30 inches (76 cm) high and up to 5 feet (1.5 m) across. The siting of the nest varies—sometimes it is hidden farther inside the rain forest than at other times. It is not uncommon for several females to nest in a similar area. The hatchlings then emerge from their eggs after a period of six weeks. Females may become more aggressive when nesting, although not all of them appear to display strong maternal protective instincts—some simply abandon their nests after they have finished laying their eggs.

Black caimans are known to undertake seasonal movements. Early 19th-century naturalists visiting the Amazon first described the way in which these crocodilians preferred the relative shallows of the flooded forests during the wet season, only returning to the main river channels during the dry season when the water level was much lower. In some parts of the lower Amazon where river flow can fall dramatically, they would even estivate by burrowing into mud, while large numbers congregated in remaining areas of water.

Hunting of the caimans was common at that time of the year, notably around Marajó Island. The island is located at the mouth of the Amazon, and cattle ranchers there were desperate to protect their stock from attacks by these crocodilians.

A butterfly seems unaware of the potential danger from the powerful jaws of this black caiman basking on a log in Peru.

Unusual Fishing Technique

Fish figure prominently in the diet of the black caiman, with larger individuals hunting the big freshwater catfish found in the waters of the Amazon basin. Black caimans have developed a very efficient way of catching and stunning fish, using movements of their powerful tail. They advance quietly toward their target. They lash out with the tail; then almost imperceptibly, using the tail like a hand, they bring the fish around to their mouth. They hunt at night, and the distinctive smacking sound made by the caiman's tail on the surface of the water indicates that these crocodilians are present in the area.

Blue-Tongued Skink

Tiliqua scincoides

The slow-moving blue-tongued skinks in the genus Tiliqua *are probably the most familiar of all Australian lizards. As their common name suggests, they have bright blue tongues.*

Common name Blue-tongued skink

Scientific name *Tiliqua scincoides*

Family Scincidae

Suborder Sauria

Order Squamata

Size Up to 24 in (60 cm)

Key features Head large; lower eyelid movable without a transparent window; body stout and flattened across the back from side to side; scales smooth; limbs relatively short but well developed; color varies according to subspecies, usually gray to tan or silver background with bands extending onto the sides; some have a dark streak running from the eye to the top of the ear opening; tongue blue

Habits Diurnal ground dweller, usually slow moving; spends a lot of time basking and foraging for food

Breeding Live-bearer; female gives birth to up to 25 young; gestation period about 110 days

Diet Omnivorous; eats insects, fruit, berries, flowers, even carrion

Habitat Temperate forests, subhumid forests, grassland, suburban gardens

Distribution Eastern and northern Australia and Irian Jaya (Indonesia)

Status Common

Similar species Centralian blue-tongued skink, *Tiliqua multifasciata*; western blue-tongued skink, *T. occipitalis*; blotched blue-tongued skink, *T. nigrolutea*; pygmy blue-tongued skink, *T. adelaidensis*; New Guinea blue-tongued skink, *T. gigas*

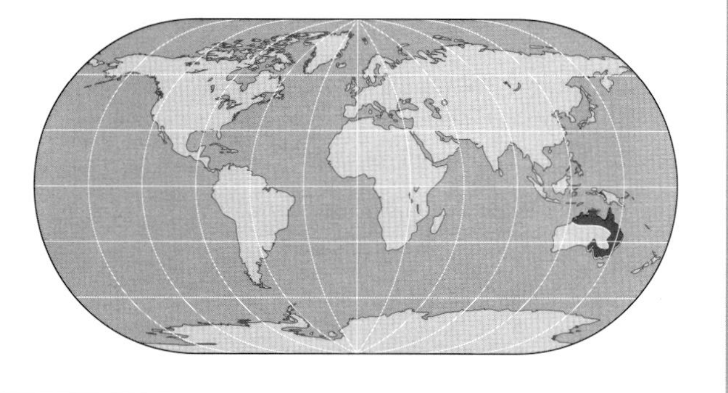

THERE ARE FOUR SUBSPECIES of *Tiliqua scincoides*. The eastern (or common) blue-tongued skink, *T. s. scincoides*, is the nominate form. It usually has a gray to tan background coloration with darker bands extending onto its flanks. Its forelimbs are plain. A dark temporal streak runs from the eye to the top of the ear opening. The northern blue-tongued skink, *T. s. intermedia*, is the largest member of the genus at 24 inches (60 cm) long. It is paler than the nominate form, the dark bands on its flanks are more orange, and its grayish-tan head lacks the temporal streak. The Tanimbar blue-tongued skink, *T. s. chimaerea*, is banded with golden brown and silver on the flanks. Its back is silver banded with gray. A fourth subspecies has yet to be officially classified but is referred to as the Irian Jaya form. Its head colors are similar to those of the northern blue-tongued skink, while the bold bands on its back are similar to those of the eastern form. It has a longer tail, and its forelimbs are dark brown with cream flecks.

When threatened, the blue-tongued skink puffs up its body, sticks out its long blue tongue, and hisses. The bright tongue contrasts with the pink mouth lining and acts as a warning to predators.

Terrestrial Habitats

Blue-tongued skinks inhabit Australia, various Indonesian islands, and New Guinea. Although location and habitat vary depending on subspecies, they are all diurnal and have adapted to a terrestrial lifestyle. They often shelter in animal burrows, hollow logs, leaf litter, or rock crevices. The range of the eastern blue-tongued skink is southeastern Australia up along the east coast to the tip of Cape York Peninsula. It includes semiscrub, woodland, and coastal habitats. The northern blue-tongued skink inhabits drier open woodlands and

subhumid forests. It ranges from Western Australia along the northern coast through Northern Territory and into Queensland. The smaller-sized subspecies *T. s. chimaerea* inhabits subhumid and tropical forests on the Tanimbar Islands off the northern coast of Australia. The Irian Jaya blue-tongued skink is found in drier tropical regions of southern coastal New Guinea where climate and vegetation are similar to adjacent northern Australia.

Activity periods for blue-tongued skinks vary with habitat and temperature. Individuals in northern parts tend to be active mostly in the early morning and later afternoon, seeking burrows in which to spend the hottest part of the day. Blue-tongued skinks are ectotherms. They bask to raise their body temperature to 95° F (35°C). They are unusual among skinks in that they pant to cool down. During this process they open their mouth when inhaling and close it when exhaling (unlike other

Suburban Skinks

More than any other blue-tongued skink, the eastern (or common) form, *T. s. scincoides*, thrives in urban areas and reaches the suburbs of most cities within its range. It has adapted to large gardens with plenty of shelter, making its home in rockeries, pipes, and cavities under houses. Lawns and paths make useful basking sites. Although the skink has become used to human activity, life in suburbia is not without its hazards. A number of adults, especially males, tend to be killed by dogs and motor vehicles in spring (the mating season) when they move around more frequently. Later in the year most deaths occur among young that are preyed on by cats. In addition, chemicals used to control snail and slug populations in gardens have killed numbers of blue-tongued skinks.

Research has shown that the skinks use corridors of dense vegetation to move between retreat sites. Females, especially gravid individuals, are highly sedentary and often use the same site for 10 or more years. Their habit of returning to the same site, combined with adequate food supplies in the form of snails, slugs, caterpillars, fruits, and plants, and a long life span mean that populations of adult blue-tongued skinks in suburbia are stable.

panting lizards, which hold the mouth open all the time). During the dry season when food and water are limited, they conserve energy levels by staying in their burrows and allowing their body temperature to fall, which lowers their metabolic rate.

Body Features

The blue-tongued skink is among the largest in the skink family. Its body is long and flattened with small legs, each with short-clawed digits. The tail is about 50 to 60 percent of the head and body length, tapering sharply at the tip. When moving in thick vegetation, the skink folds its hind limbs back along the tail and uses lateral undulation. The limbs and claws are not strong enough for the skink to dig its own burrow, so it uses those of mammals. The skink's head is quite large. It has prominent ear openings with small scales in front. It has strong jaws and a wide gape. The teeth at the front of the mouth are smaller than those farther along the jaw and are used to pick off insects and to bite plant material. The tongue is, of course, blue. It is also large and broad.

Blue-tongued skinks are opportunistic feeders. They rely on prey that is slower than themselves. Insects, fruit, flowers, foliage, fungi, snails, eggs, small vertebrates, and carrion are all eaten. When feeding on insects and snails, the skink captures its prey and crushes it in its jaws before swallowing. Consumption of other foods involves the use of the tongue and jaws.

Startling Blue Tongue

In the wild the main predators of the blue-tongued skink are large birds such as brown falcons and laughing kookaburras. Dingoes, monitor lizards, eastern brown snakes, red-bellied black snakes, and the Mulga snake also attack it. In suburban gardens dogs and cats are additional hazards. Unable to rely on speed to escape, the blue-tongued skink inflates its body, hisses loudly, and opens its mouth. At the same time, it thrusts out its broad, blue tongue that contrasts with the pinkish-red lining of its mouth, giving the impression that it is deadly. If necessary, it will deliver a bite using its strong jaws. Because of this defensive behavior blue-tongued skinks were once thought to be venomous, and many were killed by humans.

↑ ***The Centralian blue-tongued skink,* Tiliqua multifasciata, *is slightly smaller than* T. scincoides. *Its coloration enables it to remain concealed against the arid red sand of its native habitat in the Australian interior.***

Breeding Behavior

Male blue-tongued skinks have a longer, more slender body than the females. The difference is apparent in newborns and adults but harder to spot in partly grown specimens. The skinks tend

to be solitary for the majority of the year. In most parts of their range they either hibernate or experience slightly cooler conditions for six to 12 weeks depending on temperatures. After emerging in spring (September to October), males become more active. They increase the size of their home range as they look for females. As in other species of skink, male-to-male encounters can result in combat and serious injury.

When meeting a receptive female, the male grips her neck or shoulder in his powerful jaws, aligns his body with hers, and scratches her back with a hind leg. With these movements she becomes passive and raises her tail to allow mating to take place. Mating lasts for two to four minutes. If the female is unreceptive, she may try to escape, causing serious damage to her neck or shoulder from the male's jaws. After mating, males often rub their cloaca along the ground—this behavior may be a form of scent marking. Observers have reported males guarding mated females for periods of up to 25 days after copulation.

Mated females spend more time basking, which increases their chances of predation. As the time of birth approaches, the size and number of young cause the female to rest with her hind legs and tail raised. The young are born approximately 110 days after mating. While developing, they have been nourished by a placentalike organ. During birth they rupture the embryonic membrane and consume the yolk sac and membrane before dispersing.

There is no parental care. The female eats any undeveloped ova that emerge at the same time. The number of young produced tends to increase with the female's size and age. Female fertility is low—in many parts of their range they may not breed every year, especially if there is a shortage of food. The young skinks reach adulthood at three years old.

Live-Bearing—the Pros and Cons

All species of blue-tongued skinks give birth to live young. This method of reproduction protects the young from some environmental hazards. In temperate areas the female blue-tongued skink is able to thermoregulate by choosing and maintaining positions with higher temperatures, thereby increasing her body temperature and that of the developing young. Eggs buried in the ground would not be able to reach the same temperature for long enough, and development would be halted. Because the young develop inside the female's body, they can feed and accumulate reserves before the onset of winter hibernation.

It is thought that viviparity (live-bearing) has evolved in blue-tongued skinks and other large species because the adults are less vulnerable to predation, which means that the young have a greater chance of survival. Another theory is that they do not depend on speed to escape from enemies or to obtain food, and females are not inconvenienced by the additional weight and bulk of the developing young. However, there are disadvantages. If the female blue-tongued skink is killed while gravid, all the young die too. Also, multiple clutches do not occur—in fact, most female blue-tongued skinks only breed every second year.

The young of* Tiliqua scincoides *are born live rather than hatching from an egg. These two are just five days old. Females reproduce every second year and can give birth to up to 25 young.

Boas and Relatives

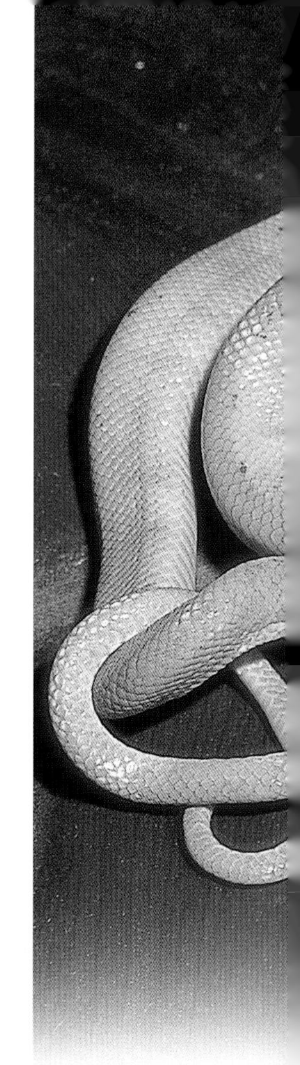

The family Boidae contains some of the best-known, sometimes notorious snakes. Not all boas are huge man-crushing monsters, however, and most are fairly average in size. Even the largest ones are harmless to humans in all but the most exceptional circumstances.

The family includes two of the world's largest snakes: the anaconda, *Eunectes murinus*, and the common boa, *Boa constrictor*. The anaconda is a massive, heavy-bodied snake that leads a semiaquatic lifestyle. Juveniles are fairly active on land; but as they grow, their body mass dictates their preferred habitat of overgrown swamps. In this environment the water supports their great weight, and the mat of vegetation hides them and allows them to ambush their prey. Stories of gigantic anacondas and boa constrictors abound, especially in accounts written by 19th-century explorers in the region. However, all these tales should be taken with more than just a pinch of salt!

Characteristics

Boas have a pelvic girdle and the vestiges of hind limbs in the form of claws or spurs on either side of the cloaca. Males usually have longer, more conspicuous spurs than females. Scales may be small and smooth, as in the boa constrictor, or heavily keeled, as in the rough-scaled sand boa, *Eryx conicus*. Similarly, the head may be covered with many small, granular scales, or larger, platelike scales as in the more advanced snakes. They have a functional left lung (a primitive feature), but their skull is highly flexible (a more advanced feature). The boas, therefore, together with the pythons and some smaller families, represent a link between the primitive and the advanced snakes. Most boas give birth to live young.

Common name Boas **Order** Squamata

Family Boidae 2 subfamilies
Subfamily Boinae—5 or 7 genera, about 27 species of typical boas, including the common boa, *Boa constrictor*; Dumeril's boa, *Boa (Acrantophis) dumerili*; Pacific ground boa, *Candoia carinata*; Emerald tree boa, *Corallus caninus*; rainbow boa, *Epicrates cenchria*; anaconda, *Eunectes murinus;* Madagascar tree boa, *Boa (Sanzinia) mandrita*
Subfamily Erycinae—2 or 3 genera, 14 species of burrowing boas, including the rubber boa, *Charina bottae*; rosy boa, *Charina trivirgata*; African sand boa, *Eryx colubrinus;* Calabar ground boa, *Charina reinhardtii*

Family Bolyeriidae—2 genera, 2 species of Mascarene boas, including the Round Island keel-scaled boa, *Casarea dussumieri;* Round Island boa, *Bolyeria multocarinata*
Family Tropidophiidae—tentatively divided into 3 subfamilies, with 5 genera, 25 species of dwarf boas or wood snakes, including the Cuban wood snake, *Tropidophis melanurus*; banana boas, *Ungaliophis, Exiliboa, Trachyboa, Xenophidion*

Death by Constriction

All boas are powerful constrictors, killing their prey by coiling around it and steadily increasing the pressure every time the victim breathes out. Eventually, they prevent it from breathing altogether and slacken their coils so that they can search for the head, using their tongue. Having found the head, swallowing begins. Depending on the size of the prey, it may take anything from a couple of minutes to an hour or more to swallow it. Large prey creates an equally large bulge in the snake's stomach, presenting it with a problem if it needs to move quickly. Recently fed boas are, therefore, vulnerable to attack from predators, and—as a last resort—they will regurgitate their prey in order to make good their escape.

The family is divided into two subfamilies: the Boinae, or "true" boas, and the Erycinae, which could be called the "burrowing" boas, but they have no universally accepted common name. Many members of the Boinae have heat-sensitive pits around their mouths, but erycine boas do not—with their burrowing lifestyles the pits would quickly become clogged with sand or soil.

⊖ *Its long, slender body and prehensile tail make the Amazon tree boa,* **Corallus hortulanus**, *perfectly designed for life in the trees.*

The Boinae

The Boinae includes the boa constrictor and the anaconda, both of which are well-known South American species. The anaconda—sometimes called the green anaconda—has a smaller relative, the yellow anaconda, *Eunectes notaeus*, which grows to about 6.5 feet (2m) in length as opposed to a potential 18 to 32 feet (5–10 m) in the case of the green anaconda. A third species, De Schauensee's anaconda, *E. deschauenseei*, which is similar to the yellow anaconda, is sometimes recognized.

Other members of the Boinae include six (or seven) species of tree boas, *Corallus*. Compared with other boas, they have slender bodies and are often compressed from side to side. This girderlike shape increases their rigidity when they have to bridge the gaps between branches. All *Corallus* species have long, curved teeth for catching and holding onto birds, which are their main prey. They also have a characteristic way of coiling over horizontal branches when resting. Tree boas have large and very obvious heat pits, and appear to hunt almost entirely by using them. (Captive tree boas have been observed having great difficulty locating dead prey unless it is first warmed up.) There has been confusion recently over the naming of the forms of the Amazon tree boa in the genus *Corallus*, and some authorities only recognize four species. Cropan's boa, *C. cropanii* from southeastern Brazil, is very rare and known only from three specimens.

The genus *Epicrates* contains one species, the rainbow boa, *E. cenchria,* which is widespread throughout South America, where it occurs in a number of different geographical races or subspecies, and there are nine other species scattered throughout the islands of the West Indies. They vary greatly in shape and, especially, in size. The largest is the Cuban boa, *E. angulifer*, which routinely grows to 9 feet (2.7 m) but can occasionally grow to over 13 feet (4 m) long and is heavy bodied. Some of the smaller species only reach less than 39 inches (1 m).

The well-named Haitian vine boa, *E. gracilis*, is extremely slender and arboreal. Many of these island dwellers are under threat because their habitat has been largely destroyed by introduced goats or cleared for agriculture. Introduced predators such as mongooses and cats have killed large numbers of vine boas. The Mona Island boa, *E. moensis*, is among the most threatened because it only occurs on the small island of Mona off Puerto Rico, where feral cats have reached epidemic proportions. Although the larger species eat mammals, including rats, bats, and small domestic animals, the smaller ones specialize in eating *Anolis* lizards, of which there are many living on the West Indian islands. The lizards often sleep on the tips of branches at night, and the agile boas catch them by hanging down or reaching out and plucking them from their roosts.

Three species of Madagascan boas used to be placed in two separate genera—*Sanzinia* with one species and *Acrantophis* with two. More recently they have been moved into the genus *Boa* along with the common boa because they share many characteristics—despite the great distance between their ranges. One of the species, the Madagascar tree boa, *B. mandrita*, is arboreal and lives in a wide range of forested habitats throughout the island. It is olive-brown or green in color with irregular markings of lighter and darker shades. The other two species, Dumeril's boa, *B. dumerili*, and the Madagascan ground boa, *B. madagascariensis*, live on the ground, where they are well camouflaged against the leaf litter and other forest debris. All species grow to about 6.5 feet (2 m) and feed on birds and mammals.

The common boa, although well known to the general public, has been poorly studied in the wild. It has a huge range from Mexico to Argentina, and a number of forms or subspecies are recognized. Its habits vary from place to place: For example, it may be highly arboreal in forest habitats but largely terrestrial in the semidesert regions of western Mexico.

Three species of Pacific boas make up the genus *Candoia*. These small- to medium-sized snakes live on the large island of New Guinea and on several smaller island groups in the same region, so they are isolated geographically from other members of the family. The smallest species is called the viper boa, *C. aspera*, and is stout bodied and very much a ground-dwelling species. The Pacific ground boa, *C. carinata*, is slightly more elongated and occurs in a range of different habitats, while the Pacific tree boa, *C. bibroni*, is very elongated and lives in trees. By having such different lifestyles, they probably avoid direct competition with each other in places where more than one species occur.

Some tree boas are brightly colored, but Cooke's tree boa,* Corallus enhydris cookii*, seen here in Trinidad, is more subtle.

The Erycinae

With 14 species the subfamily Erycinae is smaller than the Boinae. All its members burrow to some degree, but between them they inhabit a wide range of habitats. They are short, stout snakes with blunt snouts and short tails. The tails of some species end abruptly, and they may sometimes use them as "false heads," waving them above their coils in the hope of deflecting an attack while the real head remains protected in the coils. While a number of the true boas have heat pits, they are absent in the erycine boas.

Two species are North American: The rubber boa, *Charina bottae*, lives in montane habitats often associated with conifer forests, and the rosy boa, *C. trivirgata*, lives in rocky, desert, and semidesert areas. The latter grows to about 39 inches (1 m), but the rubber boa is the smallest family member, typically about 12 inches (30 cm) long. The Calabar ground boa, *C. reinhardtii*, is a West African relative sometimes placed in the genus *Calabaria*. It is unusual among boas in laying eggs. It is a forest species that spends most of its time buried in the ground.

The 11 species of sand boas make up the genus *Eryx*, although some authorities place some of them in the genus *Gongylophis*. They are short, stout snakes with flattened heads, blunt snouts, and cylindrical bodies. As in other burrowing snakes, the eyes are positioned toward the top of the head so that they can ambush prey while remaining partially buried. The upper jaw overhangs the lower one so that the upper snout forces a passage through the soil or sand without the risk of it entering the mouth. Species that live in sandy habitats "swim"

(↑) ***Not all boas are giants: The Panamanian dwarf boa,*** **Ungaliophis panamensis*****, reaches just 30 inches (76 cm) long as an adult. Pale brown in color with grayish and rust tones, it is a rare and interesting snake.***

through the substrate, while those that live in more compacted soil form tunnels, often coming to the surface beneath a stone slab. They are found from southeastern Europe and North and East Africa through the Middle East and the Arabian Peninsula to India and Sri Lanka. All eat small mammals, and some of the smaller species eat lizards, especially when they are young. All species give birth to live young, with the possible exception of the Arabian sand boa, *E. jayakari*, which reportedly lays eggs.

Boa Relatives

Two other small families are closely associated with the boas, and their members are usually referred to as "boas," even if this is not strictly correct. They are the Bolyeriidae, or Mascarene boas, and the Tropidophiidae, or dwarf boas (sometimes called wood snakes). There are two Mascarene boas, both from Round Island in the Indian Ocean, although one of them, the Round Island boa, *Bolyeria multocarinata*, is probably extinct. It was last seen in 1975 and has not been found since—a victim of habitat destruction. The surviving species is the Round Island keel-scaled boa, *Casarea dussumieri*.

The 21 species of dwarf boas are mainly West Indian, being especially well represented on Cuba, although there are three species in Central America and another three in South America. Their left lung is greatly reduced in size, and females of some species lack vestigial hind limbs, although they are always present in males. They are secretive nocturnal snakes, usually found under stones, logs, and forest debris. Some species are extremely rare. The two members of the genus *Ungaliophis* are known as "banana boas" because they are occasionally transported around the world in shipments of bananas from Central America. All species give birth to live young.

Boomslang *Dispholidus typus*

Boomslang is an Afrikaans word meaning "tree snake." In fact, the boomslang is a snake of bushes and shrubs rather than tall trees. It moves gracefully through branches using its large eyes to seek out prey.

Common name Boomslang

Scientific name *Dispholidus typus*

Subfamily Colubrinae

Family Colubridae

Suborder Serpentes

Order Squamata

Length 5 ft (1.5 m) to 6.5 ft (2 m)

Key features Elongated body with a deep head and narrow neck; snout is pointed, and upper profile is steeply angled; eyes very large with round pupils; eyes of juveniles are bright emerald green in color; bodies of juveniles gray-brown, adults variously colored

Habits Arboreal

Breeding Egg layer with clutches of 10–14 eggs (occasionally more) laid in hollow tree trunks or leaf litter; eggs hatch after 60–90 days

Diet Lizards (especially chameleons) and birds; occasionally small mammals and frogs

Habitat Grassland with scattered trees and woods; open woodland

Distribution Africa south of the Sahara, missing only from montane grasslands, deserts, and closed forests

Status Common

Similar species Two other genera of African tree snakes, *Thrasops* and *Rhamnophis*, are similar in size and coloration, and also have large eyes; none of them is dangerous to humans, however, although they have enlarged rear fangs and may bite

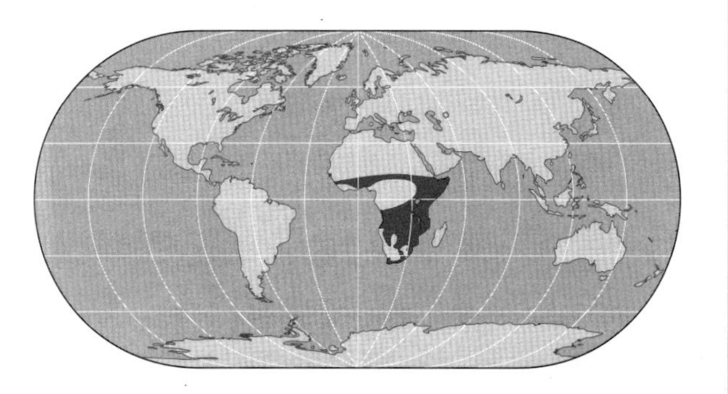

THE DISTRIBUTION OF BOOMSLANGS is determined to a large extent by the presence of their favorite prey—chameleons. Along the south coast of South Africa and in the Cape region the dwarf chameleons in the genus *Bradypodion* predominate, but farther north the flap-necked chameleon, *Chamaeleo dilepis*, is the dominant species. Boomslangs also take birds, mainly young nestlings.

Boomslangs rarely come down to the ground except to move from one group of bushes to another or (in the case of the female) to lay eggs, usually in a rotting tree stump. Boomslangs are active during the day, in a constant search for the chameleons, lizards, and small birds on which they feed. Their eyes are huge, perhaps larger relative to their body size than those of any other snake. Their vision is excellent and helps in the search for food. The eyes of juveniles are especially startling—a bright emerald green color.

Fast-Acting Venom

While foraging, boomslangs move in a controlled fashion, hardly disturbing the branches as they glide effortlessly along. When potential prey is spotted, they stop suddenly, raise their head slightly, and gather themselves for a strike by making a series of bends along the length of the body. They strike by straightening out the bends. Immediately, they get a firm grip on the prey to reduce the chances of it falling from the branches or running away. They push it to the back of their mouth and begin to chew. As they chew, venom is forced into the wound made by their long, grooved rear fangs. The venom is fast

Closeup of the head of a juvenile boomslang. The eyes in juveniles are bright emerald green in color, and the snout in both juveniles and adults is steeply angled.

acting; and as soon as the prey stops struggling, the snake rearranges it until it is facing down its throat and begins swallowing.

The boomslang is one of only a very few colubrid snakes that can kill humans. It delivers small quantities (about 1 mg) of its potent venom via the grooved rear fangs. Bites are rare in the normal course of events, however, because boomslangs invariably prefer to avoid confrontation by hiding or slipping rapidly away. If cornered, they will try to intimidate rather than bite by inflating their throat to about twice its normal diameter. Since the skin between the scales is often brightly colored, this can result in an impressive display.

As a last resort, they may strike. Because their gape is so wide, the enlarged rear fangs may puncture the victim's skin even in a quick stab, but chewing usually needs to occur before a significant amount of venom is injected. Human casualties therefore usually happen when people try to catch or handle the snakes.

Gripping firmly on its victim's body, a boomslang prepares to enjoy its favorite meal of a chameleon. Boomslangs also prey on lizards and frogs.

The venom contains a very potent hemotoxin that prevents blood from clotting. There is often a delay in the symptoms, which may not develop for 24 to 48 hours after the bite.

Color Variations

Boomslang color variation is complicated: Juveniles are colored differently than adults (this is known as ontogenetic color change), males and females are also colored differently (known as sexual dimorphism), and males occur in several different color forms (polymorphism). None of these phenomena is unique to boomslangs, but it is unusual for all to occur in the same species.

Boomslangs are among just a few snakes in which the males and females are colored differently. As juveniles both sexes look the same, with a gray-brown back, white underside, and yellow throat. By the time they are about 36 inches (90 cm) in length, the colors begin to change. Adult females are olive-brown with a white underside, but males are variable. They can be uniform brown or black, leaf green, powder blue, or even brick red. Their body scales usually have black borders, giving a checkered or sometimes a barred appearance. Some are solid black above and yellow beneath. There is some correlation between colors and geography, but more than one form can occur side by side in places. Many examples have brightly colored skin between their scales, and when threatened, they expose them by inflating their throat.

With its long, slender body, **Dispholidus typus** *is well adapted for an arboreal lifestyle. It spends most of its time among the branches of bushes and trees; and unlike most other snakes, boomslangs even mate in trees.*

Is It Venomous?

In 1957 the herpetologist Karl Schmidt was bitten on the thumb by a boomslang. Unconcerned at first, he died 24 hours later. This was the first documented case of a fatal bite by a member of the Colubridae—a family that until then was thought to consist entirely of nonvenomous snakes or at least snakes whose venom was not dangerous to humans. The following year another fatal bite was recorded, this time from a savanna vine or twig snake, *Thelotornis capensis*. Since then several other "safe" colubrids have been responsible for serious bites, sometimes resulting in fatalities, and the scientific community has had to revise its opinions.

It is hard to draw a line between venomous and nonvenomous snakes. Venom is basically modified saliva containing enzymes of various types that damage living tissue. To be harmful to humans, venom must be potent or fast acting (or both), and the snake must have a way of delivering it in significant quantity. Most colubrids have a gland known as Duvernoy's gland that produces secretions used in swallowing and digestion. Even if the Duvernoy's gland produces venom, most species have no means of storing it in significant quantities nor of injecting it into their prey—ducts from the glands simply empty into their mouth and dribble into the wound.

Venom delivery is made easier in some species by the presence of enlarged fangs toward the back of the mouth. The enlarged fangs sometimes have grooves along their length; if they do, the venom is drawn up by capillary action. Species of this type are rarely considered dangerous to humans because they need to get a firm grip before the enlarged fangs can be used. In addition, it takes time for the venom to find its way into the wound. Even then the venom is probably so mild that the effects are not dangerous.

A few species of colubrids, including the boomslangs, have a more sophisticated arrangement. They have a capsule around the Duvernoy's glands in which venom can accumulate. In addition, muscles at the angle of the jaw that are responsible for closing the mouth can be used to compress the gland and force out the venom. Their grooved fangs are positioned farther forward than in other rear-fanged snakes, so they can be brought into play more easily. The combination of potent venom, a storage system, muscles to force it out, and large fangs makes these species potentially dangerous to humans.

As well as the boomslang in Africa, other species that have these characteristics are the twig snake, *Thelotornis capensis* (also in Africa), the Asian keelback snake, *Rhabdophis tigrinus* (Southeast Asia), certain cat snakes, including Blanding's tree snake, *Boiga blandingi* (Africa), the mangrove snake, *B. dendrophila* (Southeast Asia), and possibly other *Boiga* species.

Kirtland's tree snake, *Thelotornis kirtlandii* from Africa, and the tropical racers, *Alsophis* and *Philodryas*, may also be dangerous to humans, but there are no recorded deaths. There may be others—many rear-fanged snakes can give a painful bite with mild symptoms of envenomation at the site of the bite, especially if they are allowed to chew. It is inadvisable to allow any colubrid snake to chew on your finger or any other part of the body.

↑ ***In Transvaal Snake Park a snake handler shows the large, deeply grooved rear fangs of a boomslang. The venom capsule can be seen in the roof of the mouth below the eye.***

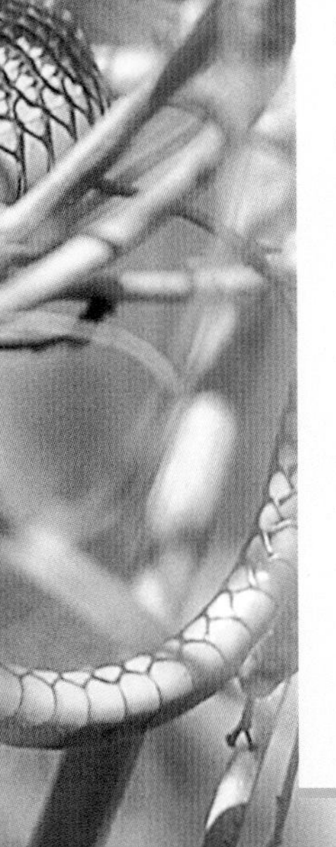

Brown House Snake

Lamprophis fuliginosus

Common name Brown house snake

Scientific name *Lamprophis fuliginosus*

Subfamily Boodontinae

Family Colubridae

Suborder Serpentes

Order Squamata

Length 35 in (90 cm) to 4.9 ft (1.5 m)

Key features Body slender but muscular with smooth scales; narrow head only slightly wider than neck; eyes small with vertical pupils; color varies from pale orange-brown through warm medium brown to dark olive or black; 4 cream lines on the head—1 pair of lines runs from the snout through the eye and stops on the neck, another is lower down, running along the edge of the upper jaw; underside is pearly white and highly iridescent

Habits Nocturnal; terrestrial

Breeding Female lays clutches of 5–16 eggs; eggs hatch after 60–90 days

Diet Mainly rodents, but birds, lizards, and frogs are sometimes eaten

Habitat Open grassland, arid rocky scrubland, cultivated ground, and around human dwellings

Distribution Most of Africa south of the Sahara; also found in a small area of North Africa in western Morocco

Status Common

Similar species Other house snakes can be confused with this species; youngsters are sometimes mistaken for venomous species such as cobras and mambas; but the cream lines on its head are characteristic

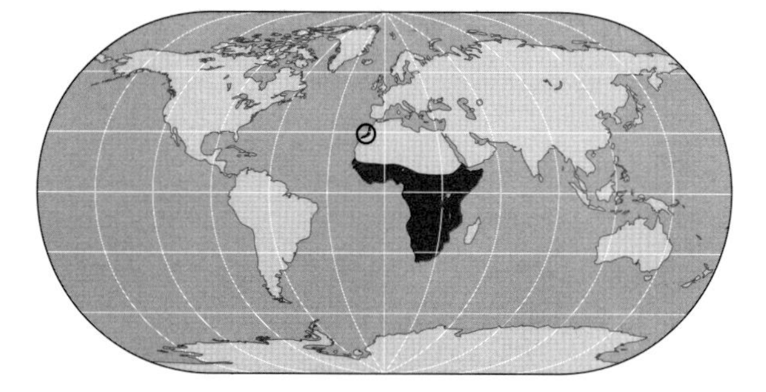

The brown house snake is one of the most common, widespread, and familiar of all African snakes and is found throughout the continent in a variety of situations.

Adaptability is the key to the brown house snake's success. Attracted by the presence of rodents, it is willing and able to make a living around farms and human dwellings. It also lives in the humid savannas of West and southern Africa, the plains and acacia woodlands of East Africa, and the arid landscapes of Namibia and Namaqualand. Throughout Africa it is absent only from dense forests and large expanses of sand desert, such as the Sahara.

Large Appetite

The brown house snake is a generalist. Although it eats mainly rodents up to the size of adult rats, it will also feed on other small mammals such as shrews and bats, as well as lizards and even frogs. Lizards are the most common prey in some of the more arid regions inhabited by the species, because they are the most readily available. Individual house snakes do not appear to have any preferences and will eat most prey species that come their way. Furthermore, their capacity is huge, perhaps even larger relative to their body size than some boas and pythons—house snakes will take on prey that seems far too large for them to subdue and swallow. They are extremely efficient constrictors, throwing a series of coils around their prey and quickly squeezing the breath out of it before swallowing it headfirst.

Because of their diet and their appetite, they are of great benefit to farmers, and some encourage them onto their land. Unfortunately, other communities consider them dangerous due to their long front fangs, and they are often killed needlessly. In fact, house snakes are

The desert form of the brown house snake, **Lamprophis fuliginosus**, *is lighter in color with yellow stripes behind the eyes. Its eyes are larger than those of its counterparts elsewhere. This snake was seen lying coiled in the sun in Namaqualand, South Africa.*

Relic Populations

The brown house snake has a small outlying population in North Africa, in the Souss Valley at the foot of the Anti-Atlas Mountains of southwestern Morocco—many hundreds of miles from its nearest neighbors in West Africa or elsewhere. Populations such as this are called relic (or relict) populations. They occur when species that were far more widespread in former times become fragmented owing to habitat changes. These changes can be caused by rising sea levels, for example, when populations can become stranded on islands that were formerly mountaintops, or, as in the brown house snake, when a desert encroaches.

The Sahara Desert has expanded in relatively recent times and is still expanding today. The brown house snake had a much wider range in former times when the area was more hospitable. Its distribution would have been continuous throughout West Africa and perhaps elsewhere; it may still be present in a few isolated oases in the Sahara. Interestingly, the brown house snake's distribution is paralleled by two other widely distributed African snakes: the common egg-eater, *Dasypeltis scabra*, and the puff adder, *Bitis arietans*. Both these species have a basically sub-Saharan distribution with small isolated populations in the same part of southwestern Morocco.

usually gentle and make good pets, but they may attempt to bite furiously when first captured. The arrangement of their teeth is quite unusual because the first five or six on the upper jaw (or maxilla) are elongated and curved. Behind them there is a small gap before the row of more normally sized teeth. Their long teeth gave them their former scientific name, *Boaedon*, meaning "boa teeth."

Efficient Breeders

As well as being efficient feeders, brown house snakes are efficient breeders. Females grow larger than males, to between 35 and 59 inches (90–150 cm), while males reach just 30 to 35 inches (76–90 cm). They can weigh up to 18 ounces (510 g), but males rarely exceed 9 ounces (255 g). Females are therefore able to lay large clutches of up to 16 eggs. They hatch after 60 to 90 days, depending on temperature.

Although small, most hatchlings are capable of tackling large prey such as nestling rodents. Females can lay several clutches in a single breeding season if they feed well, with clutches appearing every one to two months (at least in captivity). In this way a female can easily lay 30 or more eggs in a year. If she gets a chance to mate between clutches, she will do so; but in the absence of a male she will fertilize subsequent clutches with sperm stored from a previous mating.

The ability to store sperm is an advantage in animals with a limited opportunity to move around in search of mates, and that are sometimes scattered over a wide area. Another African colubrid, the common egg-eater, *Dasypeltis scabra*, can also lay several clutches a year from a single mating, and the ability may be more widespread than we realize.

Color Variations

The color of brown house snakes may vary from place to place and even within populations. This has led to the naming of several species and subspecies over the years, but none of them is widely accepted as separate, because variability within populations can often be greater than

The olive house snake,* Lamprophis inornatus, *is related to the house snake. The slight bulge in its body suggests that it has just swallowed a lizard.

the differences between them. One species, *Lamprophis arabicus* from the southwestern corner of the Arabian Peninsula, is sometimes regarded as a subspecies despite its geographic separation. Another, *L. lineatus*, is a resurrection of an old name for the brown house snake and may be valid. In this form the pale lines running from snout to neck continue on along the snake's flanks.

Desert forms of *L. fuliginosus* tend to be lighter in color and have larger eyes. In Namibia this form also has yellow stripes on its face instead of the usual cream and is sometimes regarded as a separate species, *L. mentalis*. Forms from East Africa may be darker than average, sometimes dark olive or almost black. They tend to be smaller than other forms, and the hatchlings are especially tiny. They possibly feed on young lizards initially, since they are hardly capable of swallowing even newborn rodents despite their huge gape. Other variants are less consistent and include animals with indistinct blotches on their neck and flanks, often more obvious in juveniles than adults.

The House Snake Relatives

House snake classification is extremely complex because of the variability of the brown house snake and the division of some of its forms into separate species, some of which are not widely recognized. There are between 11 and 15 species, some with very restricted ranges in southern Africa. Some of the more distinctive forms are described below.

The Aurora house snake, *Lamprophis aurora* from South Africa, is an especially beautiful species with an olive- or lime-green body and an orange line down the center of its back. Juveniles are more brightly marked than adults. Their preferred prey seems to be nestling mice, but the juveniles probably eat only small lizards, sometimes searching for them in termite nests.

The olive house snake, *Lamprophis inornatus*, is another South African species. It is plain olive-green in color without markings. Its habits are similar to those of the brown house snake, but it prefers damper habitats. It is an efficient hunter of rodents, but it also eats lizards, including the rather spiky girdle-tails, *Cordylus* species.

Fisk's house snake, *Lamprophis fiskii*, is extremely rare but very attractive. Its head and body are dusky yellow with two parallel rows of alternating dark-brown blotches. It comes from dry habitats and probably eats lizards. *Lamprophis fuscus*, the yellow-bellied house snake, also from South Africa, is similarly rare and poorly known.

The spotted house snake, *Lamprophis guttatus,* and the Swazi rock snake, *L. swazicus*, are similar to each other and are both brown. They live in cracks in rocks or under slabs of flaking rock and emerge at night to hunt geckos and other small lizards. *L. olivaceus* is a dark-brown or black species, similar to the brown house snake but with red eyes and without the pale stripes on its head. It occurs in West and Central Africa. Like *L. inornatus,* it also goes by the common name of the olive house snake, so the two species are sometimes confused.

The Seychelles house snake, *Lamprophis geometricus*, is the odd snake out. It lives on the Seychelles Islands in the Indian Ocean. It is another predominantly brown species, but its relationship with mainland *Lamprophis* species is unclear, and there is a strong possibility that it will be reclassified at some time in the future. Other house snakes come from Ethiopia (*L. abyssinicus* and *L. erlangeri*) and Central Africa (*L. virgatus*).

Lamprophis aurora, ***the Aurora house snake, prefers damp habitats such as grasslands, moist savanna, and lowland forest. It has a distinctive orange line down its back.***

Indian python *(Python molurus molurus)*

Common name Burmese python, Indian python (depending on subspecies), Indian rock python

Scientific name *Python molurus*

Family Pythonidae

Suborder Serpentes

Order Squamata

Length Maximum about 20 ft (over 6 m)

Key features Very heavy bodied, especially old adults; markings consist of a pattern of large, irregular, but interlocking blotches on a paler background; blotches dark brown on a lighter brown to tan background (Burmese python) or midbrown on a gray background (Indian python); dark arrowhead mark always present on top of head between the eyes; heat pits in rostral scales and some of the labial scales

Habits Live in dense forests and frequently climb into trees; elsewhere prefer riversides and often enter the water

Breeding Egg layers with large clutches of eggs brooded by the female

Diet Mammals up to the size of deer

Habitat Rain forests, clearings, plantations, and riversides

Distribution India, Sri Lanka (*P. m. molurus*); Myanmar (Burma), Thailand, South China, and Vietnam (*P. m. bivittatus*)

Status Common where suitable habitat remains but rare in places where the land has been extensively cleared; IUCN Endangered (Indian subspecies)

Similar species In places its range overlaps that of the reticulated python, *Python reticulatus*, but the 2 species are not easily confused

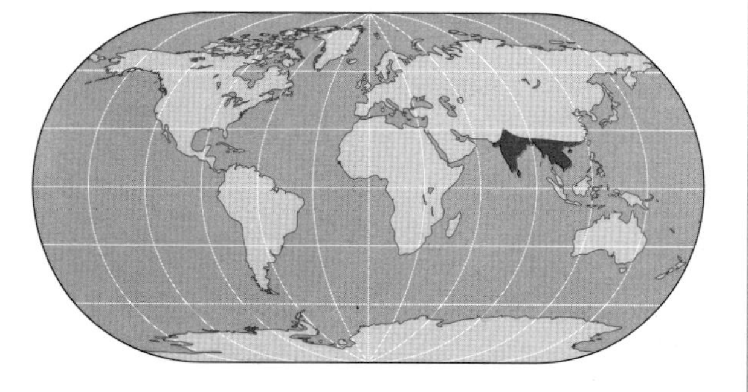

Burmese and Indian Pythons

Python molurus

The Burmese and Indian pythons are among the longest snakes in the world. They are frequent exhibits in zoos because they have a calm temperament and adapt well to captivity.

DESPITE THEIR GIGANTIC SIZE, the Indian (and Burmese) pythons are the favorite of pet keepers, snake charmers, and exotic dancers. Although juveniles can be aggressive, adults are remarkably tolerant of being handled and rarely attempt to bite, which is just as well.

In addition to the normal wild form several color variations have arisen from captive-bred colonies. They are selectively bred to give the python-buying public a wide choice of colors, including an albino form known as the "golden" python. Of the other giant snakes only the boa constrictor has attained such a degree of popularity as this species; the others are all too aggressive, or too rare, or both.

Feeding Habits

As expected, snakes that reach such large body sizes have equally large appetites. Indian pythons apparently lie in wait for their prey, often climbing among the branches of trees or hiding in hollow limbs or trunks. Banyan (fig) trees are often favored as hiding places when their fruit is ripening because many animals, especially deer, visit them to eat the fallen fruit. At other times the snake lives in trees that are used as roosts by egrets and herons.

Although they move around at night and may find prey (including domestic animals) during their travels, they also ambush prey during the day. As with all pythons, the snake seizes its prey in its mouth and immediately throws several coils of its powerful body around it, gradually increasing the pressure until the victim suffocates.

An albino Burmese python—the golden python—alongside a more usual color form. Color variations have been selectively bred to appeal to pet owners.

An Indian python swallows a deer in Mudumalai Wildlife Sanctuary in southern India. Pythons sometimes choke on the deer's antlers.

Burmese and Indian pythons eat a wide variety of prey, including birds and other reptiles, but they seem to favor large mammals. They include adult spotted deer and barking deer. A 15-foot (4.5-m) python that ate a hog deer with antlers 12 inches (30 cm) long choked while it was trying to swallow it.

There are plenty of records of Indian pythons eating monkeys, including one of a specimen that escaped from its cage at a zoo in Liverpool, England, and ate a monkey complete with its collar and chain.

Other prey includes jackals, porcupines, hares, and domestic goats and fowl. One python that was killed and slit open contained three ducks, and another had eaten five ducks, four chickens, and a pigeon. Large pythons have even killed and eaten leopards. Encounters with tigers, on the other hand, usually end badly for the snake.

Digestion

After eating a large meal, an Indian python may be unable to move easily because of the large bulge in its stomach. Snakes that have been forced to move shortly after feeding have damaged themselves by piercing their skin with horns, antlers, or porcupine quills.

Staying quietly in one place is the obvious choice. The snake slowly and carefully hauls itself to a safe retreat and lies in a more or less torpid state until it has digested the meal. If necessary, it arranges itself so that the bulge is positioned in the sun to raise the temperature there and speed up the digestive process. While a python will digest an average meal of a chicken in a week or less, a large meal consisting of a deer or goat may take three weeks or more depending on the temperature and the health of the snake.

Meals are often infrequent, and Indian pythons can survive for long periods without feeding, even when food is offered in captivity. There are several records of Indian pythons remaining in good health for over a year without food and then starting to feed again voluntarily. The python appears to shut down its digestive system and shrink its intestines between large meals. It cannot survive long without water, however.

Just How Big?

Like all large snakes, much controversy surrounds the maximum size of Burmese and Indian pythons. Many reports from the 19th and early 20th centuries of gigantic specimens have been discarded or revised because they were often exaggerated. The British army officer Colonel Frank Wall, writing in 1921, was one of the most reliable observers. He stated that "specimens of 18 feet (5.5 m) are not very uncommon" and recorded pythons of 18.3 feet, 18.7 feet, and several of 19 feet (5.56 m, 5.71 m, and 5.79 m). The Maharajah of Cooch Behar in Assam shot one that measured 19.2 feet (5.84 m).

From these records it is probably safe to say that the Indian python could reach 20 feet (6.1 m), but that specimens of that size would be extremely rare in the wild, especially nowadays. Captive records, however, prove that the snake can grow extremely fast. A batch of hatchlings grew from 20 inches (51 cm) to an average of 6.6 feet (2 m) at 20 months old. The fastest-growing one reached 7.7 feet (2.3 m) in that time. The smallest wild male that was sexually mature measured only 5.7 feet (1.7 m), and the smallest sexually mature female 8.5 feet (2.6 m), indicating that the species can potentially reach breeding size in about 18 months. Experiences of snake breeders confirm that to be the case if the young snakes are heavily fed. Under normal conditions they probably take three to four years to reach breeding size.

Temperature Control

The breeding habits of the Indian python in the wild are virtually unknown. In fact, none of its habits in the wild are well known. In captivity it breeds readily, and many interesting observations have been made.

Most notably, it was the first python shown to coil around its eggs and the first one in which internal temperature control (metabolic endothermy) was demonstrated. As long ago as 1841 the curator of the Jardin des Plantes in Paris, France, kept records of a female Indian python that laid a clutch of 15 eggs. She coiled around her clutch and was covered with a cloth. The curator made notes recording the temperature of the air in the cage, the air under the cloth, and the temperature between the python's coils. The temperature between the coils turned out to be well above that of the rest of the cage—it hovered around 90°F (32°C), while the cage was below 70°F (21°C).

More accurate records were taken later at London Zoo in England in 1881 and at Basel Zoo in Switzerland in 1946. In both cases the temperature between the female's coils was significantly higher than that of the cage. The observers also noticed that the snake's body went into spasms while she was incubating her eggs, and that the frequency with which she quivered depended on the temperature. When the cage temperature approached 90°F (32°C), the quivering stopped; but when the temperature fell to 79°F (26°C), the quivering increased to about 30 spasms per minute.

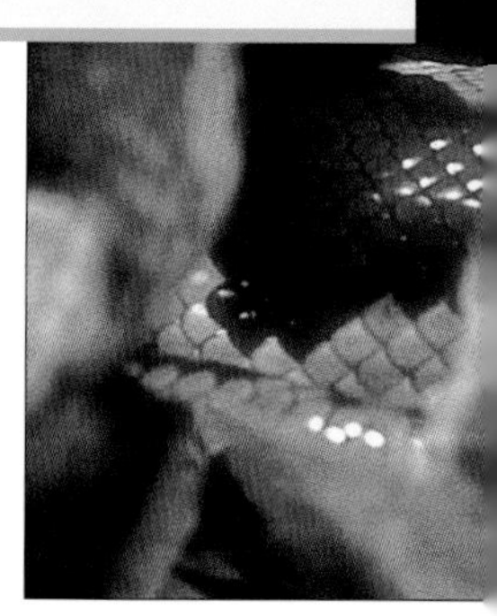

Later experiments showed that the snake was able to produce metabolic heat from her muscles by twitching. This facultative endothermy, to give it its scientific name, is a rare example of a "warm-blooded" reptile. Since the Indian python was shown to be capable of it, several other species of pythons

(notably the carpet or diamond python, *Morelia spilota*) have also been shown to brood their eggs in the same way. It may be that all female pythons can produce heat, but it is more likely that only the larger species can.

Large snakes, especially when in a coiled posture, reduce the ratio of their surface area compared to their volume and therefore slow down the rate at which heat is lost. (Small snakes would lose heat from their bodies more quickly than they could produce it.) The process of heat production in these snakes is known as "facultative" because they can turn it on or off as required. Pythons living in hot places can be be expected to produce heat only very rarely, while those that live in cooler places would find it essential.

The study of endothermy in pythons is of great interest because it may point the way toward the evolution of warm-blooded animals (birds and mammals). It also means that pythons may be able to move into colder regions where their eggs would not otherwise hatch successfully.

Indian python hatchlings. The mother will coil around her clutch of eggs to incubate them.

Common name Bushmasters (matabueys)

Scientific name *Lachesis* sp.

Subfamily Crotalinae

Family Viperidae

Suborder Serpentes

Order Squamata

Number of species 3

Length Up to 10 ft (3 m), exceptionally to over 11 ft (3. 3 m)

Key features Massive vipers; body with heavily keeled scales giving it a knobby appearance; backbone forms a prominent ridge; head large and rounded; distinct facial pits between nostrils and eyes; color usually yellow, tan, or pinkish with sooty black, roughly diamond-shaped marking down the back; some variation in marking between the 3 species

Habits Terrestrial; nocturnal

Breeding Egg layers, with clutches of 5–18 eggs; eggs hatch after about 60 days

Diet Mammals, especially spiny rats when available

Habitat Rain forests

Distribution Central and South America, with gaps between populations

Status Common to rare; population on the southeastern Brazilian coast, *Lachesis muta rhombeata,* is endangered

Similar species The 3 species are similar to each other but unlike any other snakes

Venom Extremely toxic; 75 percent of bite victims die even with medical attention

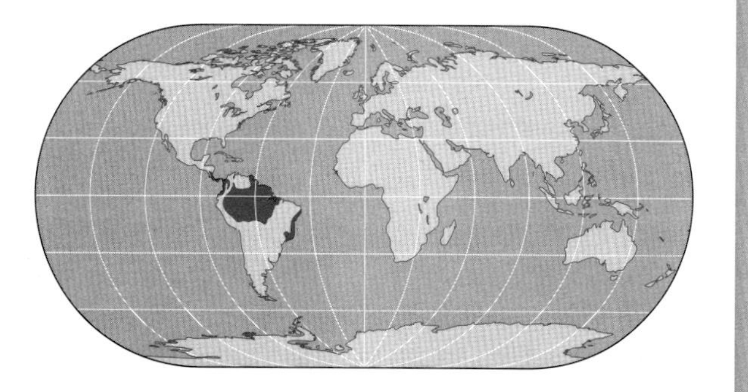

Bushmasters *Lachesis* sp.

Bushmasters are awesome snakes. They are the largest venomous snakes in the Americas and the longest vipers to be found anywhere.

Bushmasters live in undisturbed rain forests and quickly disappear from areas that have been developed for agriculture. However, individuals may linger on for some time, presenting a danger to workers. Many such bushmasters meet their end at the hands of machete-wielding farmers. There are three species of bushmasters, *Lachesis melanocephala, L. muta, and L. stenophrys.*

Collectively, bushmasters occur from eastern Nicaragua through Central America to Bolivia and southeastern Brazil. However, there is a large break in the distribution between the Amazon Basin population, *L. muta muta,* and the populations from the Brazilian coastal forests, *L. m. rhombeata.* This particular bushmaster is very rare, since forests along the Brazilian coast are being cleared to provide agricultural land for a rapidly expanding human population. Several other rare snakes are found in the same region.

The Three Species

The black-headed bushmaster, *Lachesis melanocephala*, lives in southwestern Costa Rica in a fairly small region with a high level of rainfall. The top of its head is black, and it is more aggressive than the other two species. Differences between the other two species are based largely on scale counts. Both have calmer dispositions, but individuals can be quick to strike and put on a dramatic show that includes inflating the neck, hissing, and vibrating the tail. The latter gives bushmasters one of their many Latin American names of *cascabela muda*, meaning "mute rattlesnake." In some places the snake is called *matabuey* ("bullock killer"), a reference to the potency of its venom.

Bushmasters are lethargic snakes, moving rarely and feeding infrequently. Their hunting method is to find a place where rodents are likely to visit, such as the forest floor beneath a fruiting tree. They probably do this by identifying their prey's scent trails.

Sit and Wait

Once they have found the right place, they take up a strategic position and form a coil, with their head in the center and their neck bent into an "s"-shape. Then they simply wait for a meal to come within striking range. Observations of the snakes in the wild have shown that this occurs very rarely, and a bushmaster may have to stake out a patch of forest floor for several days or even weeks before it is successful. If it has caught nothing after a week or two, it will move 5 to 20 yards (4–18 m) away to try its luck at a new ambush spot. Its meals tend to be large ones, however, weighing 50 to 70 percent of the bushmaster's own weight.

Once it has fed, it moves to a hidden place in which to digest its meal. Because it uses so little energy, a bushmaster may be able to survive on six to ten meals each year. Breeding females, however, need to eat more often than males because their bodies divert a sizable proportion of their food to produce eggs.

Reproduction

The three bushmaster species are unique in being the only American pit vipers to lay eggs (although several Old World species do so). They lay between five and 18 eggs that hatch after about 60 days. The newly hatched bushmasters feed on mammals right from the beginning and are among the few venomous snakes that eat only mammals throughout their lives—the other three being the taipans and the black mamba.

Lachesis stenophrys ***blends in perfectly among leaf litter on a rain-forest floor in Costa Rica. If disturbed, it will strike or vibrate its tail in a warning display.***

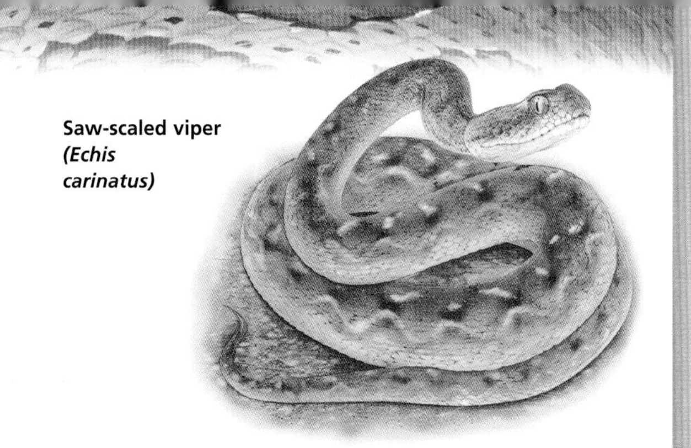

Saw-scaled viper *(Echis carinatus)*

Carpet Vipers

Echis sp.

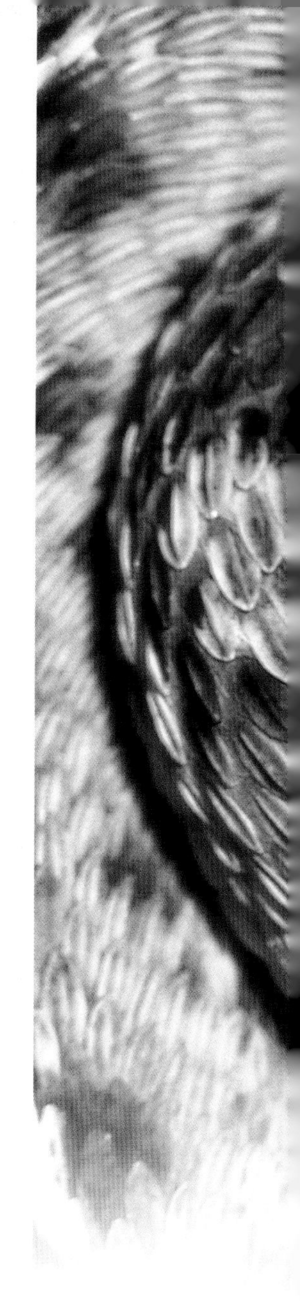

Common name Carpet vipers (saw-scaled vipers)

Scientific name *Echis* sp.

Subfamily Viperinae

Family Viperidae

Suborder Serpentes

Order Squamata

Number of species About 8

Length From 20 in (50 cm) to 36 in (91 cm)

Key features Head rounded or pear shaped, covered with many small, keeled scales; eyes large with vertical pupils; eyes visible from above; intricate patterns and serrated keels present on several rows of scales along the flanks; color usually sandy brown, yellow, or gray to match the soil on which they live

Habits Mainly nocturnal; terrestrial, although occasionally climb into low bushes

Breeding Mostly egg layers, but at least 1 African population bears live young

Diet Highly varied from invertebrates to small mammals and birds

Habitat Semideserts, often where there are rocks; absent from the most arid sandy deserts

Distribution Large and disjointed from West and Central Africa through North Africa and the Middle East to Pakistan, India, and Sri Lanka

Status Common

Similar species All carpet vipers resemble each other, making identification difficult at times

Venom Potent, preventing blood from clotting and causing extensive tissue damage; death can follow after several days without treatment

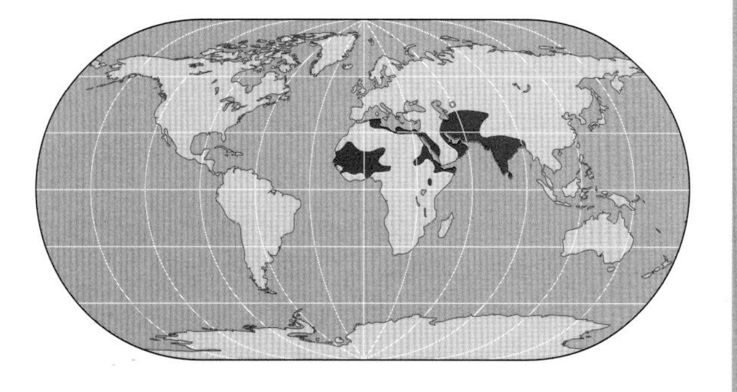

Despite their relatively unthreatening appearance, carpet vipers are the most dangerous snakes to humans over much of Africa, the Middle East, and South Asia. They kill many hundreds—possibly thousands—of people every year.

CARPET, OR SAW-SCALED, VIPERS often occur in extremely high densities in places where rural people cultivate the land. Their preferred habitat, coupled with their excellent camouflage, short temper, and destructive venom, makes them good candidates for the world's most dangerous snakes.

Carpet vipers were once thought to include just two variable species: Burton's or painted carpet viper, *Echis coloratus*, and the saw-scaled viper, *E. carinatus*. Research carried out on their venoms in 1990 showed that populations of the saw-scaled viper were different from each other; and although *E. coloratus* was unchanged, *E. carinatus* was divided into 11 separate species. Since then some have been suppressed or relegated to subspecies, and at present eight species are generally recognized. No doubt this arrangement will also be changed, which highlights the difficulties herpetologists face when trying to investigate the relationships between snakes.

Classification Conundrum

In this case the "species" involved are so similar to look at that they can be almost impossible to tell apart unless their locality is known (and then only if they come from a place where there are no overlapping species). Museum specimens that were found before the reclassification cannot be assigned to one

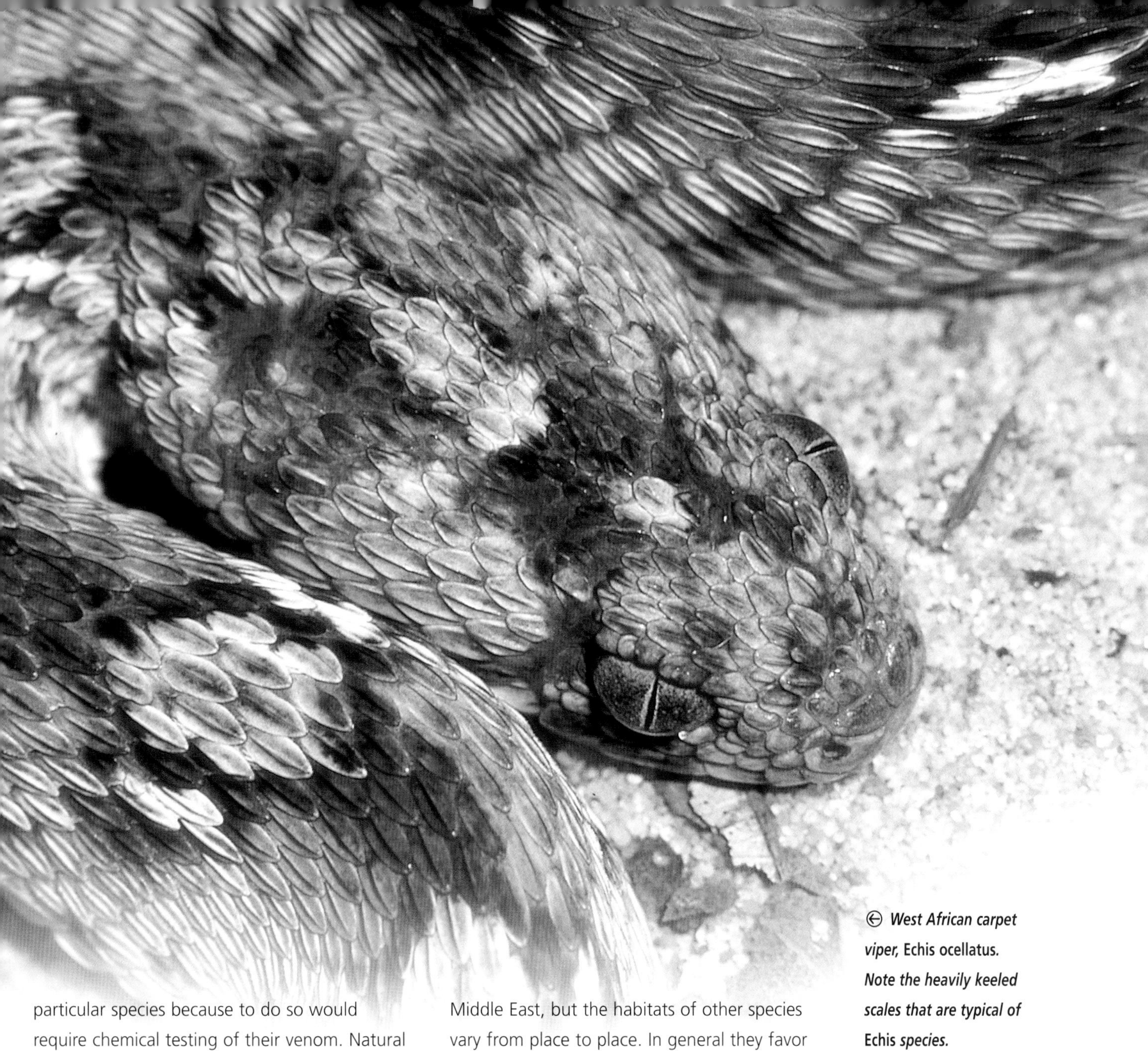

← *West African carpet viper,* Echis ocellatus. *Note the heavily keeled scales that are typical of* Echis *species.*

particular species because to do so would require chemical testing of their venom. Natural history notes made at the time when all the species were classed together are now invalid in some cases because it is impossible to know which species was involved.

Lifestyle and Habits

Despite the difficulties in assigning the snakes to particular species, all carpet, or saw-scaled, vipers have several characteristics in common. For the hottest parts of the year they are nocturnal, hiding by day under rocks, logs, beneath dense shrubs, or in vacant burrows. At other times they are active in the day and can tolerate more heat than most other snakes.

Burton's carpet viper, *Echis coloratus*, is a desert species living in the rocky deserts of the Middle East, but the habitats of other species vary from place to place. In general they favor dry, rocky, scrubby semidesert or arid grassland. They do not thrive in forests or in the most extreme sand deserts. They are absent throughout the Sahara but live around its edges, however, in the Sahel to the south and the coastal plains and plateaux to the north.

Northern populations around the Mediterranean coast are relics from a former time when the whole region was less dry, and there are isolated populations in oases and mountain ranges surrounded by desert.

Across the deserts of the Middle East, the Arabian Peninsula, Pakistan, India, and Sri Lanka saw-scaled vipers are unable to make a living where there is only sand. They are restricted to dry riverbeds (wadis), mountain foothills where rain falls, and other less arid environments. It is this preference that often brings them into contact with people who farm the lands at the edge of deserts, along with oases and flood plains. Carpet vipers have an unfortunate habit of lurking on pathways at dusk.

Although mainly terrestrial, carpet vipers sometimes climb into low bushes to escape excessively hot surfaces. They have several methods of locomotion depending on the ground. They are adept at sidewinding, which is their usual method of traveling over hot surfaces or small patches of shifting sands. They sidewind if they are disturbed and need to move quickly, but they also move by serpentine

Classifying the Carpet Vipers

At present there are several classification schemes, and anything from five to 20 species of carpet viper are recognized. There are four well-established African species and a single Asian species. Three other African species with small ranges are more of a problem, since it is difficult to ascertain whether they are full species or subspecies.

Asian species:

E. carinatus, saw-scaled viper. Widespread but previously more so. Occurs from the Middle East to India and Sri Lanka.

Established African species:

E. coloratus, Burton's carpet viper, or painted carpet viper. A distinct species and always recognized as such. Occurs in Egypt, the Arabian Peninsula, Israel, and Jordan.

E. leucogaster, white-bellied carpet viper, including the subspecies *E. l. arenicola*, which may be a separate species. North and West Africa.

E. ocellatus, Central and West Africa.

E. pyramidum, northeast African carpet viper. Occurs through East Africa (Kenya) to northeastern Africa, Sudan, and the Arabian Peninsula.

Problematical African species:

E. hughesi. Perhaps a subspecies of *E. pyramidum*. Occurs in Somalia.

E. jogeri. Perhaps a subspecies of *E. leucogaster*. Found in western and central Mali.

E. megalocephalus. Perhaps a subspecies of *E. pyramidum*. Found in Eritrea (northeastern Africa) on an offshore island.

crawling. At times they edge forward slowly using straight-line locomotion.

Defense

Carpet vipers use a form of defense that is shared by a few other vipers, such as the Sahara horned viper, *Cerastes cerastes*, and its relatives, and is mimicked by some egg-eating snakes, *Dasypeltis*. When they are cornered out in the open or uncovered under an object, the carpet vipers form a "c"-shaped or horseshoe-shaped coil. It consists of two or three sections of the body folded back on themselves so that touching sections of scales are laid in opposite directions. The head is always central and facing the threat.

The carpet vipers then move these parts of the body against each other. The scales on the flanks rub against each other to produce a loud rasping noise, likened by some authors to water dropping onto a hot plate. The sound is produced by the lateral scales (on the flanks), since their keels are serrated (hence "saw-scaled"). The sound may be amplified by the snakes' body, which they puff up with air when they are displaying. The additional air serves to tighten the skin like a drum and produces a resonating chamber.

Vibrating Coils

An angered saw-scaled viper moves its coils quickly; and as its state of excitement abates, it slows down, only to set off again if it detects some movement on the part of its opponent. The sound therefore rises to a crescendo and then drops off a number of times.

Should the aggressor come within range, the viper will strike suddenly and without warning, throwing its head forward from the coiled position and immediately recoiling in preparation for repeated strikes. The strike itself occurs so quickly that it is impossible to follow with the eye. The herpetologist Colonel Frank Wall, who had more experience of snakes in the Far East than almost anyone and was not easily impressed by venomous snakes, wrote, "This is the most vicious snake I know. Not only is it apt to bite on the smallest provocation, but strikes out without hesitation and with great malice."

Scientists believe that the production of sound by rubbing scales together is a substitute for hissing. Hissing involves the expulsion of air from the snake's body and would therefore allow water vapor to escape. In dry environments this would be disadvantageous to the snake, and so another method of sound production has evolved. The rattlesnake's rattle may have evolved for similar reasons. The other so-called "raspers" also live in dry environments, except for the harmless egg-eaters that produce sound in the same way to mimic the venomous vipers.

Strictly speaking, sounds produced by rubbing one part of the body against another are known as "stridulation." It is a term usually associated with grasshoppers—which stridulate using their legs—and crickets—which use their wings.

Feeding and Breeding

Carpet vipers are remarkable for the range of food they eat, and that is one of the reasons for their success. Population densities can be very high: Nearly 7,000 were collected in four months in an area of 2,500 square miles (6,500 sq. km) in northern Kenya. Often their most common prey consists of invertebrates, especially scorpions, centipedes, locusts, solifuges (wind scorpions), and even termites! Larger fare includes lizards along with their broken tails, frogs, snakes—including their own species—and small mammals.

Their breeding habits are something of a puzzle. Most forms, including *Echis pyramidum* and the other African species, lay eggs with clutches of four to 20. Indian and Sri Lankan saw-scaled vipers, *E. carinatus*, however, give birth to live young with three to 15 juveniles. Whether all populations of this species bear live young is uncertain.

This group of captive carpet vipers in India is about to be "milked" for their venom, which will be used in antivenom serums to treat bite victims.

Chameleons

The classification of the 131 chameleonid species is not definitive, but here they are grouped into six genera. Chameleons are unusual in appearance. They have evolved a number of special features that are ideally suited to an arboreal life and that set them apart from other lizards. They are distributed throughout the continent of Africa. A few species are also found in Spain, Portugal, the Middle East, Sri Lanka, India, and on the Seychelles Islands and Madagascar. About 50 percent of all known species live on Madagascar, including the largest, Oustalet's chameleon, *Furcifer oustaleti*, at 27 inches (69 cm) long, and the smallest, the tiny ground chameleon, *Brookesia minima* and relatives, which measure as little as 1.38 inches (3.5 cm) in length.

Habitat Types

There is a natural division of the environmental conditions in which chameleons occur throughout their range—either cool and dry or warm and humid. In tropical regions some species occur at high altitudes and are therefore subjected to temperatures similar to those in more temperate areas. In Central Africa the Senegal chameleon, *Chamaeleo senegalensis,* lives in lowland forest habitat and needs a constantly warm, humid environment. The coarse chameleon, *C. rudis,* is also from central Africa but lives in higher montane forests, where it is cooler and a little drier.

Chameleons live in three different types of habitat: highland, lowland, or forest floor. Highland habitat occurs between 2,000 and 9,900 feet (600–3,000 m), usually on mountain slopes. While daytime temperatures are similar to those at lower altitudes, night temperatures are considerably lower, resulting in a heavy morning mist that creates some humidity. The humidity evaporates when the sun warms the mountain slopes. The high-casqued chameleon, *Chamaeleo hoehnelii*, lives at about 9,250 feet (2,800 m) on the slopes of Mount Kenya, where overnight temperatures drop to freezing. Such conditions would kill most chameleons, but *C. hoehnelii* falls to the ground into leaf litter that helps keep it warm. When the first rays of the sun strike the following morning, it warms up and climbs back into the branches.

Lowland habitat is considered to be between 1,500 and 2,000 feet (500–600 m). In most areas it is a tropical rain-forest environment. Rainfall and humidity are usually high. Temperatures are fairly constant with little seasonal variation. The graceful chameleon, *C. gracilis,* is a typically equatorial species that enjoys the high humidity and heat.

Some species are found at the same altitude but in tropical scrub. These areas have a temperature similar to lowland rain forest, but rainfall is not as plentiful and

Common name Chameleons **Family** Chamaeleonidae

Family Chamaeleonidae 6 genera and about 131 species:

- Genus *Brookesia*—23 species of stump-tailed chameleons from Madagascar, including *B. stumpffi*
- Genus *Rhampholeon*—8 species of leaf chameleons from West Africa, including the western pygmy chameleon, *R. spectrum*
- Genus *Bradypodion*—21 species of live-bearing dwarf chameleons from South Africa, including *B. thamnobates* and *B. damaranum*
- Genus *Calumma*—18 species of chameleons from the more humid areas of Madagascar, including Parson's chameleon, *C. parsonii*
- Genus *Furcifer*—19 species of chameleons from the more arid regions of Madagascar (2 species on the Comoro Islands), including the panther chameleon, *F. pardalis*
- Genus *Chamaeleo*—42 species from Africa, the Middle East, Sri Lanka, India, and southern Europe, including the Namaqua chameleon, *C. namaquensis*, the veiled chameleon, *C. calyptratus*, and Jackson's chameleon, *C. jacksonii*

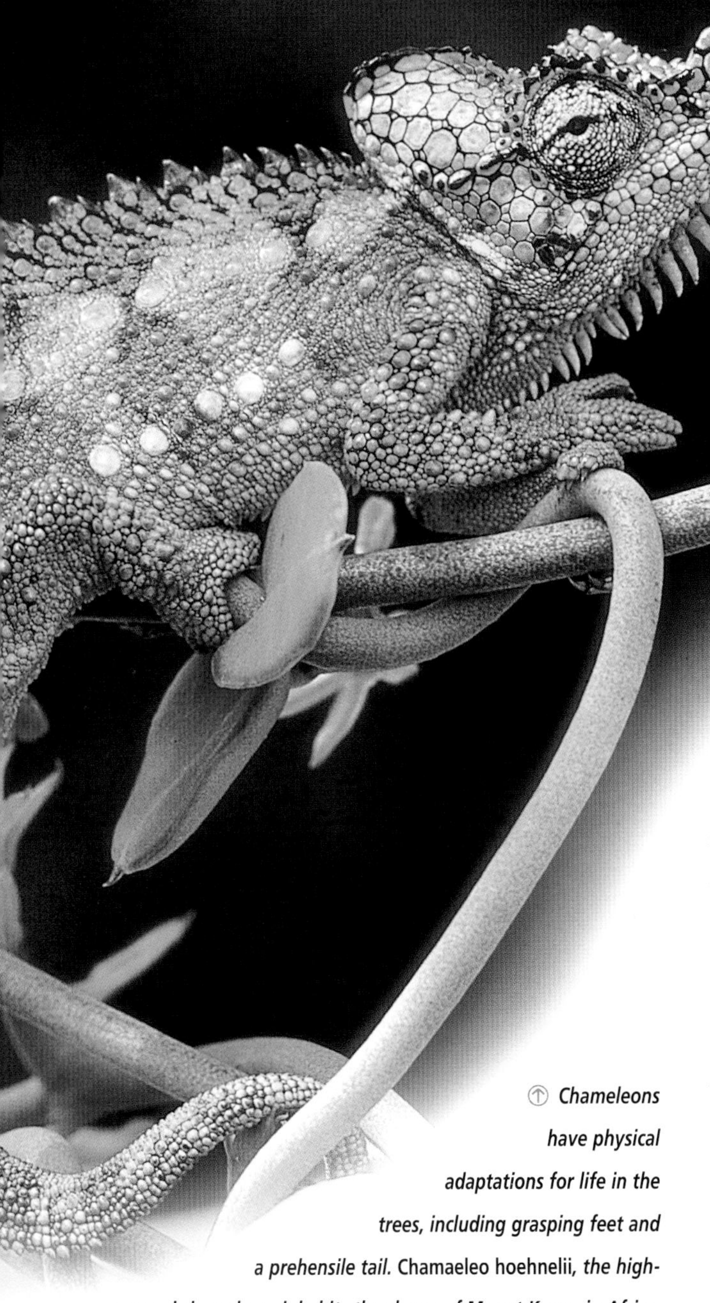

motionless. High humidity levels in these areas encourage moss to grow—it even grows on the skin of *Rhampholeon spectrum*, giving it the appearance of a dying leaf.

Chameleons have physical adaptations for life in the trees, including grasping feet and a prehensile tail. **Chamaeleo hoehnelii**, *the high-casqued chameleon, inhabits the slopes of Mount Kenya in Africa.*

tends to be seasonal. However, morning dew brings moisture that is beneficial to the chameleons. Oustalet's chameleon, *Furcifer oustaleti*, is thought to collect dew in its high casque, from where it runs down into its mouth.

Primary or untouched forest floor is the natural habitat for *Brookesia* and *Rhampholeon* species from Madagascar and West Africa respectively. Temperature and rainfall levels on the forest floor are the same as in lowland forest. Naturally decomposed leaves on the forest floor are ideal for incubating their eggs. These "stump-tailed" and leaf chameleons are difficult to detect among the leaves, being camouflaged both by color and shape. At just 1.38 inches (3.5 cm) long the tiny *Brookesia peyrierasi* looks like a thin piece of wood when

Physical Features

Most chameleons' bodies are compressed from side to side. This shape allows them to move through leaves and twigs and, viewed head-on or from above, aids concealment. The body can be flattened even more to provide a larger area for heat absorption or to fool an enemy into thinking the chameleon is larger than it is.

Like other reptiles, chameleons are covered in scales. The scales can be quite small, making the body appear smooth, or large and often tubercular, giving it a rough appearance. Scales that are all the same size are said to be homogeneous. Heterogeneous means that the scales are different sizes. Most chameleons have areas of both types on different parts of the body. The skin is sloughed (shed) periodically. Unlike in snakes, whose old skin is usually sloughed in one piece, chameleon skin cracks and dries between the scales before falling off in many pieces. The chameleon sometimes needs to help the process by scraping along branches to remove old skin, especially around the eyes, since it can impair the chameleon's vision and therefore its ability to feed.

Various physical adornments add to the unusual appearance of many chameleons. Some species have horns (anything from one to four) or other hornlike processes. The latter may not be as rigid as horns; but like horns, they are probably of some sexual significance, since they usually occur in males. *Furcifer bifidus* from Madagascar has two large rostral (nose) processes. Most species have a crest of spines or blunt tubercles along the back and underside. The crest can be exaggerated and fanlike as in the sail-fin chameleon, *Chamaeleo montium* from West Africa. Others, such as *C. hoehnelii*, have a raised casque, giving it its common name of high-casqued chameleon. (It also has a distinctive upturned snout.)

Two rather strange species are *Chamaeleo xenorhinus* and *C. carpenteri* from Central Africa. The males have a large crest on top of the head and a huge rostral process. In Pfeffer's chameleon, *C. pfefferi* from Cameroon, both sexes have a gular (throat) crest made of elongated soft "spines." In Jackson's chameleon, *C. jacksonii* from Africa and Hawaii, the males use their horns in combat—they can be interlocked to twist a rival from its branch.

Prehensile Tails

Chameleons have prehensile tails (but in *Brookesia* and *Rhampholeon* the tail is only slightly prehensile). They are particularly useful for animals living in trees. They can be coiled around twigs to provide an anchor when extending the tongue or reaching out to grasp a branch. When covering any distance, chameleons can extend the tail for balance. In many species it is coiled up during rest or sleep. Falling chameleons have been seen to use the tail as a "brake" by grasping a branch with it. On each foot opposing groups of toes have become fused to produce feet that can grasp too, often with remarkable strength.

Chameleons spend much of their time motionless, their eyes constantly searching for prey or danger. Movement is usually slow and deliberate, but certain species can move surprisingly quickly when necessary. For example, when it feels threatened, *Furcifer lateralis* from Madagascar will climb rapidly and even leap from branch to branch to avoid being caught.

Chameleons rely heavily on sight for feeding and during courtship. The flap-necked chameleon,* Chamaeleo dilepis, *shows the typically protruding eyes that can swivel independently, giving panoramic vision.

Senses

A chameleon's eyes can move independently, giving all-round vision. The eyes protrude and, except for a relatively small area, are covered with skin. Small eyes probably help the chameleon remain concealed. When prey is spotted, accurate binocular vision enables it to judge the distance. The chameleon then shoots out its extensible tongue, and the prey "sticks" to it.

The West African male sail-fin chameleons,* Chamaeleo montium, *have long rostral horns projecting straight forward along the lower snout. The females lack these horns and have only "horn buds."

It appears that chameleons have little or no sense of smell. In other reptiles smell is detected by the Jacobson's organ in the roof of the mouth. In chameleons this organ is reduced and (as far as is known) is nonfunctional. Chameleons often press their tongue against branches (a technique referred to as "tongue testing"), but the reason for this is unclear. Some scientists think it may be a means of detecting the scent of other chameleons from fecal traces using a small organ with two "lobes" on the end of the tongue. When the tongue is extended to trap prey, the end is formed into a bulbous shape; but when lapping water, the tongue is flattened, and the "lobes" can be seen as small tendrils.

Chameleons have no voice as such. In confrontations a faint hiss may be heard, but it is probably an involuntary expulsion of air as the body is flattened. A similar hiss is sometimes heard when the tongue is retracted with prey attached. The veiled chameleon, *Chamaeleo calyptratus* from Yemen and Saudi Arabia, can produce a loud, hoarse hissing sound. Hearing is practically nonexistent in chameleons. Unlike most other lizards, they have no external ear opening. They can, however, detect airborne vibrations by means of a membrane on each side of the skull. This membrane is connected with the inner ear.

Chameleon Myths

The Malagasy believe that the devil made chameleons from parts of other animals—the tail of a monkey, the skin of a crocodile, the tongue of a toad, and the horns of a rhinoceros.

During a period of prolonged drought in Central Africa a village chief sent a chameleon to the gods to ask for rain and save the people. But it moved so slowly that the message was only delivered after the people had died.

In Madagascar killing a chameleon is *fady* (taboo). They believe that if any injury is inflicted on *Brookesia* chameleons, the same will happen to the perpetrator. This has led the Malagasy to believe that chameleons have supernatural powers.

In Morocco dried chameleon is thought to act as a powerful medicine for humans, curing many illnesses.

The Use of Color

Chameleons have special skin cells (chromatophores) that contain red and yellow pigments and that can shrink or expand. Together with a reflective layer and special cells containing a dark pigment (melanophores), they cause the skin to change color, sometimes extremely quickly.

Most chameleons are various shades of green and brown, which helps conceal them in trees. There is a mistaken belief that chameleons can change to any color to match their surroundings. But color is used mostly as a form of communication, and color changes are usually related to making the chameleon more visible in order to attract a mate, threaten a rival (or predator), and in the case of females, to warn off unwelcome suitors or intruding females. Males are highly territorial and will subject any intruder to aggressive displays combined with battle coloration. Gravid females (ones that are full of eggs or young) exhibit spectacular color and pattern combinations. Female *Furcifer lateralis* display colors of white, red, yellow, light blue, black, and purple. No doubt they act as an effective warning to males, but they are no aid to concealment. In the lesser chameleon, *F. minor*, gravid females have particularly brilliant coloration (black with red, blue, and yellow scales). When compared with a male, they could easily be mistaken for a different species.

Color can be an indicator of health. A sick chameleon usually turns a sickly, yellowish shade. Color changes also relate to temperature. A cold chameleon is usually dark, often almost black. Specimens basking in the early morning sun can be dark on the exposed side (to absorb heat) with normal coloration on the other side. When it is very hot, chameleons often turn a lighter color before seeking shade among leaves. They probably gain more protection from their habit of resting motionless than from cryptic coloration. When they do move along a branch, it is usually with a jerky, swaying gait, possibly imitating the action of a leaf in the wind.

Feeding and Diet

While some chameleons eat vertebrates such as lizards, frogs, small mammals, and small birds, most enjoy a natural diet of live insects. When hungry, chameleons may stalk their prey. However, their usual method is to sit patiently and pick off any insects that come within reach of their long, extensible tongue. With this method the chameleon only has to move a minimal distance—a real advantage for a creature that lives among trees and bushes, and is unable to make a quick dash to seize prey.

The chameleon's tongue is a remarkable organ. Rather than being "sticky," it works more like a rubber sucker and needs to be moist in order to stick to the target. In some species the fully extended tongue can be almost twice the length of the body. Accelerator muscles in the tongue squeeze together to propel it from the mouth. The contractile muscles are elastic; after the tongue hits the insect, they contract to bring it back into the mouth. When not in use, the tongue is folded concertina-style in the mouth. Occasionally a chameleon might seize prey in its jaws if it appears too large to stick to the tongue or if the tongue has missed the insect once or twice and it is near enough to be grabbed.

Defense

Chameleons use various tactics when threatened. Many species have cryptic coloration, which makes hiding in trees and bushes effective. Stump-tailed chameleons combine this with "freezing" (remaining motionless until the danger has passed). Once spotted by a predator, it is difficult for the chameleon to escape. However, by jumping from one branch to another or falling to the ground, some chameleons manage to elude their attacker. Occasionally a chameleon will hiss, inflate its gular pouch, and sway with its mouth open in an attempt to fool the enemy into thinking it is larger and fiercer than it is. Some chameleons feign death by lying sideways on a branch or hanging beneath the branch with the tongue protruding.

Reproduction

Vision and color are important in helping males distinguish females of their own species. Colors and patterns can be intensified to warn off rivals (in the case of males) or to signal sexual receptivity (in females). Males are territorial and can be very aggressive during courtship. To intimidate a rival male, the short-horned chameleon, *Calumma brevicornis,* raises its occipital lobes to make itself look bigger. The flap-necked chameleon, *Chamaeleo dilepis* from Africa, also raises and wiggles its lobes.

Chameleons' tongues are often as long or longer than their body and can be shot out of the mouth at lightning speed, as shown here by a female flap-necked chameleon,* Chamaeleo dilepis *from Zanzibar.

In pursuit of a female a male flap-necked chameleon, Chamaeleo dilepis, *picks its way along the branch of a tree on the island of Zanzibar. The species shows an obvious example of sexual dichromatism.*

Identifying the sex of many species of chameleons is relatively easy in mature specimens. As a general rule, it is the male that sports some kind of ornamentation such as horns, nasal appendages, casques, crests, or tarsal spurs. Occasionally a female may have some of these features, but they tend to be smaller than those of the male. Nasal appendages are usually a male feature, but in *Calumma nasuta* from Madagascar both sexes have one. However, the male's is wider and more rounded at the end.

A number of Madagascan chameleons exhibit sexual dichromatism (the males and females have different colors and patterns). *Furcifer labordi* males are green with white stripes on the flanks, while the females have red coloration on the throat, two small, red lateral markings on the neck, and a bluish color on the flanks.

Male chameleons initiate mating by rhythmic head movements and by intensifying their color. A receptive female stays quiet and does not change color. If she does not rebuff him, he mounts and aligns their cloacae, curling his tail under hers. Mating can take from a couple of minutes to almost an hour. After mating, females change color to show they are no longer "available."

As with all reptiles, chameleons can be divided into two reproductive groups—those that lay eggs and those whose eggs stay in the body until the young hatch (live-bearers). In the latter the young are fully formed and move off almost immediately. There is no dependence on the mother after birth. Gestation varies from four to seven months depending on species and temperatures. Gravid females often seek out warmer areas and spend more time basking. The young emerge still encased in a membrane (sac), or they may already have broken out from the membrane before emerging. The two-lined chameleon, *Chamaeleo bitaeniatus*, and *Bradypodion* females move along the branches depositing "complete" youngsters that cling to the branch. In *C. ellioti* the young are contained in a sac. Some fall onto branches, while others fall to the ground. Each sac contains an amount of fluid that gathers in the lower end and seems to act as a shock absorber. Within seconds of birth the youngsters pierce the membrane, emerge, and climb into the branches.

In egg-laying species the eggs develop in the female's oviduct. About six to eight weeks after mating, the female descends to the ground and digs a hole. After laying her eggs in the hole, she fills it in and tamps down the surface to obscure signs of nesting. This helps prevent predation by other animals. Eggs of some Madagascan species experience a diapause, a period during which development is suspended. It usually coincides with cooler temperatures and results in a longer incubation time. Eggs of Parson's chameleon, *Calumma parsonii*, can take up to 21 months to hatch. Depending on seasonal variations, they can have one or more diapauses.

Chameleon eggs have leathery shells that absorb moisture throughout the incubation period and increase in size. When they are ready to hatch, the youngsters make several slits at either end of the eggs, sufficient to allow the head to emerge. Full emergence does not take place until all the yolk has been absorbed, which may take as little as one hour or as much as two days. Hatching frequently takes place during a wet period, since the ground is softer, and it is easier for the youngsters to dig their way to the surface. Most chameleon babies are usually brown—this coloration gives them effective camouflage on branches and twigs.

Cobra Family

The cobras in the family Elapidae occur mainly in the Southern Hemisphere in South America, Africa, Southeast Asia, and Australasia. There are only two species in North America and none in Europe or Central Asia, but in Australia it is the largest snake family.

Elapids are descended from colubrids. Superficially they look similar, with large scales covering the top of the narrow head and more or less cylindrical bodies. Internally they lack a functional left lung, but they have left and right oviducts. Only aquatic species have a tracheal lung (for more efficient absorption of oxygen). Most elapids have large eyes with round pupils. A few Australian species have chunky bodies and eyes with vertical pupils.

Venom Fangs

The main features that distinguish elapids from colubrids are the venom fangs at the front of the mouth. The fangs are relatively short and fit into slots on the floor of the mouth when it is closed. Each fang has a tubular canal running through the center. When the snake bites, the venom is forced through the canal. Most of the larger, more conspicuous elapids are alert, fast-moving snakes that can be unpredictable. They have a reputation for standing their ground and being more intelligent than other snakes, sizing up their enemy before launching an attack. Whether or not that is true, few snake handlers are as comfortable with cobras as they are with vipers.

Common name Cobras **Family** Elapidae

Family Elapidae 2 subfamilies

Subfamily Elapinae—17 genera, about 140 species of cobras, mambas, coral snakes, and kraits, including mambas, *Dendroaspis* spp.; Malaysian blue coral snake, *Maticora bivirgata*; American coral snakes, *Micrurus* spp.; Indian cobra, *Naja naja*; Cape cobra, *Naja nivea*; king cobra, *Ophiophagus hannah*

Subfamily Hydrophiinae—43 genera, about 165 species of Australian elapids, sea kraits, and sea snakes, including Australian tiger snakes, *Notechis scutatus* and *Notechis ater*; taipans, *Oxyuranus scutellatus* and *Notechis microlepidotus*; and yellow-lipped sea krait, *Laticauda colubrina*

Classification

The Elapidae includes several groups of well-known, even notorious snakes, such as the cobras, mambas, coral snakes, and death adders. It also includes the sea snakes and sea kraits, although they are sometimes placed in a separate family, the Hydrophiidae. While that may appear to be a logical division, there are no sound taxonomic reasons for it. In fact, scientists think that the sea snakes and sea kraits evolved from the same line of ancestors as the Australian land species after they had become separated from the rest of the family through continental drift. The two groups are therefore more closely related to each other than to elapids from other parts of the world.

For the purposes of this book the Elapidae is divided into two subfamilies. The Elapinae contains only terrestrial (in other words, nonmarine) species from the Americas, Africa, and Asia—the cobras, mambas, kraits, and coral snakes. The Hydrophiinae contains a mix of terrestrial species from Australasia together with all the marine forms, including the sea snakes and sea kraits. There are still unanswered questions regarding their family tree, however, and it is quite possible that the classification system will change at some point.

Subfamily Elapinae

In North, Central, and South America the Elapinae is represented by the coral snakes. Among this group the genus *Micruroides* has a single species, *M. euryxanthus* from Arizona, New Mexico, and north-central Mexico. The much more widespread genus *Micrurus* has 65 species. They are all brightly colored snakes, mostly with body

⊖ **Ophiophagus hannah** ***from Southeast Asia raises its body in alarm. Adult king cobras have few natural enemies, but juveniles like this one are preyed on by mongooses, civets, army ants, and giant centipedes.***

rings of white, red, and black. Most live in rain forests, but some occur in the deserts of the southwestern United States and northern Mexico. One species is semiaquatic and eats eels, and all the species are dangerous to humans. Coral snakes are semiburrowing species and feed largely on other burrowing reptiles. They are the only terrestrial elapids in North and South America.

Terrestrial elapids are well represented in Africa, however, and include several well-known genera. The mambas, genus *Dendroaspis*, are mostly tree dwelling, although the black mamba, *D. polylepis*, hunts on the ground. Mambas are among the most feared snakes in Africa, but the harm they do is insignificant compared with snakes such as the puff adder, *Bitis arietans*, and the carpet vipers, *Echis* species. The most numerous elapids in Africa are the eight species of cobras in the genus *Naja*.

The Cobra Hood

Cobras are powerful snakes with smooth scales and graceful heads. If threatened, they lift the front part of the body off the ground and spread a hood—an area of skin stretched across a series of extralong ribs. They normally keep the ribs folded along the upper part of the body but pivot them out in order to spread a hood.

Some African species are spitters—they can force a fine jet of venom through openings in the front of their fangs—including the rinkhals, *Hemachatus haemachatus*. It is unusual among elapines in having keeled scales and giving birth to live young. (Most elapines have smooth scales and lay eggs.) It specializes in eating toads and is one of several snakes that play dead when under attack.

The genus *Aspidelaps* is made up of the African coral snake, *A. lubricus*, and the shield-nosed snake, *A. scutatus*. They are small, have an enlarged scale at the tip of the snout, and spread a wide hood if aroused. The African garter snakes, *Elapsoidea* (not to be confused with the North American garter snakes), are small and have banded patterns around their bodies, especially when young. Because they are reluctant to bite even if handled, they are regarded as fairly harmless.

Four African elapids have left the terrestrial lifestyle behind: African water cobras, genus *Boulengerina,* live along the shorelines of large lakes and feed on fish, while the tree cobras, genus *Pseudohaje*, are arboreal and probably eat tree frogs. Each genus has two species.

In Asia there are several *Naja* species, including some spitters. There are also other genera, including the kraits, *Bungarus*, and the coral snakes, *Maticora*, that mainly eat other snakes. The kraits are triangular in cross-section and have a prominent backbone with a row of large scales running along it. Kraits are responsible for large numbers of snakebite fatalities, not because they are aggressive but because they are nocturnal and easily stepped on. The remaining Asian genus is *Calliophis*. These snakes are called coral snakes because of their bright coloration and their displays that involve erratic twitching and raising of the tail that are similar to those of the New World coral snakes.

Subfamily Hydrophiinae

All the Australasian elapids are placed in this subfamily. Among them they encompass a large variety of shapes, sizes, habits, and habitats. The sea snakes and sea kraits are highly specialized marine snakes. Because there are few colubrids and no vipers in Australasia, elapids have moved into their niches. Many, such as the brown snakes, *Pseudonaja*, and the whip snakes, *Demansia*, are remarkably similar in appearance and habits to the sand snakes, whip snakes, and racers of the Old and New Worlds, except that they are venomous.

Others have similar builds but are smaller and more secretive. They include the crowned snakes, *Cacophis*, which are the counterparts of other species such as the black-headed snakes, *Tantilla* species, the ringneck snakes, *Diadophis punctatus* from North America, and the

European hooded snake, *Macroprotodon cucullatus*.

The Australian copperhead, *Austrelaps superbus*, has a chunky shape reminiscent of the water snakes *Nerodia* and *Natrix*. Like them, it is found near marshes and swamps. The taipans have lifestyles similar to the king cobra and the black mamba. The 12 or so species of small coral snakes, *Simoselaps*, are all burrowers and are found on the surface only at night. They eat mainly lizards and live in dry, sandy regions. The bandy-bandy, *Vermicella annulata*, and its close relative the northern bandy-bandy, *V. multifasciata*, are both boldly banded black-and-white snakes—hence their common names. They are small and feed on other snakes, mainly blind snakes in the family Typhlopidae, which they catch in underground tunnels. When alarmed, they raise a large part of their body off the ground to form a loop, which they direct toward their opponent and can use as a club. Finally, the death adders, *Acanthophis*, are heavy-bodied snakes. Their common name connects them to the Old World adders whose appearance and lifestyle they match closely.

Ⓣ *Sea kraits such as this* Laticauda *species swimming in waters around Indonesia spend most of their time at sea on shallow tropical reefs. They come ashore during the breeding season to lay their eggs on land.*

Common name Collared lizard

Scientific name *Crotaphytus collaris*

Subfamily Crotaphytinae

Family Iguanidae

Suborder Sauria

Order Squamata

Size To 10 in (25cm); males are larger than females

Key features Head massive; 2 black rings around the neck; tail long and cylindrical; hind limbs long; body covered in small scales, giving the skin a silky texture; body color varies among populations and with the season but is typically green in males, dull green, brown, or yellowish in females, all with light spots loosely arranged into transverse lines; juveniles have distinct banding across the back that gradually fades as the animal grows

Habits Diurnal and heat loving; lives on the ground and rarely climbs, except among boulders

Breeding Egg layer with more than one clutch each year; female lays 1–13 eggs that hatch after about 45 days

Diet Large invertebrates and smaller lizards; also some vegetable material, including leaves and flowers

Habitat Hot, rocky hillsides with sparse vegetation

Distribution Western United States in desert regions

Status Common

Similar species There are other collared lizards, but their ranges do not overlap; the leopard lizards, *Gambelia*, are their only other close relatives

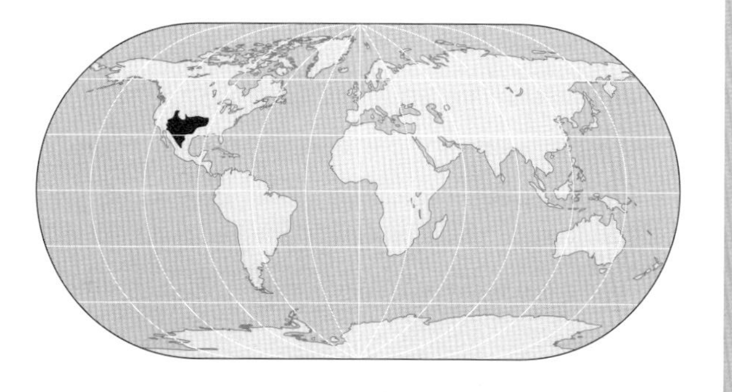

Collared Lizard

Crotaphytus collaris

Collared lizards are noted for running upright on their hind legs, making them look like miniature Tyrannosaurus rex. *They are powerful hunters and have an unusual way of waving their tail, like a cat, before grabbing at prey.*

THERE ARE EIGHT SPECIES OF COLLARED LIZARDS and three species of leopard lizards. Together they make up the subfamily Crotaphytinae, which is sometimes regarded as a full family, the Crotaphytidae. All collared lizards are broadly similar in size and shape, and all have two black collars separated by a white or light-colored one, although their general coloration varies. For instance, *C. vestigium* from Baja California and California, is rich chocolate-brown with pure white spots and bands fading to dull orange or green on its flanks. The spectacular Dickerson's collared lizard, *C. dickersonae*, is sometimes regarded as a subspecies of *C. collaris*. The males are deep blue in color.

The other half of the family, the leopard lizards, consists of three species with slightly more slender heads and bodies than the collared lizards. They prefer open habitat, such as sand and gravel flats with sparse vegetation, for example, creosote bush scrub. They shelter in burrows at night. They are pale buff or sandy brown in color and often wait for prey at the base of a small bush, where the dappled shadow helps disguise their outline. When approached by predators, leopard lizards often flatten themselves to the surface and rely on their pale camouflage colors to escape notice.

Fearless Terrorists

Collared lizards are powerful hunters—they are likely to terrorize smaller lizards from the region. They chase them down and crush them in their powerful jaws before swallowing them whole. Collared lizards also eat large invertebrates such as grasshoppers and beetles. They dominate south-facing rocky ridges and

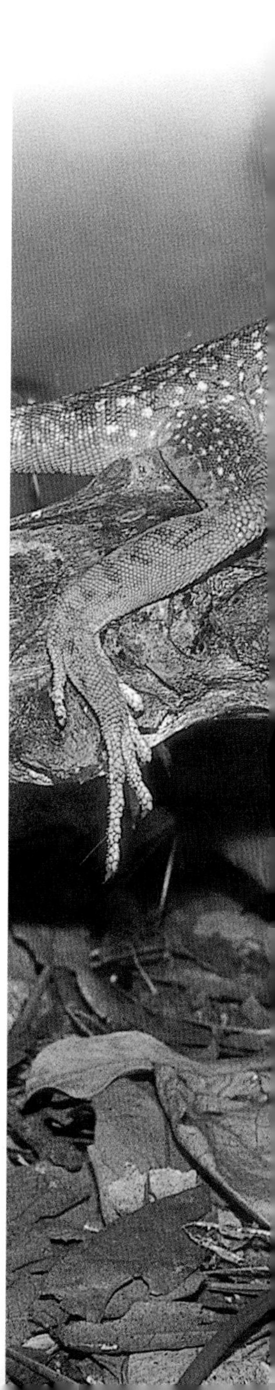

Collared lizards exhibit sexual dimorphism, as seen in this pair of Crotaphytus collaris *from Texas. The male (left) is predominantly green, while the female (right) is yellowish brown in color. Males are also larger than females.*

hillsides by taking up positions on prominent boulders. From there they can survey the large area over which they hunt. Males are especially territorial. They stand up on stiff limbs and bob their body up and down if another collared lizard approaches. They will even display to humans, standing their ground until the last minute. Then they dash off, sometimes lifting the front part of their body off the ground and taking huge strides with their hind legs. At the same time, they raise their tail as a counterweight. Their aggressive behavior has persuaded local people, especially in Mexico, that collared lizards are venomous.

Breeding Coloration

Collared lizards come into breeding condition in the spring, often in late March or early April after several months in hibernation. At that time of year the male's colors really glow once he has warmed up by basking on the hottest exposed rocks. Dominant males mate with all the females living within their territory. Once females have mated and their eggs begin to develop, they too undergo a color change. Large patches of bright orange appear on their flanks and the sides of their neck, showing that they are full of eggs (this is known as "gravid coloration"). The color probably indicates to the male that she is not receptive to more mating attempts. This prevents further attempts at courtship and saves both of them time and energy.

Gravid coloration is not unique to collared lizards: It is present in many other iguanids and in some agamid lizards and chameleons. The exact color varies from species to species, but it is invariably bright, distinctive, and easily visible from a distance. Females lay one to 13 eggs, which they bury, and which hatch after about 45 days. Each female lays two or more clutches throughout the summer provided she can find enough food to replenish her reserves. The hatchlings, the last of which emerges in late summer or early fall, lack the bright colors of adults and are mostly brown with a black collar.

Colubrids

The Colubridae is a huge "family" of snakes whose relationships are not properly understood. Because they are so diverse, it is almost impossible to describe them in general terms. They do not even have a widely accepted collective common name and are generally referred to rather clumsily as "colubrids."

Colubrids range from less than 12 inches (30 cm) to over 10 feet (3 m) long. They can be almost any color, and their coloration may be intended either to make them blend in with their surroundings or to stand out. They may be extremely slender, very stout, or somewhere in between; and they may have long or short tails. Their scales may be smooth and shiny or keeled and rough. Their eyes may be small or large with pupils that are round, vertically elliptical, or horizontal and shaped like elongated keyholes—almost any permutation is possible.

Unlike some of the more primitive groups, colubrids have no limbs or limb girdles, vestigial or otherwise. The left lung is either just a remnant (vestigial) or absent altogether, and some species have a tracheal lung.

The San Diego gopher snake,* Pituophis catenifer annectans *(subfamily Colubrinae), is a powerfully built colubrid. Its snout is slightly pointed for burrowing, and it has large, regular scales on its head.

Common name Colubrids **Family** Colubridae

The colubrid "family" contains about 1,800 species—more than all the other 17 snake families combined. It is unlikely that all these snakes come from the same ancestor, and so some time in the near future the "family" will be broken down into a number of smaller units. At the moment the only sensible way of organizing them is to divide them into subfamilies, some of which are more widely recognized than others. Some genera have not yet been assigned to any particular subfamily.

Subfamilies:

Colubrinae—A rather vague group of about 650 species with an almost global distribution; highly variable in size, shape, and lifestyle. It is likely that they will eventually be further divided into as many as 3 or 4 subfamilies or families. Species include the long-nosed vine snake, *Ahaetulla nasuta*; Asian brown snake, *Boiga irregularis*; western shovel-nosed snake, *Chionactis occipitalis*; paradise tree snake, *Chrysopelea paradisi*; western whip snake, *Coluber viridiflavus*; common egg-eater, *Dasypeltis scabra*; boomslang, *Dispholidus typus;* American ratsnake, *Elaphe obsoleta*; corn snake, *Elaphe guttata*; leopard snake, *Elaphe situla*; common king snake, *Lampropeltis getula*; milksnake, *Lampropeltis triangulum;* rough green snake, *Opheodrys aestivus*; and the ground snake, *Sonora semiannulata*

Psammophinae—About 35 species in 8 genera from Africa and Asia; slender, agile snakes that are active daytime (diurnal) hunters, including the sand and grass snakes, *Psammophis*

Homalopsinae—A well-defined group of about 35 aquatic species from southern Asia and northern Australia that live in freshwater, brackish, and seawater habitats. They are all are rear fanged, and species include the tentacled snake, *Erpeton tentaculatus*, and the Asian water snakes, *Enhydris* sp.

Boodontinae—(sometimes called Lamprophiinae). This group of over 200 colubrids from southern Africa and Madagascar might be further divided at a later date. They are mostly terrestrial but include a few arboreal and burrowing species. Species include the common slug-eater, *Duberria lutrix*; brown house snake, *Lamprophis fuliginosus*; and the leaf-nosed vine snake, *Langaha madagascariensis*

Natricinae—The ancestry of this group is uncertain, but it contains nearly 200 mainly semiaquatic colubrids. The European and American species may later be placed in separate subfamilies or families. Species include the grass snake, *Natrix natrix*; southern banded water snake, *Nerodia fasciata*; and the common garter snake, *Thamnophis sirtalis*

Pareatinae—Nearly 20 species from Southeast Asia. With one exception, they specialize in eating slugs and snails. They include the slug-eating snakes, *Pareas* sp.

Dipsadinae—20 genera of small American snakes that either live on the forest floor and eat invertebrates or climb into trees and vines, and eat slugs and snails. Some authors include them with the Xenodontinae. Species include the South American slug-eaters, *Dipsas* sp., and the blunt-headed tree snake, *Imantodes cenchoa*

Xenodontinae—A large and diverse group from the Americas. They are varied in shape and lifestyle, but many have enlarged fangs in the rear of the mouth. Species include the western hognose snake, *Heterodon nasicus,* and possibly the ringneck snake, *Diadophis punctatus*

Calamarinae—Asian reed snakes and their relatives are placed in 7 genera and total about 60 species. They are small burrowing snakes that eat soft-bodied invertebrates or skinks and include the wormlike reed snake, *Calamaria vermiformis*

Xenodermatinae—A small group of 15 species from Asia, many of which are rare and poorly known, e.g., the Javan mudsnake, *Xenodermis javanicus*

The long-nosed snake, Rhinocheilus lecontei *from North America, uses its pointed snout to burrow underground. Its diet includes small rodents, lizards, and their eggs, and small snakes.*

Females have paired oviducts. Colubrids may have one or more pairs of grooved teeth positioned toward the back of the jaw or near the front. Some species produce venom that may trickle into the mouth near the base of the teeth, or in a few cases they may use jaw muscles to force the venom into the wound made by the teeth. A few species can produce bites that are fatal to humans.

Since most of these characteristics are internal, there is no easy way of differentiating a colubrid from a variety of other superficially similar snakes, especially elapids and atractaspids. Colubrids have large scales covering the head, which should help separate them from most vipers, boas, and pythons. But there are exceptions: *Nothopsis rugosus* from Costa Rica and the western keeled snake, *Pythonodipsas carinata* from western Namibia and southwestern Angola, both have heads covered with small, fragmented scales, as in most vipers. (Some vipers, boas, and pythons have large scales on their heads too.)

The Colubrinae

Colubrids that we can regard as "typical" belong to the subfamily Colubrinae. They are among the few groups of colubrids that are found in the Americas and in the Old World. In fact, they are found nearly everywhere that colubrids occur, but there are only a few in South America and Africa and none in the southern half of Australia. One genus, the ratsnakes, *Elaphe*, is especially wide ranging and occurs in North and Central America, Europe, and Asia. Recently proposed changes would reduce the genus to five European and eastern Asian species, including the four-lined snake, *E. quatuorlineata*. Other European and Asian species would be moved to new genera: The leopard snake and the Aesculapian snake would become *Zamenis situla* and *Z. longissimus*. The North American species would be placed in a new genus, *Pantherophis*, so the corn snake would become *P. guttata* and the American ratsnake *P. obsoletus*. It is quite likely that other changes will be made in the near future, and some of the recent ones may not be accepted universally. For the time being, the older classification is used here.

Popular Pets

The ratsnakes are medium-sized muscular snakes that hunt mainly rodents. Among them the various species display almost the whole color spectrum found on snakes. They are very adaptable, making them popular pets. The corn, or red ratsnake, *Elaphe guttata* from North America, is probably the most widely kept of all serpents. Closely related species have been split away from the ratsnakes recently and placed in two small genera: *Bogertophis* (the trans-pecos ratsnake and the Baja ratsnake) and *Senticolis* with just one species, the green ratsnake, *S. triaspis*.

Another widespread genus is that of the whip snakes and racers, *Coluber*, from Europe, Africa, Asia, and North America. In North America it occurs as a single, highly variable, wide-ranging species, *C. constrictor*, with many different common names (black racer, blue racer, eastern yellow-bellied racer) depending on its distribution and appearance. Old World *Coluber* species are known as whip snakes and are far more numerous—about 35 species in total. In the Americas the nine species of coachwhip snakes, *Masticophis*, are also counterparts of the whip snakes.

Other well-known genera from North and Central America include the large, powerful gopher snakes,

Pituophis, and the brightly colored long-nosed snakes, *Rhinocheilus*. The large rear-fanged tree dwellers, *Boiga* species, live in Asia, Africa, and northern Australia. The smaller tiger snakes, *Telescopus*, are also rear fanged but of no apparent danger to humans. They are terrestrial and live in Africa and Europe. In the Americas the equivalent species are the lyre snakes, *Trimorphodon*. They are all active nocturnal hunters with large eyes and vertical pupils, and eat sleeping lizards and sometimes small mammals. Other large species include *Drymarchon corais,* the spectacular glossy-scaled indigo snake that occurs in a number of color forms from North America down into tropical America (where it is known as the cribo), and the Central American tiger snake, *Spilotes pullatus*.

The African boomslang and twig snakes, *Dispholidus typus* and *Thelotornis* species, are rear fanged and notoriously dangerous; but the green snakes and bush snakes, *Philothamnus* species, which are sometimes confused with green mambas, are inoffensive. In all, there are few colubrines south of the Sahara.

Terminology

Aglyphous teeth without a groove (some colubrids with and some without Duvernoy's glands)
Opisthoglyphous teeth grooved toward the back of the maxillary bones to introduce secretions from Duvernoy's glands
Proteroglyphous having venom-delivering teeth in the front of the mouth (there are no colubrids in this category)

Some colubrine snakes, such as the egg-eaters, *Dasypeltis* from Africa, and the flying snakes, *Chrysopelea* from South and Southeast Asia, have unusual lifestyles. There are also many less conspicuous colubrine snakes, for example, the shovel-nosed snakes, hook-nosed snakes, and leaf-nosed snakes from the deserts of the American Southwest. Their snouts are modified for burrowing, and they specialize in eating invertebrates, including scorpions and centipedes, or reptile eggs, which they can root out of the ground.

Other invertebrate eaters, such as the large genus (over 50 species) of black-headed snakes, *Tantilla*, are rear fanged; while in Asia members of the even larger genus, *Oligodon*, have enlarged teeth that they probably use to slit the

Some colubrines have unusual habits. The golden tree snake,* Chrysopelea ornata *from Southeast Asia, can glide through the air in the forests where it lives.

The aquatic natricine, Hammond's garter snake,* Thamnophis hammondi*, in Baja California. The species is sometimes called the two-striped garter snake because of the yellowish stripes on its sides.

shells of reptile eggs. This specialized method of feeding has given them the name of kukri snakes (a *kukri* is a large knife with a curved blade used in the Far East).

The Natricinae

Apart from the Colubrinae, the Natricinae is the only other subfamily with representatives in both the Old and the New World, although in the New World it is restricted to North America. Its members tend to be associated with water or, at the very least, damp places. They are slender, small to medium snakes with keeled scales. Natricine snakes are often alert, diurnal species with a large head and large eyes, and they frequently bask. Most eat frogs and fish, but a few have more specialized diets. In North America there are the water snakes, *Nerodia*, the garter snakes, *Thamnophis*, and several lesser-known genera, including the crayfish snakes, *Regina*, as well as the brown snakes, *Storeria*, and the earth snakes, *Virginia*, that eat earthworms and other soft-bodied invertebrates.

All North American species are live-bearers, but most European, African, and Asian species lay eggs. Some stay with their clutch throughout the incubation period. They include the grass and water snakes, *Natrix*—the genus that gives its name to the subfamily and that used to include the American water snakes and the Asian keelbacks, which now belong to genera such as *Xenochrophis* and *Rhabdophis*.

African natricine snakes include the marsh snakes, *Natriciteres*, that can break off their own tail when grasped (but unlike lizards, they are not able to grow a new one). In general, this subfamily has not had much success at moving south through the African continent.

The Boodontinae and the Psammophinae

By contrast, two subfamilies that have thrived in Africa are the Boodontinae (sometimes known as the Lamprophiinae) and the Psammophinae. Boodontiine snakes are nearly all terrestrial and do not have enlarged fangs. They include the house snakes, *Lamprophis*. This genus consists of one familiar and widespread species, the brown house snake, *L. fuliginosus*, and several others with more limited distributions. The Aurora house snake, *L. aurora*, is arguably one of Africa's most attractive snakes, with a mid-dorsal orange stripe on an olive-green background. Hatchlings are even more colorful—lime

green with black edges on every scale. Some species, such as Fisk's house snake, *L. fiskii* from southern Africa, are very rare, possibly because of increased urbanization and agricultural clearance in their natural habitat. *Lamprophis geometricus* is a largely overlooked species from the Seychelles Islands in the Indian Ocean. House snakes are mainly nocturnal and are powerful constrictors of lizards and small mammals. As far as we know, they all lay eggs.

The 17 species of wolf snakes, *Lycophidion*, are restricted to Africa and owe their common name to their long curved fangs that do not produce venom. They are nocturnal hunters of diurnal lizards, seeking them while they sleep in crevices. Their large teeth help them grip the scaly skin of skinks and similar species, and pull them out from their retreats. The file snakes, *Mehelya* (not to be confused with the acrochordid file snakes), are similar to wolf snakes, but they feed mainly on other snakes and other small vertebrates. They have heavily keeled scales, and their bodies are triangular in cross-section like a file.

The little slug-eating snakes, *Duberria*, are sometimes placed in the Boodontinae, but there is disagreement about their status. They live among vegetation in damp places and feed on slugs and snails. Like the brown snakes, *Storeria* from North America, there are two species, and they give birth to live young. The mole snake, *Pseudaspis cana*, also tentatively put in this subfamily, is the only member of its genus. It is widely distributed over sub-Saharan Africa and varies in color from light sandy brown to solid black (black specimens live mainly in the cooler southwestern parts of Africa). Juveniles are spotted, and some adults retain this coloration. They can grow to 6 feet (1.8 m) or more, they are common, and they are particularly useful in keeping down rodent populations. Regrettably, because of their size and their sometimes belligerent attitude, humans often kill them unnecessarily. They give birth to live young, which (very occasionally) can number up to 100.

Origins of the Name

Unlike the family names Boidae and Pythonidae for the boas and the pythons, the origins of the family name Colubridae are not immediately obvious.

In fact, it is derived from the genus *Coluber*, the whip snakes and racers (in other words, *Coluber* is the type species of the family). But where does *Coluber* come from? *Coluber* is the Latin word for snake and still survives, in slightly different forms, in several Mediterranean languages: *culebra* (Spanish); *couleuvre* (French), and *colubro* (Italian).

Madagascan Colubrids

Madagascan colubrids are traditionally assigned to the Boodontinae, but they have also been placed in a subfamily of their own, the Pseudoxyrhophiinae. One exception is *Mimophis mahfalensis*, which is related to the sand snakes, *Psammophis*, and is therefore placed in the subfamily Psammophinae. *Leioheterodon* are sometimes called Madagascan hognose snakes because of their upturned snouts. As in the American hognose snakes, *Heterodon*, they use the snout to dig up food, including frogs and toads. These large, impressive snakes are perhaps the most commonly seen colubrids on the island. Many other Madagascan species parallel the whip snakes of Europe and North America, being long and slender in

Scale Polishing

Members of the genera *Psammophis* and *Malpolon* are notable for their scale polishing. A pair of glands producing an oily substance open onto their nostrils; by rubbing the snout against the scales on their flanks and underside, they spread the substance over their entire body.

The sequence of events varies a little between the species. Starting close to the head, *Malpolon* moves its head in a zigzag pattern over its ventral surface, at the same time twisting its body slightly so that the underside is accessible. It uses first one side of its head and then the other until it has covered the entire ventral surface. Some *Psammophis* have a pattern of movements that is almost identical to that of *Malpolon*, while others have a slightly more complicated sequence. They move their head over their back in order to reach the scales on the opposite side of their body, so they use the right nostril on the left side of the body, wiping the scales back and forth before changing sides and using the other nostril. As they work their way down the body, they polish the scales on their flanks as well as their ventral surfaces.

Scale polishing probably reduces water loss from the surface of the snake by coating the scales with an oily or waxy deposit. This is substantiated by the behavior of *Malpolon*: It polishes itself at regular intervals throughout the day in warm weather, but during cooler weather it only does so immediately after shedding its skin or feeding.

shape and alert daytime hunters. Strangest of all Madagascan snakes are the three members of the genus *Langaha*. These bizarre vine snakes are long and slender with remarkable appendages to their snouts. In males the scaly "horns" come to a drawn-out point, and in females they are flattened from side to side and lobed like leaves.

In *L. madagascariensis* males can be distinguished further because they are light brown with a yellow belly and a white stripe along the flanks, while females are grayish brown with stippled markings that make them look like lichen-covered sticks. The reason for the color difference (dimorphism) and the function of the nose horns is unknown, but it may relate to camouflage. The snakes are day-active, appear to eat mainly lizards, and give birth to live young. Many other Madagascan genera are poorly known, and new species are added frequently. All are unique to the island, but the genus *Geodipsas* also has species on the African mainland.

A common sight in the Arabian Peninsula, the sand racer,* Psammophis schokari*, occurs in a variety of habitats and is active during the day. These long, thin snakes can disappear with amazing speed when disturbed or threatened.

The Psammophinae

The other mainly African subfamily is the Psammophinae, containing long, slender, fast-moving hunters of lizards. They are typically found in dry open scrub and grassland, but there are also climbing and burrowing species. They are rear fanged, and some of the larger species, such as the hissing sand snake, *Psammophis sibilans*, and the European Montpellier snake, *Malpolon monspessulanus*, can produce painful symptoms. *Psammophis* species can force their tails to break by spinning rapidly if held. In some populations over one-half of the adult snakes have been found to have broken tails. Members of the genus *Psammophylax* are known in Afrikaans as *skaapstekers*, literally "sheep stabbers." They are fairly small and produce powerful venom, but they are inoffensive and do not pose a danger to humans (or sheep!). Their breeding habits are varied: Female spotted skaapstekers coil around their eggs until they hatch, while *P. variabilis* may be live bearing or egg laying depending on where it lives.

Lizards form a large part of the diet of many colubrids. This menarana, Leioheterodon madagascariensis, *was seen digging up a lizard's nest in sandy soil in Madagascar.*

The Homalopsinae

The Homalopsinae has over 30 species. They are mostly small and consist of aquatic and semiaquatic colubrids from South Asia and coastal northern Australia. Many live in inshore waters, including mangrove forests, and they are quite distinct from other colubrids. They have a number of unique characteristics, including slitlike valvular nostrils that they are able to close to keep water out. The scales bordering their mouths close tightly for the same reason. They produce venom and have grooved rear fangs; and although the larger species can give painful bites, they are not considered dangerous to humans.

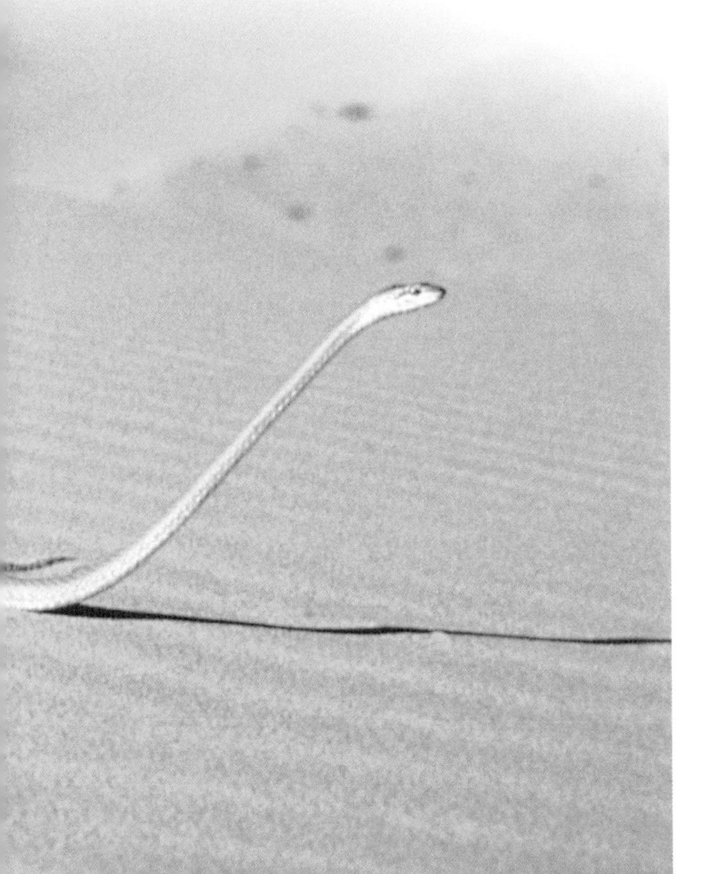

The keeled-bellied water snake, *Bitia hydroides* from around the coasts of Burma, Thailand, and peninsular Malaysia, has a small head and neck, wide body, and flattened tail, so superficially it looks like a sea snake belonging to the Elapidae. Its dorsal scales are triangular in shape and surrounded by interstitial skin. But perhaps the strangest species in the group is the tentacled snake, *Erpeton tentaculatus*, with fleshy appendages on its nose.

Other members include the bockadam, or dog-faced water snake, *Cerberus rynchops*, which moves across mud using similar locomotion to the sidewinding rattlesnake crossing sand. *Fordonia leucobalia,* the white-bellied mangrove snake, has an unusual way of hunting the crabs on which it seems to feed exclusively. It constricts them and then injects them with venom before pulling off their legs and eating them bit by bit. It is perhaps the only snake that does not swallow its prey whole.

The Calamarinae

The Calamarinae from Asia are small burrowing snakes with shiny scales. The reed snakes, *Calamaria*, are the most numerous. They are small and secretive, and poorly studied, but some species have brightly colored heads or undersides and are thought to mimic some venomous Asian coral snakes such as *Maticora bivirgata* and *M. intestinalis*. Reed snakes all probably lay eggs. Most of them eat earthworms and other soft-bodied invertebrates, but the larger species eat small lizards, especially skinks.

The Pareatinae

The Pareatinae is a small subfamily consisting of about 15 species of slug-eating snakes, *Pareas*, and the blunt-headed snake, *Aplopeltura boa*, which eats lizards and snails. *A. boa* has an extremely thin body and a large, squarish head. Its markings resemble a twig covered in lichen, making it difficult to see. Three of the *Pareas* species are sometimes placed in the genus *Internatus*. They are all snakes of forests and plantations. They are nocturnal in habit and are egg layers. Some are arboreal and have thin bodies, wide heads, and huge eyes. The rarely seen *Pareas vertebralis* has large blood-red eyes.

The Xenodermatinae

The remaining 15 or so Asian colubrids are placed in the subfamily Xenodermatinae (meaning "strange skin").This little-known cluster of species shares unusual scalation. For example, *Xenodermis javanicus* has keeled scales separated by areas of bare skin, and the rows of scales down the center of its back have knobby tubercles on them. The largest species reach about 24 inches (60 cm) long, but nothing is known of their diet or reproduction.

The Dipsadinae

In the New World the Dipsadinae contains some species that are generalists in terms of food and habitat—such as the leaf litter snakes, *Ninia*, and the genus *Rhadinea,* which includes the pinewoods snake, *R. flavilata* from southeastern North America (although most members of the subfamily are tropical species). Some have rear fangs but are so small as to be harmless to humans. Another group, including *Dipsas* and *Sibon*, are specialized slug- and snail-eaters. They are arboreal, and they track snails by following their slime trails. They use their thin lower jaw and long curved teeth to pull the soft parts of the snail out of its shell. The blunt-headed tree snake, *Imantodes*, perhaps the world's skinniest snake, cantilevers its body out over gaps between branches to pluck lizards from the leaves and twigs on which they are sleeping. The cat-eyed snakes, *Leptodeira*, eat frogs; more interestingly, they crawl around in bushes to find the egg clumps of leaf-nesting tree frogs such as *Agalychnis* and *Phyllomedusa* species. These frogs lay their eggs on leaves hanging over pools in the mistaken belief that by keeping them away from aquatic predators, they will be safe!

The Xenodontinae

The final subfamily, also mostly from the American tropics, is the Xenodontinae, a diverse group that includes some of the "false" coral snakes and others that mimic pit vipers. There are several rear-fanged genera, and the mussuranas, *Clelia* species, are powerful constrictors that eat other snakes, including venomous ones. The false water cobra, *Hydrodynastes gigas*, is a large semiaquatic species that eats fish, but most of the subfamily are terrestrial and eat lizards and frogs.

A number of species have defied attempts to place them in any particular subfamily. They include familiar species such as the American hognose snakes, *Heterodon*, and the ringneck snake, *Diadophis punctatus*, a hugely variable species found over much of North America (sometimes placed in the Xenodontinae). Other species without a home are the mud and rainbow snakes, one of which, *Farancia abacura*, lays huge clutches of eggs, sometimes numbering over 100. They are semiaquatic and eat elongated prey such as salamanders and eels.

Some snakes in the Dipsadinae specialize in particular types of prey.* Dipsas bicolor, *the banded snail-eater, eats slugs and snails. This individual is weaving its way among leaves in a Costa Rican rain forest.

Common Boa

Boa constrictor

Common name Common boa (boa constrictor)

Scientific name *Boa constrictor*

Family Boidae

Suborder Serpentes

Order Squamata

Length Up to 13 ft (4 m) but often much smaller; island forms rarely more than 6.5 ft (2 m)

Key features Head wedge shaped; a dark line runs through each eye, widening toward the angle of the mouth; background color gray, brown, tan, or pink, with a series of large rounded saddles in maroon or dark brown down the back; tail saddles may be reddish

Habits Arboreal or terrestrial; often enters the water and is a good swimmer

Breeding Bears up to 60 live young but more commonly 10–15

Diet Small mammals; birds

Habitat Very adaptable; rain forests, deciduous forests, dry scrub, and even beaches; often common around human settlements

Distribution From northwestern Mexico through Central America and into South America as far as northern Argentina; also found on some West Indian islands (St. Lucia and Dominica)

Status Generally common, but some subspecies (e.g., Argentine boa, *B. c. occidentalis*) are Endangered (CITES Appendices)

Similar species None

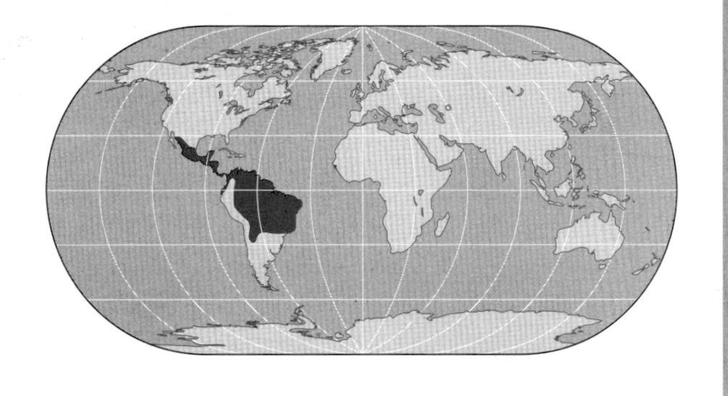

The common boa, or "boa constrictor," suffers greatly from its deadly reputation and is often persecuted by people. However, it is not dangerous to humans.

ALTHOUGH THE COMMON BOA is a massive snake by any standards, it is actually only the sixth biggest. Until quite recently its maximum size was thought to be 18.5 feet (5.6 m), from a specimen killed in Trinidad during World War II by a malaria control party. In recent years, however, doubt has been cast not so much on the size of that specimen but on its identity. It seems that the gigantic "boa constrictor" was, in fact, an anaconda.

Leaving aside this dubious record, the largest common boa came from Brazil and measured 14.8 feet (4.5 m). Common boas are also quite heavy bodied, so any snake over 10 feet (3 m) in length would be capable of eating a large dog, for instance. However, it is highly unlikely that they would take human prey.

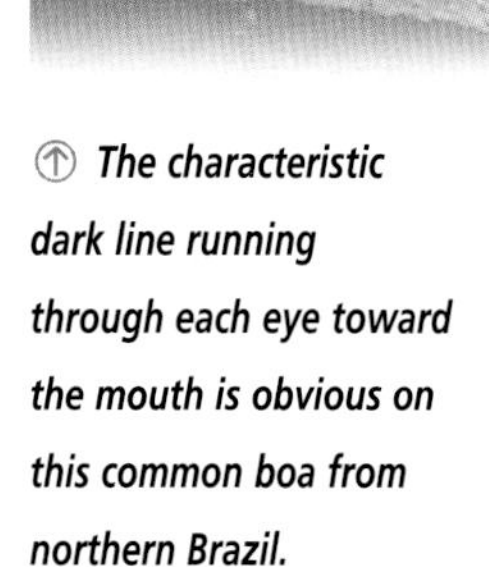

The characteristic dark line running through each eye toward the mouth is obvious on this common boa from northern Brazil.

A 19th-century engraving showing South American hunters stringing up a common boa from a tree in order to skin it. Boas are still hunted for their skins.

The pelvic spurs of this male common boa can be clearly seen. The spurs are used in mating.

Spurs

All boas, pythons, and most other primitive snakes have "spurs" in the form of small thornlike structures on either side of their cloaca. They are the remains of their limbs and are attached to the pelvic girdle that is present in these species. But what use are they?

Although the spurs are largely vestigial—they are the remains of structures that have become redundant through evolution and are in the process of disappearing altogether—males use them in courtship. When the male approaches a female, he crawls along her back. More advanced snakes often make regular twitching movements at this stage, but male boas and pythons use their spurs to scratch the female's skin. When the spurs are moved backward and forward, they seem to stimulate the female into raising her tail so that mating can take place. In species that have them, the spurs of males are nearly always significantly longer than those of the females, whose spurs may be so small that they are difficult to see. In a few species of dwarf boas females lack spurs altogether, although males have them.

Characteristics

The species is distinctive and difficult to mistake for any other snake because of its size. Young common boas, however, may be confused with several other species, including venomous pit vipers, and are often killed in the belief that they are dangerous. The basic color of a typical common boa is silvery gray with about 25 large saddles (blotches) of dark brown or reddish brown. The saddles near the tail tend to be of a richer color than those on the front end of the body. Variations in color depend to some extent on where the specimen comes from.

Up to 10 subspecies have been named, but not all of them are recognized by everyone. A common feature found on all forms of boa constrictor is the pair of wedge-shaped lines that start at the snake's snout and lead backward, through each eye, becoming broader as they reach the angle of the mouth. Another thin line starts just behind the snout and runs back along the top of the head, continuing as far as the nape of the neck. The markings on the head, body, and tail serve to break up the snake's shape and are a good example of disruptive coloration. Common boas' scales are small, and the skin has a silky texture. There is a pair of large scales, or spurs, on either side of the cloaca.

Variations

Boas living in cooler conditions tend to be darker, and the Argentine form, *Boa constrictor occidentalis*, has an overall filigree pattern in black obscuring its markings. Other forms have tan or pinkish background colors, and the colors of the saddles may be especially bright in specimens from Peru, Guyana, and Surinam, for example. They are the so-called "red-tailed" boas that are simply color forms (meaning they are the same in all other respects but color).

The form from Dominica in the West Indies is known as the clouded boa, *B. c. nebulosa*. It is darker in color than mainland forms, and the blotches on its back may be completely obscured in older snakes. The St. Lucia form, *B. c. orophias,* also has poorly defined blotches down its back, and their shape may be less regular than the blotches on typical boas. Boas from small islands off the coast of Honduras are often paler and much more pink than those from the mainland. They can change color to a limited degree, becoming lighter at night.

Boa Distribution

At first glance it seems unlikely that the common boa, *Boa constrictor*, and the three Madagascan boas, such as Dumeril's boa, *B. dumerili*, are closely related, since they come from opposite sides of the world. Continental drift, however, provides the explanation. About 100 million years ago, at a time when the early snakes were beginning to diversify, South America, Africa, and Madagascar were joined together as a Southern Hemisphere landmass known as Gondwanaland. As it broke apart and the separate pieces drifted away from each other, numbers of a boa ancestor were presumably trapped on each of the new landmasses. In South America it evolved into the common boa, and on Madagascar it evolved into three separate species. In Africa it died out, possibly due to competition with pythons. Apart from the boas, Madagascar shares a number of other animal families with South America, such as iguanas, which are also absent from the African mainland.

Habitat and Distribution

Taken as a whole, the various forms of boa constrictor have a huge range. In the north the snake occurs up the west coast of Mexico as far as Guaymas in Sinaloa, while in the south it reaches northern Argentina and Paraguay. As the habitat varies within this area, so the boa constrictor has adapted to a wide range of situations and climates. It is usually considered to be a snake of the rain forest, where it climbs into trees in search of prey such as roosting birds. It hides itself away in hollow tree trunks, caves, or crevices during the day and becomes

active at night or in the evening and early morning.

Large boas are not fast-moving snakes and often travel in a straight line (using rectilinear locomotion) when they are moving across the ground. When climbing, they do so slowly and deliberately, often using concertina locomotion to make their way up tree trunks and along branches.

Due to their great bulk adult common boas are limited to climbing large limbs of trees and often rest in the fork of two thick branches. Unlike more specialized tree boas, they are not able to span large distances between branches. Although not especially aquatic, common boas are good swimmers.

Rat Catchers

Common boas are often found around human settlements, where they perform a useful service in keeping down rat populations, sometimes at the cost of some domestic poultry. However, they are often persecuted in the belief that they are dangerous.

They also live in arid, scrubby habitats in coastal Mexico and in the more open countryside of Argentina and Paraguay, where they have few opportunities to climb at all. Climatic conditions may dictate that they become dormant for several months of the year in some places, while they are continually active in others. Those in tropical lowlands may be more active during the day, even basking in exposed places in order to raise their body temperatures.

Common boas eat lizards, mammals, and birds. Newborn young are large enough to tackle prey up to the size of mice or small rats. As they grow, the range of prey they take increases. A 3-foot (1-m) long boa will easily take prey up to the size of a domestic chicken, and adults can handle medium- to large-sized mammals. They regularly eat white-tailed deer where they occur.

Their method of hunting is typical "sit-and-wait." They often coil just inside the entrance to a hollow tree or a mammal burrow and wait for prey to come within range. At other times they home in on places where numbers of possible prey are likely to congregate, such as fruiting forest trees.

Although common boas do not have the heat-sensitive pits that characterize some other members of the family, they do have temperature-sensitive nerve endings in the scales of the head. They help them locate and pinpoint warm-blooded prey even in darkness. When suitable prey has been located and identified, the boa lunges forward at great speed with its mouth agape, ready to grasp the prey with its backward-pointing teeth. As soon as contact is made, it immediately throws several coils of its body around the victim. Constriction may take some time, during which the snake progressively tightens its coils each time the prey takes a breath, until it becomes

On Hog Island (Cayo de los Cochinos), Honduras, a lighter-colored form of the common boa is found. It reaches about 5 feet (1.5 m) long and is often pinkish gray, although it can change color and be milky white, orange, or lemon depending on the time of day.

unable to fill its lungs. When the snake senses the prey is dead, it slowly releases its coils. Swallowing starts with the head and for large prey items may take an hour or more.

Breeding

The breeding habits of common boas are hardly known in the wild, but they are widely bred in captivity. Information from both sources shows that those from tropical populations prefer to mate during the cooler parts of the year, but year-round matings have been recorded. However, populations from regions with a well-defined cold season, such as northern Argentina, probably mate in spring (but there are no field observations to confirm this).

The gestation period lasts about six months, and a litter consists of 6 to 60 young, with the older and larger females producing the largest litters. A typical medium-sized boa of about 6 feet (1.5 m) in length will usually give birth to about 10 to 15 young.

Human Adversaries

Common boas have a number of enemies, including birds of prey and carnivorous mammals. As they grow, the number of animals that can harm them diminishes, and they may even be able to turn the tables on some of them—a captured adult boa has been seen to eat an ocelot. By the time they are 6.5 feet (2 m) long, they probably have few natural enemies apart from humans. Boas are hunted for food and for their skins, and are killed by traffic. Boas are also collected for the pet trade, although most of the demand in recent years has been met by captive breeding—wild common boas are protected internationally by CITES regulations. Their main means of defense is to retreat into a hole or crevice; but if cornered, they defend themselves vigorously by hissing loudly and lunging repeatedly. A bite from a large boa can cause serious injury.

A newborn common boa emerging frm its egg sac. Each newborn snake measures 18 to 24 inches (50 to 75 cm) in length.

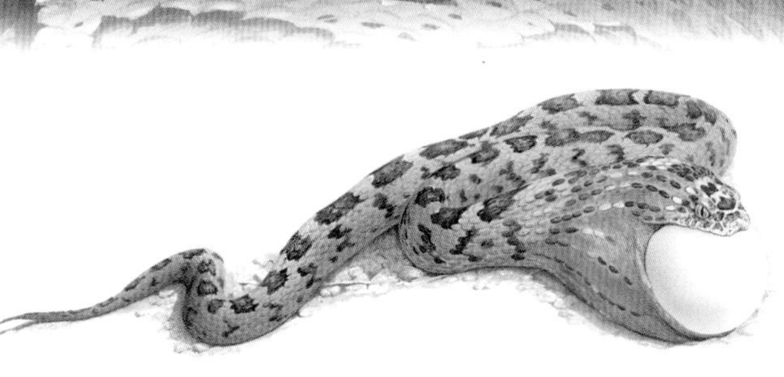

Common name Common egg-eating snake (rhombic egg-eater)

Scientific name *Dasypeltis scabra*

Subfamily Colubrinae

Family Colubridae

Suborder Serpentes

Order Squamata

Length From 31 in (80 cm) to 39 in (1 m)

Key features Slender snake with a small head and rounded snout; markings often consist of darker blotches or chevrons on a light-gray, brown, or reddish background, but some individuals are plain colored; eyes small with vertical pupils; scales are very heavily keeled and have a rough texture

Habits Mostly terrestrial, although will climb into bushes and trees to find food; active at night

Breeding Egg layer with clutches of 6–18 eggs; may lay more than 1 clutch in a single breeding season; eggs hatch after about 60 days

Diet Birds' eggs

Habitat Lives in a variety of habitats and only avoids extreme deserts and rain forests

Distribution Africa south of the Sahara; also a small relict population in North Africa, in western Morocco

Status Very common but not often seen because of its nocturnal habits

Similar species Mimics a number of different venomous snakes (including adders, of which there are several species) according to its locality; there are 5 other species in the genus, some of which are similar to the common egg-eater

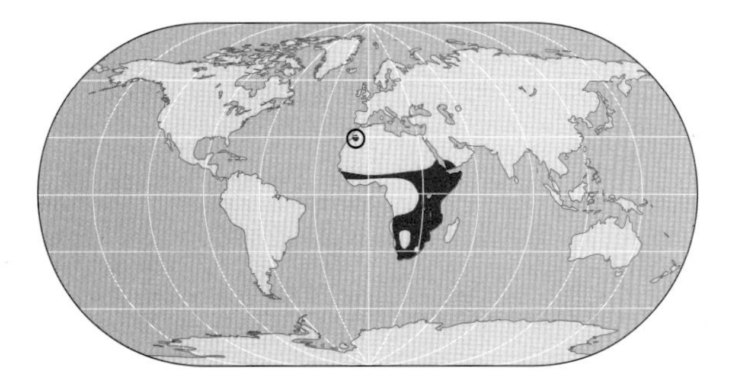

Common Egg-Eating Snake

Dasypeltis scabra

Few snakes are as specialized as the African egg-eating snake. This rather delicate-looking colubrid only eats birds' eggs, which it has to locate and swallow whole. Its abundance in some places is testament to the success of its technique.

LIVING EXCLUSIVELY ON birds' eggs can be a risky business, not least because birds breed seasonally over much of the snake's range, and because, like sensible parents everywhere, they take precautions to stop predators from stealing their offspring. As a result, the egg-eating snake has evolved specialized anatomical structures and behavior patterns in order to succeed.

The greatest hurdle for an egg-eater must be to survive the first year of its life. Hatchlings are bootlace thin, and it almost seems impossible that such small snakes could engulf birds' eggs whole. Youngsters start by feeding on the eggs of small species such as weaver birds (*Ploceus* species) and titmice (for example, *Parus fasciventer*) as hatchlings. They graduate to those of doves, pigeons, and ground-nesting game birds as they grow. Adults, with a head no bigger than a man's thumbnail, will accept chickens' eggs (although they often raid nests containing smaller eggs, too).

The outline of a recently swallowed egg can be clearly seen in this West African egg-eating snake, *Dasypeltis fasciata* in Ghana.

Egg Detectives

The snakes have a number of unique adaptations associated with their feeding habits. They have a highly developed sense of smell and can detect birds' eggs at some distance and differentiate between those that are freshly laid, which they eat, and partially developed ones, which they do not. They closely investigate each egg before swallowing, using the tongue and snout to judge its size. Mistakes can happen, however: Juveniles have died tackling eggs that are too large for them and that have become stuck in their throat.

Once the snake decides the egg is acceptable, things happen rather quickly. It may roll the egg against a solid surface so it can push against it, or it may maneuver it against a loop of its own body. Small eggs are attacked from the top and held against the ground or the bottom of the nest. The snake's mouth expands greatly to take the egg, and the snake then pushes the rest of its body over the egg.

Once the snake has engulfed the egg, it immediately pushes it back into its throat, and the egg disappears from view save for the enormous bulge it causes. Once the egg is in the throat, the snake begins to "saw" through the top of the shell, using its gular "teeth"—actually sharp, downward-pointing extensions of its vertebrae. This results in a strange rippling motion just behind the snake's head. Eventually, the gular teeth penetrate the shell, and it collapses—the cracking sound can be heard quite easily as this happens. The snake swallows the contents of the egg, but not the shell: That is regurgitated a few minutes later, a process that often seems to be more of a struggle for the snake than the swallowing itself. It uses the same rippling motion that it used to swallow the egg, but in reverse. The regurgitated

↑ ***Egg-eating snakes often mimic more harmful species, such as vipers or adders, with different species imitating the colors and markings of venomous species that occur in the same region.***

How Do they Do it?

The feeding habits of egg-eating snakes have been known for well over 150 years, but the mechanics of the operation were not properly understood until quite recently. Eating eggs that can measure up to three times the diameter of the head needs serious modifications to the egg-eater's skull and jaws.

Their teeth are so small that they are barely visible, buried inside their gums—long teeth, essential for capturing live prey, would hamper the ingestion of eggs. Their reduction in size was an evolutionary commitment to subsisting exclusively on eggs. Their skull bones are highly modified in several ways to increase the snake's gape. This also involves the expansion of the skin and internal soft tissue around the jaws, which lie in folds when the mouth is closed. (The same structures also enable the egg-eater to flatten its head and spread its jaws in imitation of a viper when threatened by predators.)

Under the chin are a number of large plates that reach across the bottom of the mouth. This is in contrast to the usual arrangement in which there is a joint, called the mental groove, running down the center of the chin, separating the scales on each side. When the egg is forced down into the snake's throat, it lodges in an area roughly beneath the 17th and the 38th vertebrae, whose lower surfaces are strengthened and modified into spines. The spines are known as gular "teeth," and they protrude into the snake's throat—those in the center actually penetrate the esophagus wall.

The snake brings pressure to bear on the eggshell by contracting powerful muscles along its neck. The first group of spines weaken the shell, after which it is pushed farther back toward the longest spines. As well as crushing the shell, the spines prevent the egg from slipping too far down the throat. The eggshell is sawn through using the longer spines, and the empty shell is forced back up into the snake's mouth, from which it is spit out.

No other snakes have such specialized feeding apparatus. Although many other species, including ratsnakes and king snakes, eat eggs, they swallow the egg whole and digest the shell. Only the Japanese ratsnake, *Elaphe climacophora*, has spines on the undersides of several vertebrae to crush eggs, but they are smaller than those of the egg-eater, and the snake swallows the shell.

eggshell has a characteristic boat shape caused by the edges around the cut being rolled inward as it collapses.

Irregular Meals

Because of the seasonal availability of their food, egg-eaters tend to gorge themselves when eggs are plentiful and can go for many months with no food whatsoever. Considering their feeding habits, it is a surprise to learn that the common egg-eater is not especially arboreal but is more at home on the ground. Of the six species in the genus, only one species, the East African egg-eater, *D. medici*, is considered to be arboreal. It is more elongated than the other species and has a longer tail. It lives in forested areas, whereas the common egg-eater and the other four species prefer open woods and scrub-covered countryside. They forage on the ground and occasionally climb into low bushes and trees when not actively searching for food. Common egg-eaters frequently turn up in termite mounds and in the nests of weaver birds, having first eaten the eggs. They appear to be completely nocturnal in habits and are often abroad on cool nights at the beginning of the bird-breeding season.

Demonstrating the amazing ability of egg-eating snakes to widen their gape to envelop eggs, this specimen from western Africa has lodged an egg in its mouth and is preparing to swallow.

Dasypeltis fasciata ***with an egg lodged in its neck. Having broken the egg, the shell has partially collapsed, and the snake is ready to regurgitate it.***

Convincing Mimics

Another fascinating part of the egg-eater's repertoire is its defensive behavior. Egg-eaters across Africa tend to be colored and marked in ways that match the venomous snakes with which they share their range and habitat. Some populations look like rhombic night adders, *Causus rhombeatus*, others look like lowland swamp adders, *Proatheris superciliaris*, and yet others look like saw-scaled vipers, *Echis* species. In parts of the western Karroo in South Africa common egg-eaters are reddish-brown in color, partially to match the red soil on which they live and, presumably, to mimic the horned adders, *Bitis caudalis*, that also live there.

Common egg-eaters often try to increase their defense by copying the behavior of the adder they mimic. In one of its most impressive displays the common egg-eater forms itself into a horseshoe-shaped series of coils, then moves the coils against each other to produce a loud rasping sound. The keels on the scales of its flanks are angled down and have minute serrations along them. As the snake coils itself, the saw teeth rub against one another. This behavior is a convincing imitation of saw-scaled

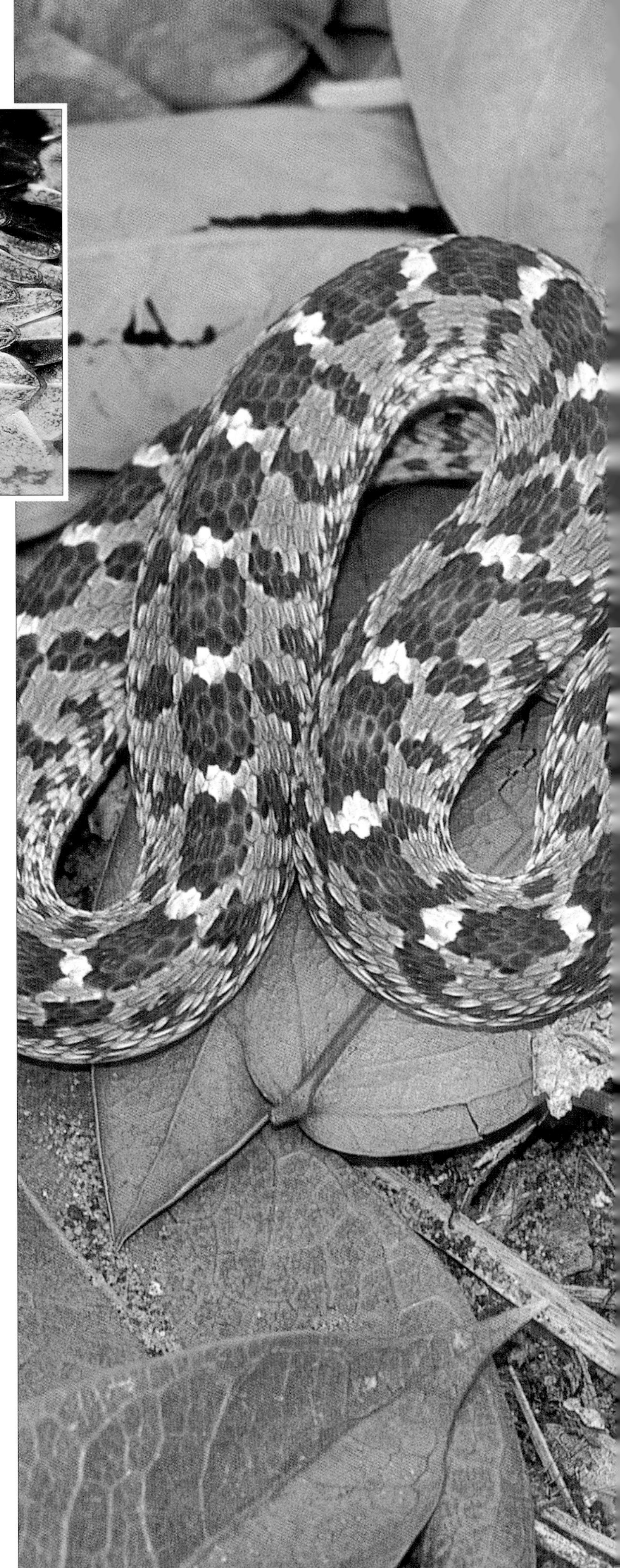

Closeup of an egg-eater's scales. To deter predators, the keeled scales can be rubbed together to produce a sound similar to that produced by African vipers.

Common egg-eaters from the Kalahari region are reddish in color to match the soil on which they live.

vipers, *Echis*, and desert horned vipers, *Cerastes*, which also occur in Africa.

To further enhance its deception, the snake flattens its head and spreads its jaws, giving it a viperlike profile. It may lunge at its aggressor, opening its mouth widely to show off the black interior. Even egg-eaters living in southern Africa where there are no saw-scaled vipers often go through this performance, perhaps because migratory birds of prey are among their main predators. Some of these birds will have had experience of the dangerous "raspers" farther north.

Breeding

Apart from eating eggs, egg-eaters also lay them. Large female common egg-eaters can reportedly produce clutches of up to 25 eggs, although six to 10 is more usual. At 82°F (28°C) they hatch in about 60 days. Hatchlings measure about 8 inches (20 cm) in length. One can only assume that a large proportion fail to find an adequate supply of tiny eggs, but some feed well enough during their first few weeks to survive the eight or nine months when there

is no food to be had. They grow quickly, and by the following season they are large enough to accept the eggs of a wider variety of birds, at which time their future is more secure.

They reach breeding size in two to three years, and at maturity the males are noticeably smaller than the females. Mating takes place in the spring, and females can continue to lay fertile eggs after a single mating, producing clutches about every six weeks throughout the summer. Egg-eaters living in tropical parts of Africa (including some populations of the common egg-eater as well as some of the other species) probably breed throughout the year or for an extended period, and their food supply is also likely to be more constant as well.

The Other Egg-Eaters

Apart from the common egg-eater, there are five other species in the genus, all from Africa. Less is known about their natural history than that of the common egg-eater, but they are all birds' egg specialists. They are the montane egg-eater, *Dasypeltis atra*, which may be black, plain brown, or brown with darker markings; the West African egg-eater, *D. fasciata*, which is pale olive in color with subtle mottling of slightly lighter and darker markings; the East African egg-eater, *D. medici*, which is dull brick-red or pink in color, often with a broad central stripe of a darker hue interrupted by small pale markings, but sometimes plain colored; the southern brown egg-eater, *D. inornata*, whose range is fairly small along the southeastern coastal woodlands of South Africa; and finally the almost unknown *D. palmarum*, most of whose range falls within Angola, herpetologically speaking one of Africa's least explored countries.

A seventh, unrelated egg-eating snake, *Elachistodon westermanni*, comes from India, but this rare and poorly known species is not closely related to the *Dasypeltis* genus, and its egg-eating habits may have evolved independently. It is not known if it feeds exclusively on birds' eggs or whether they make up just part of its diet.

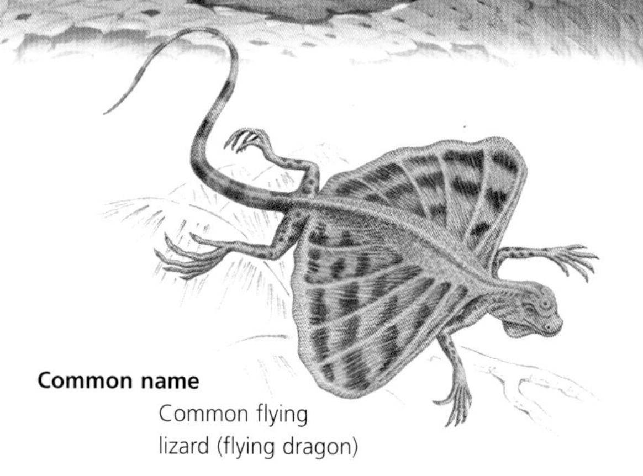

Common name Common flying lizard (flying dragon)

Scientific name *Draco volans*

Family Agamidae

Suborder Sauria

Order Squamata

Size From 6 in (15 cm) to 8 in (20 cm)

Key features Body slender with a long, thin tail and long, spindly legs; there is a flap of skin on each side of the body supported by elongated ribs; these "wings" are normally held against the side of the body but can be opened for gliding; wings are mottled orange and black; both sexes have a dewlap (piece of loose skin) on their throat that is yellow in males and blue in females

Habits Diurnal; arboreal, living on vertical tree trunks

Breeding Egg layer with clutches of 3–6 eggs buried in the ground

Diet Small insects, mainly ants and termites

Habitat Rain forests

Distribution Southeast Asia (Malaysian Peninsula, Indonesia, Borneo, and the Philippines)

Status Common

Similar species There are 28 species of flying lizard altogether, and their ranges overlap in many places; they are most easily distinguished by the color of their wings and dewlaps

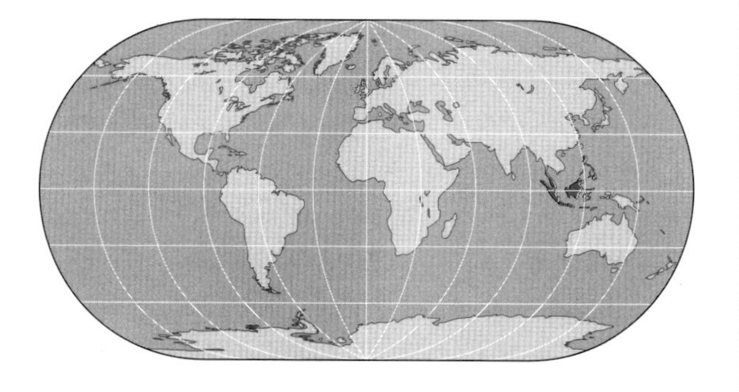

Common Flying Lizard

Draco volans

Flying lizards, of which Draco volans *is one of the most widespread species, live on large tree trunks in rain forests. They flit from one tree to another using their collapsible "wings."*

THE FLYING LIZARDS DO NOT ACTUALLY FLY—they glide between the trees. Their "wings" are formed from a membrane of skin supported by elongated ribs, usually five but sometimes six. The ribs are movable; and when the lizard is at rest, they fold down like an old-fashioned fan. During gliding, however, the first two ribs are swung forward by muscular contractions. They in turn pull the remaining ribs forward, because they are attached by ligaments. The wingspan is roughly equal to the lizard's body length.

Free-Falling

To glide, the lizard first drops from its resting place with its wings folded and its tail raised, so that it falls at a very steep angle. Almost immediately, however, it erects its wings and lowers its tail so that its rate of fall is slowed, and it glides almost horizontally. It uses its tail as a rudder to steer. Just before it lands, the lizard raises its tail again and changes the shape of its wings so that they scoop more air. These actions slow the lizard down and change its position so that its head is pointing up, in the same way as when an aircraft stalls. The lizard can then land lightly on its new tree trunk.

It runs up the tree immediately to regain the height lost during the glide. Flying lizards can glide for up to 50 yards (46 m), but their normal "flight" is much shorter, often a route between two neighboring tree trunks that are only a few yards apart. Taking the 28 species of the genus as a whole, small species can glide farther than large species; the common flying lizard, *D. volans*, is roughly intermediate in size.

↑ The common flying dragon can glide considerable distances using lateral membranes stiffened by elongated ribs. It controls its descent with a degree of accuracy by small adjustments of the tail and membranes.

Other Winged Reptiles

The only other flying lizards are the flying geckos, *Ptychozoon*, which also live in Southeast Asia, although there is evidence of limited gliding in another gecko, the wood slave, *Thecadactylus rapicauda* from Central and South America, as well as in a lacertid, the blue-tailed tree lizard, *Holaspis guentheri* from southern Africa. Flying snakes, *Chrysopelea*, are also found in Southeast Asia, as are a number of flying frogs, *Rhacophorus* species.

When at rest, the lizard holds its head away from the tree trunk, probably to give it room to raise and lower its dewlap. All flying lizards have dewlaps, but they vary in size, shape, and color according to species and sex. The common flying lizard has a triangular dewlap that is yellow in males and blue in females. The female's dewlap is smaller than the male's. Variations in color and shape of the dewlap help different species recognize one another, just as anole lizards in North, Central, and South America use their dewlaps.

Female *Draco volans* are slighter larger than males. That is not the case in every species of flying lizard, but females of all species have relatively larger wings. These probably enable them to fly even with the added burden of a clutch of eggs. In one species at least (*Draco melanopogon*) and possibly others, females have a larger head relative to body size than males. It might act as a counterweight to the eggs, which are carried in the abdomen. However, this theory begs the question of how the female lizard manages to allow for the larger head when she is not carrying eggs.

Flying lizards spend their entire lives high up in forest trees except during one activity: Females need to come down to the ground to lay their eggs. They are not agile on the forest floor and are very susceptible to predation. They lay small clutches of three to six eggs, which are elongated and spindle shaped with heavily calcified caps at each end. As far as is known, this type of egg is unique to flying lizards, but the reason for its shape is unknown. Details of incubation and the size and habits of the hatchlings are also unknown.

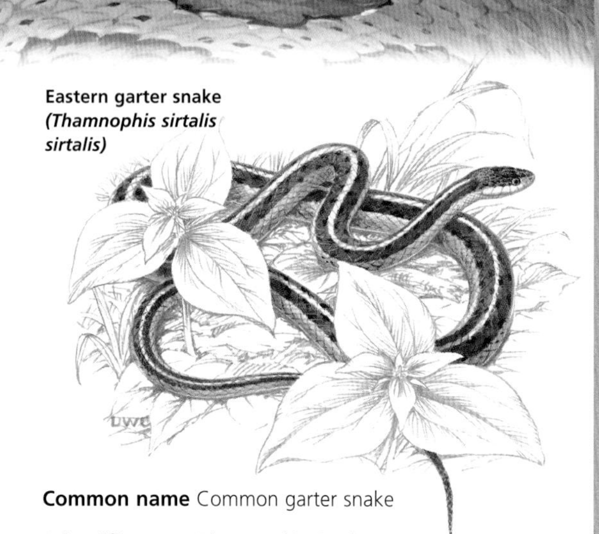
Eastern garter snake (*Thamnophis sirtalis sirtalis*)

Common name Common garter snake

Scientific name *Thamnophis sirtalis*

Subfamily Natricinae

Family Colubridae

Suborder Serpentes

Order Squamata

Length From 18 in (45 cm) to 4.25 ft (1.3 m)

Key features Body slender; head well separated from neck; large eyes with round pupils; keeled scales; pattern consists of several longitudinal stripes from neck to tail, but width and color of stripes depend on locality

Habits Semiaquatic in most places; diurnal and very alert, taking off at the slightest disturbance

Breeding Bears live young, with litters of up to 24; gestation period 60–90 days

Diet Amphibians, fish, and invertebrates such as worms

Habitat Ditches and ponds, streams in dry regions, damp meadows, marshes, parks, and gardens; from sea level to 8,000 ft (2,450 m)

Distribution North America (Atlantic to Pacific coasts and Canada to the Southwest and Southeast); absent from deserts

Status Common in most places, but some forms are rare: The San Francisco garter snake, *T. sirtalis tetrataenia,* is one of North America's rarest snakes

Similar species Garter snakes in general can be difficult to identify since most of them are striped

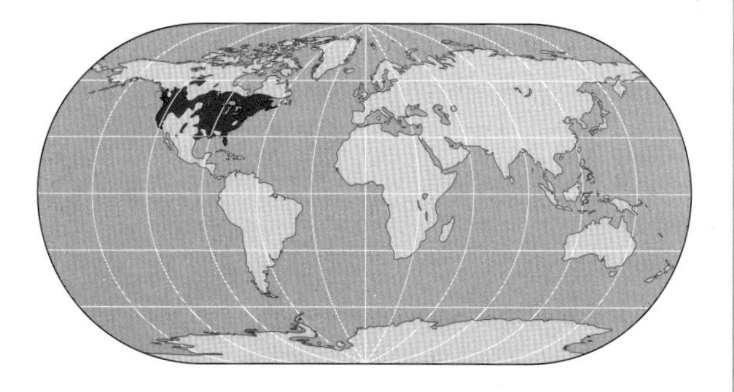

Common Garter Snake

Thamnophis sirtalis

The common garter snake is an easy species to study since it is active in the day. They are also numerous and fairly predictable in their comings and goings.

SEVERAL IMPORTANT STUDIES on snake biology have used the common garter snake as their subject, including investigations into hibernation, chemical communication, and feeding strategies. Some results are specific to garter snakes, but many have helped us understand snakes as a whole.

Because common garter snakes have a large north–south range, their behavior differs from place to place. In southern states such as Texas and Florida, for example, they are active all year around, although they may take cover for a few days at a time during cold snaps.

Suspended Animation

However, in the north, especially in parts of Canada, common garter snakes hibernate for several months every year and undergo dramatic body changes that allow them to survive temperatures close to freezing. The red-sided garter snake, *Thamnophis sirtalis parietalis*, lives in a region that has winter temperatures as low as -40°F (-40°C), with snow on the ground from September to May. During hibernation their body temperature plummets, and they live in a state approaching suspended animation, moving only rarely and extremely slowly. Their blood thickens as it begins to freeze.

Their hibernation dens can become flooded with water, and the snakes are totally submerged. If this occurs, they slow down their system and use less oxygen than usual, obtaining what little they need from the water by absorbing it through the skin. Once spring arrives, they begin to emerge, often while snow is still on the ground.

This eastern garter snake, **Thamnophis sirtalis sirtalis** *from Canada, has distinctive red stripes down its sides.*

Breeding Patterns

Common garter snakes from different regions have different breeding systems. Those in the far north (such as the red-sided garter snake) are the most studied. Their breeding season is concentrated into a few weeks or less, and competition between males is intense. Males emerge first and wait around the den site for females to emerge. They bask by day but return to the den at night to avoid the cold.

Females emerge gradually over a period of about a month. As each female emerges, she secretes pheromones that stimulate the males. These airborne scent molecules are specific to each species of garter snake, so that where several species occur close together, males do not waste their time and energy pursuing females of the wrong species.

Life in the Freezer

Animals that live in subzero temperatures have to develop behavioral or physiological means of protecting themselves from freezing. Birds and mammals manage by producing metabolic heat internally. They can maintain their body temperatures at a safe level regardless of the environment. Cold-blooded animals do not have this option, however, and have to evolve an alternative strategy. Two such strategies have been explored by reptiles. The first is "supercooling," in which the body fluids can fall below their normal freezing point because they contain substances known as cryoprotectants that act as antifreeze to prevent the formation of ice inside the animal cell.

Garter snakes have cryoprotectants, but only in small amounts, mainly in their liver and heart. They mainly rely on another system known as freezing tolerance. As the temperature drops, ice begins to form in their body fluids, and their blood thickens as it begins to freeze. In experiments garter snakes were able to survive temperatures as low as -58° F (-50°C) for up to three hours in the fall, during which the blood contained up to 40 percent ice. After 10 hours the ice content rose to 50 percent, and only about half the snakes survived. After 24 hours it rose to 70 percent, and none survived. In practice what this means is that garter snakes can survive sudden temperature drops in the fall when they are still above ground. By midwinter they are safely tucked away in their underground chambers and are unlikely to experience such severe temperature fluctuations. As would be expected, their tolerance of freezing temperatures at that time of the year is lower. Despite these "tricks," winter mortality rates among garter snakes can be high—34 to 50 percent in harsh winters.

The male detects the scent with his forked tongue and rubs his snout along her back.

Dozens of males mob each female, forming huge mating balls with the larger female at the center. Each male rubs his chin along the female's back and tries to wrap his tail under hers to copulate. Eventually one male succeeds, and the others abandon the female in search of others. Mating lasts about 15 to 20 minutes. After mating, the successful male leaves a "copulation plug" in the female's vent. This waxy gelatinous plug forms a physical barrier to other males trying to mate with the female and may also contain pheromones that inhibit other males. The successful male is free to pursue other females, safe in the knowledge that his sperm cannot be displaced by a second male.

Summer Migrations

Females leave the den site as soon as they have mated and migrate to their summer feeding sites. Males do not leave until all the females have dispersed. It seems likely that some males mate with several females, but most of them do not mate at all.

The best feeding grounds can be quite a distance from the best hibernation sites, and common garter snakes travel up to 11 miles (18 km) to reach them. They are thought to navigate by the sun, but other factors including pheromone trails can deflect them. Some individuals' travels apparently take the form of a loop that begins and ends at the hibernation den rather than retracing their route in the fall.

Developing Young

Males are free to feed throughout the summer and build up reserves for the next period of hibernation, but females stop feeding when their developing young begin to take up space in their bodies. Garter snakes are truly viviparous: The developing young obtain nourishment from their mother through her blood, which also transports embryonic waste products away from them.

The gestation period is about 60 to 90 days, and the young are born toward the end of summer. Bearing live young has huge benefits in cold climates, where females can bask in the sun to raise their body temperatures

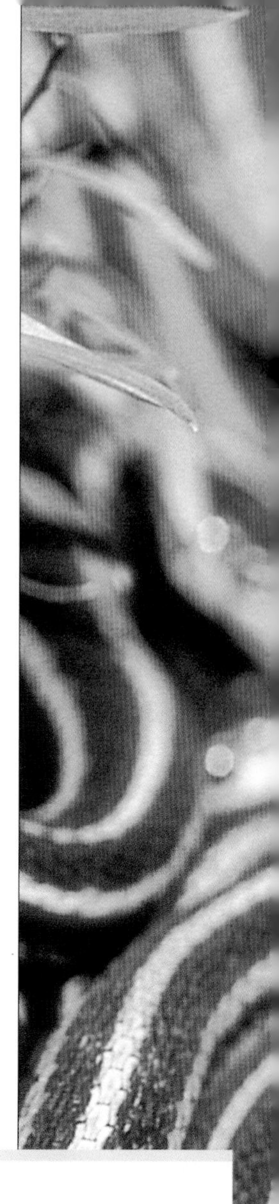

Large Congregations

Garter snakes from northern climes come together to hibernate in underground limestone cavities where frost cannot reach them. Some dens have an estimated 8,000 to 10,000 snakes in them, although populations fluctuate. Farther south, in Michigan for example, winter aggregations are smaller, with up to 150 individuals, while in the far south garter snakes do not congregate at all.

They do not always use underground crevices—abandoned ant mounds are widely used. In a two-year survey in the 1960s 11 ant nests yielded 2,019 red-bellied snakes, *Storeria occipitomaculata*, 276 smooth green snakes, *Liochlorophis vernalis*, and 131 common garter snakes. The highest number from a single mound was 299 snakes, including all three of the above species.

Snakes congregate during hibernation for three main reasons. First, there is often a shortage of good hibernation sites; this is increasingly important in areas where temperatures drop dramatically, because only the deepest crevices will provide adequate protection from the cold. Second, by clumping together, the snakes may gain an additional advantage in reduction of heat loss. Third, they can use each other for mutual insulation and reduce water loss by exposing less of their surface to the atmosphere. The last reason is especially important for small snakes, including several north American species such as the brown snake, *Storeria dekayi*, and the closely related red-bellied snake, Butler's garter snake, *Thamnophis butleri*, and common garter snakes.

A group of red-sided garter snakes,* Thamnophis sirtalis parietalis, *emerge from hibernation at the same time. Large numbers aggregate in favored places and hibernate for many months of the year.

and therefore those of the developing young. On the other hand, females that have carried their young throughout the spring and early summer have little time to build up their fat reserves. It is likely that in the north they breed every second year, but in the south they have a longer season and breed every year.

Litter sizes vary greatly in garter snakes. Most forms give birth to six to 12 young, but much larger litters are possible, and litters of up to 80 have been recorded.

Varied Diet

All garter snakes eat a variety of food items, including earthworms, leeches, fish, and amphibians. Different populations have different preferences, probably depending on availability. Some populations enter tidal pools to hunt for small marine fish, while others prey extensively on toxin-bearing California newts, which individuals from other populations avoid altogether. Scientists think that where newts are an important food source, the young snakes obtain immunity to toxins from their parents.

Garter snakes are diurnal and hunt initially by sight, but they need the stimulation of smell before they will attack their prey. Once food has been identified as edible, it is simply grabbed and swallowed. Garter snakes eat mostly food that cannot fight back, so there is no rush to subdue it before swallowing. It is not unknown for prey that has been swallowed to be found still alive if it is disgorged later. A few of the larger garter snakes, especially those from drier habitats, occasionally eat small mammals and nestling birds.

Different Forms

Because of its large range, the common garter snake shows considerable variation in form, and several subspecies are recognized. Many bear little resemblance to one another and can be difficult to identify.

The most distinctive form is the San Francisco garter snake, *T. s. tetrataenia*, a highly endangered subspecies that lives only in four small sites within San Mateo County, California.

Thamnophis sirtalis concinnus, *the red-spotted garter snake, is a beautifully colored subspecies found along the coasts of Oregon and California.*

This beautiful snake has a red head. A pair of bright-red, black-edged stripes runs along its back on either side of the white central stripe that is common to most members of the species. Since it lives in a built-up part of California, the species' best hope of survival is through the captive-breeding programs in North American and European zoos.

Neighboring subspecies, including the California red-sided garter snake, *T. s. infernalis*, and the Oregon red-spotted garter snake, *T. s. concinnus*, also have extensive areas of red in their patterns, while others, such as the northern red-sided garter snake, *T. s. parietalis*, have red spaces between their black scales, which only show when their skin is distended, either after they have eaten a large meal or when they flatten themselves in response

A San Francisco garter snake,* Thamnophis sirtalis tetrataenia, *takes to the water in California. These beautiful snakes are highly endangered.

The Relatives

There are 26 species of garter snakes altogether and another two species of *Thamnophis* that are popularly known as ribbon snakes, making 28 in all. Of these, 16 occur in the United States, including three that enter Canada and five that enter Mexico. Another 12 live in Mexico (but not the United States), and they include three that extend down into Central America. Two species, the checkered garter snake, *T. marcianus,* and the western ribbon snake, *T. proximus*, extend from the United States in the north to Costa Rica in the south, while one species, *Thamnophis valida*, only occurs in Baja California, Mexico.

to a possible predator. Some herpetologists think that this show of red may be a form of warning coloration.

Common garter snakes in the central and eastern states are less likely to have red in their pattern, but one subspecies from Florida, *T. s. similis*, has an overall bluish wash. It is also among the largest of the common garter snake subspecies. Melanistic (all-black) individuals have been found in several populations but are especially frequent in coastal areas, including those of the Great Lakes, and on islands. George Island, in Halifax harbor, Canada, has a particularly high incidence of melanistic common garter snakes as well as a number of other unusual color forms.

California king snake *(Lampropeltis getula californiae)*

Common name Common king snake (each subspecies has its own name, such as California king snake, black king snake, etc.)

Scientific name *Lampropeltis getula*

Subfamily Colubrinae

Family Colubridae

Suborder Serpentes

Order Squamata

Length 35 in (90 cm) to 5.8 ft (1.8 m)

Key features Muscular snake with an almost cylindrical body; small head hardly wider than neck; pupils round; smooth, glossy scales; markings variable but nearly always consist of a contrasting pattern of black (or dark brown) and white (or cream) in various arrangements; the Mexican subspecies is uniformly black

Habits Nocturnal; mainly terrestrial

Breeding Egg layer with clutches of up to 24 (but typically 6–12); eggs hatch after about 70 days

Diet Small mammals, lizards, and other snakes

Habitat Varied from lowland swamps to deserts

Distribution Southern half of the U.S. and adjacent parts of northern Mexico

Status Common in places

Similar species None in the area

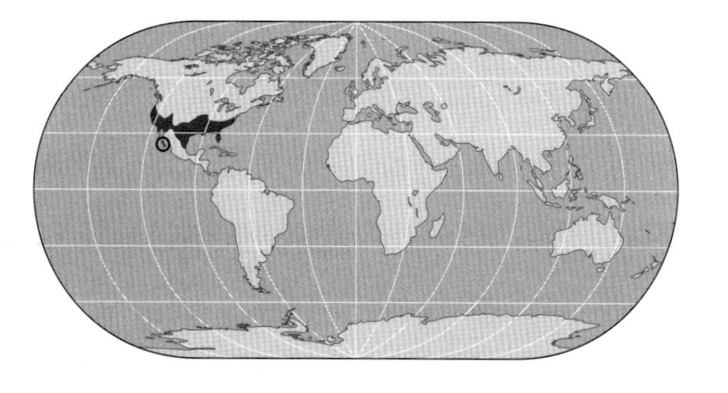

Common King Snake

Lampropeltis getula

The king snake probably gets its common name from its habit of killing and eating other snakes, including venomous ones. Snakes, however, form only part of its diet. A powerful constrictor, it tracks down small mammals and reptiles before squeezing them to death.

King snakes hunt at night, covering the ground methodically as they search for prey in burrows or rock crevices. Their roughly cylindrical shape helps them enter small spaces and burrow through loose soil, sand, or leaf litter. When they locate their prey, they throw several coils of their body around it and squeeze tightly. When it stops breathing, they swallow it. If the prey is a venomous snake, such as a rattlesnake, the king snakes bite it behind the head and coil around its upper body. In this way they immobilize it; but if they do get bitten, they are immune to the venom and can retain their vicelike grip. King snakes have been found with rattlesnakes equal to their own length inside their stomach, packed in like a concertina.

Pit vipers identify king snakes by smell and react by immediately taking up a defensive posture: They raise the thickest part of their body off the ground in an arc. They then thrust it toward the king snake while trying to make themselves look as large as possible. They may even use the raised part of their body to club the king snake.

Egg Thieves

Less ambitious items on the king snakes' menu include birds' eggs (especially those of ground-nesting species) and reptile eggs, including those of freshwater turtles. They find the latter by smell and dig them up using their snout. In places where turtles nest regularly, king snakes may travel some distance to the breeding sites at the appropriate time of year, showing that they have good memories. They also eat young

The king snake, **Lampropeltis getula,** *devours a black-headed snake,* **Tantilla** *species.*

turtles on occasion. In short, king snakes are "general purpose" snakes.

Because of their southerly distribution king snakes may be active all through the year. But populations living at high elevations, such as the desert king snakes, *L. g. splendida* from Arizona, and animals from the northern states, may undergo a period of prolonged hibernation in winter. Otherwise, their activity period depends on local conditions, and they may simply become inactive during cold snaps.

Either way they mate in the spring, and males fight for dominance, intertwining their bodies while lying on the ground and testing each other's strength. The female lays her eggs in the summer, about six to eight weeks after mating. She chooses a place that will retain some moisture. The eggs usually stick together to form a single clump, and they take about 70 days to hatch. The young measure 10 to 12 inches (25 to 30 cm) in length.

Variation

Across their geographical range king snakes have remarkable color variations. In the southeastern United States, for example, each scale is yellow or cream with black edges. Some

As its common name implies, the variable king snake,* Lampropeltis mexicana thayeri *from Mexico, occurs in different color phases. Its background color can be light gray, peach, silver, or buff.

scales have more yellow on them than others, and may be arranged into a faint pattern of bars or blotches. The speckled king snake, *L. g. holbrooki*, is the most evenly marked of this type, with each scale having a small yellow spot at its base. Moving farther west, the markings have progressively more contrast, until on the West Coast the California king snake, *L. g. californiae*, may have one of two different but equally bold patterns. They are either black or brown with wide white or cream bands encircling the body, or they are brown or black with a single white stripe running right along the back. Interestingly, both forms can hatch from a single clutch of eggs, but the striped form only occurs in part of the subspecies' range (mainly in San Diego County, where about 30 percent of king snakes are striped). One subspecies, the black king snake, *L. g. nigritus* from Sonora, Mexico, is jet black in color without any traces of paler markings.

In the West desert forms tend to be black and white, whereas forms from farmland, woodland, swamps, and coastal marshes tend to be less contrasting. They are often brown and cream or even yellow.

Night Hunters

Although they are basically nocturnal, common king snakes may be active in early morning and late afternoon in the spring and fall. However, in midsummer they rarely emerge before sunset, especially in the warmer parts of their range (such as California, Arizona, and Mexico). Their shiny skin reflects car headlights, and they are often seen crossing roads at night. During the day they hide in rotting stumps, under debris (including human trash), among rocks, and in rodent burrows. In the American Southwest they sometimes shelter at the base of old agave bushes.

The common king snake is popular among snake keepers and breeders in all its forms

Lampropeltis getula brooksi, ***the southern Florida king snake, lays a clutch of eggs. She will leave shortly after finishing, and the eggs will develop independently.***

because it adapts well to captivity and is usually placid and pleasant to handle. It breeds well in captivity—some king snakes can reach sexual maturity in just 18 months—and several color forms, such as albinos, have become commonly available through selective breeding.

Because they eat other snakes, however, common king snakes must be kept separately. Even introducing males to females at breeding times can be nerve-racking (for the male snake as well as its owner, presumably). They rarely eat each other, but an overeager female may grab a newly introduced male if she thinks she is about to be fed.

Polymorphism

Most species of snakes have fairly constant colors and patterns. Some species, however, exist side by side in two or more distinct forms. This is known as polymorphism. (To be exact, if they occur in just two forms, it should be known as dimorphism.) It is important to distinguish between regional variations, sometimes classified as subspecies, and polymorphism.

In polymorphism two or more forms occur together in the same region; and in the case of many snakes they may even hatch from the same clutch of eggs. Snakes can be polymorphic in color or in pattern. The eyelash viper, *Bothriechis schlegelii*, for example, may be colored green and brown, like lichen, or plain orange.

Other examples are the American mangrove water snake, *Nerodia fasciata compressicauda*, which can be greenish with dark blotches or plain dull orange, and the Asian crab-eating snake, *Fordonia leucobalia*, which can be plain yellow, orange, or black in color. It may even be mottled with a combination of these colors.

California king snakes provide a good example of pattern polymorphism. In most parts of their range they are black with white bands, but in some areas a proportion of them are black with a white line down the center of their back (sometimes the combinations are brown and cream, but the patterns are still the same). Other snakes with polymorphic patterns are the Sonoran ground snake, *Sonora semiannulata*, which can be plain, striped, or banded, and the leopard snake, *Elaphe situla*, which can be spotted or striped. Male boomslangs, *Dispholidus typus*, are not only different than the females (this is called sexual dimorphism) but are also highly variable themselves both in color and in pattern—a very complicated situation.

The purpose of polymorphic patterns is thought to be defense. To be successful, polymorphism relies on the principle that predators build up a mental picture or "search image" of their preferred prey species. Animals that do not match this image are often ignored, even though they may be equally good to eat. If a predator builds up an image of a striped snake, for example, the banded individuals may be overlooked. As a rule, predators would be expected to maintain a search image of whichever form was most common in the region, and the less common form would benefit by being overlooked. After a while the latter may become the most common form because of this, and predators in the area may switch their attention accordingly.

In the long term polymorphism is maintained in the population, and both forms would be expected to occur in roughly similar numbers, all other things being equal. (In the California king snake, however, the proportions are slightly skewed in favor of the banded form, so it appears that other factors may also have a bearing on the frequency of the two forms.)

Common name Common snapping turtle

Scientific name *Chelydra serpentina*

Family Chelydridae

Suborder Cryptodira

Order Testudines

Size Carapace length up to 24 in (61 cm)

Weight Up to 82 lb (37.2 kg)

Key features Head powerful; jaw hooked; barbels present on lower jaw with small tubercles on the neck and underparts; eyes prominently located near the snout; carapace brown and relatively smooth in older individuals but with a more pronounced keel in younger turtles; plastron relatively small in area with no patterning and varies from whitish to coppery brown; tail quite long with a crest running down the upper surface

Habits Relatively shy; spends long periods concealed in mud or vegetation; usually more active at night; often rests during the day, floating just under the surface with the eyes protruding

Breeding Female lays single clutch of 25–80 eggs (but may lay more than once a year); eggs hatch after minimum of 2 months

Diet Predominantly carnivorous; eats fish, amphibians, other turtles, birds, snakes, and small mammals; also eats plant matter

Habitat Occurs in virtually any type of standing or flowing fresh water, especially where there is a muddy base and vegetation

Distribution Southern parts of Canada through the United States and Central America south to Ecuador

Status Reasonably common

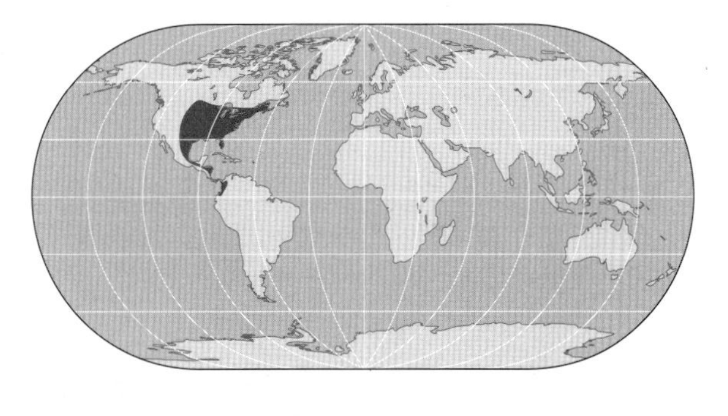

Common Snapping Turtle

Chelydra serpentina

Although it does not grow as large as its relative, the alligator snapping turtle, Macroclemys temminckii, *the snapping turtle is a voracious predator. It has powerful jaws that can inflict a painful, damaging bite.*

THERE IS SOME DISPUTE over the taxonomy of the snapping turtle. Its range extends from southern parts of Canada through the United States and Central America as far south as Ecuador. Although it is traditionally divided into four subspecies within this range, DNA studies have led to the suggestion that the population should be split into two or three distinct species, with those in southern areas being recognized as separate from the northern populations. In the east the Florida snapping turtle, *C. s. osceola*, is also considered to be a separate species by some taxonomists. It is restricted entirely to the Florida peninsula but appears to differ little from examples occurring elsewhere in North America.

Musky Females

The mating period of snapping turtles varies from April to November in different parts of their range. It is thought that females emit a pheromone from a gland in the cloaca that attracts males when they are ready to mate. The secretion has a distinctive, musklike odor. Courtship is aggressive: The male pursues the female, snapping initially at her head and legs to slow her down. If she accepts his advances, she raises her hindquarters, and both turtles then face each other for a period, while moving their head from side to side. The male then

moves around the female's body and climbs on top of her, anchoring himself with his feet. Mating lasts for about 10 minutes.

Female common snapping turtles do not need to mate every year in order to lay fertile eggs. In common with many other chelonians females are able to store viable sperm in their reproductive tract from previous matings, which will fertilize future egg clutches. As a result, the eggs laid by a female of this species can produce hatchlings of different parentages.

Nesting begins in May, peaks in June, and can continue until as late as September. A rise in air temperature above 50°F (10°C), especially in combination with some rain, triggers nesting behavior. Egg laying begins either at dusk or dawn. Larger females tend to nest earlier in the season and lay proportionately more eggs in a clutch than younger individuals laying for the first time.

Dug in a wide variety of soil conditions usually in the open and often some distance from water, the nests themselves are vulnerable to predators, including skunks and raccoons. Some snakes, notably the eastern king snake, *Lampropeltis getulus getulus*, will also eat turtle eggs readily. Often what happens is that eggs at the top of the nest are destroyed

⊖ Female snapping turtles make a bowl-shaped cavity in loose sand, loam, or plant debris. In the nest the eggs are vulnerable to predators, and this female in Ohio takes no chances and stays close by.

⊖ An average of 40 eggs are laid, and incubation lasts at least two months. The young emerge from the nest and head for water, where they hide under cover from predators.

as a result of predation, but those at the bottom survive and hatch.

The incubation period tends to vary markedly through the snapping turtle's range. Eggs hatch after just two months in warm surroundings but take much longer in northern parts, where the young may even overwinter in the nest before emerging the following spring. If the nest is shallow, however, they may be killed by frost over this period. The young hatchlings emerge under cover of darkness and are directed toward the water by the light reflecting off its surface. (Marine turtle hatchlings, notably loggerhead turtles, *Caretta caretta*, use a similar method of finding their way to the sea.)

Predators and Prey

Common snapping turtles measure about 1 inch (2.5 cm) when they hatch, and their long, flexible tail is similar in length to their shell. They may use the tail as an anchor at first, clasping onto vegetation to keep themselves from being swept away by the current once they enter the water.

They spend much of their early life concealed in these surroundings, often lying partly hidden in the streambed to avoid drawing attention to themselves. This is a particularly dangerous time for the young turtles—they face many predators, from wading birds such as herons to amphibians such as large bullfrogs, *Rana catesbeiana*.

Young snapping turtles have voracious appetites of their own, however, enabling them to grow rapidly. Males eventually reach a larger size than females. Both sexes are sexually mature once their carapace has grown to about 8 inches (20 cm) long, by which time they are between five and six years old. Their growth then slows significantly. Other changes in appearance are evident as they grow: The shell becomes relatively long compared with its width, and the tail is proportionately shorter compared with their overall size.

Their hunting habits also tend to change as they grow older, with mammals and birds more likely to fall prey to them at this stage. Muskrats, for example, may be seized and dragged under water. They are held under until they drown and are then eaten. Adult ducks may suffer a similar fate. Larger individuals have even been seen preying on smaller members of their own kind, and carrion also features in their diet. But even adult snapping turtles are not entirely safe from predators, particularly in southeastern parts of the United States, where their distribution overlaps with that of the American alligator, *Alligator mississippiensis*. These large reptiles will eat the whole turtle, including the shell, which is crushed in their powerful jaws.

Snapping turtles are also hunted on a wide scale for their meat. The equivalent of over 6,000 adults were caught commercially in Minnesota in the late 1980s, for example, and 8,000 in southern Ontario. Regulations are in force to regulate this trade, certainly in the northern part of the species' range, but trade in the subspecies *C. s. rossignonii* from Central America and *C. s. acutirostris* from South America is less well documented and could be endangering these populations.

Nasty Bite

Estimating numbers of snapping turtles in a given area is not easy because they are less inclined to enter baited traps than other turtles whose range overlaps with theirs. Handling common snapping turtles is not straightforward either, since they can be a genuine danger to the unwary. While most turtles can be safely held by the sides of their shells, common snapping turtles are able to reach around with their head and inflict a serious bite. They are highly aggressive when restrained, and the safest way to move an individual is by grasping the upper and lower ends of the carapace and holding it away from your body.

Few turtles occur in such a wide range of aquatic habitats as the common snapping turtles. They can be found anywhere from muddy pools to fast-flowing rivers and readily move across land if food becomes short, or if

water levels fall significantly. They often travel under cover of darkness, when they are naturally more active. During the day they often rest by floating just under the water's surface with their eyes protruding, keeping a watch on their surroundings. As a result, their shell develops a covering of green algae, which helps conceal their presence even more.

They rarely bask on land, however, unless plagued by leeches. In that case they are forced to dry off in order to make the parasites dehydrate and let go. Their disinclination to bask is probably related to the relatively large amount of water they lose from their body when on land as well as to their dislike of high temperatures.

During the winter, however, individuals in northern areas will hibernate in the mud on the bottom of a river or pond. They dig themselves in usually by the end of October and then emerge the following March. In areas closer to the equator the turtles may bury themselves in mud and wait for the rains to return if the pools in which they are living dry up. While common snapping turtles are essentially found only in fresh water throughout their range, on rare occasions they can be encountered in brackish areas.

↑ ***Not afraid to tackle even venomous prey, a snapping turtle eats a rattlesnake,*** **Crotalus viridis.**

Common name Wall Lizard

Scientific name *Podarcis muralis*

Family Lacertidae

Suborder Sauria

Order Squamata

Size 8.5 in (22 cm) long

Key features Graceful lizard with a narrow head, pointed snout, long tail, and relatively long limbs; coloration extremely variable, and many subspecies are recognized; some forms entirely brown with light and dark markings on their back and sides; in other places they are green (especially males) with extensive black markings; females usually have darker flanks than males and often have a line down the center of the back; markings of males are more likely to be netlike or scattered randomly

Habits Terrestrial and climbing; diurnal

Breeding Female lays clutches of 2–10 eggs that hatch after 6–11 weeks

Diet Insects and spiders

Habitat Dry, open places, including south-facing banks and rock faces, and stone walls around fields and the sides of buildings; often found in villages and the outskirts of larger towns

Distribution Europe from northern Spain and western France through Central Europe and northern Italy to the Peloponnese, Greece; also in Turkey.

Status Very common in suitable habitat

Similar species Many small lacertids from the region are similar and difficult to separate from the wall lizard and each other; locality is often the best means of identification

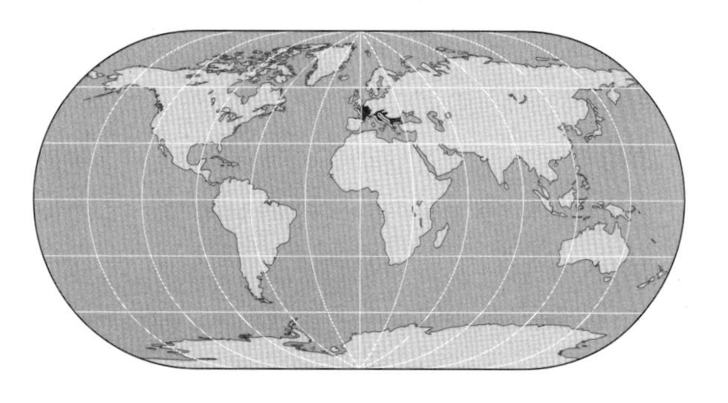

Common Wall Lizard

Podarcis muralis

In many parts of Europe the small lively lizards scurrying about on stone walls and paths, catching insects and pausing to bask occasionally, are most likely to be common wall lizards.

COMMON WALL LIZARDS ARE GENERALISTS and opportunists, and can occur in large numbers—up to 600 per acre (1,500 per ha) in good habitats, often around human dwellings. In some regions wall lizards are hardly ever seen away from houses and are replaced in open countryside by other related but more specialized types of wall lizards. They undoubtedly benefit from the flies and other pests that are attracted to villages and farmyards. They can be seen alternately basking and chasing food, occasionally taking cover in a crevice and slowly moving around during the day to follow the direction of the sun.

Males are territorial but not aggressively so, defending a territory of about 30 square yards (25 sq. m). Females lay from one to three clutches of eggs each year depending on their locality and the abundance of food. A typical clutch consists of about six eggs that are buried, often under a flat stone. They take six to 11 weeks to hatch, and the offspring mature in two to three years.

Lizard Neighbors

In Central Europe the wall lizard has little or no competition. Although it shares the region with the viviparous lizard, *Lacerta vivipara*, the latter prefers moister habitats, and the two are rarely seen side by side. Farther south it overlaps with the ranges of numerous other small lacertids. In Spain and Portugal, for instance, it occurs with the Spanish wall lizard, *Podarcis hispanica*, which is a little more delicately built and is more likely to be seen away from human dwellings on vertical rock faces. Over Italy, Sicily, Sardinia, and Corsica the Italian wall lizard, *P. sicula*, usually has more bright green on its back, often with a dark line down the center. This species occurs on many Adriatic islands, and there are 48 recognized subspecies! In the Balkan region there are a number of wall lizards, including the Balkan wall lizard, *P. taurica,* and Erhard's wall lizard, *P. erhardii*, but here the common wall lizard is again most likely to be seen around human dwellings.

The wall lizards of the Mediterranean islands are easier to identify because few islands have more than one species—the common wall lizard is absent. In the Balearics, for example, the large islands of Majorca and Minorca are home to Lilford's wall lizard, *P. lilfordi*, which occurs in a number of color forms, including several melanistic (all-black) populations on offshore islets. Some of these forms, or subspecies, include a large proportion of vegetable material in their diet. The other two Balearic islands are Ibiza and Formentera, and the wall lizard here is *P. pityusensis*, an attractive species that is often bluish in color: One form from the tiny conical islet of Vedra, *P. p. vedrae*, is especially brilliant.

Wall lizard populations on some of the uninhabited islets are unbelievably numerous. The Ibiza wall lizard can reach densities of over 12,000 per acre (30,000 or more per ha), while Lilford's wall lizard easily exceeds this at nearly 18,000 per acre (44,000 per ha) in places. On some islands it is impossible to walk without causing waves of lizards to move away in front of your feet.

Lacerta or *Podarcis*?

Some small European lacertids are included in the genus *Lacerta,* whereas others are in *Podarcis*. Previously they were all included in *Lacerta,* but changes in their classification in the 1980s and 1990s separated them on the basis of their internal anatomy. In practice, the more delicately built species with a slightly flattened body are *Podarcis* species. Those with a more robust body and deeper head are *Lacerta*, which also includes a number of larger species such as the European sand lizard, *L. agilis*, the eyed lizard, *L. lepida*, and the green lizards (of which *L. viridis* is one). There are no large *Podarcis* species.

The coloration of Podarcis muralis varies according to location. There are thought to be about 20 subspecies of this common wall lizard.

Copperhead

Agkistrodon contortrix

The copperhead is a common snake of the forested hills of the American Southeast, but because of its excellent camouflage coloring it is often overlooked.

Common name Copperhead

Scientific name *Agkistrodon contortrix*

Subfamily Crotalinae

Family Viperidae

Suborder Serpentes

Order Squamata

Length From 24 in (61 cm) to 4 ft (1.2 m)

Key features Head triangular with large scales covering the top and a prominent facial pit; a number of broad, rich reddish-brown, chestnut, or coppery bars cross the body and become narrower toward the midline, like a bow tie, but the two sides often fail to meet perfectly, so they are staggered; background color is tan, and despite its name, the head is also tan color

Habits Terrestrial; nocturnal or diurnal according to season

Breeding Live-bearer with litters of 4–14; gestation period about 83–150 days

Diet Small mammals, lizards, amphibians, and invertebrates

Habitat Well-drained, lightly wooded places, including rocky hillsides and gardens, often near streams and ponds

Distribution Southeastern United States and adjacent parts of northeastern Mexico

Status Common

Similar species Some water snakes in the region are superficially similar, but their habitat is different, and they lack the facial pit

Venom Mildly dangerous; copperheads are placid and do not bite unless provoked

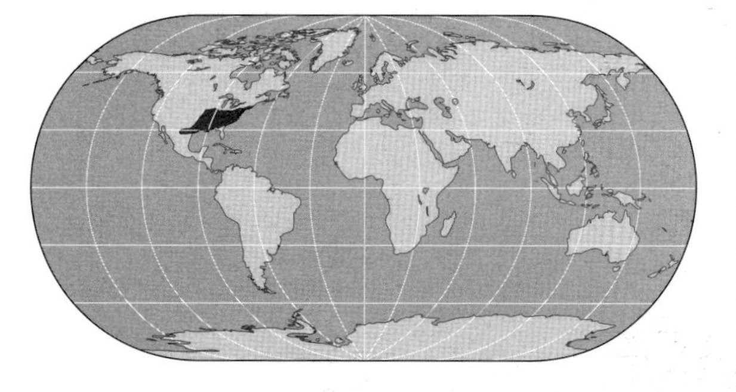

COPPERHEADS PREFER WOODLANDS with clearings in which to bask and rock outcrops in which to hide and hibernate. In the northern parts of their range the outcrops are even more important because the snakes may hibernate for up to six months of the year. Their other requirement is for water or a damp area nearby and open fields or meadows in which to hunt. In the absence of natural outcrops copperheads will live around drystone walls, woodpiles, and tumbledown buildings. In some places the woods may become inundated with water, and they tolerate swampy conditions.

They are also found throughout southern Texas in scrub and among the yucca and mesquite bushes of the Chihuahan desert in northern Mexico, but usually along river courses or near springs. Much of this region has been extensively settled with subsequent habitat change. Copperheads may remain near human settlements provided some natural areas are left. They may be nature reserves, neglected farmland, or narrow gorges that are unsuitable for development. If such places are not available, the snakes become isolated into small colonies that eventually die out.

Perfect Camouflage

It says a lot for their camouflage that although they are often numerous along the edges of human habitations and can come into contact with people in parks and even gardens, they are rarely noticed. Their markings, which appear so bold when seen against an unnatural background, blend beautifully in color and shape when they are resting among the dead leaves of their natural habitat. The disruptive coloration makes it very difficult to pick out the complete form of the snake. It only works,

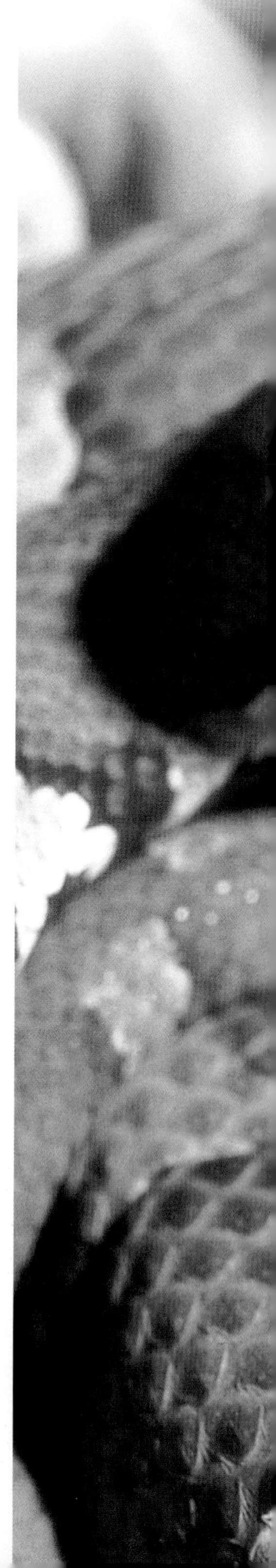

however, because the snakes tend to stay motionless. Even if they are suddenly exposed, they remain in a flat coil unless they are provoked. Even then they will turn and crawl away, avoiding confrontation.

The Human "Copperheads"

During the American Civil War (1861–65) southern sympathizers living in the north were called "Copperheads." The name was a play on words because not only was the South the place where copperhead snakes were most common, but there was also a suggestion of hidden danger, even treachery, in the name. Groups of Copperheads called themselves "Sons of Liberty" or "Knights of the Golden Circle" and identified themselves to each other by surreptitiously showing a copper coin with the head uppermost. Toward the end of the war, when events started to swing against the South, the Copperheads gradually melted away.

The only time they are likely to bite is if they are accidentally stepped on or deliberately molested. Then they will vibrate their tail rapidly, often rustling the dead leaves among which they are resting, and strike ferociously. Copperhead venom is typical of vipers, acting on blood corpuscles and blood vessels, causing hemorrhaging, pain and swelling to the bitten area. Yet bites are rarely serious, and effects usually disappear after a day or two.

Unusual Prey

Copperheads are unusual among pit vipers in taking cold-blooded as well as warm-blooded prey. Alongside the list of predictable prey items such as rodents, birds, lizards, and small snakes are surprise items such as frogs, salamanders, and insects and their larvae. They are surprising because they are cold-blooded, and the copperhead's heat pits have evolved to detect warm prey. Cicadas figure highly in the diet of some populations, both as larvae and as newly emerged

Like other pit vipers, the copperhead has heat-sensitive facial pits between its nostrils and eyes to help seek out its prey.

adults. In short, copperheads are not fussy eaters.

Juveniles are about 8 inches (20 cm) long at birth and probably eat mainly frogs and lizards at first. The tip of their tail is yellow or yellowish green. When they sense a likely meal nearby, they raise it above their coils and wave it slowly back and forth to attract the victim's attention and lure it within striking range. The tail loses its conspicuous color as the snake grows and switches its diet mainly to mice—it is no longer needed as a lure.

Shared Dens

Hibernation is an important part of the copperheads' annual cycle, especially in the northern part of their range. They need a suitable place to escape the worst of the cold and, ideally, a basking place nearby where they can benefit from the last warm days of the year and take advantage of early spring sunshine the following year.

Rock outcrops with deep fissures and south-facing ledges are perfect, and large numbers may gravitate there in the fall. They are often shared with other species of snakes, notably timber rattlesnakes, *Crotalus horridus*, but also with black ratsnakes, *Elaphe o. obsoleta*, and milksnakes, *Lampropeltis triangulum*. Mild spring days may bring groups of two or more species to the surface to "lie out" or bask together. Once the danger of frosts has passed, the snakes disperse, probably returning to home ranges with which they are familiar.

There are folktales of copperheads and black ratsnakes warning rattlesnakes that enemies are approaching. In some places copperheads are sometimes even called "rattlesnake pilots" (and the black ratsnake is sometimes called the pilot black snake). These small, alert species sense danger before the large, heavy-bodied rattlers and take cover first, with the rattlesnakes moving away afterward.

Mating Rituals

Copperheads give birth to live young. They mate in the spring shortly after they emerge from hibernation. Males go through a combat ritual before mating, and females sometimes

↑ ***Young cottonmouths,* Agkistrodon piscivorus, *open their mouths in a threatening behavior. The behavior shows the white interior of the mouth that gives them their common name.***

initiate mock combats with males, presumably to judge their fitness as a good mating partner. If the male does not respond, they break off and move away.

Females that have mated are likely to mate again with one or more additional males, and the young in a single litter may have several fathers. This system is probably widespread in snakes, including the adders, but it has not been studied in more than a handful of species.

The young are born in August, September, or later depending to some extent on the weather. Litters range from one to 20 young, but between four and seven are more usual. There are regional variations, with females from northern populations on average having larger litters than those from the south. The markings of the young are rather dull.

After giving birth, the female remains with her young until they have shed their skins, which usually takes place when they are about one week old. Again, such parental care is not often reported for live-bearing snakes but may be more widespread than has been assumed.

The Relatives

The genus *Agkistrodon* used to include a number of Asian species until fairly recently, when they were placed in a new genus, *Gloydius. Agkistrodon* now contains three species, all from North America. The copperheads themselves are divided into five geographical races, or subspecies, all of which differ slightly in their markings, but which have areas where they intergrade (merge gradually) with each other.

The other members of the genus is the cottonmouth, *A. piscivorus*, and the cantil, *A. bilineatus*, from northern Mexico south into Central America. Of these the cantil is the closest relative to the copperheads and has much more in common with them than the cottonmouth does. Distribution of the cantil follows the coast on both the Pacific and Atlantic sides. It lives in dry deciduous forests, grasslands, and scrub but is not found in the more humid, rain-forested interior of the region. It has a reputation for being aggressive and irascible, striking repeatedly. Like the copperhead, it is banded but much darker overall. Four striking white lines radiate from the tip of the snout, two follow the ridges along the top of the head, and the other two follow the jawline. Juveniles have a yellow tip to their tail like those of the copperheads.

The cottonmouth, also called the water moccasin, is a semiaquatic species from much the same area as the copperheads, but including Florida. Its habitats are low-lying swamps, wetlands, and lake edges. It often basks on emergent logs or stumps and can be reluctant to move away when disturbed. Instead, it opens its mouth widely and gapes, displaying the white interior that gives it its common name.

Cottonmouths are remarkable for their diet: In addition to mammals, birds, frogs, and lizards it may include young alligators, freshwater turtles, and carrion.

Corn Snake

Elaphe guttata

The corn snake is bred in greater numbers than any other species. It is undoubtedly among the most familiar snakes to pet keepers over much of North America and Europe.

Common name Corn snake (red ratsnake)

Scientific name *Elaphe guttata*

Subfamily Colubrinae

Family Colubridae

Suborder Serpentes

Order Squamata

Length From 43 in (110 cm) to 5.8 ft (1.8 m)

Key features Slender but muscular snake; head narrow; scales weakly keeled; eyes moderately large; pupils round; pattern consists of black-edged, deep-red to orange saddles on a background of gray, silver, or yellow; there is nearly always an arrow-shaped marking between the eyes; underneath it is black and white, often arranged in a checkered pattern

Habits Basically terrestrial but climbs well and may also spend time below ground

Breeding Egg layer with clutches of 5–25 eggs; eggs hatch after about 65 days

Diet Mostly small mammals (including bats) and occasional birds, but it sometimes takes frogs and lizards when young

Habitat Open woods, hillsides, clearings in forests, and forest edges; often attracted to human settlements, especially farm buildings

Distribution Eastern United States to southern and northern Mexico

Status Common

Similar species Some milksnakes, such as the eastern milksnake, *Lampropeltis triangulum triangulum*, are similar but have smooth scales; young ratsnakes, *Elaphe obsoleta*, are blotched but lack the arrowhead mark

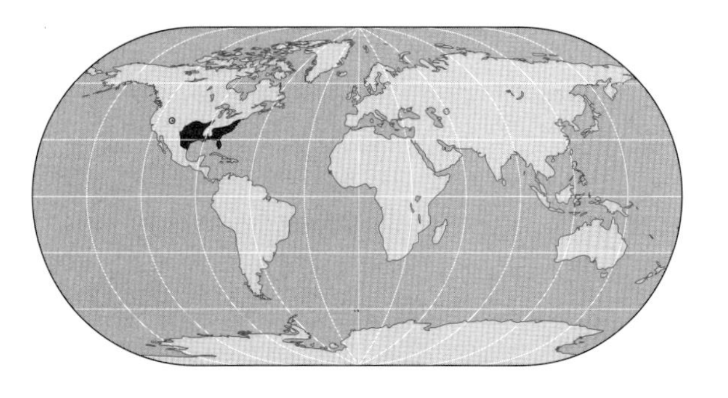

THE CORN SNAKE'S GREAT popularity among breeders is due to its (usually) docile nature and its adaptability. Corn snakes thrive under the sometimes inappropriate conditions provided by well-meaning but inexperienced reptile keepers. In their natural habitat corn snakes live in a wide range of habitats and frequently turn up in the vicinity of houses and barns—no doubt in search of mice and rats attracted by discarded and stored foodstuffs. (This habit has earned them the alternative common name of red ratsnakes.) They are powerful constrictors, and adults tackle prey up to the size of full-grown rats. They also raid birds' nests, including those inside bird boxes.

False Rattlers

Corn snakes are active at night during the summer, but in early morning and late evening when the weather is cooler. They are rather slow, deliberate crawlers and will often "freeze" if discovered in the open, relying on their disruptive markings to help them melt into the background. If approached, they often vibrate the tip of their tail rapidly. If they happen to be resting among dead leaves, this can cause a rustling or buzzing noise similar to that made by some rattlesnakes.

They hibernate from October to March in northern parts, but in the south they may be active throughout the year, emerging from their retreats on warm days. They mate in spring, and the female lays her eggs, which can number up to 30, in June, July, or August. Northern forms tend to lay fewer, larger eggs. The hatchlings typically measure 8 to 12 inches (20 to 30 cm), but those of the northern subspecies are larger. Males reach breeding size in two years, but females usually take three.

Selective breeding has resulted in an amelanistic form of the corn snake, in which all traces of black pigment are missing.

The corn snake's natural coloration varies depending on location. The most colorful examples are those from South Carolina, where a typical example will have rich red saddles with wide black borders on a straw-colored background. Farther south in parts of Florida the background color is often silvery gray, and the saddles are paler red, sometimes orange, with no black borders. In the Florida Keys the subspecies *E. g. rosacea* has little or no black pigment, and the whole snake has a reddish-orange wash. The population farther west, known as the Great Plains ratsnake, however, has no red on it but has brown blotches on a grayish background.

↑ Snow corn snakes have no black or colored pigmentation and only a hint of a pattern. The pinkish coloration comes from their blood.

Designer Snakes

The number of naturally occurring color forms, or subspecies, is nothing compared with the array of "sports" produced by selective breeding in captivity. Because the corn snake is such a prolific and easy animal to breed, it was inevitable that occasional mutations would occur sooner or later.

The first was a form in which all the black pigment (melanin) was missing. These amelanistic corn snakes have all the red markings in the right places but no other markings, and their eyes are pink. The next mutation to be produced was the opposite—snakes in which all the black areas were present as normal, but in which the red pigment was absent. These are anerythristic corn snakes, not as colorful as the wild type or the amelanistic form, but interesting in their own right. More importantly, they represent a step toward producing corn snakes without any pigment at all by crossing them with amelanistic individuals. The resulting form, which takes two generations to create, was the first man-made corn snake variety and is called the snow corn snake.

Other slight variations appear from time to time among breeders' "crops," such as specimens in which the blotches are replaced with a single straight or zigzag stripe down the back. They have been propagated by selective breeding, and some dealers list a dozen or more corn snake "sports," often with fanciful names, no doubt created simply to enhance their desirability among snake-keepers.

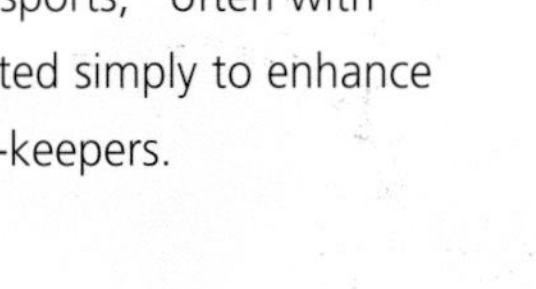

CROCODILIANS

The ancestry of today's crocodiles stretches back over 265 million years, making them one of the oldest vertebrate groups on the planet. In common with the turtles they survived the major upheaval that resulted in the disappearance of the dinosaurs about 65 million years ago. By that time the direct ancestors of today's crocodilians had already developed.

Some of the early crocodiles are believed to have been much larger than those existing now. Remains of *Deinosuchus* have been unearthed in various parts of the United States. Its skull measured over 6.5 feet (2 m). Based on the head to body ratios of modern crocodilians, it could easily have grown to over 50 feet (15.2 m) long and may have weighed as much as 6 tons (6,096 kg). These crocodilians probably preyed on duck-billed dinosaurs, since fossilized remains of both species have been found together.

Even the direct ancestors of today's species are known to have been larger in the past, as shown by the fossilized remains of caimans. The largest-known example yet unearthed is *Caiman neivensis*, which occurred in the area of present-day Colombia and grew up to 30 feet (9.1 m) in length, dwarfing modern species.

While the majority of crocodilians today inhabit fresh water, there were many crocodilians in the past that roamed the world's oceans and were fearsome predators in the sea. Today just two species are regularly encountered in the marine environment: The saltwater crocodile, *Crocodylus porosus*, occurs in the Indo-Pacific region, and the American crocodile, *Crocodylus acutus,* in the Caribbean region. A number of others, including the Nile crocodile, *C. niloticus*, the American alligator, *Alligator mississippiensis*, and the common caiman, *Caiman crocodilus*, can be found in the sea occasionally.

Crocodiles that live in marine habitats face the inevitable challenge of avoiding the risk of dehydration, which could be fatal. Their thick skin helps prevent loss of water by osmosis, however, and there are also highly effective salt glands located at the back of the mouth that excrete sodium chloride from the body. They are present in all crocodiles, not just those that range out to sea, which suggests that their ancestors originated from the marine environment.

Some crocodilians in the past appear to have been primarily terrestrial. It is assumed that they died out

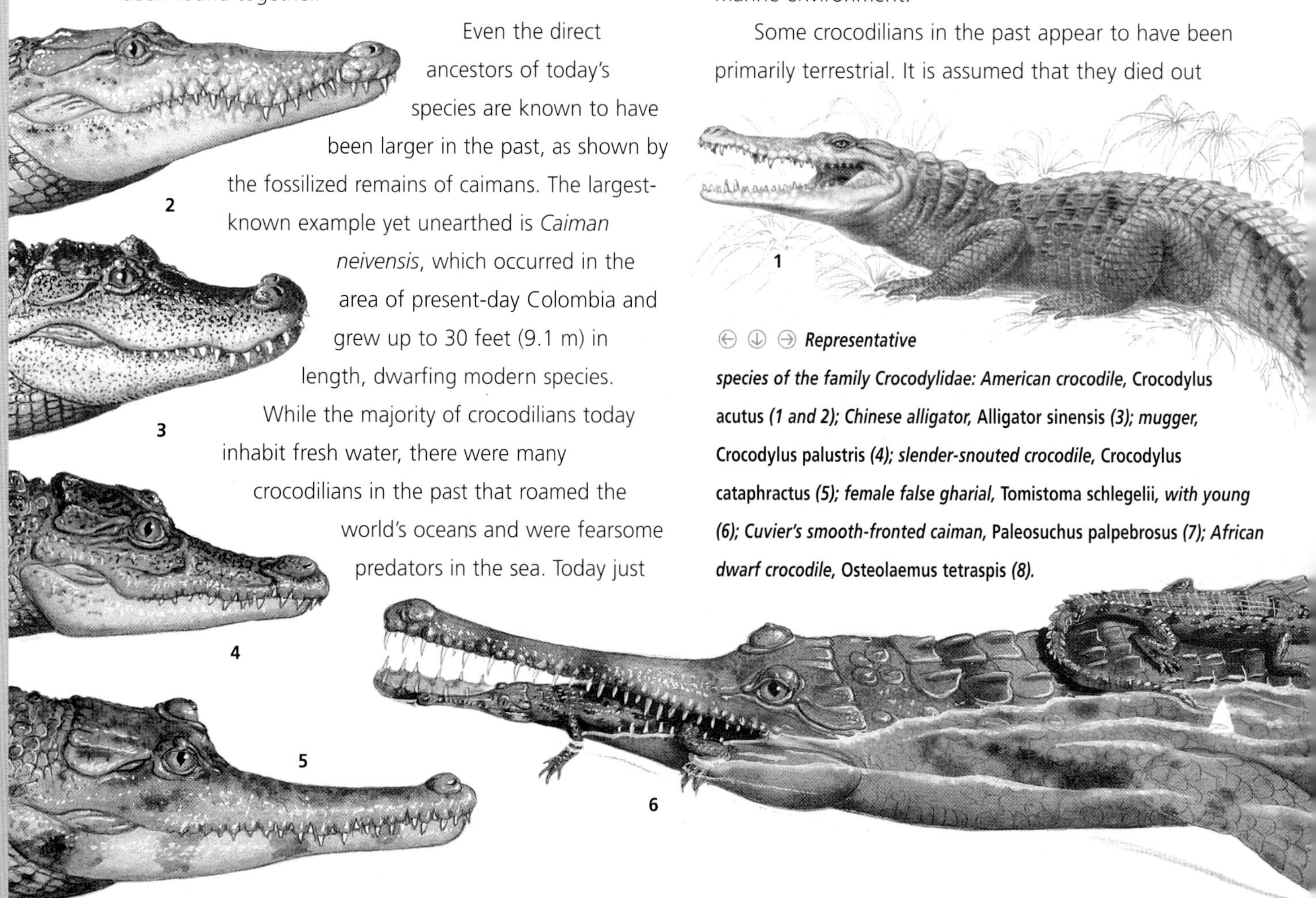

Representative species of the family Crocodylidae: American crocodile, **Crocodylus acutus** ***(1 and 2); Chinese alligator,*** **Alligator sinensis** ***(3); mugger,*** **Crocodylus palustris** ***(4); slender-snouted crocodile,*** **Crocodylus cataphractus** ***(5); female false gharial,*** **Tomistoma schlegelii,** ***with young (6); Cuvier's smooth-fronted caiman,*** **Paleosuchus palpebrosus** ***(7); African dwarf crocodile,*** **Osteolaemus tetraspis** ***(8).***

largely because they were no match for the mammalian hunters that colonized the land after the end of the Cretaceous Period about 65 million years ago. The New Caledonia crocodile, *Quinkana fortirostrum*, was the last survivor of this group, evolving at a stage before the island for which it is named separated from the much larger landmass of Australia. Growing to approximately 6.5 feet (2 m), these crocodiles had powerful limb muscles to help them move on land. They lived right through until the first native people reached the islands about 2,000 years ago, but they had disappeared before the arrival of the Europeans.

In the past crocodilians also had a much wider distribution than they do today, and remains have even been unearthed in Europe. Although there are no surviving representatives of the family there today, they are found on all other continents. The basic appearance of crocodilians has changed relatively little over a long period of time—recognizable ancestors of the American alligator existed for over 5 million years in the same area of the United States where the species is still found.

Predatory Lifestyles

The appearance of the snout of a crocodilian gives a clear indication of its feeding habits. Those with long, narrow jaws with relatively small, sharp teeth such as Johnston's crocodile, *C. johnstoni* from Australia, are primarily fish-eaters. Others with broader jaws, for example, the saltwater crocodile, *C. porosus* (also from Australia), prey predominantly on mammals. In some cases they can catch creatures that are almost as large as themselves.

The fearsome teeth in the jaws of crocodiles are regularly replaced throughout their lives, although the

Who's Who among the Crocodilians?

Order Crocodylia

Family Crocodylidae

Subfamily Alligatorinae: 4 genera, 7 species of alligators and caimans—the American alligator, *Alligator mississippiensis*; Chinese alligator, *A. sinensis*; common caiman, *Caiman crocodilus*; broad-snouted caiman, *C. latirostris*; Cuvier's smooth-fronted caiman, *Paleosuchus palpebrosus*; Schneider's smooth-fronted caiman, *P. trigonatus*; black caiman, *Melanosuchus niger*

Subfamily Crocodylinae: 3 genera, 14 species of crocodiles—the American crocodile, *Crocodylus acutus*; Central American crocodile, *C. moreletii*; Colombian crocodile, *C. intermedius*; Cuban crocodile, *C. rhombifer*; Nile crocodile, *C. niloticus*; African slender-snouted crocodile, *C. cataphractus*; saltwater crocodile, *C. porosus*; Johnston's crocodile, *C. johnstoni*; mugger, *C. palustris*; New Guinea crocodile, *C. novaeguineae*; Philippine crocodile, *C. mindorensis*; Siamese crocodile, *C. siamensis*; African dwarf crocodile, *Osteolaemus tetraspis*; false gharial, *Tomistoma schlegelii*

Subfamily Gavialinae: 1 genus, 1 species, the gharial, *Gavialis gangeticus*

Total: 1 family, 8 genera, 22 species

rate of replacement slows down with age. It is actually the difference in the arrangement of the teeth that distinguishes alligators from crocodiles. In alligators the lower set of teeth is concealed when the mouth is closed, whereas in true crocodiles the fourth tooth in the lower jaw remains visible when the jaws are closed, sliding into a notch rather than being concealed in a pit.

The tremendous power in the jaws of large crocodilians is used to deadly effect to seize and incapacitate mammalian prey, dragging the victim under water. Crocodiles usually kill their prey by drowning. The crushing force that can be inflicted by the jaws of a large crocodile is equivalent to about 11 tons (13 tonnes). Yet, remarkably, the muscles that open the jaws are very weak. A simple rubber band applied around the jaws of a young crocodile less than 6.5 feet (2 m) in length will be enough to keep it from moving these muscles.

Their teeth are not equipped to chew food, so it is not uncommon for crocodiles to work together to dismantle a carcass. One individual holds onto the body while the other tears off a chunk, usually by spinning around under water. The crocodiles then raise their head and proceed to gulp the whole piece of food down into their gullet.

Many large crocodilians rely on their ability to ambush prey as their main hunting strategy, lurking close to the water's edge where mammals and birds come to drink. The design of the crocodile's body is ideally suited to this type of hunting, since it can rest with just the top of the ears and eyes positioned out of the water. It can also raise the nostrils located on the tip of the snout above the water in order to breathe. Since they have no lips, however, water flows into the crocodilian's mouth when it is submerged. A special flap at the back of the mouth stops the crocodile from swallowing the water.

Crocodilians occur in a wide range of habitats. Some favor marshland areas, while others are restricted to rivers. The small African dwarf crocodile, *Osteolaemus tetraspis*, lives in forested areas, where it inhabits small waterways. As its name suggests, it is the smallest of all contemporary crocodiles, reaching a maximum length of approximately 6.5 feet (2 m).

Caimans are distinguished by having broad, blunt snouts, and the teeth in the lower jaw lie inside the mouth when closed. This broad-nosed caiman,* Caiman latirostris, *is feeding on a catfish in Argentina.

Reproduction

All crocodilians reproduce by means of eggs. Since males grow at a faster rate than females, they mature earlier, usually at about seven years old, by which stage they have reached a length of about 6 feet (1.8 m). Females are unlikely to lay for the first time until they are about nine years of age. While most crocodilians construct an elaborate nest mound, the American crocodile, *C. acutus*, is unusual—females often simply bury their eggs in a hole dug in the sand. In many cases the female guards the nest throughout the incubation period, occasionally with assistance from the male, and helps the young into the

water once they emerge. Just as in the case of many chelonians (turtles and tortoises), the incubation temperature determines the sex of the offspring. The actual temperature parameters vary significantly, however, among different species.

Crocodilian Senses

Since crocodilians tend to hunt at night, their eyes have typical vertical, slit-shaped pupils that maximize the amount of light entering them. They also have a reflective layer called the *tapetum lucidum* at the back of the retina. It acts like a mirror, reflecting back the available light and helping provide a clearer image. As a result, censuses of crocodiles in an area are usually carried out at night by flashlight, because this causes their eyes to glow in the dark and makes counting them straightforward. However, this technique has the disadvantage that it gives little indication of an individual's size.

Crocodiles can also see well during the daytime, and they have color vision to help them. They also have a protective transparent shield, known as the nictitating membrane, that covers the eyes under water without impairing their vision.

If the water is muddy, crocodiles can use their keen sense of smell. They also have very acute hearing, which not only helps them hunt but is also important for communication. For example, a female is able to detect her young calling from within the nest once they are ready to hatch. Adult crocodiles keep in touch with each other by bellowing and by slapping their head down on the water's surface. This noise carries some distance, so it can be detected by other crocodiles in the area.

Common name Eastern box turtle

Scientific name *Terrapene carolina*

Family Emydidae

Suborder Cryptodira

Order Testudines

Size Carapace up to 8 in (20 cm) in length

Weight Approximately 2.2 lb (1 kg)

Key features Carapace relatively domed, usually brownish in color often with variable markings; body predominantly brown with yellow and orange markings particularly on the chin and front legs (depending to some extent on the subspecies and the individual); plastron relatively plain with distinctive hinged flaps front and back, allowing the turtle to seal itself into its shell completely; males generally have reddish irises, but those of females are brownish

Habits Spends much of its time on land but usually remains close to water; may immerse itself for long periods, especially during dry spells

Breeding Female lays 3–8 eggs in a clutch, sometimes more than once in a season; eggs hatch after 9–18 weeks

Diet A wide variety of invertebrates as well as smaller vertebrates and carrion; also feeds on vegetable matter and fruit

Habitat Most likely to be encountered in open areas of woodland; sometimes also occurs in marshy areas

Distribution Eastern United States to northern Mexico

Status Has declined in various parts of its range

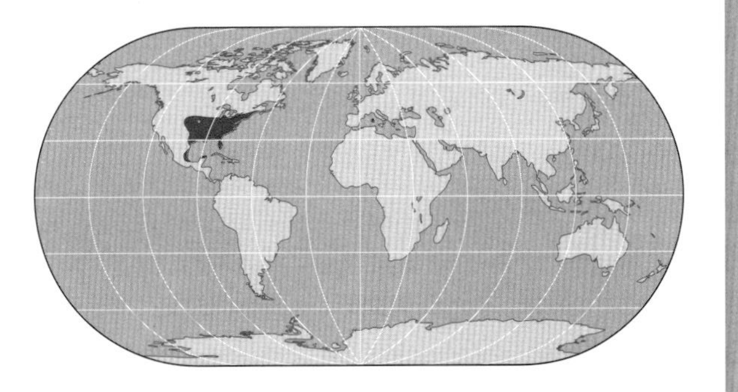

Eastern Box Turtle

Terrapene carolina

Eastern box turtles are among the most well-known turtles and have even appeared on a U.S. postage stamp. Their appearance is variable, and some individuals are thought to live for over 130 years.

THE DOMED CARAPACE of the eastern box turtle is a clue that these turtles spend much of their time on land—more aquatic species have a more streamlined shape. The turtles range widely across eastern parts of the United States from Maine to Florida and south to northern Mexico.

The species is divided into six subspecies that differ in size, markings, and shell shape. The largest race is the Gulf Coast box turtle, *Terrapene carolina major*, which also displays flaring of the hind marginal shields. Its shell is dull in color compared with that of the northern nominate race, *T. c. carolina*. Both these subspecies have four toes on each hind foot. The other two subspecies in the United States are *T. c. triunguis* from Missouri south to Alabama and Texas, and *T. c. bauri* from Florida. Both have three rather than four toes on each of the hind feet. However, *T. c. bauri* can be distinguished by the light pattern of lines radiating across the carapace and two stripes on each side of the head. The carapace of the Mexican races, *T. c. yucatana* and *T. c. mexicana*, is particularly domed—in the latter it is yellow with dark spots.

Box turtles have two hinges on the plastron, allowing them to close up like a box when threatened. This is the Florida box turtle, Terrapene carolina bauri.

Hibernation Sites

In northern parts of their range eastern box turtles hibernate in a variety of different places and often use the same site each year. They may prefer to hibernate under water in mud or bury down into soil or under vegetation. Some sites are shared by several box turtles, and they will continue burrowing if the winter proves severe, sometimes digging down distances as great as 24 inches (60 cm). On occasion they

"Old 1844"

In the past it was something of a tradition for people to carve their initials or dates into the shells of eastern box turtles, particularly in northeastern parts of the United States. Since studies have shown that these turtles are sedentary, researchers have tried to correlate the initials carved on shells of individual turtles with local parish records to come up with a novel way of estimating their age. This led to the discovery of the Hope Valley turtle, also known as "Old 1844," which is believed to be the oldest living vertebrate recorded in the United States.

This particular box turtle had two sets of initials, one of which was E. B. K., and two dates etched on its plastron. It was discovered that in 1844 there was a 19-year-old farmhand called Edward Barber Kenyon working on the land where the turtle was found. The fact that there was no distortion of the carving suggested that the reptile was fully grown then and would therefore have been at least 20 years old at the time. Although it proved impossible to match the second set of initials, G. V. B., alongside the date of July 11, 1860, it was discovered that two families (the Bitgoods and the Bigwoods) had owned the land during that period. It could have been one of their family members who found the turtle again. These two pieces of evidence suggested that Old 1844 was therefore about 138 years old at the date when these enquiries were made.

may also move from one site to another during mild spells, but this can prove fatal if the weather takes an unexpected turn for the worse—the turtles are left stranded and defenseless against the frost.

Those that survive over the winter usually emerge during April, and mating begins soon after. Their domed shells mean that males have to balance themselves at a semivertical angle when mating. It can be very dangerous for them if they fall over and are unable to right themselves. In most cases, however, they can use their powerful neck to flip their body over if they should fall while mating.

Females lay their eggs typically between May and July, digging the nest site under cover of darkness. The incubation period depends greatly on the temperature and can range from nine weeks to 18 weeks.

⊖ A closeup of the eastern box turtle shows bright yellow markings on the chin and head. Males have red irises, while those of females are usually brown.

Common name Eastern glass lizard (Florida glass lizard)

Scientific name *Ophisaurus ventralis*

Subfamily Anguinae

Family Anguidae

Suborder Sauria

Order Squamata

Size 39 in (99 cm)

Key features A stiff, legless lizard with eyelids and external ear openings; a groove along its side marks the change from grayish-brown flanks with white bars to the plain, off-white underside; the back is plain brown in color; side of the head is marked with dark-edged, whitish bars, but they may disappear with age

Habits Diurnal; terrestrial

Breeding Female lays 8–17 eggs that hatch after 8–9 weeks

Diet Invertebrates, especially slugs, snails, and earthworms

Habitat Grassland, open woods, fields, and parks

Distribution Southeastern United Sates (North Carolina, the whole of Florida, eastern Louisiana)

Status Common

Similar species The ranges of 3 other glass lizards—the slender, the island, and the mimic (*Ophisaurus attenuatus*, *O. compressus*, and *O. mimicus*)—overlap the range of the eastern glass lizard; the first 2 usually have some black striping along their back or flanks, while the mimic glass lizard is much smaller, about 15 in (38 cm)

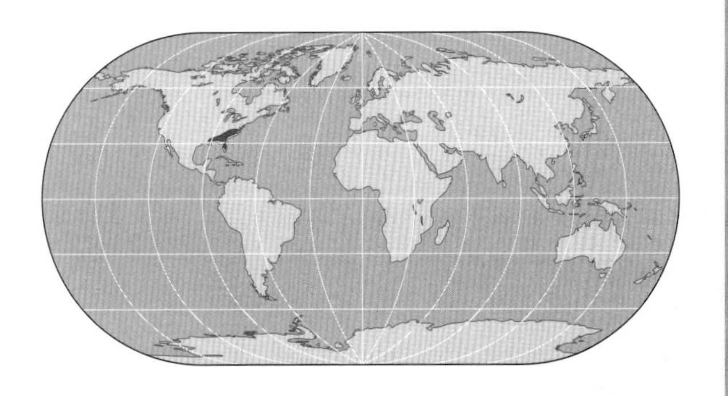

Eastern Glass Lizard

Ophisaurus ventralis

Glass lizards are well named: If their tail is held or attacked, it can shatter into several pieces, each of which continues to move independently.

GLASS LIZARDS SHOULD BE EASY TO DISTINGUISH from snakes because they have movable eyelids and external ear openings, whereas snakes do not. The head and body are only half as long as the tail (showing that it is a terrestrial rather than a burrowing species), while a snake's tail is generally about one-fifth of its total length. Glass lizards also have a distinct groove along each side of the body that allows it to expand when the lizard is distended with food or eggs.

A characteristic shared by all four North American glass lizards is the ability to lose and regrow the tail. A new, shorter tail will grow in its place, and adults with complete and original tails are rare. (The tail of the island glass lizard, *O. compressus*, does not have fracture planes like the other three species, and its tail is therefore not shed as easily.) The only other means of defense for glass lizards is to wriggle furiously and empty the contents of their cloacal glands, smearing and spraying them over their enemy.

Eastern glass lizards are usually found sheltering under objects lying on the ground, often pieces of board, tin, old carpet, or sacking. They are active during the day, especially early in the morning when the humidity is high, and their prey—slugs, snails, and earthworms—are most easily found.

Guarding the Eggs

Eastern glass lizards usually mate in the spring and lay eight to 17 eggs in May, June, or July. The eggs are laid in hollows in the ground, often under a log or flat rock. The female

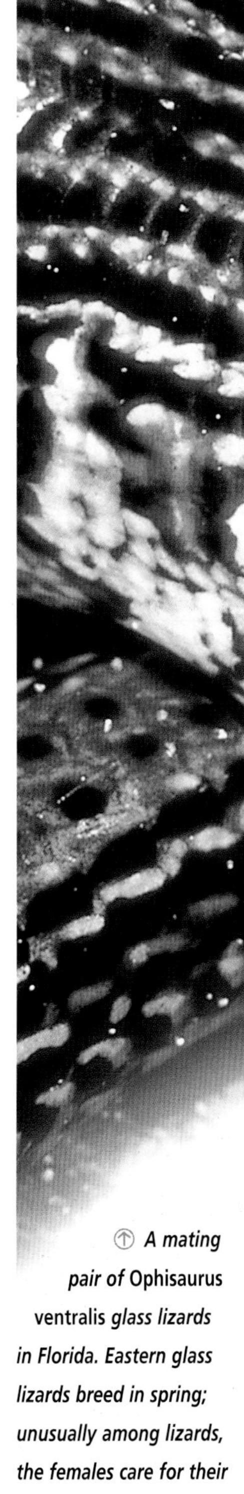

A mating pair of* Ophisaurus ventralis *glass lizards in Florida. Eastern glass lizards breed in spring; unusually among lizards, the females care for their eggs by coiling around them until they hatch.

Lateral Folds

All the glass lizards and many other members of the family (except the legless lizards, *Anniella*, and the slow worm, *Anguis fragilis*) have folds of skin along the sides of the body from the front legs to the back legs. The folded area is covered with small, soft scales, making it very flexible. The lizards need the fold because they have osteoderms (small bones) under their other scales. The bones make them so stiff that without the fold the lizards would not be able to expand their body to breathe, to take in large meals, or in the case of females, to hold developing eggs or young.

remains coiled around the eggs to guard them until they hatch during August and September.

Egg brooding in lizards is very rare. The only other species in which females are known to brood their eggs are the oviparous (egg-laying) members of the skink genus *Eumeces,* such as the prairie skink, *E. septentrionalis*, and the five-lined skink, *E. fasciatus*. Coincidentally, several of these species live in the same region as the glass lizards. The other members of the genus are viviparous (live-bearers). Some authorities think that egg brooding is the first evolutionary step toward viviparity.

Rare Relative

The mimic glass lizard, *O. mimicus*, was described to science only in 1987. It is the smallest glass lizard in North America, rarely exceeding 15 inches (38 cm) in total length, and is light tan to golden brown in color with several stripes. It is found in pine grasslands of northern Florida, coastal North Carolina, Georgia, and Mississippi.

Common name Emerald tree boa

Scientific name *Corallus caninus*

Family Boidae

Suborder Serpentes

Order Squamata

Length From 5 ft (1.5 m) to 6.5 ft (2 m)

Key features Adults bright green with narrow white crossbars; sometimes yellow underneath; newborn young are bright yellow or red; scales covering the head are small and granular; lips bear prominent heat pits; eyes have vertical pupils

Habits Completely arboreal; rarely, if ever, comes down to the ground; when resting during the day, it drapes its coils over horizontal boughs in a characteristic way

Breeding Bears live young, with up to 15 in a litter

Diet Birds and small mammals, which it catches at night

Habitat Lowland rain forests up to 3,000 ft (900 m)

Distribution South America (mainly within the Amazon Basin)

Status Common but hard to find; habitat destruction is its biggest threat

Similar species Green tree python, *Morelia viridis*, in Australasia; within its range it could possibly be confused with several green pit vipers

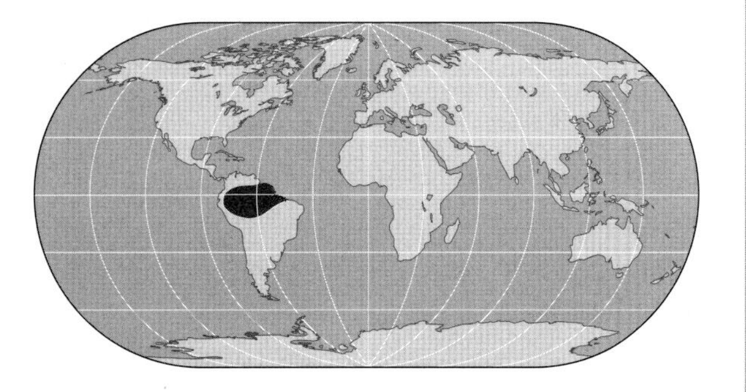

Emerald Tree Boa

Corallus caninus

This beautiful snake lives only in the humid tropical forests in the high rainfall areas of the Amazonian lowlands. It is highly specialized for life in the canopy and rarely, if ever, comes down to the ground.

THE EMERALD TREE BOAS ARE undoubtedly some of the most distinctive and colorful members of the family Boidae. As their common name suggests, they are bright green. They have widely spaced, narrow white bars across their back. Sometimes the bars are continuous from one flank to the other, but usually they are broken or staggered along the center of the back. They may be gray near the head, and individuals from some regions have black borders to the white crossbars.

The head is wide and covered in numerous small scales. Two large bulges on the back of the head behind the eyes are muscles that help in catching prey. The teeth are long and curved backward. The snakes have a formidable bite; and once the prey has been grasped, there is little chance of it escaping. All these features relate to the snakes' lifestyle and diet.

Lazy Days

Emerald tree boas live in tall rain-forest trees mainly in the Amazon Basin, where the high temperatures and year-round humidity are essential for their survival. They spend the day motionless, their coils draped over a branch. At night they may move around to find a good hunting place, or they may stay where they are and unwind a couple of coils, allowing their head to hang down, with one or two bends in the neck. They can remain in this position for hours on end—all night if necessary.

If a suitable prey animal moves into range, the snake strikes by straightening its neck and

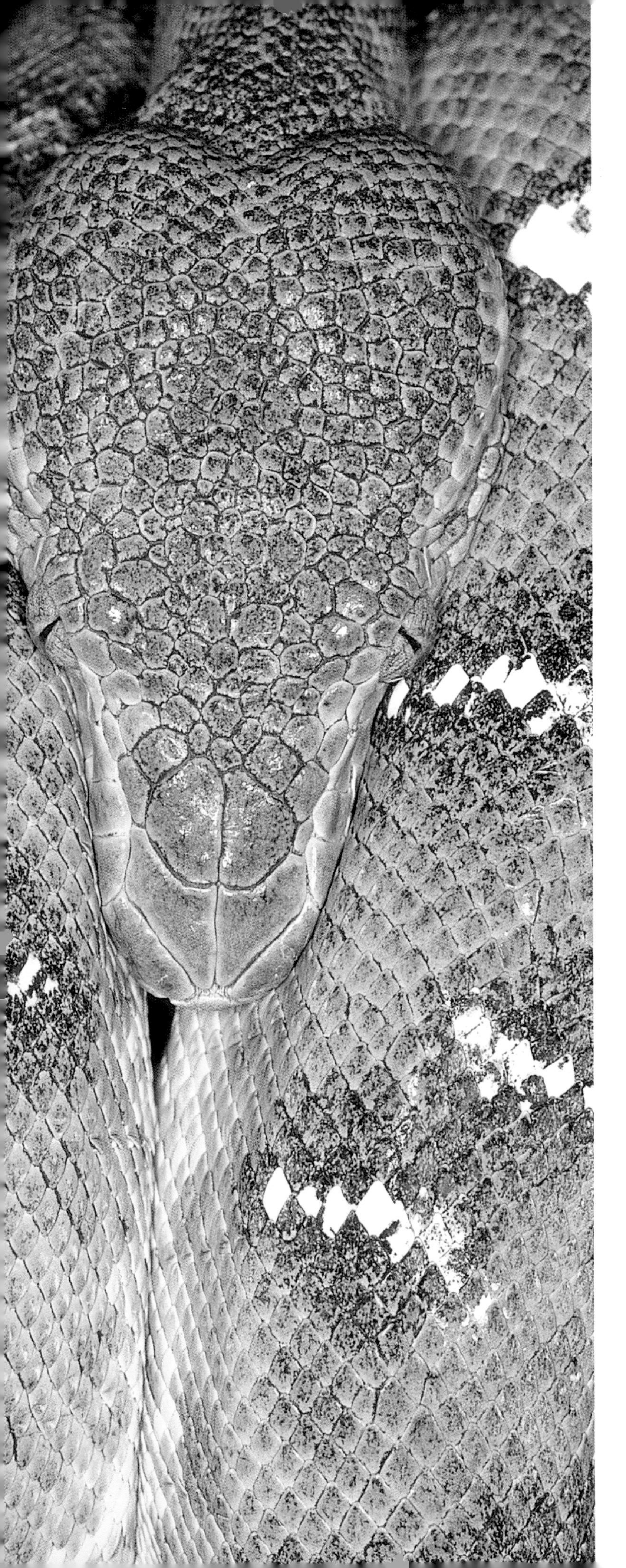

grasping the animal in its wide mouth. The massive muscles on top of the head help the snake clamp the prey firmly in its mouth while—still hanging from a branch—it throws a coil or two of its body around it. The extremely long, backward-pointing teeth enable it to get a good grip on mammals with thick fur or on the plumage of birds.

Because the snake hunts in trees, it only gets one chance to capture its prey. Once the prey has stopped struggling, the snake begins to swallow, still with its head hanging down. This is a very specialized method of feeding, and the emerald tree boas seem incapable of swallowing while in any other position.

Sensing Prey

As in all members of the genus *Corallus*, the emerald boas have very deep, prominent heat-sensitive pits. They are situated between the scales bordering the mouth (the labial scales) on both the upper and lower jaws. The snakes seem to rely almost exclusively on these sense organs when locating and striking at prey. If there are other heat sources in their cage, captive snakes are often confused, and their heat pits do not function well—having sensed the proximity of food by smell, they strike at the heater instead. Their eyes are small and probably play little part in the hunting process.

In common with the other tree boas, their body shape is long and slender, quite unlike that of ground-dwelling boas. It is also flattened from side to side and is girder shaped in cross-section, allowing the snakes to stretch out the front part of their body to span the gap between branches when they are moving through the forest canopy. The tail is relatively long and prehensile. The ventral scales are narrow and join the scales on the flanks at a sharp angle, forming a ridge along both sides.

Heat pits for locating prey line the upper and lower lips of the emerald tree boa and are more numerous in this snake than in other boids. After a large meal it does not feed for a long time and rests for weeks.

These features help them grip irregular surfaces, especially bark, and can also be seen in other climbing snakes such as some of the North American ratsnakes, *Elaphe*. When climbing stout branches, they move slowly with a concertina type of locomotion. When they come to rest, they slowly arrange their coils over a horizontal bough, hanging down on each side with their head in the center. They will remain like this for hours, or even days, on end. The green tree pythons, *Morelia viridis* from New Guinea and Australia, which are similar in appearance, also rest in this position.

Reproduction

Like other boas, this species gives birth to live young. Mating takes place in response to a slight temperature drop, perhaps triggered by the beginning of the rainy season. Details of courtship and mating in the wild are lacking. In captivity, however, the mating couple coil together and allow their tails to hang down during copulation. The gestation period lasts about six to eight months, during which time the female will actively seek out warm basking positions to maintain her body temperature at about 86°F (30°C). Litter size ranges from two in young females to 15 in older, larger ones.

The young measure about 8 to 12 inches (20 to 30 cm) long and come in a variety of colors. The most common color is orange, but they may also be bright coral red, yellow, green, or brown. Young of different colors may be present in a single litter, and nobody knows why the juveniles' colors should differ so much from that of the adults. In each case the white markings along the back are present at birth.

As the youngster grows, its color changes slowly, so that by the time the snake is about one year old, it is green, like the adult. Males probably reach breeding size in three to four years, while females take a year longer.

Looking at this group of juvenile brown emerald tree boas shortly after birth, it is hard to believe they are related to the green adults.

Convergent Evolution

When unrelated organisms look and behave like one another, they may be showing signs of a phenomenon known as "convergent evolution." There are many examples throughout the plant and animal kingdoms (such as agave and aloe plants in North America and southern Africa, and bats and birds), but none are such close counterparts as the emerald boa and the green tree python (*Morelia viridis*).

These two species live on opposite sides of the world—South America and Australasia—but resemble each other so closely that they are almost identical to the untrained eye. Both grow to about the same size, and both are flattened from side to side. Both species are green with white dorsal markings, and both coil characteristically over horizontal branches. Even more remarkably, both species produce young that are different in color than the adults—orange, red, or yellow in the case of the emerald boa, and yellow or red in the case of the green tree python.

Evolution converges when two species develop parallel adaptations due to similar lifestyles and habitats. The emerald tree boa and the green tree python both live in trees in humid tropical forests and feed on birds and tree-dwelling mammals. Both have developed similar physical adaptations relating to climbing, feeding behavior, and camouflage.

Other examples of convergent evolution in snakes include the sidewinder of North America, *Crotalus cerastes* (a small species of rattlesnake), and the horned adder of southern Africa, *Bitis caudalis*, (a true viper). In lizards the plumed basilisk, *Basiliscus plumifrons* from Central and South America, closely resembles the Asian water dragon, *Physignathus cocincinus*.

Common name European glass lizard (Pallas's glass lizard, scheltopusik)

Scientific name *Ophisaurus apodus*

Subfamily Anguinae

Family Anguidae

Suborder Sauria

Order Squamata

Size Up to 4.6 ft (1.4 m)

Key features Effectively legless, although reduced legs are present in the form of small, flipperlike flaps of skin on either side of the cloaca; body thick; tail accounts for about two-thirds of the total length; it has eyelids and small external ear openings; scales arranged in regular rows across and down the body; a fold of skin runs along the sides; color uniform brown, paler on the underside; juveniles are gray with irregular brown blotches and crossbars

Habits Terrestrial; diurnal

Breeding Female lays 6–10 eggs that hatch after 45–55 days

Diet Mostly invertebrates, especially snails, and occasional small vertebrates such as mice

Habitat Dry, rocky hillsides, sparse woodlands, fields, and meadows

Distribution Southeast Europe, the Caucasus, and part of the Crimean Peninsula

Status Very common in suitable habitat

Similar species Unlikely to be confused with any other reptile in the region

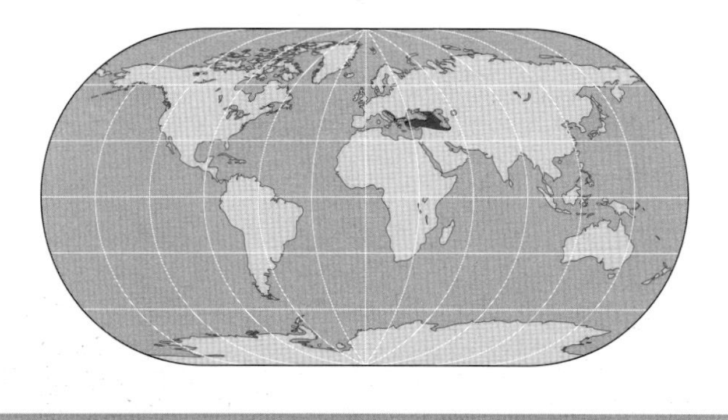

European Glass Lizard

Ophisaurus apodus

The European glass lizard is usually found in fairly dry habitats, often on rocky hillsides with some cover. It can sometimes be found sunning itself in meadows.

THE EUROPEAN GLASS LIZARD was first recognized as a legless lizard rather than a snake as long ago as 1775, when it was named *Lacerta apodus* by the German zoologist Peter Pallas. It is the largest legless lizard in the world, and is not only long but also stocky. Its body can be as thick as a human wrist.

Foul-Smelling Fluid

Adults can give a powerful bite if they are restrained, but they are normally harmless. When captured, their first line of defense is to defecate copiously over their captor. At the same time, they twist, spin, and hiss, ensuring that the obnoxious fluid is well distributed. If this does not have the desired effect (it usually does!), they may cause their tail to break off. But this is a last resort, and they are more reluctant than the American glass lizards to do this. They can regenerate a new tail if the old one is lost, but it does not grow back very well and is little more than a stump. Like the slow worm, *Anguis fragilis*, the European glass lizard has a short body and a proportionately long tail suited to its terrestrial (as opposed to burrowing) lifestyle.

It is an active species and is often quite conspicuous. In places where the grass is kept short by grazing animals, for example, individual glass lizards can even be seen from a passing car, lying stretched out in the turf. They usually remain still when they are first approached by a human but make off rapidly when the intruder reaches a distance of about 6 feet (1.8 m) away.

When foraging, they move slowly by lateral undulation and raise their head slightly from time to time, looking for possible food. They

The most obvious feature of the European glass lizard is its apparent lack of limbs, but there are two very small stumps on either side of the cloaca. It also has the lateral fold along its body that is distinctive of many anguid species.

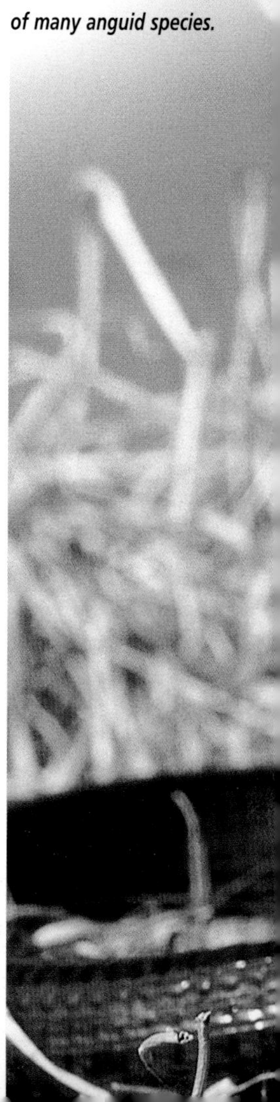

also crawl up into the lower branches of bushes in this way. They may "paddle" their reduced hind limbs ineffectually when creeping slowly forward. When they hunt, they approach slow-moving prey such as snails with a rush; more agile prey (such as a grasshopper) is stalked more carefully—the lizard edges forward every time the grasshopper moves or is distracted.

Breeding

European glass lizards breed in the spring, usually from March to April. Nothing is known of their courtship. The female lays six to 12 eggs in an underground chamber (usually under a flat rock) about 10 weeks after mating and typically coils around them. She will chase off small animals that try to approach the nest. Egg guarding seems widespread in *Ophisaurus* and is practiced by all five species whose reproductive habits are known. The eggs hatch after 45 to 55 days, and the young glass lizards are about 4 to 5 inches (10–13 cm) long. They are gray with wavy dark-brown crossbands, most of which end at the lateral fold, and their scales are more strongly keeled than those of the adults. They grow slowly, taking over four years to reach sexual maturity. When they shed their skin, it comes away piecemeal in rings. Some scientists report that related species, such as the North African glass lizard, *O. koellikeri*, only shed their skin very rarely with gaps of up to two years between each shedding.

Common Names

The alternative common names for the European glass lizard are Pallas's glass lizard and scheltopusik. Peter Pallas was a German zoologist who became a professor at St. Petersburg University in Russia. He first described and named the species in 1775. Scheltopusik also has Russian connections: It comes from the Russian word *zheltopuzik*, which means "yellow belly."

European Pond Turtle

Emys orbicularis

Common name European pond turtle

Scientific name *Emys orbicularis*

Family Emydidae

Suborder Cryptodira

Order Testudines

Size Carapace from 5 in (13 cm) to 7 in (18 cm) long

Weight Approximately 1.8 lb (0.8 kg)

Key features Carapace relatively low and flattened; coloration varies but usually consists of a dark background with yellow markings in the form of streaks or spots; head coloration blackish with yellow spots; tail in males longer than in females; males also have a slightly concave plastron and red rather than yellow eyes; females larger on average than males

Habits Semiaquatic; will emerge to bask, but behavior varies throughout its wide range

Breeding Mating takes place under water; female lays clutch of 3–16 eggs (with an average of 9) on land; eggs usually hatch after about 70 days

Diet Aquatic invertebrates and fish; adults tend to eat more plant matter

Habitat Relatively sluggish stretches of water as well as ponds

Distribution Europe from Lithuania and Poland in the north across most of southern Europe and into North Africa and west to Turkmenistan

Status Declining in many areas because of pollution and habitat modification; Lower Risk (IUCN)

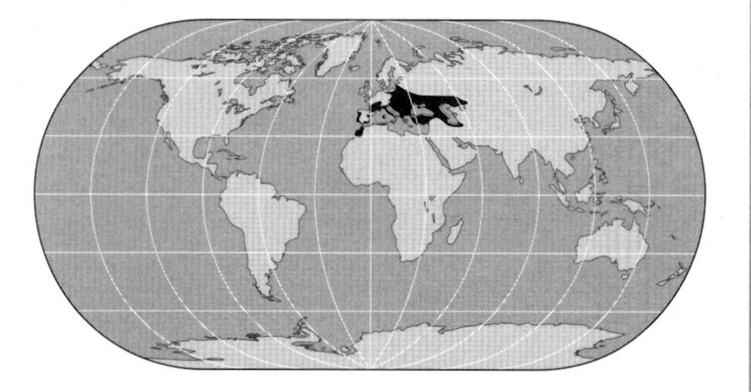

Although its appearance varies over its large range, the European pond turtle is often identifiable by the bright yellow or gold speckling on its dark carapace and skin.

THE EUROPEAN POND TURTLE is the only freshwater turtle to be found throughout much of Europe. These chelonians have a more restricted range than they did in the geological past, however. They once reached Scandinavia and parts of England, but they died out there about 5,000 years ago. Several attempts to reintroduce them to the British Isles in Victorian times failed, indicating that climatic changes were probably responsible for their demise.

Their distribution still extends over a wide area ranging from Lithuania and Poland in the north and then across most of southern Europe and into North Africa along the Mediterranean shore from Morocco to Tunisia. Unfortunately, European pond turtles are becoming much rarer in many areas, particularly in agricultural regions of France and Italy, as a result of loss of habitat, including the concreting of drainage ditches. They are adaptable, however, and in some areas they have colonized artificial stretches of water.

European pond turtles favor waters where there is a muddy base combined with plenty of aquatic vegetation, which provides them with plenty of retreats. They prefer slow-flowing or even still areas of water and are sometimes found in marshland. On a few occasions they have been recorded in brackish areas as well.

Multiple Subspecies

Owing to the fact that they occur over such a wide area, there are marked variations in the appearance of the turtles throughout their range. Taxonomists tend to recognize 13 different subspecies grouped across five geographical ranges. There are also noticeable

Conservation Success

A detailed study of European pond turtles in Spain has confirmed that the species has declined markedly—in many cases because of the effects of water pollution. Existing populations soon become isolated and gradually decline in number. Then they fail to breed, usually as a consequence of habitat changes along the waterways where the turtles live.

However, careful conservation measures can lead to a significant increase in numbers where suitable protected habitat is available. In parts of Catalonia in the northeast of Spain conservation efforts increased the local population of pond turtles from barely 40 to over 300 during the course of about 10 years.

differences among the groups in terms of breeding habits. Breeding times extend from March through June but start later in northern latitudes. It is thought that, at least in some parts of their range, these turtles vocalize when seeking a mate, uttering brief calls. Males chase after a potential mate, bumping her shell and biting at her legs from behind to slow her down before climbing on top of her carapace. Mating occurs under water, and the female lays her eggs on land between four and six weeks later. Clutches consist of about three to 16 eggs, the average being nine—larger females produce bigger clutches. The eggs themselves have leathery rather than hard shells.

The incubation temperature is critical to the development of the hatchlings and determines what sex they will be. (This is known as temperature-dependent sex determination, or TDSD, and occurs in other chelonians, notably the green turtle, *Chelonia mydas*.) Temperatures from 75 to 82.5°F (24–28°C) produce only male offspring, but a temperature of 86°F (30°C) gives rise to almost entirely female hatchlings. Because of this sensitivity to temperature global warming may have a significant effect on populations of these and other turtles affected by TDSD in parts of their range.

The young emerge from the nest site between August and October, although hatching takes longer in cooler conditions, and in northern areas they may overwinter underground. When they hatch, the carapace is approximately 1 inch (2.5 cm) long and clearly marked with a central keel and two lateral keels on each side. The young mature quite rapidly and in some cases can breed successfully by the time they are just three years old.

Lifestyle varies significantly depending on distribution. In northern areas the pond turtles hibernate over the winter months, burying themselves in the bottom of the stream or other stretch of water to protect themselves from the cold. Although the water at the surface may freeze, the ice is unlikely to extend down into the muddy bottom. They sometimes bury down as deep as 6 inches (15 cm) to escape being frozen to death.

Around the Mediterranean region, however, European pond turtles often estivate in a similar way during the heat of the summer, when food often becomes scarce.

European pond turtles are found in slow-moving or still water with or without vegetation. They feed on fish, invertebrates, amphibians and their larvae, and plants.

Common name Eyelash viper (eyelash pit viper and many other local names)

Scientific name *Bothriechis schlegelii*

Subfamily Crotalinae

Family Viperidae

Suborder Serpentes

Order Squamata

Length From 20 in (50 cm) to 30 in (76 cm)

Key features Cluster of raised scales above each eye; body fairly slender for a viper; tail prehensile; color highly variable but basically gray-green or golden yellow; color of pupil matches that of body

Habits Tree dwelling, although it sometimes descends to the ground

Breeding Live-bearer with litters of 6– 24 small young; gestation period about 140–168 days

Diet Frogs, lizards, birds, and small mammals

Habitat Rain forests; in Central America it is a lowland and foothill species living to about 4,500 ft (1,400 m), but in South America it is found at higher elevations up to 8,000 ft (2,400 m)

Distribution Central America and northern South America

Status Common in suitable habitats

Similar species Other tree pit vipers in the region, but none have the raised scales above the eyes

Venom Bites quite common, and a small proportion are fatal

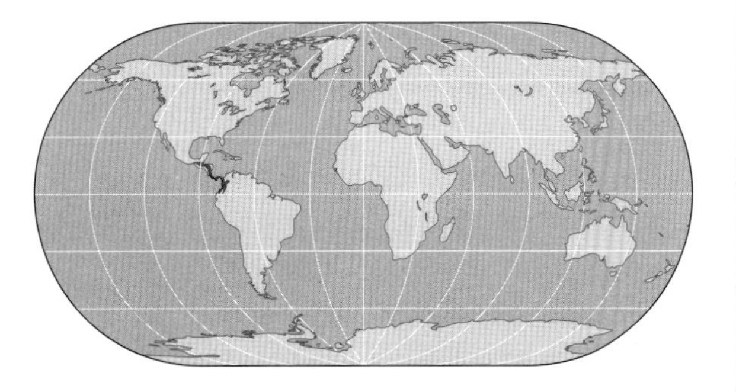

Eyelash Viper

Bothriechis schlegelii

The eyelash viper is a beautiful, dainty, but dangerous snake that ambushes its prey from a coiled position on a branch, leaf, or flower. Its common name refers to the distinctive cluster of raised scales above each eye.

Eyelash vipers are often found at about 3 to 5 feet (1–1.5 m) above the ground. At that height there is a danger of them biting the human face, upper body, or hands. In Costa Rica about 100 people are bitten by the species each year; and although their venom is not as strong as that of some other pit vipers, its effects can be disfiguring, and a significant number of bites result in fatalities.

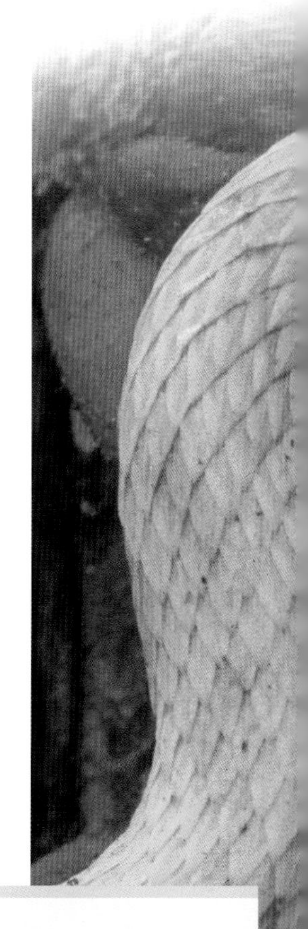

Leaf Ambush

The prey of eyelash vipers includes small mammals such as opossums, birds, lizards, and tree frogs. Some of them visit forest flowers, such as lobster claws, *Heliconia* species, for nectar, and the vipers may coil nearby waiting for their next meal to arrive. Even though the vipers are normally nocturnal, they strike opportunistically at hummingbirds that visit the flowers at frequent intervals throughout the day. The young eyelash vipers, which are very

Other Family Members

The genus *Bothriechis* contains seven species. Four of them—*B. aurifer, B. bicolor, B. marchi, and B. rowleyi*—have fairly limited ranges in Central America. The other three (*B. lateralis, B. nigroviridis, and B. schlegelii*) are more widespread. All are attractive, mostly green in color, and arboreal (tree dwelling). However, the young of *B. lateralis* are dull brown in color and spend the first six months of their lives on the forest floor, eating small lizards and frogs that live in leaf litter. As they grow, their color gradually changes until by the time they are two years old, they are bright bluish green. The local name for this colorful species is *lora*, the Spanish word for "parrot."

small, have pale tips to their tails, which they use as lures to attract frogs and small lizards.

Despite its venomous nature, the image of the eyelash viper (especially the golden form) has become something of a symbol in Costa Rica today, when ecotourism and a welcome reevaluation of biodiversity are a growing force throughout the world. The image is used to epitomize the rain forest or "jungle" in the same way as the toucan and the red-eyed leaf frog have been exploited.

A female eyelash viper with newborn young. The young include several color forms.

The golden-yellow form of the eyelash viper,* Bothriechis schlegelii. *The raised scales above its eye can be clearly seen.

Multicolored Young

Eyelash vipers occur in two basic color forms—gray-green and golden yellow—but both of them are subject to a great deal of variation. The mottled green form often has two rows of reddish spots along its back interspersed with smaller beige or pink ones. Specimens from South America are often suffused in black speckles, while those from Ecuador may have a continuous white line along their flanks. Snakes with any of these variations are well camouflaged when they are resting among lichen-covered branches.

The golden form is the other extreme. It is called the *oropel*, which is Spanish for tinsel (literally "gold skin"). It may also be uniform pale pinkish-brown, salmon, or deep yellow in color. The purpose of this coloration is hard to imagine, but there are several parallels: Young emerald tree boas, *Corallus caninus*, and green tree pythons, *Morelia viridis*, may be yellow or orange, as may tree vipers, *Trimeresurus* species from the Philippines, and bush vipers, *Atheris* species from Africa. Some scientists think that the bright colors may enable the snakes to mimic fruits or flowers and so escape notice (or possibly attract prey). The plain yellow color form is probably controlled by a simple recessive gene, and young of both types may be present in a single litter depending on the genetic makeup of the parents. The large eyes are always the same color as the body.

Five-Lined Skink

Eumeces fasciatus

Five-lined skinks can be seen in woods where there are plenty of logs, stumps, rock piles, and leaf litter. The brightly colored juveniles are more distinctive than the adults, whose colors fade with age.

Common name Five-lined skink

Scientific name *Eumeces fasciatus*

Family Scincidae

Suborder Sauria

Order Squamata

Size Up to 8 in (20 cm)

Key features Body slender and elongated; body color tan, bronze, or grayish olive-green with pale stripes; juveniles have 5 longitudinal bright-cream to yellow stripes on a black background; tail is blue in juveniles and some females but fades to gray in adult males; head wedge shaped; ear opening distinct; limbs short, each bearing five digits with claws; scales smooth

Habits Diurnal; terrestrial; may climb onto tree stumps to bask and look for insects; also burrows under rocks

Breeding Egg layer; clutch containing 4–15 eggs laid in a nest dug in moist soil; eggs hatch after 33–35 days

Diet Insects, spiders, earthworms, crustaceans, and small lizards

Habitat Humid woods with leaf litter and tree stumps; may also be seen around human habitations

Distribution Southern New England to northern Florida west to Texas, Kansas, Wisconsin, and southern Ontario; isolated groups may occur farther west

Status Common but listed as being of special concern in some parts of its range, e.g., Iowa

Similar species *Eumeces inexpectatus* and *E. laticeps* have similar colors and longitudinal stripes

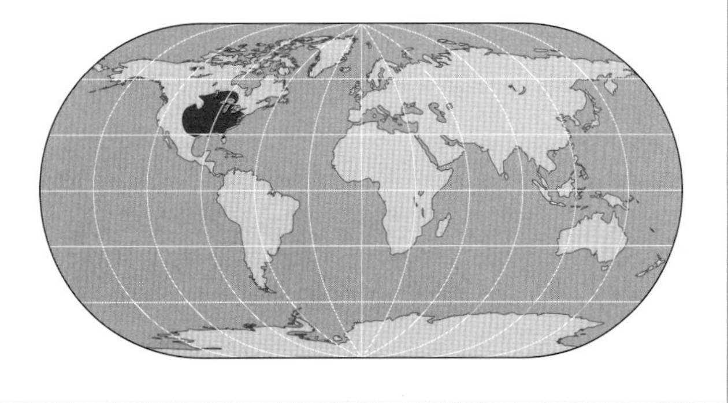

THE FIVE-LINED SKINK IS PROBABLY the most common member of the genus found in the United States. Although its habitat varies, it prefers moist areas. It lives in wooded and partly wooded places as well as disturbed environments such as forest edges and cleared areas. It favors sites with wood and brush piles, stumps, logs, buildings, and outcrops, all of which provide shelter and basking places.

Five-lined skinks are diurnal and mainly terrestrial creatures, although they will climb onto stumps and small rotting trees to bask and search for insects. During the hottest part of the day in midsummer they often take refuge under rocks or logs. They are also accomplished burrowers and excavate dugouts under rocks. At night or when hibernating, they seek shelter in rotting logs, rock crevices, and sawdust piles.

The cream stripes and bright-blue tail indicate that this five-lined skink in Florida is a juvenile. The long, tapering tail can be broken off in the presence of a predator.

They have a wedge-shaped head and a long, slender, cylindrical body. Males are slightly larger than females and grow up to 8 inches (20 cm) long, of which about 60 percent is the tail. The limbs are small but powerful with five clawed digits on each. The species can be distinguished from similar species by a middle row of enlarged scales under the tail and 26 to 30 longitudinal rows of scales around the center of the body. The skin is supported by small bones called osteoderms that lie beneath each scale, giving it greater strength.

Diet and Predators

Small invertebrates such as spiders, crickets, grasshoppers, beetles, millipedes, and caterpillars make up most of the skinks' diet. Snails and small vertebrates, including smaller lizards and newborn small mammals, are also

eaten. The skinks are often seen climbing on houses and are regarded as beneficial, since they eat a number of insect pests. They crush food in their strong jaws prior to swallowing it.

The skinks rely on speed to escape from predators, including snakes, crows, hawks, racoons, and foxes. However, if it is grabbed, the skink can break off its tail voluntarily, a process known as autotomy. While the skink runs for shelter, the predator is distracted by the disconnected tail, which continues to twitch. The skink regenerates a new tail, but it is usually not as long as the original.

Bright Youngsters

Juvenile five-lined skinks are attractive animals with five narrow, longitudinal cream to yellow stripes running along the back from the snout to the tail. The background body color is black, and the tail is blue. A light-colored "v" shape on the head merges with the mid-dorsal stripe. The tail color dulls with age and turns gray, although some females may retain some of the blue coloration. Body coloration also changes to a tan, bronze, or grayish olive-green with pale stripes. In old males only faint traces of stripes may remain.

Male five-lined skinks will attack other males and smaller lizards. However, they do not attack lizards with blue tails. This enables both the adults and the juveniles to feed on different sizes of food in the same area and reduces the risk of juveniles being killed by aggressive, mature adults.

Reproduction

In spring the snout and jaws of mature males develop a reddish-orange coloration. Mating occurs between mid-May and the end of June, and females lay a clutch of up to 15 eggs some four to five weeks later. They prefer secluded nest sites under cover such as logs, boards, rocks, or partially decayed stumps.

They also prefer areas where the soil has a higher moisture content. The eggs absorb moisture from the soil, which enables them to swell. Incubation time varies from 33 to 55 days depending on temperature. During this time the female coils around the eggs, feeding on any passing insects and exhibiting defensive biting behavior toward small predators.

She also regulates the temperature of the eggs by moving them up or down in the nest site; if there is a danger of it flooding or the eggs becoming too moist, she moves them to safety. After the eggs hatch, the female plays no further part in looking after the young.

Common name Frilled lizard

Scientific name *Chlamydosaurus kingii*

Family Agamidae

Suborder Sauria

Order Squamata

Size From 24 in (61 cm) to 36 in (91 cm)

Key features A large lizard with a triangular head and a wide, circular frill around its neck; frill is normally folded along its neck and chest when at rest but raised when the lizard is alarmed; body usually brown; frill can range from brown to black, the latter having a red center

Habits Arboreal; diurnal

Breeding Egg layer with 8–14 eggs per clutch; eggs hatch after 54–92 days

Diet Insects and small vertebrates, including other lizards

Habitat Open woodland

Distribution North Australia and southern New Guinea

Status Common

Similar species None

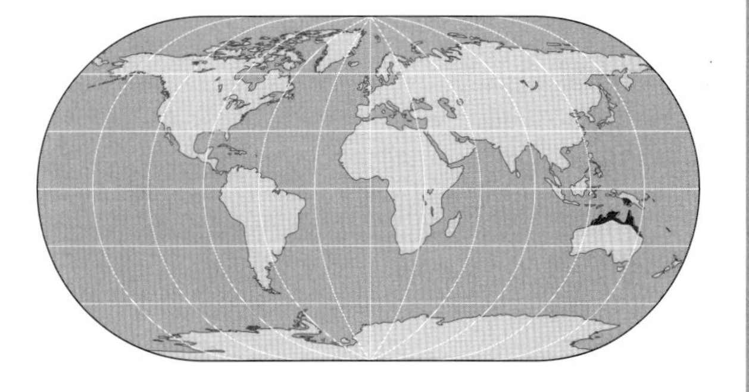

Frilled Lizard

Chlamydosaurus kingii

The frilled lizard is a spectacular dragon lizard from Australia. Its defensive display is unique in the reptile world.

WHEN AT REST, THE FRILLED LIZARD has a large flap of skin folded along the side of its upper body. When it feels threatened, it erects the flap to form an elaborate frill that completely encircles its head like a ruff. The frill is supported on long, slender bones radiating from the hyoid bones, which originate in the lizard's mouth and support the tongue. The extensions act like the struts of an umbrella. As the frill extends, the lizard opens its mouth widely to show the bright yellow interior; the wider it opens its mouth, the more the frill opens out. At the same time, it straightens its front limbs so that its head is held well off the ground. It can even stand up on its hind limbs to make itself look bigger than it really is.

The display may be reinforced by hissing and jumping at its enemy before turning and fleeing. It raises its front limbs off the ground as it goes, speeding along on its two long hind limbs until it reaches the safety of a tree. It is a good climber and clambers quickly out of reach. Frilled lizards also use their frills during encounters with individuals of the same species, and males may use them in territorial displays. Males also fight during the breeding season, biting and lashing each other with their tails.

Daily Activity

Their behavior pattern involves basking in the early morning by clinging vertically to a tree trunk and orienting their body to soak up as much heat as possible. Then during the hottest part of the day they move into the shade, often climbing higher into the crown of the tree where there is some breeze. They rarely venture out of the tree except in the early morning and late afternoon when, having spotted potential prey, they sprint across the ground to capture it before hurrying back to the safety of the tree.

For such a large lizard the frilled lizard includes a surprising number of small prey such as ants and termites in its diet. This is simply because they are the most numerous prey species in the arid grassland and sparse woodlands in which it lives. Frilled lizards also eat larger invertebrates, small mammals, and other lizards when they get the chance.

They have very sharp eyes, and their hunting is done almost entirely by sight. Frilled lizards thrive immediately after the bush fires that sweep across their dry habitat with some regularity. Once the undergrowth of dead grasses has burned off, their prey is much easier to see from their elevated perches.

Frilled lizards are famous for their dramatic defensive displays and have even been adopted as the reptile emblem of Australia. At rest, the frill acts as camouflage, allowing the lizard to blend in with the bark of a tree.

Frilled Lizard History

The frilled lizard was one of the first lizards to be described from Australia in 1825.

Its generic name, *Chlamydosaurus*, means "cloaked lizard," and its specific name, *kingii*, honors Rear Admiral Philip King, who commanded British naval ships that explored Australia in the 19th century. It is the only member of its genus.

Breeding takes place during the southern summer (November to April), which also coincides with the rainy season, when food is most abundant. Females lay eight to 14 eggs in a clutch, although numbers vary among populations living in different places, perhaps depending on food supply. Females typically bury their eggs in a patch of open sandy soil, where the nest will receive sunlight for most of the day.

Common name Gaboon viper (Gaboon adder)

Scientific name *Bitis gabonica*

Subfamily Viperinae

Family Viperidae

Suborder Serpentes

Order Squamata

Length From 4 ft (1.2 m) to 6.6 ft (2 m)

Key features The longest and heaviest African viper with a massive girth and enormous, spade-shaped head; 1 pair of small hornlike scales on its snout, which are larger in individuals from West Africa (subspecies *B. g. rhinoceros*); its body is instantly recognizable with a pattern of geometrically arranged triangles, rectangles, and diamonds

Habits Terrestrial; active in the evening and at night

Breeding Live-bearer with litters of up to 60; gestation period about 90–120 days

Diet Small mammals

Habitat Tropical rain forest and open woodland

Distribution Central and West Africa with a few isolated populations in East and southeastern Africa

Status Common in suitable habitat

Similar species The rhinoceros viper, *Bitis nasicornis*, also has a pattern of brightly colored shapes

Venom Not particularly potent but injected in large quantities is easily enough to kill a human; bites very rare, however, because it is not often encountered and is apparently unlikely to bite even when stepped on

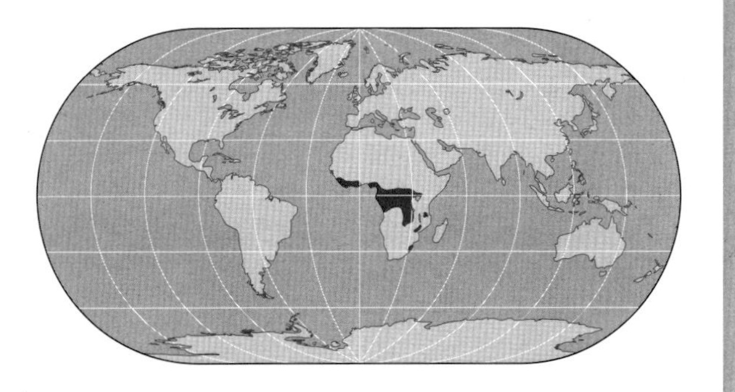

Gaboon Viper

Bitis gabonica

The Gaboon viper is Africa's most easily identified snake. Its gaudy coloration helps it blend in perfectly when it is resting among the dead leaves of its rain-forest habitat. Its common name is a corruption of Gabon, the place where it was first found.

GABOON VIPERS DISPLAY disruptive coloration, a type of camouflage in which the animal does not imitate its surroundings realistically but breaks up its outline with irregular shapes and patches of color—for example, the zebra. Predators and prey have a mental "search image" that they use to identify potential meals or enemies. If what they see does not match their search image, they tend to ignore it. Disruptive coloration is an ideal trick for predators such as the Gaboon viper that ambush their prey.

Like many ambushers, the Gaboon viper's heavy body keeps it anchored to the ground as it strikes, so it has enough grip to force its fangs deep into its prey. On a large Gaboon viper measuring 6 feet (1.8 m) the fangs can be 2 inches (5 cm) long, making them the longest of any venomous snake. They allow the viper to inject venom into the body of a rat or small mammal where it will act most quickly.

Gaboon vipers are more inclined to hold onto bitten prey than many other vipers, but large prey items are released and tracked down later. Their prey can include small antelopes and brush-tailed porcupines. Unlike the puff adders, however, Gaboon vipers seem to feed almost entirely on mammals, but birds and frogs may also be taken.

The main habitat of the Gaboon vipers is forests, but they avoid the densest growth where food is scarce. Instead, they prefer to live in clearings and forest edges. They sometimes occur around plantations.

Slow and Heavy

A 6-foot (1.8-m) Gaboon viper weighs up to 22 pounds (10 kg), but captive individuals that do

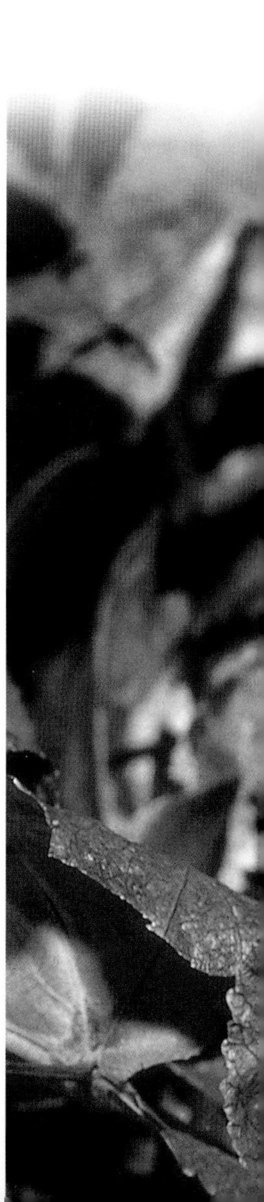

little but eat can weigh much more than that. The only other viper that is close to the Gaboon viper in size is the eastern diamondback rattlesnake, *Crotalus adamanteus*, which may be longer but is not as stout. The bushmasters, *Lachesis*, although considerably longer, are nowhere near as bulky.

Its great bulk prevents the Gaboon viper from making any but the most leisurely journey, and it is likely that most do not stray far from favored hunting areas. Juveniles are more active than adults and venture out into nearby grasslands and even villages and farms in search of mice and rats.

Little is known about their breeding habits, but males battle, and only the dominant male mates. Females in East and southern Africa have litters of up to 30 or 40, while those in West Africa have larger litters of up to 60.

Subspecies

The Gaboon vipers are divided into two easily recognized subspecies. Animals from East Africa have a pair of small horns at the tip of the snout. They form the subspecies *B. gabonica gabonica*. Those from West Africa have very prominent horns and form the subspecies *B. g. rhinoceros*.

However, they should not be confused with another species, the rhinoceros viper from West Africa, whose scientific name is *Bitis nasicornis* (meaning "nose-horned"). That species is also stunningly attractive, perhaps even more dramatic in coloration than the Gaboon viper, but not as large. As their name suggests, they have a cluster of large hornlike scales on the snout. They are easily distinguished from Gaboon vipers because the top of their head has a large sooty-black arrowhead marking, while the head of Gaboon vipers is buff in color with a narrow line down the center. The rhinoceros vipers grow to about 36 inches (90 cm) and are forest dwellers. They have a liking for damp and swampy habitats, giving rise to their alternative common name of riverjacks. However, they also live in dry forest areas, wooded slopes, and cocoa plantations. Bites from this species are rare, but the effects can be serious.

A female Gaboon viper,* Bitis gabonica rhinocerus *from West Africa, lies superbly camouflaged among forest debris. The very prominent horns typical of the subspecies can be seen on the end of its snout.

Galápagos Giant Tortoise

Geochelone nigra

Common name Galápagos giant tortoise

Scientific name *Geochelone nigra*

Family Testudinidae

Suborder Cryptodira

Order Testudines

Size Carapace from 29 in (74 cm) to over 4 ft (1.2 m) in length depending on subspecies

Weight Approximately 500 lb (227 kg)

Key features Large, bulky tortoise; shell shape varies depending on subspecies; neck often long; carapace and plastron are a uniform dull shade of brown; males have longer, thicker tails than females and often have a more yellowish area on the lower jaw and throat

Habits Seeks out sun in the morning, basking before setting off to feed; usually inactive in the latter part of the afternoon, sometimes wallowing in a muddy hollow; quite agile despite its large size

Breeding Female lays clutch of 2–10 eggs, occasionally up to 16; eggs hatch usually after 3–4 months

Diet A wide range of vegetation and fruit; can even eat the spiny shoots of the prickly pear cactus

Habitat Generally prefers upland areas

Distribution Restricted to the Galápagos Islands

Status Endangered, critically in some cases (IUCN)

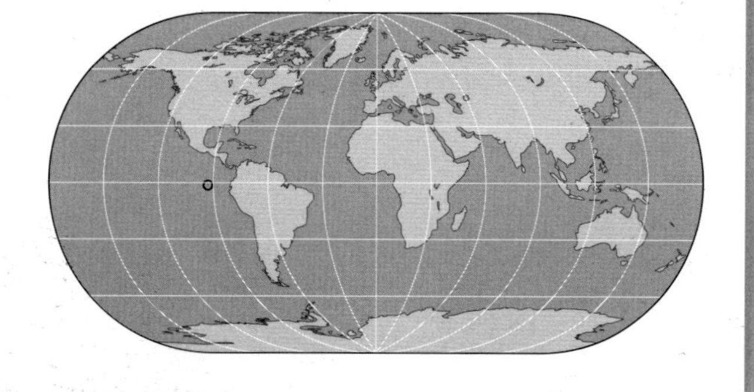

The giant tortoises of the Galápagos Islands have played an important part in the field of biological science. They provided one of the pieces of evidence used by Charles Darwin to support his theory of evolution, which stunned the world in the 19th century.

The British naturalist and explorer Charles Darwin visited the Galápagos Islands as the zoologist aboard *H.M.S. Beagle* in 1835. The population of giant tortoises, *Geochelone nigra*, on these volcanic islands had already been known to European whaling ships for many years. Crews regularly used to take the giant tortoises on board, and the creatures were able to survive for up to 14 months without food before being butchered and eaten.

Ships' Mutton

The numbers removed are quite staggering. Ships' logs reveal that a total of 115,000 were taken off the islands between 1811 and 1844, and the trade was well established by the time Darwin visited. The meat of the tortoises was even named "Galápagos mutton." They were not just used for food, however—their fat was distilled into oil along with their eggs. This plundering had a damaging effect on the tortoise populations, not all of which survived.

It is thought that there were 14 distinct races, or subspecies, on the islands. Only 11 are still there today—one of them, *Geochelone nigra abingdoni,* is facing imminent extinction and has been reduced to a single individual.

After such heavy persecution it is remarkable that it was not until 1876 that the first of these tortoises died out. It used to live on Charles Island and was flat backed with a very shiny shell. The next to disappear was the undescribed Barrington Island race, which was certainly extinct by 1890. Subsequently,

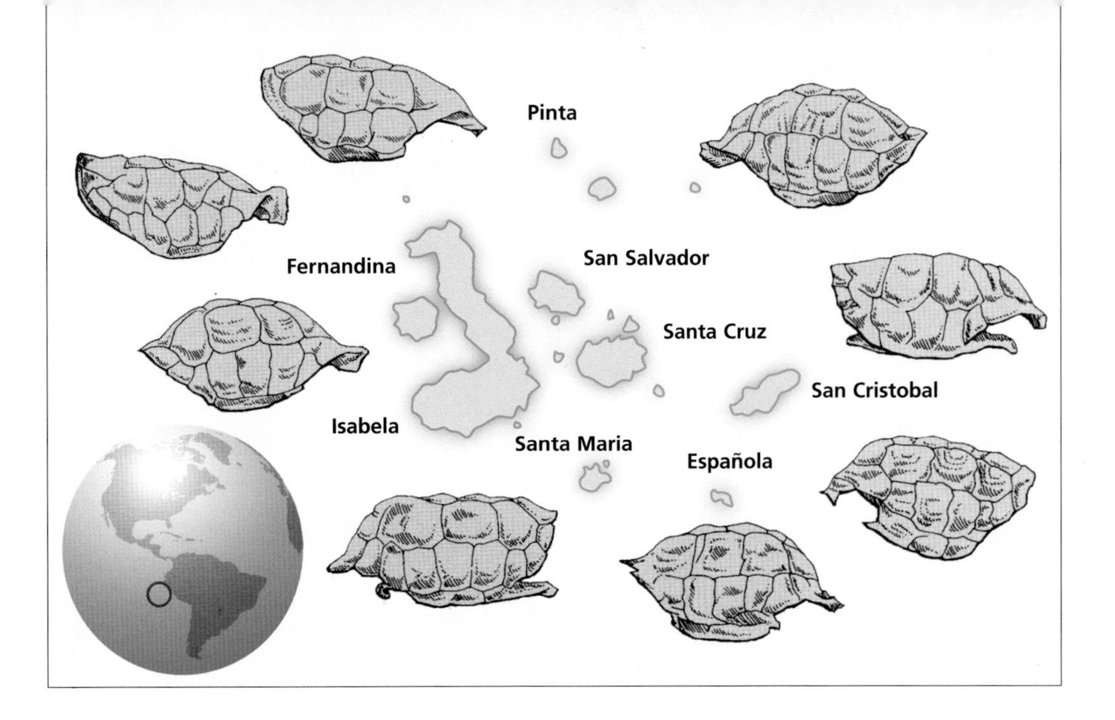

Several races of the Galápagos giant tortoise live isolated from each other on different islands in the Galápagos. The correlation between the different conditions on each island and the marked divergence in shell shape helped Charles Darwin formulate his ideas on evolution.

the Narborough Island tortoise with its characteristically flared marginal shields located at the back of the carapace had also disappeared by 1906.

Where Did They Come From?

The Galápagos Islands themselves consist of a group of 20 islands located approximately 600 miles (1,000 km) off the coast of Ecuador in northern South America. They are volcanic in origin, which means that there was no indigenous wildlife on the islands.

It is thought that the ancestors of the giant tortoises drifted south from Central America through the Pacific Ocean and were ultimately washed up on the beaches. They managed to colonize and spread

The Galápagos giant tortoise is the largest and heaviest living tortoise. This old male roams across the landscape at a speed of just 0.16 miles per hour (0.3 km/h).

Lonesome George

The population of giant tortoises on Abingdon Island, *Geochelone nigra abingdoni*, was particularly vulnerable to trade in tortoise meat in the 18th and 19th centuries because of its location—it was the first island in the archipelago to be encountered by whalers heading down toward Antarctic waters. Surprisingly, the population just managed to survive this era until the introduction of three goats in 1958. They stripped the tortoises' habitat, and by 1971 there was thought to be only one surviving male on the island. Named "Lonesome George" by staff at the Charles Darwin Scientific Research Station on Santa Cruz Island where he now resides, this individual is estimated to be between 50 and 80 years old.

Sadly, in spite of intensive searching on Abingdon Island for more than 30 years, it has proven impossible to find George a mate, and he shows no interest in breeding with females of any other subspecies. He is likely to be the last of his line, although he is only middle-aged at present. With a potential life expectancy approaching 200 years, there is the possibility that some of today's oldest tortoises on the islands were already resident there as hatchlings at the time of Charles Darwin's visit in 1835.

across the islands, or (more probably) there were several such strandings. Recent studies involving mitochondrial DNA, which is used for tracking ancestries, have revealed that the oldest group of tortoises can be found on Española. A second wave of colonization established populations on southern Isabela, Volcan Darwin, and Volcan Alcedo, followed by a third on other islands within the group.

It may seem unlikely that tortoises could drift through the ocean and end up being washed onto such tiny specks of land. Yet they are not the only reptilian colonists from the American mainland on these islands. Marine iguanas, *Amblyrhynchus cristatus*, thought to be descended from the green iguana, *Iguana iguana*, also live there. It is not uncommon even today for tortoises to be carried out to sea on floodwaters—they are, in fact, well equipped to

survive in these surroundings, remaining afloat with little effort and bobbing along on the waves. They do not need to eat, since their body stores fat that can be metabolized. It is possible that the early reptiles made the crossing not in the sea itself but by floating on debris, a process known as rafting.

What is particularly significant is that there would not necessarily have been any need for more than one tortoise to have reached the Galápagos Islands in the first instance to begin the colonization process. A mature female that had bred previously would not have needed to mate again in these new surroundings in order to lay fertile eggs, since she could still have been carrying viable sperm in her reproductive tract. Based on studies of modern chelonians, we know that a female may well be able to lay regularly for as long as four years without a male, producing many tens of offspring during this period. The young would have hatched in a relatively safe environment, so the population probably started to increase in number quite rapidly.

As night falls over the Alcedo Volcano on Isabela Island, a group of giant tortoises wallow in a pool to cool down and gain relief from ticks and mosquitoes.

Food and water can be scarce. The Galápagos giant tortoise eats prickly pear cacti and fruits, bromeliads, water ferns, leaves, and grass. It can store large amounts of water, enabling it to survive the long dry season.

It is impossible to calculate how many of these island giants there were, but some people suggest there could have been hundreds of thousands of them at one stage. The biggest problem they would have faced would have been the ability to find enough food in a limited range. As a result, they evolved physical adaptations to reflect their environment, a fact that excited Darwin during his visit.

Three Types of Tortoise

Zoologists now tend to classify the different races of Galápagos giant tortoises in three separate categories based on these physical adaptations. First, there are those known as saddlebacks, a reference to the way in which the shell at the front of the carapace is raised above the neck. These individuals tend to be found on islands where conditions are relatively arid. This adaptation enables them to browse on taller plants that are less affected by periods of drought than vegetation growing at ground level. Their neck is long as well, and their limbs are elongated, so they can use their height to maximum effect.

Remarkably, the plants on the islands have responded to the browsing behavior of the tortoises. Where saddleback populations occurred—for example, on Abingdon Island—the prickly pear cactus, *Opuntia*, which is one of the tortoise's main foods, altered its shape too. It developed a tougher outer casing and adopted a more treelike growth pattern, making it harder for the tortoises to feed on it.

On islands where saddlebacks are not present, the prickly pear's shape is unaffected.

Second, there are Galápagos giant tortoises with a more typically dome-shaped appearance and a short neck. They tend to be encountered on upland areas of the islands where grazing conditions are generally good. The third group consists of tortoises that display intermediate characteristics between these two extremes.

The legacy of the seafarers' visits to the islands is still apparent today and represents an ongoing threat to the survival of various populations, in spite of the fact that hunting has ceased and the tortoises are fully protected. Today's problems revolve largely around the other creatures that were brought on the ships and were frequently abandoned on the islands. They too have thrived, often to the detriment of the tortoises. In order to supplement their rations, sailors often left goats on remote outposts, quite oblivious to the environmental problems they would cause. On the Galápagos Islands goats have competed with the tortoises for food and can destroy their nests as well. Pigs are also a hazard, since they will actively seek out nests using their keen sense of smell.

Saving the Tortoises

A number of different programs are being undertaken to conserve and increase the populations of giant tortoises, not least by controlling the introduction of mammals, including black rats, which are another legacy of sailing ships that visited the islands. Intensive captive-breeding programs are also underway both in the Galápagos and abroad. Zoological collections around the world are cooperating for this purpose.

The success achieved can be remarkable, as shown by the work of the Darwin Foundation, which has hatched over 2,500 young tortoises in barely 40 years. This has provided a major boost to a population estimated at no more than 10,000 individuals in total. The tortoises are kept safely in pens until they are three years old and large enough to be no longer at risk of predation by endemic birds of prey. It may take a quarter of a century for these tortoises to become sexually mature.

Even today, however, unexpected hazards can occur and threaten populations of the tortoises, as in October 1998, when lava from the erupting Cerro Azul volcano started to flow toward a group of tortoises. An airlift was organized, but in spite of these efforts, a small number of tortoises died from being trapped in the lava or the resulting fires nearby. They included a member of the critically endangered subspecies *G. n. guntheri,* whose total population is at most about 100 individuals. Shifting the gigantic tortoises is a difficult and costly exercise. Helicopters were used to ferry them to the coast from 5 miles (8 km) or so inland, from where they could be taken in boats to the safety of the breeding center.

Reproduction

Male Galápagos giant tortoises often make a roaring sound when mating. Courtship itself is a fairly brutal process, with the larger males battering at the shells

The sailors have gone, but today tourists still flock to see the wildlife on the islands, including the giant tortoises on Santa Cruz.

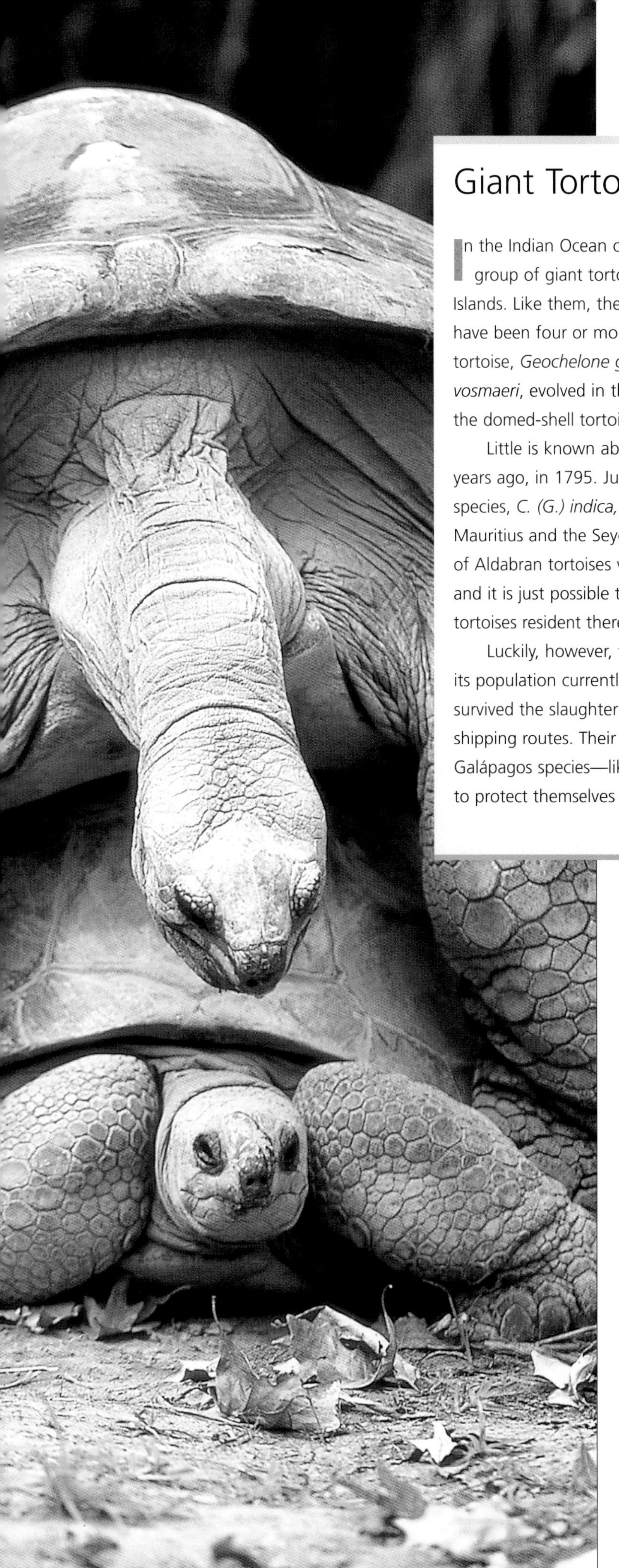

Giant Tortoises Elsewhere

In the Indian Ocean off the southeast coast of Africa a completely separate group of giant tortoises suffered a fate similar to those on the Galápagos Islands. Like them, they were captured in large numbers by mariners. There may have been four or more species, of which only one survives today—the Aldabran tortoise, *Geochelone gigantea*. Interestingly, a saddleback form, *Cylindraspis (G.) vosmaeri*, evolved in the region—on the island of Rodriguez—and was larger than the domed-shell tortoises, *C. (G.) peltastes*, which were also found there.

Little is known about these two species, which became extinct about 200 years ago, in 1795. Just prior to this the last surviving example of the Reunion species, *C. (G.) indica,* had died out as well. Native populations also existed on Mauritius and the Seychelles, but information about them is sketchy. A number of Aldabran tortoises were taken to the Seychelles and Mauritius in the 1800s, and it is just possible that they might have interbred with any surviving native tortoises resident there.

Luckily, however, the future of the Aldabran species itself seems secure, with its population currently consisting of over 100,000 individuals. Its ancestors survived the slaughter essentially because of their remote location away from shipping routes. Their behavior is similar in some respects to that of the Galápagos species—like them, they wallow in mud to cool their bodies and to protect themselves against mosquito bites.

Mating in large reptiles can be a dangerous business. This is a mating pair of captive Aldabran tortoises,* Geochelone gigantea, *in the Seychelles.

of the females and pinning them down by clambering on top.

Females lay their eggs between July and December, but this varies on different islands. Although clutches contain relatively few eggs—usually from just two to 10 but occasionally up to 16—females may lay more than once during this period. The eggs themselves have hard shells and are spherical in shape. The incubation period lasts on average between three and four months but can extend up to seven months. The young tortoises emerge during the wet season when there is fresh grass and other vegetation available for them to eat. The carapace of a newly hatched individual measures about 2.4 inches (6 cm) in length.

Common name Garden lizard (changeable lizard, red-headed lizard, bloodsucker)

Scientific name *Calotes versicolor*

Family Agamidae

Suborder Sauria

Order Squamata

Size To 20 in (51 cm)

Key features Body quite deep; limbs long, head triangular; snout pointed; there is a crest of enlarged scales on the nape of its neck; males are larger than females and have higher crests; light brown to buff in color with no markings on the body or with faint dark bands; males develop a red throat and chest during the breeding season

Habits Diurnal; mainly arboreal but sometimes ventures onto the ground

Breeding Egg layer; female may keep eggs in oviduct to lay later

Diet Insects and other invertebrates

Habitat Open forests, field edges, gardens, and parks

Distribution The Middle East (Iran and Afghanistan) through India and Sri Lanka, Indochina, South China, Hong Kong, and Sumatra; also several islands, including Andaman, Mauritius, and the Maldives (where it is probably introduced); also introduced to Oman

Status Very common

Similar species Other *Calotes* species and related genera are found throughout the region, but *C. versicolor* is by far the most common and the one most likely to be seen

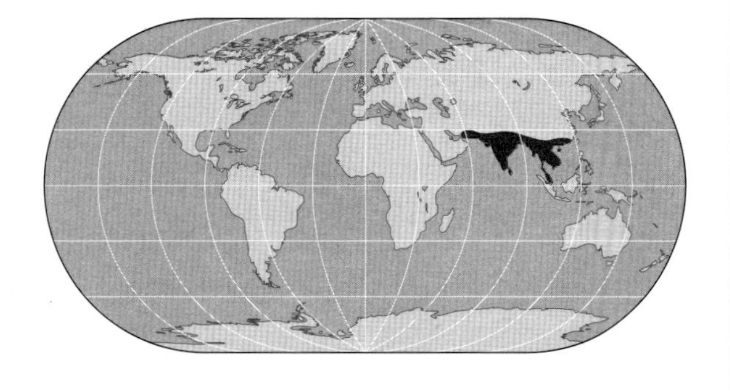

Garden Lizard

Calotes versicolor

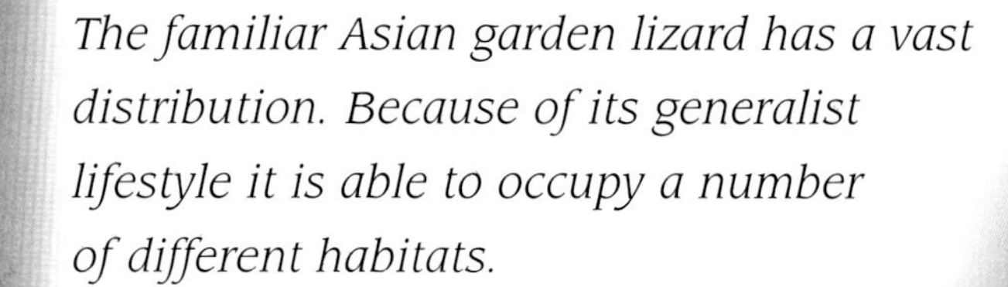

The familiar Asian garden lizard has a vast distribution. Because of its generalist lifestyle it is able to occupy a number of different habitats.

In contrast to some other Asian agamids whose tiny ranges can be measured in hundreds of square yards, the garden lizard is found almost everywhere throughout the warmer parts of southern and Southeast Asia. It occurs from the Himalayan foothills to the steamy plains.

Blood-Red Throats

The garden lizard is a conspicuous diurnal species, often resting vertically on tree trunks and in bushes even near busy roads. It gets one of its common names, "bloodsucker," from the breeding coloration of the males. They take on a bright red flush to their heads, throats, and shoulders. They also develop jet-black patches on their neck and cheeks. At this time they are very aggressive toward each other, and their colors intensify or fade as they fight according to whether they are winning or losing—defeated males lose their bright colors and sulk.

Courting males raise their bodies as high as possible as they approach the female, who usually stays hidden. They extend a pouch in their throat (the "gular" pouch) while nodding their head slowly up and down. As they do so, they open and shut their mouth rapidly.

Because this species is so common, it has been studied quite thoroughly. Some of the results are very interesting, and they may apply to other lizards as well as the garden lizard.

Females have a very adaptable breeding cycle. At the beginning of the breeding season, which varies according to locality, females have good fat reserves. They use the stored energy to produce their first clutch of eggs. Subsequent clutches are produced from food eaten by the female at the time they are being formed. By

↑ ***Calotes calotes*, the common green forest lizard from Sri Lanka and southern India, is a close relative of the garden lizard. It relies on its green coloration to protect it from predators.**

then the food supply has increased, and they can eat enough to keep themselves healthy as well as diverting energy to egg production.

Delayed Production

The ability to store energy is not their only trick. Females can store sperm in their oviducts for the whole breeding season if necessary. This allows them to lay more clutches without needing to mate again. What's more, females with eggs developing in their oviducts can put off laying them for up to six months if the conditions are not right, for example, if the weather turns cold and wet. Some scientists believe that egg retention in the oviduct may be a stepping stone toward viviparity (live birth) in lizards. There is a problem, however. If the eggs continued to grow in the oviduct, they would not be able to absorb enough oxygen to develop healthily, and the embryos would die. Furthermore, as the eggs absorbed moisture from the walls of the oviduct, they would continue to swell until they were too large to be laid. By reducing their body temperature to 73°F (23°C), females can arrest the development of their embryos.

Another adaptation is that both males and females only develop gametes (sperm or eggs) in the presence of the opposite sex. In experiments garden lizards were kept in isolation and in groups containing only the same sex. In both cases eggs and sperm failed to develop. In groups in which members of one sex had their testes or ovaries removed, sperm and eggs also failed to develop in the members of the opposite sex. This shows that they need to sense (probably through hormone release) that they are in a position to mate before expending energy on producing gametes.

Switching Genera

There are 21 species of *Calotes* altogether, but none are as widespread and adaptable as *C. versicolor*. A number of species that used to be included in the genus have been moved to another genus, *Bronchocela*. They are agile, bright green species that live in leafy shrubs. As in *Calotes*, the males have a low, serrated crest of pointed scales. *B. cristatella* is the most common species and is found throughout the Malaysian Peninsula and on Borneo. Its range in Singapore is thought to be declining because of competition with the garden lizard, which is better at adapting to man-made habitats.

Geckos

All geckos used to be classified as a single family, the Gekkonidae, but scientists now recognize at least three families: The Gekkonidae contains the "typical" geckos; the Eublepharidae, the eyelid geckos; and the Diplodactylidae, the so-called "southern geckos." A fourth family, the flap-footed lizards (Pygopodidae) is closely related to the geckos and is also included here, as is the Dibamidae, a family of very uncertain relationships.

The Gekkonidae

The most numerous family is the Gekkonidae. With about 910 species in 75 genera it ranks second in number of species only to the skinks (Scincidae). Many of them share certain characteristics, and generally speaking, geckos are among the more conservative in appearance families of reptiles and therefore easily recognizable.

Common name Geckos **Order** Squamata

The geckos are divided into three families. Two other families, the Pygopodidae (flap-footed lizards) and the Dibamidae (blind lizards), are allied to them.

Family Gekkonidae—75 genera, about 910 species of typical geckos and American dwarf geckos, sometimes divided into 2 subfamilies
- **Subfamily Gekkoninae**—about 70 genera, 770 species of typical geckos, including the tokay, *Gekko gecko*, the Moorish gecko, *Tarentola mauretanica*, and the day geckos, *Phelsuma* sp.
- **Subfamily Sphaerodactylinae**—5 genera, 140 species of American dwarf geckos, including the ashy gecko, *Sphaerodactylus elegans*, and the striped day gecko, *Gonatodes vittatus*

Family Eublepharidae—6 genera, 22 species of eyelid geckos
- **Subfamily Aleuroscalobotinae**—1 species, the cat gecko, *Aleuroscalabotes felinus* from Southeast Asia
- **Subfamily Eublepharinae**—5 genera, 21 species with a global distribution, including the western banded gecko, *Coleonyx variegatus*, and the leopard gecko, *Eublepharis macularius*

Family Diplodactylidae—14 genera, about 115 species of "southern" geckos, including the Australian leaf-tailed geckos, *Saltuarius* sp., the golden-tailed gecko, *Diplodactylus taenicauda*, and the live-bearing gecko, *Rhacodactylus trachyrhynchus* from New Caledonia

Family Pygopodidae—7 genera, 35 species of flap-footed lizards from Australia and southern New Guinea in 2 subfamilies
- **Subfamily Pygopodinae**—3 genera, 20 species, including the black-headed scaly-foot, *Pygopus nigriceps*
- **Subfamily Lialisinae**—4 genera, 15 species, including Burton's snake lizard, *Lialis burtoni*

Family Dibamidae—2 genera, 10 species of poorly studied "blind" lizards from Southeast Asia and Mexico

Gekkonid, or "typical," geckos have no eyelids. The eyelids have become fused and transparent (like those of snakes) and are shed along with the rest of the skin. Most species are nocturnal, so their eyes are large, and they often use their tongues to wipe them clean. In bright light the pupils of the eyes of nocturnal species close down to vertical slits with a number of small pinhole openings. Their irises are often intricately marked and can be quite beautiful. Diurnal species, which include all the small species in the subfamily Sphaerodactylinae, members of *Phelsuma* in the subfamily Gekkoninae, and several other genera as well, account for perhaps one-tenth of all geckos. They have smaller eyes and round pupils.

Typical geckos have an almost worldwide distribution and are absent only from the Arctic and Antarctic regions. They are successful colonizers of islands and archipelagos throughout the world's oceans and may be the only lizards present on some of the smaller ones. A tendency to wander has also resulted in the accidental introduction of many species to places where they are not native. Some species are so closely allied to man-made habitats that they are rarely found anywhere else.

Members of the Gekkonidae range in size from tiny creatures such as the American *Sphaerodactylus* species,

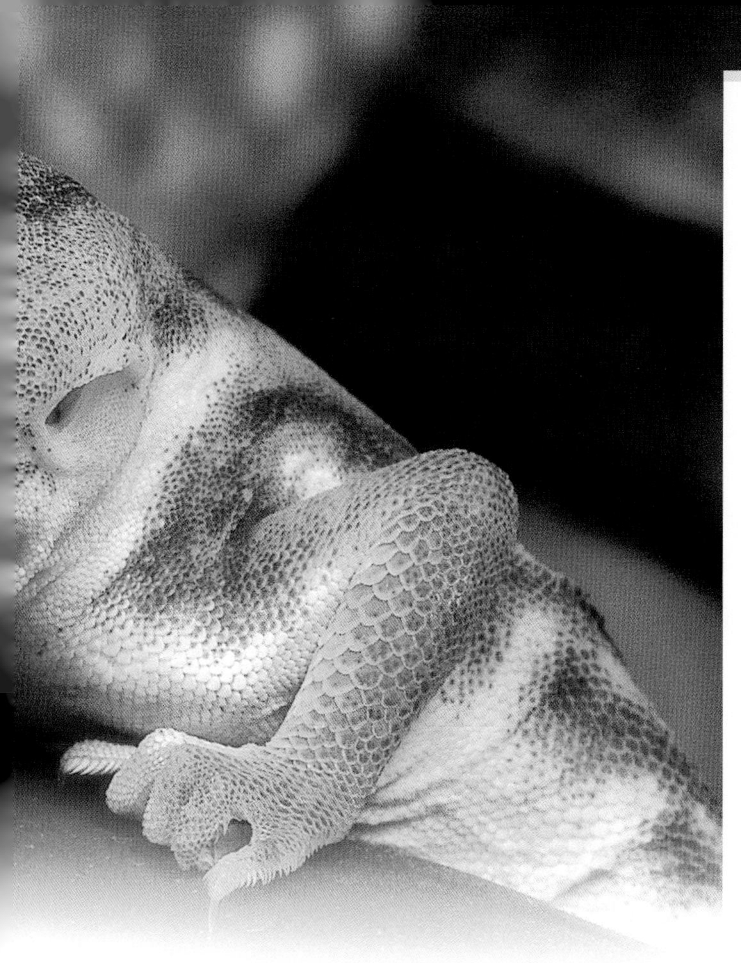

The frog-eyed gecko, Teratoscincus microlepis from Asia, belongs to the family Gekkonidae. It is unusual among gekkonids in having large overlapping scales.

at about 2 inches (5 cm) long, to the giant gecko from New Caledonia, *Rhacodactylus leachianus,* which grows to 13 inches (33 cm) from its snout to the end of its body. Its tail is short and stubby, and adds only a couple more inches to the total length. A more common species, the large, pugnacious tokay, *Gekko gecko*, grows to just over 13 inches (33 cm) in total.

Subdued Appearance

Nearly all gekkonids have small granular scales, but some also have larger tubercular scales scattered among the small ones. A few species, notably the ground-dwelling frog-eyed geckos in the genus *Teratoscincus* (six species from Asia) and the fish-scale geckos, *Geckolepis* (five species from Madagascar), have larger overlapping scales.

Nocturnal geckos tend to be dull in color, often gray or brown with indistinct, mottled markings. Many have a limited ability to change shade, becoming paler at night. The tokay is an exception: It is gray-blue in color with spots of orange or blood-red covering its entire body. Diurnal geckos are often colorful, the most attractive being the day geckos, *Phelsuma* from Madagascar and other islands in the Indian Ocean. Most are bright green with red markings, while some are bluish-green or

How Do They Stick and Move?

Geckos are spectacular climbers, but establishing exactly how their toes stick has not been straightforward. Unlike treefrogs, geckos do not grip by suction—if they did, they would be able to stick to wet surfaces, which they cannot. Nor do they have minute hairs on their feet ending in tiny hooks for gripping small irregularities in surfaces—if they did, a piece of perfectly clean glass would defeat them, and it does not. In fact, the smoother the surface, the better they stick. It now seems likely that they use "surface energy," or "surface adhesion." This force is the same one that creates surface tension on water, for instance, which is the reason why a pond skater can live on the water's surface without sinking.

The physics of this phenomenon are complex, but experiments have shown that the gecko relies on the molecular attraction across closely positioned surfaces. The gecko's toe pads are made up of special structures called lamellae that form rows across each toe pad. Each lamella consists of thousands of hairlike cells known as setae, and each seta is again divided into even smaller branches. Each of the branches ends in a slightly swollen, or spatulate, structure, of which there can be up to one billion altogether. Their cumulative surface energy is more than enough to support the gecko.

With all these setae acting together to allow the gecko to cling to a smooth surface, it could easily become stuck. It gets around this problem by curling up the ends of its toes when it wants to move its foot—in effect "unpeeling'" them from the surface. Of course, all this happens so quickly that it is usually impossible to follow with the human eye.

Geckos are nature's supreme climbers thanks to the enlarged adhesive toe pads containing thousands of microscopic setae.

Versatile Tails

Geckos are often prone to discarding their tails if they are grasped: Finding an individual that does not have a regrown tail is almost impossible in some populations, although species that use their tails to store fat or that have very short, stumplike tails are less inclined to sacrifice them.

There is more to a gecko's tail than this, however. Many species, such as several naked-toed geckos, *Cyrtodactylus* species from Asia, the leopard gecko, *Eublepharis macularius*, and the Australian velvet geckos, *Oedura* species, have tails that are banded in black and white. They serve to distract predators, which then attack the tail rather than a more vulnerable part of the gecko's body. Many species, especially those from arid regions, have thick, carrot-shaped tails that are used to store food. A few, notably the Madagascan and Australian leaf-tailed geckos, *Uroplatus* and *Phyllurus* respectively, have tails that are flattened and leaflike. These geckos press them against the tree trunks on which they rest to eliminate shadows. The Asian flying, or parachute, geckos, *Ptychozoon*, have a scalloped fringe running around the edge of the tail that increases air resistance and helps them glide.

Other activities associated with geckos' tails include the production of sound or substances. The wonder geckos, *Teratoscincus* species, and the viper gecko, *Teratolepis fasciata*, produce a sound by moving their tail and thereby rubbing together the large scales covering it, a process known as stridulation (which is more common in insects such as crickets and grasshoppers). Some *Diplodactylus* species produce a sticky substance in glands in the tail; if they are in danger, they deter predators by spraying them with it. Staying in Australia, the knob-tailed geckos, *Nephrurus* species, have small knobs on the end of the tail that are well endowed with nerve endings. Scientists believe that they may be important in chemical communication, thermoregulation, or defense.

turquoise. Some of the dwarf geckos, *Lygodactylus* species from Africa, are also colorful with yellow or black-and-white heads.

Sticky Toe Pads

A number of different species are known as "house" geckos. One of their more notable characteristics is the presence of adhesive pads at the ends of their toes. They vary slightly in shape but have expanded surfaces that enable the geckos to run up any type of vertical surface, rough or smooth, and even to rest upside down on ceilings. The exact design of the toes and their pads varies from genus to genus, and the scientific names often reflect this. So we have, for instance, *Hemidactylus*, meaning "half toe," *Sphaerodactylus*, meaning "ball toe," and *Phyllodactylus*, meaning "leaflike toe."

Other species do not have adhesive pads, however. They are divided into the ground geckos that do not climb at all, such as *Chondrodactylus* ("grain toe") and *Pachydactylus* ("thick toe"), and the species that climb into trees and bushes and use claws rather than pads. They include members of the genera *Cyrtodactylus* ("curved toe") and *Gymnodactylus* ("naked toe").

Ground-dwelling geckos are especially common in deserts (where there is often nothing

Displaying a brightly colored tail, as in the Thai bow-fingered gecko,* Cyrtodactylus peguensis, *can be a ruse to divert attention away from more vulnerable areas of the body.

The aptly named fat-tailed gecko,* Hemitheconyx caudicinctus *from Africa, uses its thick, stumpy tail as a means of storing fat to see it through times when food is scarce.

↑ ***Duvaucel's gecko,* Hoplodactylus duvaucelii, *belongs to the family Diplodactylidae and is endemic to New Zealand. It supplements its diet of insects with pollen and nectar. Here it is feeding on flax nectar.***

for them to climb anyway), and they typically live in burrows or rock crevices. Many have plump tails in which they store fat, and some have large heads with powerful jaws for crushing hard-bodied insects such as beetles. In Africa two species of web-footed geckos in the genus *Palmatogecko* have webbed feet to help them run across loose sand without sinking. Barking geckos, *Ptenopus* species from Africa, have fringes of hairlike scales around the edges of their feet for the same reason. All ground geckos are nocturnal.

Of the two subfamilies the Gekkoninae contains all the medium to large species from the Old World, such as the tokay, *Gekko gecko*, the day geckos, *Phelsuma* species, and the house geckos, *Hemidactylus* species and others, as well as some from the New World, such as the turnip-tailed gecko, *Thecadactylus rapicauda*. All members of this subfamily lay hard-shelled eggs, usually in pairs but occasionally singly or in threes. The small, diurnal sphaerodactyline geckos are restricted to the New World, mainly to Central and South America. In many species males are more brightly colored than females. In the striped gecko, *Gonatodes vittatus* from northern South America, their color varies so much that the two sexes look as though they belong to different species. These tiny geckos also lay hard-shelled eggs, but the clutches consist of a single egg.

The Diplodactylidae

The Diplodactylidae are sometimes known as "southern" geckos because they are found only in Australia, New Zealand, and New Caledonia. There are about 115 species in 14 genera, and they include several unusual species. Southern geckos produce two offspring, like most typical geckos, but they may either lay soft-shelled eggs or give

⊖ *Six species of giant geckos in the genus* Rhacodactylus *are endemic to New Caledonia.* Rhacodactylus ciliatus *has a row of enlarged, spinelike scales above the eye and along the neck.*

birth to live young. The New Zealand geckos belonging to the genera *Hoplodactylus* and *Naultinus,* with 10 and eight species respectively, are all live-bearers, as is a single species from New Caledonia, *Rhacodactylus trachyrhynchus*. Five other species of *Rhacodactylus* (all from New Caledonia) are egg layers.

Live-bearing species often have long gestation periods: The spotted sticky-toed gecko, *Hoplodactylus maculatus*, has a 14–month pregnancy and therefore only breeds every second year. Even so, it only gives birth to two young. It therefore has the lowest reproductive potential of any lizard, indicating that its life span is long.

Australian species include the leaf-tailed geckos, *Phyllurus*, and nine species of knob-tailed geckos, *Nephrurus*. As their common name suggests, their tails (which are very short in some species) end in a small knob that is well equipped with nerve endings and may be used in communication or thermoregulation.

Southern geckos live in a variety of habitats. Many are desert dwellers, while others are arboreal. They include many of the velvet geckos, *Oedura*, which are often habitat specific—some live under the bark of fallen trees, others inhabit standing trees, and yet others make their home under flakes of rocks.

The fringe-toed velvet gecko, *Oedura filicipoda*, has a fringe of scales around its toes to enhance its climbing abilities. It has a prehensile tail that helps it clamber about in vegetation. The same tail adaptation is also found in the jeweled gecko, *Diplodactylus elderi* from Australia, and the New Zealand day geckos, *Naultinus* species, as well as three species of *Eurydactylodes* from New Caledonia. In addition, the fringe-toed gecko, the New Caledonian geckos, *Rhacodactylus*, and the pad-tailed gecko, *Pseudothecadactylus australis*, have adhesive pads on the tips of their prehensile tails to supplement those found on their toes.

The Eublepharidae

The Eublepharidae is the most primitive of the gecko families. All of its members have movable eyelids—they are often referred to as the eyelid geckos—and none of them have sticky toe pads. They are all ground dwellers except for the cat gecko, *Aleuroscalabotes felinus* from Malaysia. This unusual species is semiarboreal and has a prehensile tail that it holds in a coiled position as it moves slowly along twigs.

Other eyelid geckos live in North and Central America (*Coleonyx* species), in Africa (*Hemitheconyx* and *Holodactylus* species), in the Middle East (*Eublepharis*), and in Japan and eastern China (*Goniurosaurus*). There are 22 species altogether, and their scattered distribution indicates that they were once more widespread. The species that remain are relicts of a much larger family that had an extensive range.

Eyelid geckos lay their eggs in pairs (but occasionally singly or in threes). Like the southern geckos, they have a pliable shell that absorbs water during its development. Species from North America, Africa, and the Middle East live in dry habitats, those from Central America and Asia live in rain forests, while the species from Japan and China live in cool, moist habitats, especially caves. One characteristic of the species that have been studied to date is that they defecate in one place, as though marking a territory.

The Dibamidae

The biology of this family of "blind" lizards is practically unknown. It contains 10 species, of which nine are in the genus *Dibamus*, living in the forests of Southeast Asia; the other species, *Anelytropsis papillosus*, lives in Mexico. All are limbless, although the males have small vestigial (reduced) hind limbs in the form of scaly flaps like those of some of the Australian flap-footed lizards. Their eyes are also vestigial and are covered by a single scale, and they have no external ear openings. They burrow in soil or live in or under rotting logs and grow to a maximum of about 10 inches (25 cm) including their tails. Nothing is known about their eating habits or breeding behavior, although females of the species studied so far lay clutches consisting of a single egg.

Scientists are undecided on where blind lizards come in the scheme of things, and some believe them to be more closely related to the worm lizards (suborder Amphisbaenia) than to the true lizards in the suborder Sauria.

Geckos on Islands

There are few islands where geckos of some sort do not live; on some islands they are the only reptiles. Many of the islands have never been attached to the mainland, so the geckos must have arrived by sea—clinging to drifting debris torn up during a storm on the mainland or possibly as eggs hidden behind the bark of a tree trunk. Geckos are suited to spreading in this way because their eggs are well protected against desiccation. They can survive long journeys before hatching out sometime after reaching their destination. In addition, they often use communal egg-laying sites, with several females laying in the same place. A single landfall can therefore provide enough individuals for a colony to get started. A number of species, including several island species, are parthenogenetic (meaning they are all-female species that can reproduce without males); this is obviously an advantage, since it takes just a single individual to start a colony.

Many island species are widespread. The common house gecko, *Hemidactylus frenatus*, for example, is found throughout the tropical and subtropical world. Others, however, obviously arrived tens of thousands of years earlier and evolved into new forms that were not found anywhere else. On the island of Socotra off the Arabian Peninsula, for instance, there are 18 species of geckos, of which 15 are endemic.

Common name Gharial

Scientific name *Gavialis gangeticus*

Subfamily Gavialinae

Family Crocodylidae

Order Crocodylia

Size Up to 21 feet (6.4 m) overall

Weight Large males may weigh up to 1 ton (1,016 kg)

Key features Body relatively slim with an elongated, narrow snout; in males snout has a knob at its tip; teeth narrow, sharp, and interlocking; color usually olive-green with dark bands across the body; rear feet heavily webbed

Habits Highly aquatic

Breeding Female lays clutch of 28–43 eggs; eggs hatch after 65 or 80 days

Diet Predominantly fish; older individuals may also catch birds; reputedly scavenges on the cremated remains of humans

Habitat Rivers

Distribution Northern parts of the Indian subcontinent

Status Endangered (IUCN); listed on CITES Appendix I

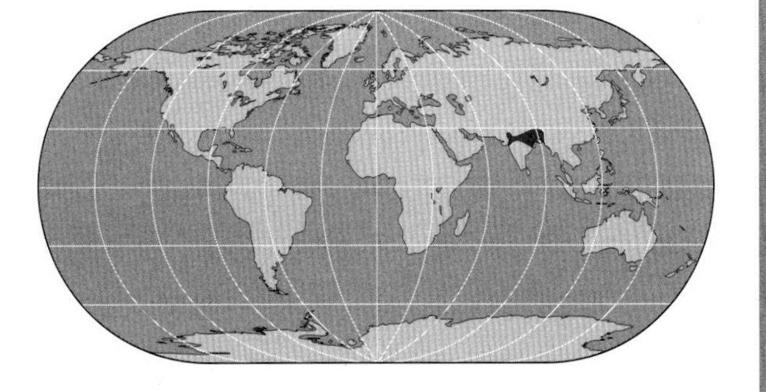

Gharial

Gavialis gangeticus

Known since ancient times, the gharial is the only living species of the subfamily Gavialinae. It is highly aquatic and well adapted for life in the deep waters of the great rivers of the Indian subcontinent.

THE UNUSUAL NAME of these distinctive crocodilians comes from the Indian word *ghara*. It describes the swollen handle on a type of cooking pot popular in the north of the country, which is said to be the same shape as the snout of a mature male gharial.

The gharial's range has contracted significantly over recent years. It is the last in the lineage of a much larger group that first emerged in the fossil record about 20 million years ago. These early gharials ranged across the Americas as well as in Africa and Asia.

Today's species is also much smaller in size than its ancestors. They could grow to at least 56 feet (17 m) long. Nevertheless, present-day gharials can reach 21 feet (6.4 m) long from the snout to the tip of the tail. Despite their size, they pose no threat to people. They are adept hunters, however, and can seize even quite large fish using the sharp teeth in their narrow jaws. They then raise their head out of the water and deftly reposition the fish as necessary, making it easy to swallow. Waterbirds may occasionally be caught as well.

Threatened Species

Over the course of the past century gharials have faced a number of threats to their survival. They used to be heavily hunted for sport. Then they began to be killed for their skins, which were used in the manufacture of leather goods. More recently, the growth in human populations has also contributed to the gharial's decline. The swollen snout of the males has acquired an ill-founded reputation as an aphrodisiac. As a result, they have been heavily hunted compared with females, creating a serious imbalance in the surviving population.

↓ The gharial's needlelike teeth are ideal for seizing slippery fish. Lying submerged, it snatches fish as they go by, but it has to raise its snout above the surface when swallowing to avoid taking in too much water.

Gharial eggs, which are the largest eggs laid by any crocodilian and weigh about 5.6 oz (160 g), are also considered to be a rare delicacy. Nests may be dug up and destroyed for this reason.

Changes in habitat, such as the construction of a dam at Kalagarh on the Ramganga River, have also had a serious effect on the numbers of these crocodilians. Gharials prefer fast-flowing rivers, but the blocking of the river flow by the dam caused flooding over a wide area and effectively destroyed their breeding habitat.

Ranching (which involves taking eggs from nests in the wild and hatching them in incubators) and captive-breeding programs are both being used by Indian zoologists to increase gharial numbers. Their efforts have been relatively successful, with the gharial population increasing from a low point of about 250 individuals in 1974 to as many as 2,500 by 2004. One of the major release sites for the young gharials is the National Chambal Sanctuary, where there is plenty of suitable habitat in which they can safely increase in numbers. The river system in this area extends for a distance of over 372 miles (598 km), providing an important potential refuge for these crocodilians. However, outside India in other parts of their former range there are believed to be fewer than 200 gharials surviving in the wild.

The Misjudged Gharial

Gharials were persecuted because they fed on fish; any of these crocodilians that had the misfortune to become entangled in fishermen's nets were likely to be killed as a consequence. Ironically, however, as the number of gharials fell, so did fish stocks. This apparent discrepancy was resolved once it became clear that gharials actually preferred to prey on fish that were themselves predatory by nature; by keeping their numbers in check, the gharials ensured that there were more fish for local people to catch.

Gharials also suffered persecution because it was not uncommon for individuals to be found with bracelets and other similar items of jewelry in their stomach. The belief grew up that these crocodilians attacked people. In fact, these objects are deposited in rivers along with the remains from cremations. The gharials swallow them to aid their digestion and possibly also to help with buoyancy. Most crocodilians simply swallow stones for this purpose.

Giant Leaf-Tailed Gecko

Uroplatus fimbriatus

Madagascar is home to many unique and unusual animals, but none is more bizarre to look at than the giant leaf-tailed gecko, which has been described as an "animated gargoyle."

Common name Giant leaf-tailed gecko

Scientific name *Uroplatus fimbriatus*

Subfamily Gekkoninae

Family Gekkonidae

Suborder Sauria

Order Squamata

Size Up to 12 in (30 cm) long and therefore one of the largest geckos in the world

Key features Body large and flattened; head large and triangular; eyes cream in color, massive and bulging with intricate markings; tail flattened and leaflike; toes have expanded pads for climbing and clinging; a frill of skin present around the lower jaw and along the flanks; coloration plain or mottled gray or brown, but it can change from light to dark possibly in response to temperature

Habits Strictly arboreal and nocturnal

Breeding Female lays 2 hard-shelled eggs; eggs hatch after about 77–84 days

Diet Invertebrates and possibly smaller lizards

Habitat Rain forests

Distribution Eastern Madagascar

Status Common in suitable habitat but hard to find

Similar species *Uroplatus henkeli* is very similar but has a more strongly patterned back, is slightly smaller, and has a different eye color; other species of *Uroplatus* are significantly smaller and are more easily distinguished from this species

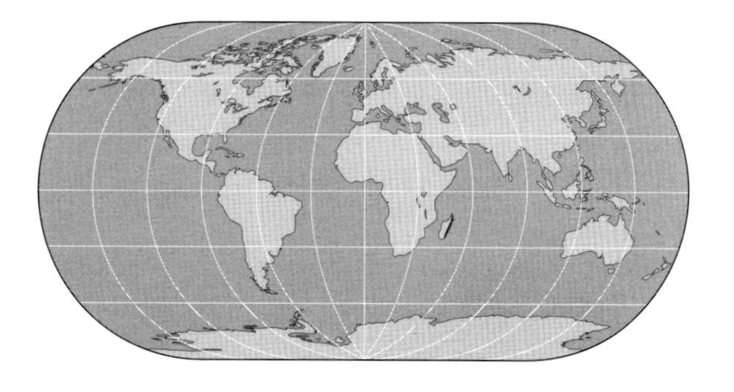

LIKE THE NORTHERN LEAF-TAILED GECKO from Australia, *Saltuarius cornutus*, the giant leaf-tailed gecko from Madagascar relies heavily on crypsis, or camouflage, to escape the notice of predators. Its colors and markings imitate the mottled rough bark of forest trees. It spends its days clinging to the trunk of a tree with its head pointing downward and its hind legs stretched out at an angle of about 45 degrees. Although it often rests in an exposed position, it can be almost impossible to see. It stays completely still even when approached at close quarters.

As well as blending in with the background, the gecko avoids detection by pressing itself close to the surface. Its flattened tail and the frilly fringes of skin along its jaws and flanks prevent shadows from forming. If it thinks it has been discovered, however, it pushes itself away from the trunk and opens its mouth wide, revealing its bright red tongue. Such intimidating displays and the gecko's strange appearance cause local people to fear it, believing it to be a devil named *taha-fisaka*.

Night Forays

The geckos come to life after dark and move away from their daytime retreats in search of food. In common with some other leaf-tailed geckos, their markings develop more contrast at night. Their movements are slow and deliberate; but if startled, they can move quite quickly.

Because of their size they can tackle most invertebrates and probably small vertebrates as well. Their method of catching prey is to pause with their limbs gathered beneath their body, leaning forward slightly at the same time. They study their prey from this position before

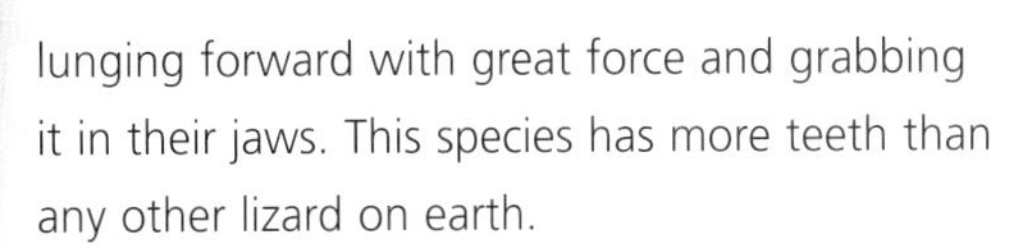

lunging forward with great force and grabbing it in their jaws. This species has more teeth than any other lizard on earth.

After a night's activity a gecko will often return to the exact place where it rested the day before—there are even reports that by resting in the same place, the geckos can create bare patches on tree trunks that are otherwise covered in lichens.

Individuals are rarely found close together in nature, and these geckos are thought to be territorial. They wave their tail when they meet another member of their species and may vocalize, but such interactions have only been casually observed and are not fully understood. During courtship the male vibrates his tail rapidly before mating. Females lay a pair of spherical, hard-shelled eggs about four to five weeks after mating and bury them in leaf litter on the forest floor. At a constant 80°F (26°C) the eggs hatch after 77 to 84 days.

A disturbed giant leaf-tailed gecko is an intimidating sight with its wide gape and brightly colored mouth. These geckos are among the largest in the world.

The Relatives

There are 11 species in the genus *Uroplatus* from Madagascar and probably more yet to be discovered. Some are newly described—*U. pietschmanni* as recently as 2004. Several species are almost as large as the giant leaf-tailed gecko, but there are a number of smaller species. *Uroplatus phantasticus*, for example, grows to about 4 inches (10 cm) long. Its tail is shaped like a shriveling dead leaf, and it has a prominent ridge along its back and small spines over its eyes. The lined leaf-tailed gecko, *U. lineatus*, is more slender than other species and is tan to yellow in color with numerous fine lines running along its body. It is thought to live in bamboo forests—its color and markings provide effective camouflage. The mossy flat-tailed gecko, *U. sikorae*, is possibly the most effectively camouflaged of all species, however. This medium-sized species is marked with patches of scales in various colors mimicking the patches of green, white, and gray lichens and mosses that grow all over trees and branches in the rain forests. Even from a few inches away it can be almost impossible to make out this gecko's outline.

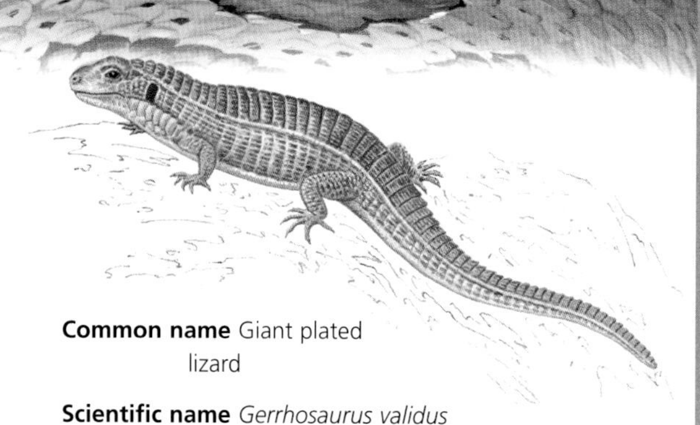

Common name Giant plated lizard

Scientific name *Gerrhosaurus validus*

Family Gerrhosauridae

Suborder Sauria

Order Squamata

Size 14 in (36 cm) long

Key features A large lizard with a flattened body and head; tail long, often thickened at the base; legs well developed but short; body scales rectangular and arranged in regular rows; an obvious fold present down each side of the body; adults are dark brown, but each scale on the head and back has a small yellow spot, creating a speckled appearance; some have a pair of cream stripes on the back; juveniles are dark brown with larger yellow spots on their back and crossbars on their flanks

Habits Diurnal and terrestrial

Breeding Female lays clutch of 2–5 eggs in rock crevice

Diet Invertebrates, small vertebrates, and some vegetable material

Habitat Grassland, mainly on rocky slopes or well-vegetated rock outcrops

Distribution Eastern subspecies, *G. v. validus*, occurs from northeast South Africa through Zimbabwe and Mozambique to Malawi and Zambia; western subspecies, *G. v. maltzahni*, lives in Namibia

Status Common in suitable habitat

Similar species The rough-scaled plated lizard, *Gerrhosaurus major*, is the other large species, but it has heavily keeled scales and is pale brown in color, sometimes with a stripe down either side of its back

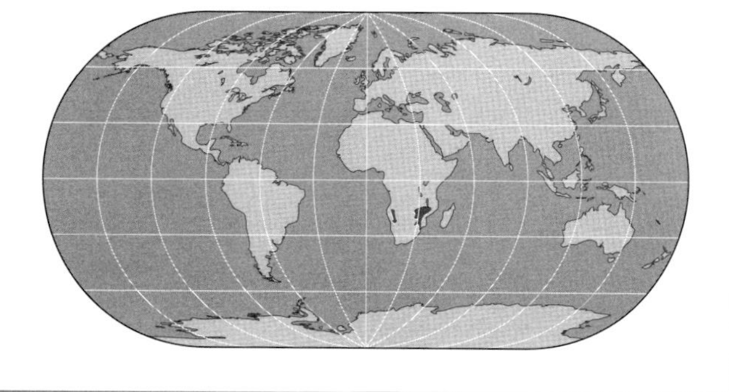

Giant Plated Lizard

Gerrhosaurus validus

The largest member of its genus, the giant plated lizard is shy and difficult to approach. When disturbed, it squeezes itself into a crevice and inflates its body.

THE GIANT PLATED LIZARD is large and bulky. It is heavily armored with rows of scales and underlying bony plates (osteoderms). Its body is roughly cylindrical in cross-section, and its limbs, although fully formed, are relatively small. It is not swift or agile, and it relies on its armor plating for defense.

The lizard usually makes a burrow between two slabs of rocks by digging out accumulated soil or leaf litter and never strays far from this retreat. If it is chased into its crack, it will inflate its body, making it difficult to dislodge. It usually looks for food in the immediate vicinity, using its front feet to scrape away soil at the bases of bushes or around rocks. Its main source of food is insects, but it also takes some vegetable material such as flowers, fruit, and berries. Small lizards and even baby tortoises are also eaten. It is truly an omnivorous lizard.

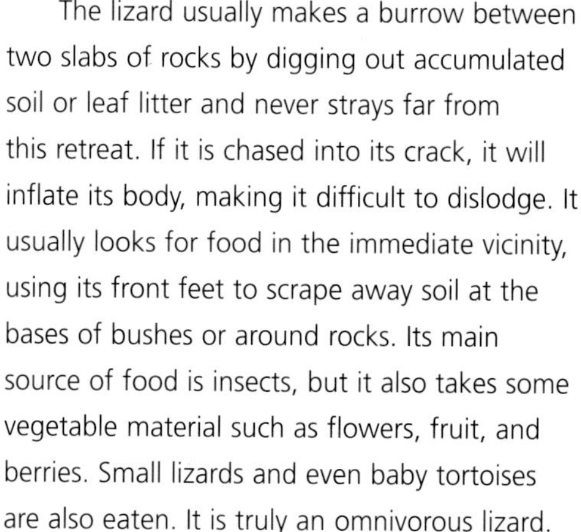

Detail of the giant plated lizard, showing the regular scales, flattened head, and dirty-white throat.

Living and Breeding

Giant plated lizards inhabit dry, grassy hillsides with rock piles or the upper slopes of vegetated *koppjes* (rock outcrops). They live in loose colonies, but they are not especially sociable. The sexes look the same, although males often have better-developed pores on the undersides of the thighs. In the breeding season their chin, throat, and sides of the head become tinged with purple. Nothing is known of their courtship or mating behavior. Females lay two to five (usually four) large oval eggs in soil-filled rock crevices in midsummer (November to December). The hatchlings appear at the end of summer and measure just over 6 inches (15 cm) in total length.

When basking, plated lizards often raise their limbs but keep their underside in contact

The Relatives

There are five other species of plated lizards. The rough-scaled plated lizard, *Gerrhosaurus major,* is also a large species. It is likely to be found in grasslands, but it hides in rock crevices and occasionally old termite mounds. It is usually light brown in color; specimens from the southern part of Africa often have dark centers to their scales and can also have a pale stripe down each side of the body, but those from the drier northern latitudes are often uniform brown. Like the giant plated lizard, this species is omnivorous. The Kalahari plated lizard, *G. multilineatus,* is similar, but its scales have light centers and darker edges, giving an overall checkered appearance. It lives in burrows beneath shrubs in the sandy soil of the Kalahari Desert.

The other three species are smaller. The yellow-throated plated lizard, *G. flavigularis*, is a handsome species. Its body is a rich brown color with a wide, bright-yellow stripe down each side of the back. The male's cheeks and throat are usually orange or yellow, becoming brighter in the breeding season, but males from some areas have blue throats. The black-lined plated lizard, *G. nigrolineatus*, is rather similar but larger and not nearly as graceful. The Namaqua plated lizard, *G. typicus*, is the smallest species at about 8 inches (20 cm) long. It is buff with dark speckling on its back and light spots on its flanks. The undersides of males turn bright orange-red in the breeding season. This is the rarest species as well as the one with the smallest range.

with the ground. They sometimes do this when moving down gentle slopes, sliding on their smooth belly scales. This is an intermediate step toward limb reduction and gives us a clue as to how limblessness may have evolved in the related seps lizards, *Tetradactylus* species.

The giant plated lizard's favored habitat is rocky, open grassland. Like other members of its family, it uses the cracks and crevices between the rocks to hide, and its body and head are suitably flattened.

Common name Gila monster (Aztec lizard)

Scientific name *Heloderma suspectum*

Family Helodermatidae

Suborder Sauria

Order Squamata

Size From 13 in (33 cm) to 22 in (56 cm)

Key features Head rounded and bears a patch of light-colored scales; nose blunt; neck short; body heavy with short, powerful limbs and long claws; tail short and fat; scales beadlike; eyelids movable; camouflage colors of black, orange, yellow, and pink on the body; has 2 elongated cloacal scales

Habits Active by day, at dusk, or at night depending on season and temperatures; spends much of the time in burrows or in shaded areas

Breeding Female lays 1 clutch of up to 12 eggs in late summer; eggs hatch 10 months later

Diet Small mammals, eggs of birds and reptiles, insects

Habitat Dry grassland, deserts, and foothills of mountains

Distribution Southwestern United States and northwestern Mexico

Status Vulnerable (IUCN); listed in CITES Appendix II

Similar species *Heloderma horridum*

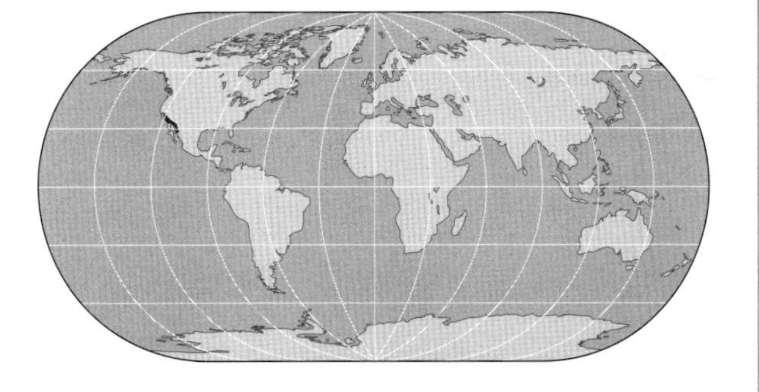

Gila Monster

Heloderma suspectum

Surrounded by myths and superstitions, the Gila monster is one of just two venomous lizards. Although potent, its venom has hardly ever been known to kill a human.

THE GILA MONSTER is named after the Gila basin in Arizona where numbers of the lizard are plentiful. It is sometimes referred to as the Aztec lizard, since it has featured in paintings by the Aztecs. Although it has a wide distribution in southern Nevada, southeastern California, southwestern New Mexico, and Arizona, numbers are concentrated in small pockets. There are two subspecies: *Heloderma suspectum suspectum*, the Gila monster, and *H. s. cinctum*, the banded Gila monster, which is slightly smaller and whose coloration contains lighter-colored bands. Its range includes southwest Utah and southern Nevada.

The habitat of Gila monsters varies from desert grassland, Mohave and Sonoran desert scrub, to Sonoran thorn scrub. They can be found on lower mountain slopes in arid and semiarid areas and also on adjacent plains and occasionally irrigated areas. They inhabit canyon bottoms—deep ravines with streams that may dry up for part of the year. In parts of Arizona the Gila monster's range extends into oak woodland and in Sonora onto beaches.

Seeking Shade

Within its habitat it seeks shelter under rocks, in dense thickets, and in wood-rat nests. It also digs burrows as well as making use of those belonging to other animals. Although Gila monsters are adapted to an extremely dry habitat, their optimum temperature is only about 86°F (30°C), which is considerably lower than other desert lizards. To avoid the high daytime temperatures during the summer, Gila monsters tend to be active at dawn and dusk. They spend the rest of the day in burrows, often dug using their powerful limbs and long

claws, or under rocks and shrubs. In Arizona Gila monsters spend 98 percent of their active season underground, and in Utah they live for 95 percent of the time in burrows.

During winter when temperatures fall below 50°F (10°C), they hibernate in burrows that have a south-facing entrance. On sunny days they wake and emerge to bask at the entrance. During the rainy season they become active at dusk and nocturnal; but after emerging from hibernation when temperatures are still relatively low, they are diurnal.

The Gila monster has a more rounded head, a shorter tail, and is a smaller species overall than its close relative the Mexican beaded lizard, *H. horridum*. Its long claws are useful when climbing trees, which it frequently does in the rainy season to escape the threat of torrential rain flooding its burrows. In spring it eats insects, but in June and July it changes to small mammals and birds. It can live for several months without food, although loss of weight shows mainly in the tail, which can lose 20 percent of its girth in one month without food.

Distinctively patterned in colors of light yellow, orange, pink, and black, the Gila monster is hard to spot against a dark background or in dappled shade.

Camouflage and Warning Coloration

Coloration of Gila monsters consists of irregular bands and blotches of black, orange, yellow, and pink. Younger specimens have more extensive lighter areas. Unusually for lizards these colors act both as camouflage and as a warning to enemies. Since the Gila monster is active primarily at dawn and dusk, the bands of color are difficult to see in the dappled shade when it moves or shelters beneath creosote bushes and other shrubs. Against a dark background the black markings blend in, and the light markings resemble gravel. Its nose is black, providing excellent camouflage when peering from burrows.

When it moves away from vegetation, its bright body markings become warning coloration, advertising its toxicity. In response to a threat the Gila monster will hiss and gape, revealing the pink venom glands that contrast with the dark lining of the mouth—yet another warning signal.

Beaded Lizard Venom

Venom in reptiles is usually associated with snakes. However, the two species of beaded lizard (the Gila monster and the Mexican beaded lizard) are the only venomous lizards. At first scientists debated whether or not they were venomous and gave the Gila monster the name *Heloderma suspectum*, since at the time it was only suspected to be venomous.

Beaded lizards have 10 teeth in each jaw. When compared with snakes, their venom-delivery mechanism is rather primitive. There is a gland on each side of the jaw with ducts next to the points where the teeth emerge from the jaw. When the animal bites, venom is expelled from the glands some distance from the teeth. The venom flows along a mucous fold between the lip and the lower jaw before reaching the front surface of the teeth. This is an inefficient method compared with the stabbing or biting stroke of vipers and cobras. Instead, Gila monsters must grip the prey or enemy tightly with both jaws and hang on to allow time for the venom to flow into the wound. Its jaws are very strong and difficult to disengage.

The poison produced is a neurotoxin that causes swelling, dizziness, drowsiness, vomiting, palpitations, swollen tongue, paralysis, labored breathing, and a fall in blood pressure. Some people unfortunate enough to have been bitten may experience just one or two symptoms. The swelling and pain that accompany a bite are due to the way in which the venom is injected. The lizard uses its vicelike grip to hold on, and chews with a sideways action of the teeth. It is possible for the elongated, inwardly curved, sharp teeth to break off and remain embedded in the victim. Teeth lost in this way are difficult to detect even using X-rays. Tissue destruction at the site of bites indicates that the venom also contains certain enzymes that play a role in digestion.

Gila venom is classified as sublethal, since there have been relatively few human deaths from it. Exhaustive studies have concluded that only eight to 10 people have ever died from beaded lizard bites. (It is interesting to note that all of them had consumed varying quantities of alcohol.) In the mid 1990s, as a result of studies on beaded lizard venom, pharmaceutical companies began experimenting with new treatments for diabetics based on elements of the venom. Even more recently the venom was found to have memory-enhancing properties, but more research will need to be done on this. Had the Gila monsters not been given protection, such medical advances would not have been possible.

A flaccid tail is an indication of poor condition in a Gila monster. As with other desert creatures, most of the moisture it needs is obtained from its food.

Reproduction

In the mating season Gila monsters have a structured social system in which dominance is established by male-to-male combat. Having spent much of the cooler months hibernating in burrows, they feed voraciously to regain body weight as soon as they emerge. Males become highly territorial in April, and wrestling matches take place. They frequently bite each other but are immune to the venom. Mating occurs in late spring and early summer. In late summer

females lay three to 12 elongated, leathery eggs, which they bury in a sunny spot near a stream at a depth of about 5 to 6 inches (13–15 cm). The eggs overwinter and hatch out about 10 months later.

Endangered Gila Monsters

Gila monsters live in small groups each with a home range of several acres. Although slow moving, they can travel several hundred yards a day. Much of their habitat has been reduced by human encroachment or destroyed by agriculture and industry. Deliberate killing through fear, superstition, or ignorance has depleted numbers further. Many Gila monsters have been collected for the reptile trade, and some have gone to institutions and serious herpetologists for captive breeding programs.

These lizards have enlarged lungs, which means that they need extra biotin (part of the vitamin B complex) to turn oxygen into carbon dioxide. In the wild this presents no problems, since fertilized eggs containing biotin form part of their diet; but in captivity many have been fed solely on unfertilized hens' eggs that lack biotin, with disastrous results.

Gila monster enjoy a degree of protection. In Arizona it is forbidden to keep them, but the law is not always enforced. They are listed in CITES Appendix II, and pressure is being applied to upgrade the listing to CITES Appendix I to restrict the trade in them still further.

Gila monsters grip their prey very tightly in their jaws. Since most of their prey is small and defenseless, venom is not usually needed. Here a Gila monster feeds on a young rodent.

Grass Snake

Natrix natrix

The European grass snake has a problem—it has to find a way to survive in the cool parts of northern Europe where winters can be long and cold, and the summer breeding season is relatively short.

Common name Grass snake (ringed snake, collared snake)

Scientific name *Natrix natrix*

Subfamily Natricinae

Family Colubridae

Suborder Serpentes

Order Squamata

Length 47 in (1.2 m) to 6.6 ft (2 m)

Key features Slender when young, but females become more stocky as they grow; head well separated from the neck; eyes moderately large with round pupils; scales keeled; body usually olive-brown or olive-green in color with a yellow or white crescent bordered by black behind the head; upper lips also yellowish and marked with a number of black bars; in some places grass snakes can be entirely black, black with white specks, or checkered black and gray

Habits Semiaquatic, swimming on the surface and diving occasionally

Breeding Egg layer with typical clutches of 8–40 eggs; eggs hatch after 42–70 days

Diet Frogs, toads, newts, and fish

Habitat Usually found around fresh water, including ponds, lakes, canals, and slow-moving rivers

Distribution Most of Europe, parts of North Africa and the Middle East, and a large part of Central Asia

Status Common but becoming less so in heavily populated areas

Similar species Two other European members of the genus are similar but smaller. An Asian species, *Natrix megalocephala*, has a slightly larger head. Young Aesculapian snakes, *Elaphe longissima*, also have yellow patches behind their heads

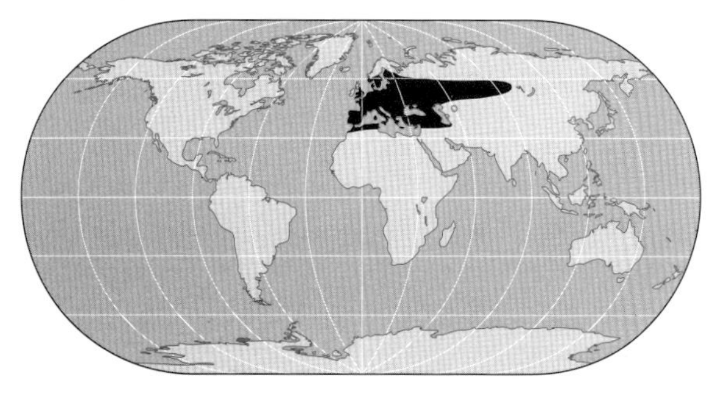

LIKE ALL REPTILES, THE grass snake depends on outside heat sources for its body functions such as digestion and locomotion to operate. Living in a cool climate makes this difficult, but the grass snake manages to live and breed by being active during the day, hibernating for up to six months, and basking when necessary.

Most snake species that live in cool regions give birth to live young. They overcome the conditions by basking on sunny days, thereby exposing their developing embryos to warmer temperatures and speeding up their development. Egg-laying snakes such as the grass snakes, however, must leave their eggs in a suitable place and trust to the weather. Snake eggs develop best at a constant temperature of about 82°F (28°C). If it is too cold, development slows down or stops altogether. In the relatively short northern summer any delay in hatching can be fatal.

Grass snakes have to choose their egg-laying sites carefully. They lay eggs in piles of leaves or even in piles of seaweed on the upper shore, where the heat produced by decomposition keeps the eggs warm. Since many grass snake habitats are near fields and farms, they may use compost and manure heaps. In days gone by, when baking ovens were common in houses, grass snakes often laid in holes in the walls behind the ovens.

Communal Egg Sites

Because suitable egg-laying sites can be scarce, females use them year after year, and several may lay their eggs in the same place. The record number laid together goes to a mass of 3,500 to 4,000 eggs laid every year in the shavings cellar of a saw mill in Mecklenburg,

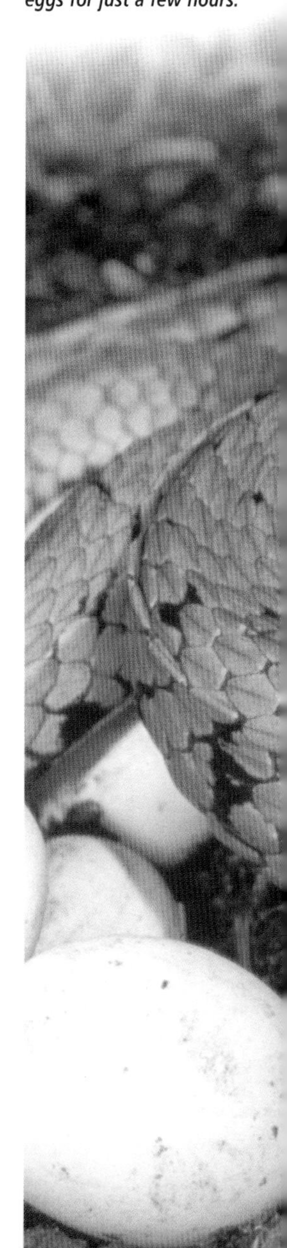

A female grass snake coils around the clutch of eggs she has just laid. She will remain with the eggs for just a few hours.

Conservation

Having recognized that one of the grass snake's major problems is finding a suitable egg-laying site, conservationists in Germany came up with a conservation plan. They are replacing traditional but diminishing farmyard manure heaps with purpose-built heaps. After one large heap was constructed, 62 eggs were found the next summer. After only two years the egg total rose to 400.

Recommended materials for the purpose-built heaps are those normally found naturally in the surrounding area: dung heaps near farms, leaves in parks and woods, compost heaps in gardens, and piles of reeds in reed beds. The heaps have to be maintained by adding to them or rebuilding them every year, preferably in April and May before the eggs are laid. Problems associated with the plan include the pollution of the surrounding area, so the site has to be chosen with care—not everyone wants a large pile of steaming, decomposing organic material at the bottom of their garden!

Germany, up until the 1960s. Since the average clutch size is thought to be about 30 eggs, this number represents the accumulated production of over 100 females. Unfortunately, many egg-laying sites of that type are no longer available to grass snakes, since the countryside is becoming "cleaned up," and farming practices are being modernized. As a result, grass snakes are much less common today than in the days when every village and farm pond had its resident population.

Populations of grass snakes that live farther south in the warmer climate of the Mediterranean region do not have the same problems with their eggs. They use more natural egg-laying sites, for example, hollow tree trunks, mammal burrows, and chambers under rocks and logs.

In northern Europe grass snakes hibernate for several months over the winter. They emerge in March or April: These are the months when they are most easily found. They often lie about near their hibernaculum in small groups, basking to raise their body temperature.

Mating Balls

Mating takes place shortly after hibernation. The male finds a female by following her scent trail. He crawls over her, rubbing his chin against her back. This is often accompanied by jerky movements. He wraps his tail under hers; and once their cloacae are next to each other, copulation takes place. Where there is a high density of males, they may compete with each other for females. Unlike the adders, they do not engage in individual combat but push and shove each other until one male—usually the biggest—displaces the others. While this is taking place, the males may form a "mating ball" with the female at the center. The largest mating ball recorded—in Sweden—contained 22 males, although three to eight are more common.

Once the female has mated, she is no longer attractive to males (unlike female adders, which often mate several times with different males). She moves off to search for a suitable egg-laying site. The eggs are laid in June or July, and clutches usually contain between 10 and 30 eggs, although in Germany a clutch of 105 has been recorded. Hatching time varies depending on temperature: Under ideal conditions they hatch in 42 to 45 days, but they may take as little as 35 days or as many as 70 days. The hatchlings measure about 5 to 8 inches (13 to 20 cm). Longer incubation times tend to produce larger young, although it is essential that the young are hatched by the fall before the weather turns too cold.

After laying, females spend the rest of the year foraging for food to replace the weight lost through breeding and to store up fat for the next hibernation period. Males feed as soon as the mating season is over—moving around within a well-defined home range of 7 to 296 acres (3–120 ha). Each snake moves between 10 and 300 yards (0.9–275 m) a day. During the course of a year it may travel a total of 2.5 miles (4 km).

Diet of Amphibians

The main food eaten by grass snakes is frogs and toads, which they swallow live and head-first. Some populations also eat newts, while others avoid them, perhaps because they are distasteful. Fish

Family Likenesses

The genus *Natrix* used to be much larger and included the North American water snakes that are now placed in the genus *Nerodia*, as well as several Asian species now called *Rhabdophis*. At present there are just four species in the genus.

The viperine snake, *Natrix maura*, lives in southwestern Europe and gets its common name from a zigzag stripe down the back of some individuals, although many are blotched instead. The dice snake, *Natrix tessellata*, is from southeastern Europe, overlapping with the viperine snake in a small area of northwestern Italy. The ranges of both species also overlap that of the grass snake, but they are easily told apart: The two southern species are smaller than the grass snake, and their eyes and nostrils are positioned toward the top of their head. Their markings are different, and in particular, they lack the yellow or white collar of the grass snake. Both species are egg layers; but coming from a warmer region, they do not specifically seek out warm places in which to lay their clutches. A fourth species, the big-headed grass snake, *Natrix megalocephala*, lives in the northeastern coastal region of the Black Sea. It is similar to the grass snake except for its proportionately large head.

⊖ Grass snakes are good swimmers and can stay under water for up to half an hour. This individual is swimming among duckweed in an English pond.

Ⓘ ***A grass snake,* Natrix natrix, *moves along a moss-covered rock and comes head to head with a toad* (Bufo bufo)*.***

make up a small proportion of their diet, and occasionally they take other prey, including mice, shrews, voles, nestling birds, and invertebrates such as slugs.

Grass snakes belong to the subfamily Natricinae, the members of which are known for their aquatic habits. Like the grass snakes, their main requirement is a constant supply of frogs and toads. Since these amphibians are more or less restricted to damp places, it follows that the grass snakes are too. They are not as closely tied to water as some of their close relatives, however; and when moving from one area to another (while searching for a place to hibernate or to lay their eggs, for instance), they travel along hedges and through sparse woodlands. In the south of their range, or where they live at higher altitudes, they are more closely associated with open water and rarely stray far from it.

Grass snakes are commonly found around ponds and lakes, in drainage ditches and canals, and in damp and boggy places. Unfortunately, many of these countryside features are disappearing as farming practices change and urban development spreads. Garden ponds can become a temporary refuge for grass snakes, but the food supply in them is often limited.

Grass snakes swim well (*natrix* means "water snake" in Latin and comes from the verb *natare*, meaning to swim). They often hunt in still or slow-moving water, swimming around the edges of ponds and lakes with their body at

or just below the surface and their head raised. If disturbed in water, they dive to the bottom and can remain under for up to 30 minutes. When disturbed on land, they make off at great speed. If they feel cornered, they turn and flatten their head, hiss, and strike. They strike with a closed mouth, however, and rarely bite.

Defensive Odor

Their main line of defense becomes all too apparent when they are picked up: They empty the copious foul-smelling contents of their anal gland and usually manage to smear it over their captor by writhing and thrashing their tail around. This usually has the desired effect; but if not, the grass snake may resort to its last line of defense—feigning death. The snake goes limp and allows its mouth to gape open and its tongue to hang out. If placed on the ground or if it goes into its "death mode" prior to being picked up, it will lie upside down for several minutes, righting itself only when it thinks it is safe. The occurrence of such behavior seems to depend on the location: In some places over one-third of all grass snakes play dead, whereas elsewhere none adopt this strategy.

Grass snakes are masters at playacting. With its jaws agape and lying absolutely still, this one is feigning death, presumably to deceive a predator.

Playing Possum

The grass snake is one of several snakes that play dead when threatened. Another well-known example is the hognose snake, and there are also several lizards that do it. Many other animals that play dead are cryptic (disguised) species, such as walkingsticks. They may gain some protection by dropping to the ground and not moving, since some of their predators have trouble seeing prey that is motionless. However, this does not explain the snakes' behavior. Many of the species that would prey on grass snakes, such as badgers, hedgehogs, crows, and birds of prey, will happily take dead as well as live prey. One can only presume that some predators are put off by the appearance and foul smell of what might appear to be carrion that is long past its sell-by date.

Common name Green Iguana (common iguana)

Scientific name *Iguana iguana*

Subfamily Iguaninae

Family Iguanidae

Suborder Sauria

Order Squamata

Size Males to 6.6 ft (2 m), females to 4.8 ft (1.4 m)

Key features Very large green or greenish lizards; adults have crest of tooth-shaped scales along the back and the first third of the tail; tail has broad, dark bands around it; limbs long; each toe is also long and has claws for grasping; large males develop a flap of skin (dewlap) under the chin (in females it is smaller); a single, very large smooth scale present on each side of the head below the eardrum in both males and females

Habits Arboreal; diurnal

Breeding Egg layers with large clutches of 9–71 eggs; eggs hatch after 65–115 days

Diet Mostly vegetation, especially leaves

Habitat Forests, especially rain forests but also dry deciduous forests in some places

Distribution Central and South America (northwestern Mexico in the north to Ecuador, northern Bolivia, Paraguay, and southern Brazil to the south); also on some West Indian islands and introduced to Florida

Status Common in places but under pressure from humans in others

Similar species The closely related *Iguana delicatissima* from the Lesser Antilles (West Indies)

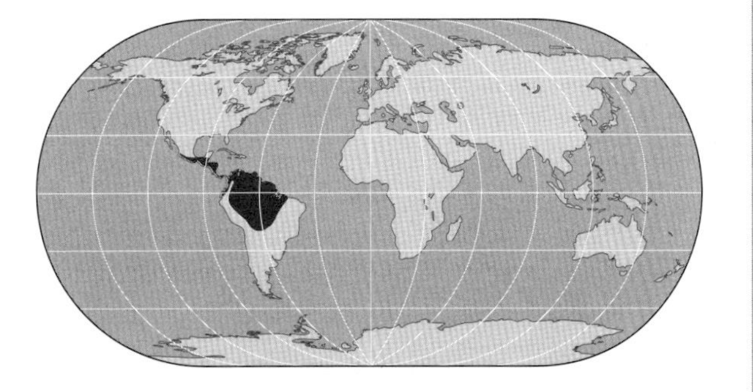

Green Iguana

Iguana iguana

The green iguana is an instantly recognizable lizard from Central and South America. It is a dinosaurlike lizard of impressive proportions that lives in the highest tree canopies.

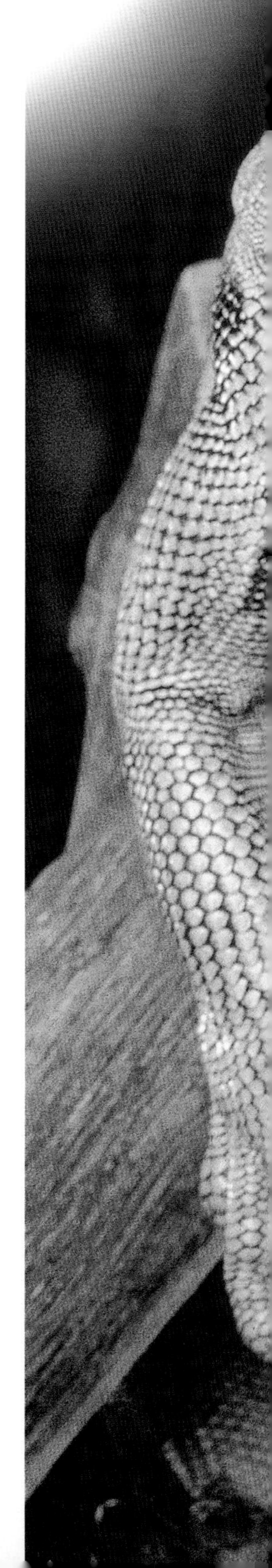

ALTHOUGH THE GREEN IGUANA is not tied to the water to the same extent as some other iguanids, such as the plumed basilisk, *Basiliscus plumifrons*, it is most often seen along the edges of rivers. Since it is easier to see into the tops of tall trees that are isolated at the edge of rivers than those that are surrounded by other trees, it is possible that green iguanas are everywhere in the forest but that we only see them near rivers. Iguanas often use water as an escape route and are good swimmers.

Life Cycle

Iguanas go through several life phases. Starting with the eggs, female iguanas lay them in burrows that they dig themselves. The burrows are usually between 3 and 6 feet (0.9–1.8 m) in length and 12 to 24 inches (30–60 cm) below the surface. They end in a chamber that is large enough to allow the female to turn around. Where populations are dense or where nest sites are scarce, a number of females (sometimes as many as eight) may dig interconnecting tunnels with several entrance holes and egg-laying chambers.

The female green iguana lays a clutch of nine to 71 eggs. In western Costa Rica an average clutch numbers about 35 eggs (it may vary in other localities). Where egg-laying sites are scarce, females may remain near their nests and defend them from other females, which may otherwise dig up the eggs in the course of their own excavations.

The eggs hatch after 65 to 115 days, and clutches often hatch simultaneously. The mass hatching may limit predation, because predators can only eat so many young iguanas at one time. If the eggs hatched over an extended

period, the predators would be able to pick off the young lizards as they emerged. Also, large numbers of young iguanas together are more likely to spot potential danger or may even intimidate predators.

Even before they leave their nest chamber, the juveniles eat some soil, which is thought to provide their gut with bacteria that aid digestion. After emerging, they move off into low shrubbery, still maintaining contact with each other. Contact involves tongue licking, rubbing their chins over each other, and nipping each other with their jaws, and is similar to the grooming behavior seen in many social birds and mammals.

The group of young iguanas may forage on the ground at first, but they always sleep on branches, often in small groups. After a few days they move into low shrubs and gradually move farther up into the canopy. By feeding below the adults and larger juveniles, they eat their feces (either deliberately or inadvertently),

← Green iguanas have small, granular scales and long claws on their fingers and toes to help them climb. The prominent dewlap shows that this individual is a male.

Tasty Cousin

The genus *Iguana* has one other member, *I. delicatissima*, which means "delicious iguana." Before it became rare, it was valued as food on the Lesser Antilles islands. It is almost identical to the green iguana, but it is slightly smaller and lacks the enlarged scales on its jowls. It was once abundant on every island from Anguilla to Martinique, but habitat destruction, harvesting for food, and introduced predators such as dogs, cats, and mongooses have finished it off on St. Kitts, Nevis, and Antigua. Numbers on the other islands have been reduced to critical levels. To raise awareness of the situation, the Anguilla Post Office issued a set of stamps featuring the species in 1997.

↑ Young green iguanas like this juvenile from Costa Rica are more brightly colored than the adults. Young iguanas mature after about two years; and as they grow, the bright green color fades.

which inoculate their gut with the bacteria needed to help them break down plant material containing cellulose.

Young iguanas have many enemies, including a variety of snakes, birds of prey, and birds such as anis and toucans. A number of small opportunistic mammals, such as coatis and kinkajous, also prey on them. As the iguanas grow, some smaller predators are no longer a problem; but they can attract the attention of larger ones such as crocodiles, caimans, and wild cats. Assuming they survive, they reach breeding size in two or three years. Females and young adult males maintain their green coloration, but they are rarely as bright as the hatchlings. Older males are often gray or tan in color and may turn orange at the height of the breeding season.

Social Behavior and Courtship

A typical colony consists of a large dominant male, a few smaller but also mature males, several subadult males, and four to six females. The large male maintains his dominance by perching at the top of large, prominent rainforest trees and rarely comes down to the ground. He displays to neighboring males at regular intervals by lowering his large throat flap (dewlap) and nodding his head vigorously.

If close encounters between two large males occur, they raise themselves to their full height, compress their bodies, and lower their dewlaps to make themselves look as large as possible. They circle each other, hissing all the time. If neither is intimidated, they may begin to fight using their long, whiplike tails to thrash each other and bite their opponent's neck until

Spiny-Tailed Iguanas

In Central America the spiny-tailed iguanas, *Ctenosaura* species, are the terrestrial counterparts of the green iguana and are also herbivorous. There are 14 species in the genus altogether, some with small ranges, including the small islands in the Gulf of California, although the most widespread species, *C. similis*, is found from Mexico to Panama. This species is as large as the green iguana but is less colorful, being pale gray with darker markings. Its tail is strongly banded and has rings of thorny scales around it. Spiny-tailed iguanas dwell among rocks and are often common in forest clearings and around human settlements, where they dig extensive tunnels. Juveniles are bright green and can easily be mistaken for young green iguanas.

The clubtail iguana, Ctenosaura quinquecarinata *from Central America, is a lesser-known relative of the green iguana. Like the other members of its genus, it can be distinguished by the thorny scales around its tail.*

one turns dark and retreats. The dominant male mates with all the females in his territory.

Courtship can last for two weeks or more before the pair finally mate. A male approaches a receptive female from behind and vibrates his head. The female moves her tail to one side, inviting the male to mount her. He continues to bob his head while he climbs on top of her, grips her neck in his jaws, and twists his tail beneath hers so that mating can take place. Small males living nearby look like females, and the large male may not drive them away; in fact, they sometimes try to mate with receptive females while the large male is otherwise occupied.

Once mating is over, females with mature eggs come down from the trees and move to places where they can dig nests. The sites are often open, sandy patches along the riverbank or clearings in the forest. Females may need to travel up to 1.5 miles (2.4 km) to find a suitable egg-laying site.

A significant predator of iguana eggs is the American sunbeam snake, *Loxocemus bicolor*. Although the snake eats a variety of food, including small mammals, frogs, and lizards, it specializes in digging up reptile nests and eating the eggs. In western Costa Rica where it is fairly common it enters iguana nest tunnels in search of the lizards or digs them out with its pointed snout. First, it slits the iguana eggs to make them collapse, then pushes them against a loop of its body and swallows them whole. One snake that was examined contained 23 green iguana eggs in its stomach, and another had eaten 32 eggs of a spiny-tailed iguana, *Ctenosaura* species.

Large adults are fairly safe from predators (except humans) because of their sheer size. Iguanas basking near water often drop from their branches and swim to safety or disappear under the surface. Females from Barro Colorado Island in Panama even swim to a small, sandy island to lay their eggs.

If they are captured, green iguanas struggle frantically, using their long claws to scratch and tear at their captor and their whiplike tail to thrash it. Handling an angry iguana is no easy matter. Having said that, they are no match for a determined human, and iguanas are routinely sold as food on the streets of Mexico, Guyana, and other parts of Latin America. To reduce the effect on wild populations, iguana "farms" have been set up to provide a supply of animals for both the food and the pet markets.

Green Tree Python *Morelia viridis*

Common name Green tree python

Scientific name *Morelia viridis* (formerly *Chondropython viridis*)

Family Pythonidae

Suborder Serpentes

Order Squamata

Length Average 5.2 ft (1.6 m) to 5.9 ft (1.8 m); maximum 7.2 ft (2.2 m)

Key features Adults bright green with white markings along dorsal midline; juveniles yellow or orange; top of head covered with many small scales; conspicuous heat pits in the rostral scales, the first few upper labial scales, and in most of the lower labial scales

Habits Tree dweller but may crawl on the ground at night in search of new hunting possibilities; drapes coils over horizontal branches

Breeding Egg layer with typical clutch of about 10 eggs; female coils around the eggs and remains with them throughout incubation period

Diet Adults feed on mammals and some birds; juveniles eat lizards and possibly frogs

Habitat Forests; garden trees and hedges on the outskirts of towns

Distribution New Guinea from sea level to 6,560 ft (2,000 m); also found on many small islands off the mainland, several Indonesian island groups, and at the very tip of Cape York, Queensland, Australia

Status Common in suitable habitat

Similar species The emerald tree boa, *Corallus caninus* from South America

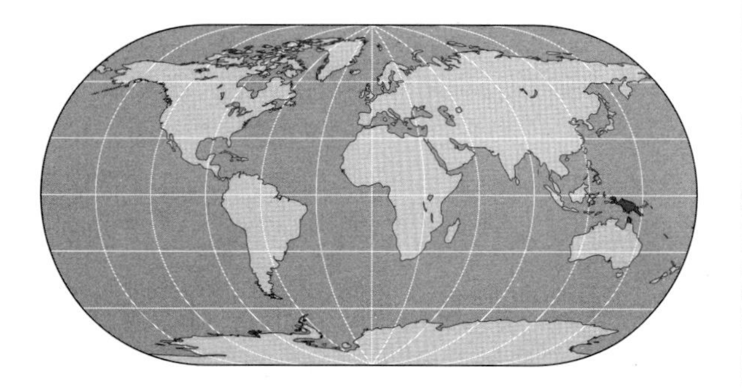

The beautiful bright-green tree python is thoroughly at home among the rain-forest trees and bushes in the Australasian tropics.

Like the emerald boa, *Corallus caninus* from South America, which it closely parallels, the green tree python has evolved to live only in trees. All its activities—catching and eating its prey, mating, and laying its eggs—take place in the forest canopy. Its coloration helps it remain hidden during the day when it drapes itself over a thin horizontal branch.

Its body is slightly flattened from side to side, and its spine forms a ridge, making the body slightly triangular in cross-section. Compared with other pythons, it is slender with a lightweight body ideally suited to crawling along narrow branches. It rarely needs to come down to the ground and then perhaps only when it feels the need to hunt in a different group of trees or when searching for a mate.

Although it can use its muscular body to bridge the gaps between branches and is able to move from one tree to another in this way, larger gaps in the canopy may necessitate coming down to ground level. It is often spotted crossing roads that cut through the forests, creating wide swathes that the python cannot span. If the roads were not there, the python might not descend to the ground at all.

Blue Mutation

There is variation in the markings, with some individuals having a broken white line along the center of their back, while others are more uniform. In some cases there are patches of blue scales concentrated along the dorsal midline, and in a few rare cases the whole snake may be blue. This is thought to be caused by a mutation in which a layer of oil-

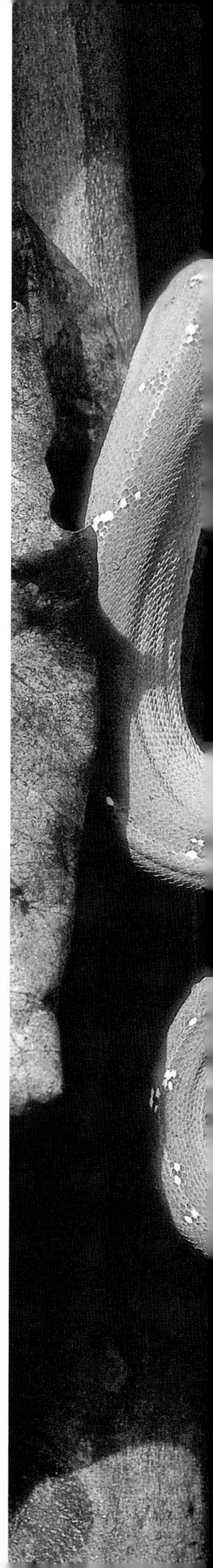

bearing cells (xanthophores) is missing. Since this layer combines with the blue pigment in the scales to form the typical green color, the absence of the layer allows the blue pigment to be seen without the yellow filter. (Holding a piece of yellow cellophane up to the blue sky produces the same effect.) There is a theory that females are more likely to have this mutation than males, although it has yet to be proved (and there is no obvious reason why it should be so).

This blue snake is, in fact, a green tree python from New Guinea. The blue variant is extremely rare. Not surprisingly, it is highly prized by breeders and collectors.

The most remarkable aspect of the green tree python's coloration, however, is the way in which juveniles differ from adults. When they first hatch, young green tree pythons are either brilliant sulfur yellow or orange. Both color forms can emerge from the same clutch of eggs (although clutches belonging to Australian pythons have only produced yellow hatchlings so far).

Both color forms have dorsal markings made up of a dark brown line, sometimes broken up into short sections or even individual blotches. Patches of white scales may be enclosed within this line or row of spots. A dark line extends from the snout to the angle of the jaw, passing through the center of the eye. The iris also has a horizontal dark line running through it so that the eye stripe is not interrupted. This helps break up the outline of the eye, which might otherwise give away the snake's presence and is a fine example of disruptive coloration.

Green tree pythons are well adapted to life in the trees. Once their coils are intricately draped around a branch, they are very hard to remove.

Color Change

Whether yellow or orange, juvenile green tree pythons are among the most strikingly marked of any reptile. They keep their juvenile colors for nine to 15 months, when they change to the all-green, adult pattern. Changing can occur in a matter of days, or it might take up to a year to complete. Typically, it takes two to three months from start to finish, and it occurs independently of skin shedding—the snake is not noticeably greener immediately after it sheds its skin, but changes almost imperceptibly day by day. A similar transition takes place in the emerald tree boa, *Corallus caninus*.

Progressively Larger Prey

In parallel with the color change the green tree python also undergoes a change in diet as it grows. Whereas juveniles feed mainly on lizards, especially the skinks and geckos that are abundant in the region, adults feed almost exclusively on mammals. Perhaps surprisingly and contrary to what early naturalists assumed, they eat relatively few birds. Juveniles increase their chances of catching lizards (and perhaps small frogs) by using their tails as lures. The tip of the tail is marked in black and white. When the animal is hungry or senses prey nearby, it lowers its tail between its coils and twitches it in a series of jerky spasms.

Although it will do this speculatively even if there is no prey within sight, the frequency of the spasms increases if suitable prey approaches. If the lure is successful and the lizard comes within striking range, the snake will grasp it, using its long, sharp teeth to penetrate the skin and get a good grip.

Ⓣ ***In New Guinea a female green tree python coils protectively around her eggs and new hatchlings.***

Ⓓ ***In a few weeks this red hatchling from New Guinea will gradually turn green. Other hatchlings may be bright yellow.***

At about the same time as the juvenile snake begins to change color, the tip of its tail loses the contrasting black-and-white bands, and it uses it as a lure less and less often. Adults may lure occasionally, however, and captive green pythons are known to lure when their owner enters the room, having obviously learned to associate him or her with food.

Eggs in the Treetops

Like all members of the family, green tree pythons lay eggs. A typical clutch numbers about 10, but extremes of six to 30 are known.

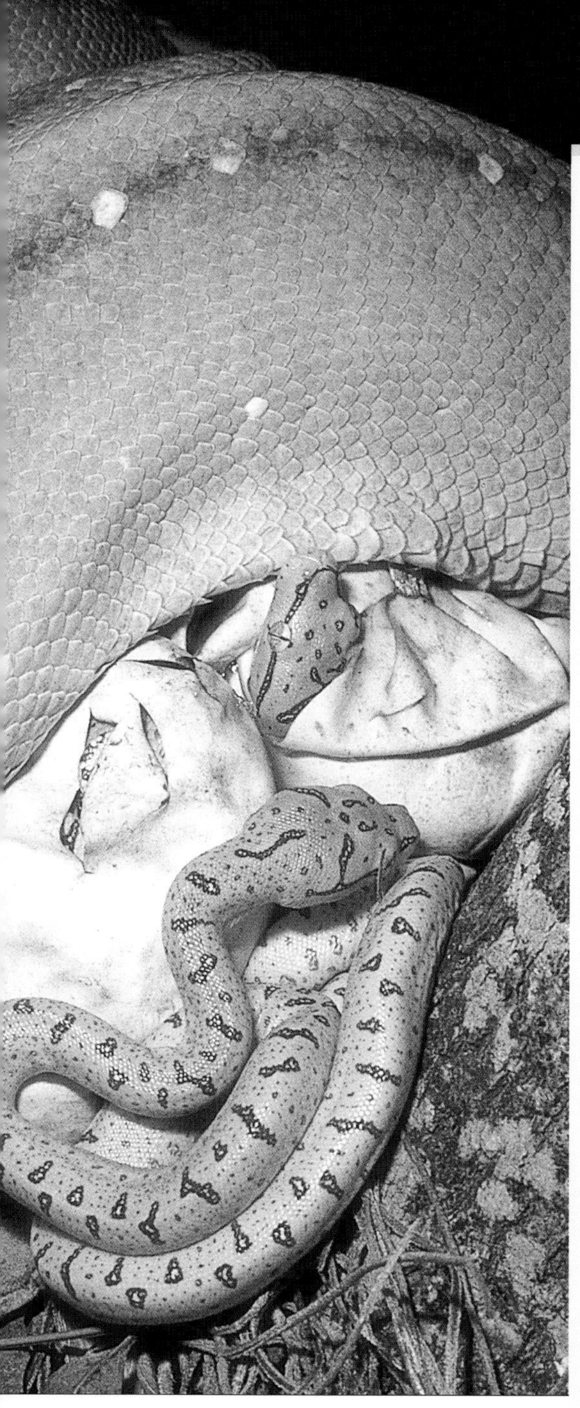

All these records are from captive individuals—naturally occurring clutches of green tree python eggs are unknown.

Laying eggs presents difficulties in the case of tree-dwelling snakes that do not like to come down to the ground, and it seems likely that the females lay their clutches among mosses and decaying vegetation trapped in large epiphytic plants or in tree holes. In captivity they have been observed coiling around the eggs and remaining with them until they hatch, about 42 to 60 days later. The brightly colored hatchlings are about 12 inches (30 cm) long.

Fruit, Flower, or Snake?

The dramatic color change that juvenile green tree pythons (and emerald boas) undergo when they are about one year old presents scientists with a puzzle that, so far, they have not solved. Since we have to assume that any behavioral or physiological characteristic increases an organism's chances of survival through natural selection, the snakes must benefit in some way from the color changes. One of the more interesting theories (and one that has been shown by limited experiments to have some merit) is that the young snakes are imitating flowers. There are other snakes that are bright yellow, including the eyelash viper, *Bothriechis schlegelii* from Central America, and several pit vipers from the Philippines.

Apart from their coloration, all these species have one other thing in common: They all live in trees and bushes. The vipers do not change color as they age, but then they do not grow as large as green tree pythons—even as adults they are still within the size range of juvenile pythons.

In experiments researchers showed that birds feeding around bushes would approach the snakes without concern. The snakes did not necessarily attract them, but nor did they frighten them off. Bearing in mind that the young pythons coil up into a tight ball when they are resting, they may even look like ripe yellow or red fruits. Perhaps the birds are so used to seeing brightly colored flowers and fruits that they simply ignore the snakes. The problem with this theory, of course, is that young green tree pythons do not often eat birds (as far as we can tell). It is possible that lizards are attracted to flowers that, in turn, attract insects such as butterflies and beetles. Nobody knows for certain.

There are plenty of other species of snakes in which juveniles are colored differently than the adults. Examples include the North American ratsnake, *Elaphe obsoleta*, and a related European species, the four-lined snake, *Elaphe quatuorlineata*. In *E. obsoleta* juveniles are blotched, but adults are more uniform in color (black, yellow, or orange depending on the subspecies). In *E. quatuorlineata* boldly blotched markings give way to a pattern of four longitudinal stripes as the snake matures.

An explanation that is frequently put forward for the color changes is that as the snake grows, it moves from one habitat to another. The colors and markings that help it survive in one place are not ideally suited to another. This may or may not be correct, but it is unlikely to explain color change in green tree pythons because both juveniles and adults live in similar locations.

Common name Green turtle

Scientific name *Chelonia mydas*

Family Cheloniidae

Suborder Cryptodira

Order Testudines

Size Carapace usually over 36 in (91 cm) long

Weight Up to 352 lb (160 kg) when adult

Key features Head relatively small with a prominent pair of scales in front of the eyes; jaw is serrated along its edges; distinctive differences in appearance between Atlantic and Pacific populations, the latter having a significantly darker plastron; carapace dome shaped and appears green; limbs paddlelike, usually with a single claw on each one

Habits A marine turtle occurring in coastal areas rather than roaming across the open ocean; some populations bask during the day

Breeding Female lays about 115 eggs per clutch on average 3–5 times during a season; interval between laying is usually 2–3 years; eggs hatch after about 65 days

Diet Small individuals feed on small crustaceans and similar creatures; larger individuals are entirely herbivorous, eating sea grass and marine algae

Habitat Coastal areas, bays, and shallow water in tropical and temperate seas

Distribution Pacific and Atlantic Oceans

Status Endangered (IUCN); listed on CITES Appendix I

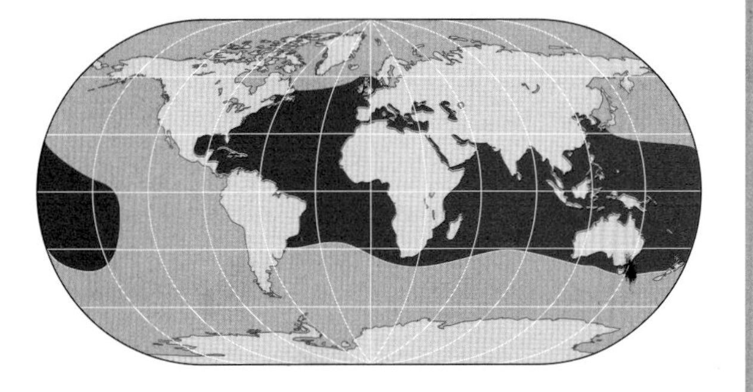

Green Turtle

Chelonia mydas

The green turtle was once common in the warm oceans of the world. Unfortunately, it has become increasingly rare in some areas, where it suffers from commercial exploitation and is at risk of extinction.

THE NAME GREEN TURTLE COMES FROM the color of the fat beneath the shell in these marine turtles. It varies among individuals: In some there may be light stripes radiating across the carapace with yellow and brown areas apparent as well. The Pacific population is often described as black sea turtles because they typically have a much darker, grayish, or even black carapace. Their shells also appear more domed than those of their Atlantic relatives.

Green turtles are probably the slowest growing of all vertebrates thanks largely to their diet. Although initially they hunt small crustaceans, they soon become vegetarian, feeding mainly on sea grass and marine algae once their carapace measures approximately 8 inches (20 cm) long. In areas where sea grass is prevalent, regular grazing by the turtles encourages the plant's growth. This is important to the turtles' own well-being, since fresh shoots are more nutritious. As in herbivorous mammals, the green turtles rely on a beneficial population of microbes in their intestinal tract to aid the digestion of plant matter.

Dietary Adaptations

Green turtles are quite adaptable in their feeding habits, and their diet varies significantly throughout their range. In northern Australia, for example, they have learned to snap the leaves off mangrove trees, adding to their feeding options. The serrations on the sides of their jaws help them tear off mouthfuls that can be swallowed easily, and they will sometimes even pluck leaves growing above the water's surface.

In areas where sea grass is not present, notably around Hawaii and on Australia's Great

↑ A green turtle feeds on marine algae growing on coral in waters off the coast of Malaysia. Its other favorite food is sea grass.

← The beautiful markings of the carapace can be seen clearly on this green turtle from Malaysia. The green effect comes from the layer of fat beneath the shell.

Barrier Reef, the lack of sea grass may explain why green turtles there are so slow to reach sexual maturity. Studies involving Hawaiian green turtles revealed that some were mature by 11 years old, but others were not capable of breeding for the first time until they were 59. In the case of some green turtles living off the coast of Queensland, Australia, sexual maturity may not occur until the turtles are 70 years old. Growth and therefore breeding capability relate largely to the availability of food. Ultimately,

Temperature-Dependent Sex Determination

It was in green turtles that researchers first discovered the phenomenon of temperature-dependent sex determination (TDSD). It is now known to apply quite widely to many chelonians that lack sex chromosomes to regulate gender. In such cases the temperature at which the eggs are incubated is significant in determining the sex of the hatchlings. Clutches exposed to higher temperatures contain female offspring, while those that hatch under cooler conditions are likely to produce mainly males. However, the exact details vary according to species, and TDSD does not apply to all species.

however, the green turtle can grow larger than any other member of its family. Those in the Atlantic are bigger on average than their Pacific counterparts.

Sand Pits

When they come ashore to nest, females dig a pit for their eggs, using their hind flippers. This laborious task usually takes about two hours. A nesting female must do this roughly every two weeks during the breeding season, since she lays from two to seven clutches. As in other marine turtles, the eggs have relatively rubbery shells and are laid with mucus around them. This stops them from being damaged as they fall on top of each other in the pit. The female uses her flippers to cover the eggs with sand.

Because the eggs are laid above the high-tide line, they are free from the risk of flooding and should hatch about 65 days later. Hatchling green turtles are about 2 inches (5 cm) long and weigh roughly 0.9 ounces (25g). Their carapace is dark above and lighter below. Young green turtles that hatch on Hawaiian beaches face relatively few predators. Seabirds do not seriously affect their numbers, nor do large fish waiting offshore in the ocean.

Basking

A very unusual behavior pattern has been observed in green turtles found in the northwestern area of the Hawaiian archipelago. Individuals haul themselves out of the water onto the beaches of isolated islands to bask during the day. Why they do this is unclear, but it may be a way of avoiding attacks from tiger sharks, *Galeocerdo cuvier*, which are prevalent in the area. Basking is almost completely unknown in other marine turtles.

Once established in an area, the turtles are unlikely to leave except to nest. They also seem to be particular about their feeding preferences. In Hawaiian waters green turtles regularly look for just nine out of over 400 species of marine algae growing in the region.

Using her hind flippers and becoming covered in sand in the process, a nesting female green turtle digs a pit for her eggs on a beach on Ascension Island.

Populations around the World

Green turtles are widely distributed throughout temperate and tropical seas, and have been seen off the North American coast as far north as Massachusetts.

In the past green turtles were heavily hunted as a source of food, but in many parts of their range they are now strictly protected. Nicaragua shut its turtle-processing plants in 1976, a move that has probably also helped the important Tortuguero population in neighboring Costa Rica—today turtles of all ages can be seen feeding together there.

Although it is nearly 30 years since green turtles were given legal protection in Hawaii, their numbers have not increased dramatically. This is partly because of changes in their habitat that have reduced the amount of food available. In addition, a significant number are affected by skin tumors known as fibropapillomas, which may be linked with harmful environmental conditions. In the vicinity of the island of Honokowai over 90 percent of the green turtles are suffering with these viral growths on their bodies.

The green turtle population in the Atlantic has fallen dramatically in some areas, and the species has become extinct on the Cayman Islands and Bermuda. Breeding populations occur on the mainland right around the Caribbean from the coast of Florida to Costa Rica and down to Surinam in South America as well as on islands throughout the region.

Success Story

One of the most significant breeding colonies can be found on Ascension Island. The turtles nesting there have been tracked back to the coast of Brazil, which means that they must migrate over 1,600 miles (1,000 km) to their nesting grounds. Individuals within the Ascension Island breeding population represent some of the largest living examples of the species, possibly because they have not been subjected to heavy predation.

Even so, they would have been dwarfed by some of the monsters recorded from centuries ago. Particularly large specimens weighing as much as 1,000 pounds (454 kg) have been recorded from the Cedar Key region of Florida.

It is hoped that turtle farms such as this one on Grand Cayman Island will prevent the hunting of green turtles in the wild.

Ranching

One of the practical ways of safeguarding wild populations of green turtles is by ranching. This entails collecting a percentage of the eggs laid by the turtles and hatching them artificially. The young hatchlings are reared in captivity and ultimately used to meet the demand for soup and other by-products.

By legitimizing trade in this way, it is hoped that wild turtles will be left alone. However, a number of welfare issues have arisen surrounding programs of this type, especially with regard to the turtles' growth rates. Young green turtles raised in this way grow at a much faster rate than in the wild. This affects their physical appearance, causing the carapace to appear more domed.

Fears have also been expressed that these programs simply encourage trade and could provide cover for the illegal collection of wild turtles, which are killed and sold as if they were reared in captivity. International trade in marine turtles (and various associated products ranging from soup to shells that may be sold to unsuspecting tourists) remains illegal under the Conservation in International Trade in Endangered Species (CITES) treaty.

Hawksbill Turtle

Eretmochelys imbricata

The name imbricata *describes the overlapping plates on the hawksbill's upper shell. Unfortunately, it is the beautiful carapace, known as "tortoiseshell," that has led to widespread hunting of this small sea turtle.*

Common name Hawksbill turtle

Scientific name *Eretmochelys imbricata*

Family Cheloniidae

Suborder Cryptodira

Order Testudines

Size Carapace up to 36 in (91 cm) long

Weight Up to 150 lb (68 kg)

Key features Head narrow and distinctive with a hawklike bill; 2 pairs of scales in front of the eyes; carapace elliptical with an attractive blend of yellow or orange mixed with brown, but coloration is highly individual; scutes overlap behind each other on the carapace; 2 claws present on each flipper; serrations present on side of carapace

Habits Often encountered looking for prey on coral reefs

Breeding Female usually lays up to 160 eggs in a season but breeds on average only every 2–4 years; eggs hatch after 58–75 days

Diet Invertebrates, mainly sponges, but also squid and shrimp

Habitat Mainly tropical waters

Distribution Occurs in the Atlantic and in parts of the Indian and Pacific oceans

Status Endangered (IUCN); listed on CITES Appendix I

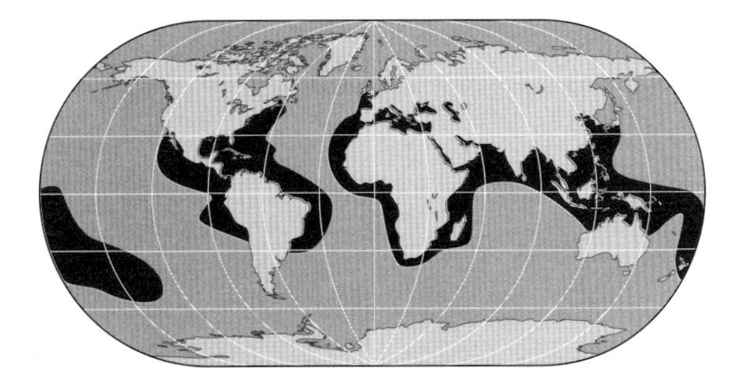

RECENT SURVEYS SUGGEST THAT numbers of the hawksbill turtle have declined more seriously and rapidly in recent years than was previously thought. This is true especially throughout the Caribbean and the western Atlantic Ocean in spite of the widespread protection given to the species. In the past huge numbers of these turtles were caught. Their shells were used to create "tortoiseshell" objects ranging from tea caddies to hair combs, which were highly fashionable in the late 19th and early 20th centuries. The scutes that extend over the carapace are the most highly prized because of the amber mottling that is apparent once the upper shell of the turtle is polished. In Japan, where the use of tortoiseshell has been elevated to an art form, this material is known as *bekko*.

Deadly Sponges

The hawksbill turtles get their name from the shape of their narrow mouthparts. Their "v"-shaped lower jaw is used to reach and pluck sponges from inaccessible areas of a reef. Some species of sponge protect themselves by toxins in their bodies, but they do not seem to harm the hawksbill turtle. Bizarrely, the sponge toxins remain potent, a fact confirmed by cases in which humans died from eating the flesh of hawksbill turtles that had eaten the sponges.

The turtles depend on sponges as their major food source. Any deterioration in the conditions on a reef will cause a decline in the number of sponges and leave the turtles vulnerable to starvation.

Hawksbill turtles are relatively small and agile, and can therefore reach nesting beaches

Hitching a Ride

Sea turtles in general carry a number of other creatures on their shell, particularly barnacles and various types of coral. Hawksbills from the Atlantic region often have Columbus crabs, *Planes minutus*, attached to the rear of their body. They are probably acquired when the hatchling turtles are feeding in the sargassum, where young crabs are frequently found.

In contrast, it is normally adult crabs that are seen on mature turtles. They help keep the reptile's body clean, feeding on algae and other creatures that may also attach themselves. If the crabs did not keep these organisms in check, they could make swimming much harder for the reptile by reducing its streamlined profile.

over reefs that would exclude larger, heavier species. They lay their eggs quite high up on the beach, often in sites partially concealed by vegetation. Once on land, the nesting process can be completed in an hour, although it often takes longer. Females lay a total of up to 160 eggs, returning to the same beach to lay at intervals of roughly 14 days about four or five times during the season. The number of eggs laid is determined by the size of the female, with larger turtles laying more eggs.

Dangerous Detritus

The hatchlings have a carapace length of just over 1.5 in (3.8 cm), and they generally weigh less than 0.7 ounces (20 g). They head into areas of sargassum at first and are vulnerable to floating detritus, including pieces of styrofoam, which can sometimes lodge in their digestive tracts with fatal consequences. Assuming they survive, the young turtles head back to reef areas once they have grown to 8 inches (20 cm) long—they often retreat under rocky overhangs and similar safe places.

The carapace alters in shape as young hawksbills mature, taking on a more elongated outline. The serrations running down the sides toward the rear of the carapace may have a protective function, but they shrink as the turtle grows older. Their ultimate disappearance in adults is regarded as a sign of old age.

Many hawksbill turtles do not seem to travel long distances to their nesting grounds, but there are exceptions, discovered as a result of tagging. One female caught and marked in the Torres Strait region of northern Australia was caught again 11 months later approximately 2,240 miles (3,600 km) away in the Solomon Islands, where she came on land to nest.

A hawksbill turtle swims near the Virgin Islands in the Caribbean. The closeup shows the beaklike mouthparts that give the turtle its common name.

Iguanas

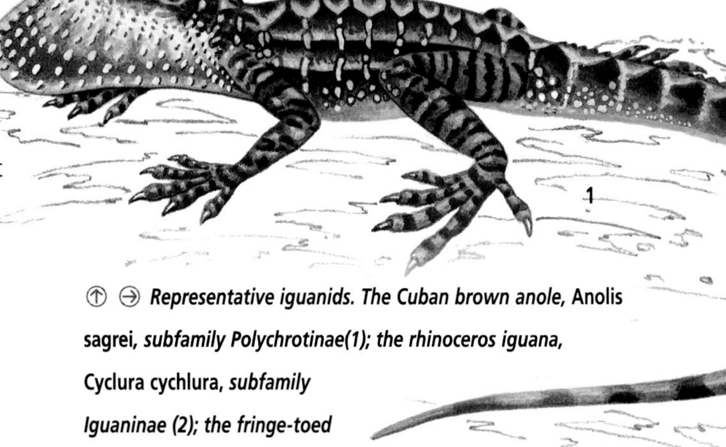

Ⓣ ⊖ ***Representative iguanids. The Cuban brown anole,* Anolis sagrei, *subfamily Polychrotinae(1); the rhinoceros iguana,* Cyclura cychlura, *subfamily Iguaninae (2); the fringe-toed lizard,* Uma notata, *subfamily Phrynosomatinae (3).***

Current understanding is that there are 892 species of iguanas. Some scientists believe that they should be divided into smaller units and that each unit should be regarded as a separate family. Others consider them to be subfamilies, which is how they are treated here.

The iguanids have an interesting distribution in South, Central, and North America and in Madagascar. Their distribution is similar to that of the boas and is evidence of continental drift. About 100 million years ago South America, Africa, and Madagascar formed a large southern landmass called Gondwanaland. When Gondwanaland broke apart, iguanas were trapped on each of the newly formed, smaller landmasses. They died out in Africa and were replaced by members of a similar family, the Agamidae, which are their ecological counterparts. However, on Madagascar they survived in the form of two genera: *Oplurus* (with six species) and *Chalarodon* (with just one species, the four-eyed lizard, *C. madagascarensis*). Other groups spread west by a process known as rafting. Ancestral stocks from South or Central America, perhaps similar to the spiny-tailed iguanas that live there today, were swept out to sea and drifted on the ocean currents. Nearly all of them would have died long before they reached land, but a lucky few made it to pastures new—the Galápagos Islands—where three large species and seven small species live. Remarkably, others continued their voyage as far as the Fijian Islands and Tonga, where there are two large and colorful species, *Brachylophus fasciatus* and *B. vitiensis*. Others colonized several Caribbean islands; they included a few large species such as the rhinoceros iguanas, *Cyclura* species, and a multitude of small, agile species belonging to the genus *Anolis* and related genera.

Once the land connection between South and North America was made, iguanids spread northward along the isthmus of Panama and into Mexico and the United States, where small iguanids in the subfamily Phrynosomatinae (horned lizards and side-blotched

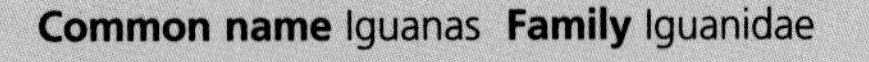

Common name Iguanas **Family** Iguanidae

Family Iguanidae
Although there is some dispute over the classification of iguanas, they are usually divided into eight subfamilies (some of which are considered full families by some authorities) and about 892 species iin 44 genera:

Subfamily Corytophaninae—3 genera, 9 species of medium-sized basilisks and related species from Central and South America, including the plumed basilisk, *Basiliscus plumifrons*
Subfamily Crotaphytinae—2 genera, 10 species of medium-sized collared and leopard lizards from North America, including *Crotaphytus collaris*
Subfamily Hoplocercinae—3 genera, 10 species from Central and South America
Subfamily Iguaninae—8 genera, 36 species of large iguanas from Central and South America, incorporating the Galápagos Islands, Cuba, and the West Indies; species include the green iguana, *Iguana iguana*, the marine iguana, *Amblyrhynchus cristatus*, and the Galápagos land iguana, *Conolophus subcristatus*
Subfamily Oplurinae—2 genera, 7 species of Madagascan iguanas, including the Madagascar collared lizard, *Oplurus cuvieri*
Subfamily Phrynosomatinae—9 genera, about 124 species of horned lizards, fence lizards, and side-blotched lizards from North and Central America, including the side-blotched lizard, *Uta stansburiana,* the coast horned lizard, *Phrynosoma coronatum*, and the western fence lizard, *Sceloporus occidentalis*
Subfamily Polychrotinae—8 genera, 394 species of anoles and relatives from North, Central, and South America (especially the Caribbean region), including the green anole, *Anolis carolinensis*
Subfamily Tropidurinae—9 genera, 302 species of lava lizards and relatives from South America, Galápagos, and the West Indies, including the Galápagos lava lizards, *Microlophus* sp.

lizards) came to dominate, especially in the desert regions of the Southwest.

Some have the ability to discard part of their tail if it is grasped, a trick known as caudal autotomy. Having said that, three species of *Uracentron* from South America have thick, heavily armored tails that cannot be autotomized. These lizards live in forest canopies and sleep in tree holes, and they can use their tails to block out predators from the openings.

Sexual dimorphism is widespread across the family. Males are often larger than females, sometimes massively so. For example, in the marine iguanas, *Amblyrhynchus cristatus* from the Galápagos Islands, males can weigh up to four times more than females. Males are often more likely to have crests on their backs and heads; or if both sexes have them, those of the males are larger. In basilisks only males have crests on their heads, while in green iguanas, for example, females have reduced crests.

Form and Function

It is difficult to define a "typical" iguanid. We are inclined to think of the large, spectacular species such as the green iguana, *Iguana iguana*, the Galápagos land iguana, *Conolophus subcristatus*, and the marine iguana, *Amblyrhynchus cristatus*, with stocky bodies and large, saw-tooth crests along their backs. In fact, they make up just a tiny minority of iguanids.

Most are quite small with slender bodies without crests or other ornamentation. Most have relatively large, keeled scales, but some have small, granular ones. The members of the family must value their limbs, since there is no tendency for them to become small or disappear altogether as in many other families—many iguanids are long legged and very agile. Fast-running species can lift their front feet off the ground as they gather speed and run on the hind legs only (known as bipedal locomotion). The plumed basilisk, *Basiliscus plumifrons* from Central America, can even run across water, a talent that has locally earned it the name "Jesus Christ lizard."

Iguanids typically have long tails and often use them for counterbalance, curling them up in the air as they run.

Color and Communication

Nearly all iguanas communicate visually, and males often have elaborate display routines that can include head bobbing, pushups, or the lowering of dewlaps. To this end, some male iguanids are very colorful, and many have bright patches to enhance their displays: The anoles have brightly colored dewlaps, for example, that act as "flags" when they flick them up and down. Female *Anolis* species often have dewlaps too, but those of males are larger and more showy. This probably reflects the fact that females can be territorial in some species but not as aggressively so as males. Dominant male collared lizards, *Crotaphytus collaris*, Peruvian swifts, *Liolaemus* species, and lava lizards, *Microlophis* species (to name just a few), often glow with color during the breeding season, while subordinate males, females, and juveniles are much duller by comparison.

Male iguanids can increase their visibility further by standing stiff legged and bobbing up and down to attract attention, often displaying bright patches of color that are hidden when they are at rest. Some of them pose on prominent rocks or tree stumps where they can be seen many yards away, a habit that gives the fence lizards their common name. Males of the species have bright patches on their throats and chests, which are only visible when they raise their bodies off the ground. They become cryptic (disguised) when at rest to avoid attracting the attention of predators.

However, the markings on some species are designed to attract the attention of predators. Several species, such as the zebra-tailed lizard, *Callisaurus draconoides*, the Texas earless lizard, *Holbrookia texanus*, and the fringe-toed lizards, *Uma* species, have black bands on the underside of their tails, which are otherwise white. They raise their tails and wave them from side to side when they see a predator, which is thought to be a signal telling the predator that it has been spotted. It is intended to persuade the predator that the lizard is not worth stalking and, at the same time, saves the lizard from using energy in running away.

Coloration can also send other signals. Females of several species develop orange areas on their flanks when they are carrying eggs. The color tells males that they have already mated and avoids both male and female wasting time and energy in courtship and rebuff.

General Behavior

The great majority of iguanids are diurnal. They typically bask in exposed places to gain body heat before going off to forage for food or look for a mate. Some live in the deep shade of forest floors and in the understory, however, and experiments have shown that those species operate below their preferred body temperatures for much of the time.

A few are unusual. The *Uracentron* species live high up in the forest canopy and are therefore rarely seen. They seem to exist in colonies containing a single dominant male, a number of smaller, subordinate males, and females and juveniles of varying ages. They move through the canopy in large groups. Another atypical species is *Uranoscodon superciliosus* from the Amazon Basin in Brazil and neighboring countries. It lives at the water's edge along rivercourses, moving out into the forest when the rivers flood. It forages among the flotsam and jetsam in search of living and dead invertebrates. In the humid environment algae grows on its scales, which enhances its camouflage. In fact, it relies so heavily on camouflage that it rarely moves, even when approached closely. It is therefore known locally as the "blind lizard," implying that it is unaware of danger.

The marine iguana, *Amblyrhynchus cristatus*, is unique not just among iguanas but among lizards as a whole because it feeds in the sea. However, there are other iguanids that live along the shore, feeding on small isopods known as slaters. Like the marine iguana, these species (the side-blotched lizards in the genus *Uta*) gather excess salt in salt glands situated near their nostrils.

At the other extreme some iguanids live in the most arid deserts. The fringe-toed lizards, for example, live among dunes in the American Southwest. As their name suggests, they have long, pointed scales around the edges of their toes, forming a fringe that helps them run across loose, windblown sand. They have additional adaptations in the form of a countersunk lower jaw and flaps over their ear openings that prevent sand from entering their ears and mouth when burrowing. They also have extremely sensitive hearing and typically stand on the surface of the sand with their head cocked to one side listening for sounds of insects moving beneath the surface. Then they dive headfirst into the sand to catch them. When chased by predators, they run rapidly across the surface, often lifting their front feet from the ground before plunging into the sand.

Diet and Feeding

Among them the iguanas consume just about the whole range of potential food. The larger species, such as the desert iguana, *Dipsosaurus dorsalis*, the rhinoceros iguanas, *Cyclura* species, and the chuckwallas, *Sauromalus* species, are completely herbivorous. They often feed on sparse desert vegetation, including cactus leaves and fruit and desert flowers. By contrast, the

⊖ Sceloporus malachiticus *is known as the green spiny lizard. It is found at higher altitudes in Central America. Some members of the genus give birth to live young and are among the few iguanids to do so.*

marine iguana eats only seaweed. Medium-sized species such as the collared lizards, *Crotaphrytus* species, are voracious predators. They eat large insects and spiders, and they are especially fond of smaller lizards.

Most small iguanids are insectivores, eating a wide variety of small prey. The horned lizards, *Phrynosoma* species from the American Southwest and Mexico, feed almost exclusively on ants and eat huge quantities at a single sitting.

Foraging behavior also varies. The herbivorous species forage actively, most small and medium-sized species feed opportunistically, while others adopt a "sit-and-wait" technique. Horned lizards station themselves alongside ant trails, picking off the ants as they pass by.

Habitat

Iguanids have moved into almost every available habitat within the region they occupy. Many North American species live in deserts, adapting well to intense heat and scarce food. They are fast, agile hunters, seizing every opportunity to chase down and capture small prey or developing special techniques for exploiting prey animals that tend to gather in one place.

Others exist on seemingly unpalatable cactus pads, twigs, and dried-up vegetation, eating constantly in order to accumulate what little nourishment they can. Iguanas from dry habitats often live in burrows, or they can be "sand swimmers," paralleling species in other families, such as the sandfish, *Scinus scinus* (Scincidae), and the wedge- and shovel-snouted lizards from the Namib Desert (in the family Lacertidae).

Iguanids have successfully colonized many islands, and are the dominant lizard family over most of the West Indies, on the many small islands in the Gulf of California, and on the Galápagos Islands. Some species exploit the seashore as a source of vegetable or animal food, while others hang around seabird colonies feeding on spilled fish or the flies that are attracted to such places.

On many small islands that would otherwise be barren, seabirds provide the only ecological input by fishing surrounding waters and returning to shore with their harvest. Species such as the side-blotched lizards, *Uta* species, have been quick to exploit the service provided by the seabirds.

Tropical forests are the habitats that are the richest in species, especially of anoles and related iguanids. In any given area it is not unusual to find three or four separate species "sharing" the resources by occupying slightly different microhabitats: There may be a terrestrial, an arboreal, and a shrub-dwelling species, for example, or they may vary in size and therefore prefer different prey. They may even prey on each other.

↑ ***In North America a clutch of eggs laid by a northern fence lizard,* Sceloporus undulatus hyacinthinus, *begins to hatch out. The eggs are laid in loose soil or sand and take from 60 to 80 days to incubate.***

← ***A spiny-tailed lizard,* Uracentron azureum, *from French Guiana. Like other members of its genus, it has a heavily armored tail that cannot be broken off.***

Breeding

Most iguanids are egg layers, but a few give birth to live young. Both types can be present even within the same genus, as in the horned lizards, *Phrynosoma*, and the South American swifts, *Liolaemus*. In both cases there is a correlation with habitat—species from warm lowlands tend to lay eggs, while those from higher, cooler altitudes give birth. Live-bearing seems to have evolved several times within the spiny lizards, *Sceloporus*. Within the genus there is also a tendency for those species living in cool montane areas to give birth, a good example being Yarrow's spiny lizard, *Sceloporus jarrovi*. However, other live-bearers, such as the crevice spiny lizard, *S. poinsettii*, and the Central American green spiny lizard, *S. malachiticus*, live in warmer habitats.

Many iguanids have complex breeding systems in which a single dominant male controls territory containing several adult females (and sometimes also younger, subordinate males and juveniles). Opportunities for research into lizard behavior are good, especially where colonies can be observed easily in habitats such as deserts and open scrub, and some interesting studies have resulted in recent years.

The competition for good territory and the females it contains leads to a selective pressure that favors large, more powerful, and more colorful males. Males of some species operate a "lek" type of system more commonly associated with birds, in which males congregate in well-defined areas and battle for the right to mate with nearby females. Alternative strategies have also evolved in which smaller and less assertive males "steal" matings using a variety of "dirty tricks" to gain access to the females.

Common name Indian cobra (spectacled cobra)

Scientific name *Naja naja*

Subfamily Elapinae

Family Elapidae

Suborder Serpentes

Order Squamata

Length From 4 ft (1.2 m) to 5.5 ft (1.7 m)

Key features Body cylindrical with smooth scales and large shields on the head; usually brown, sometimes black, with a white or cream "spectacle" marking on the back of its neck only visible when it spreads its hood

Habits Usually day-active but also active in the evening, especially in places where there is human activity

Breeding Egg layer with clutches of 12–22 eggs; larger clutches are known but are unusual; eggs hatch after about 8–12 weeks

Diet Frogs, toads, snakes, mice, rats, and birds; appears to have no preference

Habitat A generalist living in forests, open grasslands, fields, and gardens; often attracted to human settlements

Distribution Indian subcontinent from the southern Himalayas to Sri Lanka

Status Becoming rare in places; in India considered endangered locally

Similar species Other cobras in neighboring regions

Venom Very potent; easily capable of killing an adult human

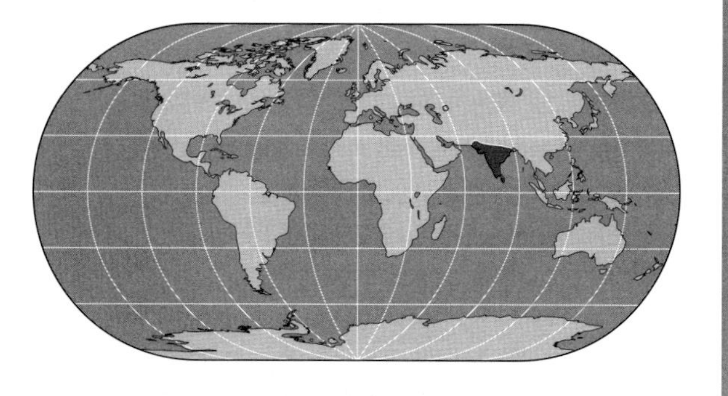

Indian Cobra

Naja naja

A rearing cobra with its hood spread is one of the quintessential images of reptiles in the Far East. The Indian cobra, with a spectacle marking on its hood, is the one that is familiar to most people.

OTHER MEMBERS of the cobra family spread a hood in order to intimidate, but its development is most advanced in the genus *Naja*, especially the nine Asian species. The hood is formed from elongated ribs that are moved outward, stretching the skin between them. The first three vertebrae (the atlas, the axis, and the third vertebra) have no ribs attached to them, but the next 27 vertebrae have ribs that are involved in the formation of the hood. These ribs are less bowed than the others and are longer, the longest ones being the ninth or tenth, with the ribs on either side of them becoming progressively shorter.

When "at rest," these ribs are folded back and downward, and are barely visible. When the cobra spreads its hood, they are moved forward and pulled back until they are at a right angle to the snake's spine. The overlying skin is only loosely attached to the ribs and does not interfere with their movement. However, when the hood is extended, the skin is pulled tight across the ribs to expose the pale markings between the scales, and the whole structure takes on the shape of a shallow spoon.

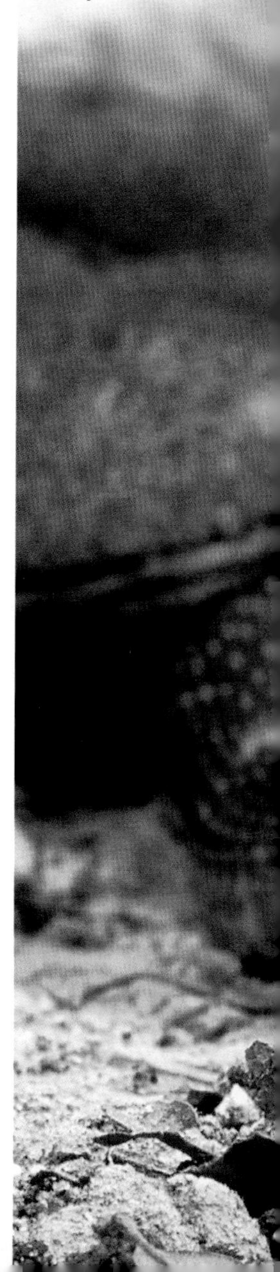

When alarmed, the Indian cobra spreads its hood and raises its body off the ground. It is then in an ideal position to strike at its attacker if necessary.

Spectacle Marking

When erecting its hood, the snake also raises the front part of its body off the ground and holds it upright. At the same time, it turns its head at right angles to the hood. It may face its enemy or turn its back on it to show off the spectacle marking on the back of its neck.

Other Asian cobras have markings on their hoods, such as an eyespot, or "ocelli," in the case of the monocled cobra, *Naja kaouthia*. Some species have bold spots and bands on the underside of their throat that serve the same

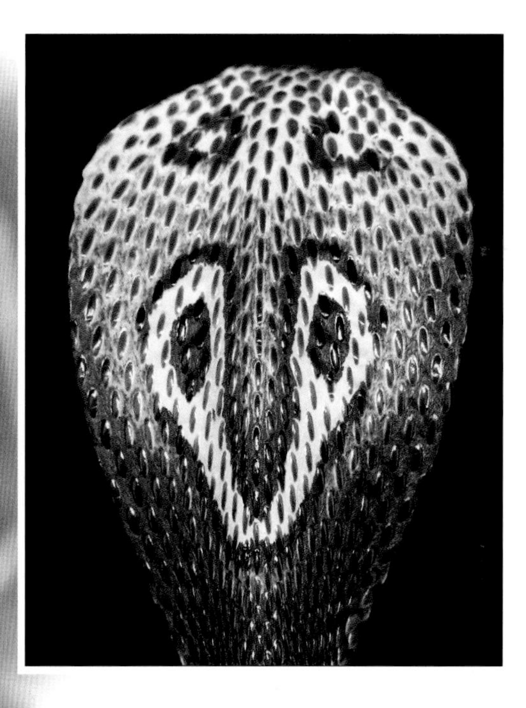

⬆ ***The back of the head of this cobra,* Naja naja *from Pakistan, displays a clear eyelike, or spectacle, pattern designed to intimidate possible predators.***

Indian Cobra Relatives

Until quite recently all the Asian cobras were classified under a single widespread species—the "Asiatic" cobra—but they have since been divided into nine distinct species. Another two new species have been added more recently. There are eight spitting cobras:

Naja atra, the black or Chinese cobra;
Naja kaouthia, the monocled cobra;
Naja philippinensis, the Philippine cobra;
Naja sagittifera, the Andaman cobra;
Naja samarensis, the Visayan or southeastern Philippine cobra;
Naja siamensis, the Indochinese spitting cobra;
Naja sputatrix, the south Indonesian spitting cobra; and
Naja sumatrana, the equatorial spitting cobra.

Two species do not have modified fangs and therefore do not spit. They are *Naja naja*, the Indian, or spectacled cobra, and *Naja oxiana*, the Central Asian cobra. An 11th species, *Naja mandalayensis*, has been described recently from Myanmar, but nobody knows yet whether or not it spits.

purpose. Once its hood is erected, the cobra sways gently backward and forward while hissing continuously. It may make half-hearted strikes by flopping forward but quickly recovers its upright position. Its maximum striking range is the same as the length of its body that is off the ground. Predators such as mongooses (and meerkats in Africa) exploit this to induce the snake to strike. They then dart in to pounce on the snake's neck before it has time to recover.

Snake Charmers

Skillful snake charmers can judge the distance to perfection, often allowing the snake to strike inches away from their hands. Such characters, with wicker basket and wooden flute, are a familiar sight in places where cobras are common. Although much of the mystique can be explained rationally, they demonstrate a unique interaction between man and snake.

Contrary to popular belief, the snakes only rarely have their mouths sewn up, and adult Indian cobras are surprisingly easy to tame. Indeed, they often need a sharp slap on the top of their head before they will rear up, unless they are recently captured. Even then, some snake collectors in the Far East are happy to catch wild cobras by hand, simply picking them up and bagging them. By contrast, juveniles are often aggressive and quick to strike.

Indian snake charmers may soon be a thing of the past thanks to a government ban on the possession of many species of snakes.

Spitting Cobras

Whatever the reason for the evolution of venom, some snakes rely heavily on it for defense. Circumstantial evidence for that conclusion is the extremely toxic nature of some venoms, which are far more powerful than they need to be to kill the snake's prey. More direct evidence comes from the evolution of methods of squirting or spitting venom at enemies, seen in the spitting cobras of Africa and southern Asia. The spitting cobras are the rinkhals, *Hemachatus haemachatus* from southern Africa, two members of the genus *Naja* from Africa, and another seven species in the same genus from Asia.

These species spit by forcing venom through small openings in the front of their fangs, causing it to squirt out at high speed in a fine spray. They can do this because their fangs are modified with small, forward-pointing openings at the front. The snakes can spray venom with a fair degree of accuracy for several yards, often aiming at the victim's eyes, where it causes immediate and intense pain. If not treated, it can cause permanent blindness in humans. It is worth pointing out that spitting cobras also bite, and that the spitting behavior is reserved for repelling predators at a distance.

Living with Humans

Indian cobras live in one of the most crowded parts of the world, yet they exist side by side with humans. They do this by adapting to a wide range of conditions. Probably forest snakes originally, they are equally at home in plantations, farms, gardens, and even buildings such as warehouses, timber yards, and houses. No doubt they are attracted by the abundance of mice and rats that are in turn attracted by the lure of food. In some places the cobras are revered by Buddhists and Hindus, and left in peace. They also climb trees occasionally in search of nestling birds or their eggs, and they take to the water willingly, sometimes even swimming out to sea.

Indian cobras are not fussy eaters. Although rats and mice probably make up the bulk of their diet, especially in urban surroundings, they also eat toads, other snakes,

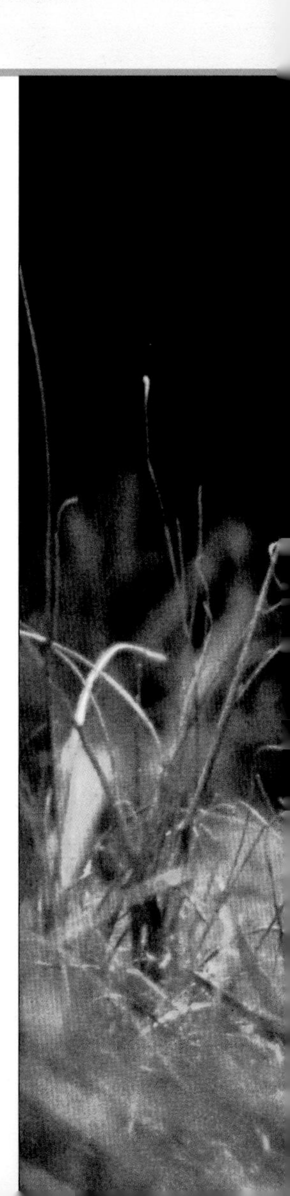

lizards, and birds. Surprising examples have included a Bengal monitor lizard, *Varanus bengalensis* of 24 inches (60 cm), that was eaten by a 4-foot (1.2-m) cobra, and a 6-foot (1.8-m) Indian ratsnake, *Ptyas mucosus*, that was eaten by a slightly shorter cobra. They sometimes eat domestic fowl and their chicks, as well as their eggs.

In a reversal of fortune chickens frequently kill cobras by stamping on them, and similar tactics are used by pigs, cows, and wild animals such as deer. Monitor lizards eat large numbers of snakes, including cobras, as do mongooses, although the enthusiasm with which they single out cobras for special treatment has probably been exaggerated.

Aggressive Youngsters

Male and female Indian cobras look similar, but the males have longer fangs. They breed at the start of the rainy season, which lasts for about three months, and there is evidence that the pair stay together throughout the breeding season. The female lays 12 to 22 eggs, although an exceptional clutch of 36 fertile and nine infertile eggs has been recorded. She lays them in a termite mound, rodent burrow, or other cavity, sometimes pushing a pile of leaves or vegetation around them to form a rough nest. Two to three months later the hatchlings emerge. They are 7 to 11 inches (18–28 cm) long but are more aggressive than the adults.

A black-necked spitting cobra, Naja nigricollis nigricollis *from Africa, launches a shot of venom at its attacker. The venom of this species can cause permanent blindness.*

Jackson's Chameleon

Chamaeleo jacksonii

Jackson's chameleons are found on the African continent, predominantly on the highlands of Kenya and Tanzania. The three-horned males look like miniature Triceratops.

Common name Jackson's chameleon

Scientific name *Chamaeleo jacksonii*

Family Chamaeleonidae

Suborder Sauria

Order Squamata

Size 12 in (30 cm) long

Key features 3 horns present on the head—2 at eye level (orbital), the 3rd located on the tip of the snout (rostral) and curving upward; dorsal crest of prominent tubercles gives the impression of a saw blade; female's horns much reduced or absent; basic coloration green; small crest to the rear of the head is outlined by conical scales

Habits Solitary, each with its own territory; individuals from middle elevations hold their body perpendicular to sun's rays to warm up in the morning; color changes to yellow when it becomes too warm; moves into deep foliage for shade and to begin feeding

Breeding Live-bearer; female produces 1 or 2 clutches each year; each clutch contains up to 35 young; gestation period about 6–9 months

Diet Insects, particularly grasshoppers, butterflies, katydids, spiders, and flies

Habitat High altitudes; found at elevations of 8,000 ft (2,440 m) that have high rainfall and distinct wet and dry seasons leading to fluctuations in temperature and humidity; common in primary and secondary forest

Distribution Mountains of Kenya and Tanzania (East Africa); introduced to Hawaii

Status Common

Similar species *Chamaeleo johnstonii* (although this species is an egg layer)

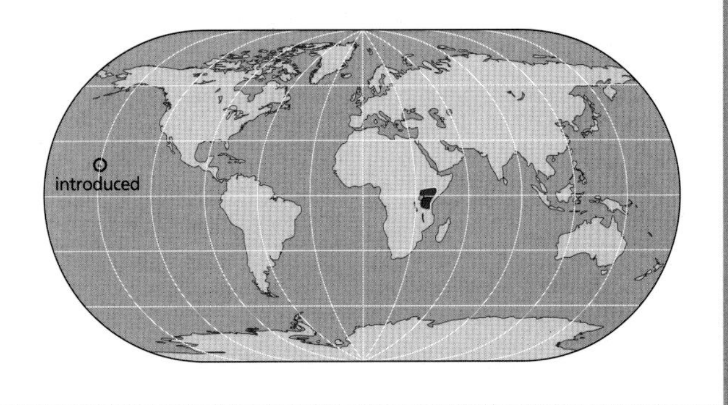

Of all the three-horned chameleons, Jackson's is the best known. There are three subspecies—*C. j. jacksonii*, *C. j. merumantana*, and *C. j. xantholophus*. The main differences between the subspecies relate to size, with *C. j. xantholophus* being the largest and *C. j. merumantana* the smallest. There is also variation in the females' horns (or lack of them) depending on subspecies. *Chamaeleo j. xantholophus* is often described as the most attractive of the three forms. It has a yellow crest and yellow on the ridge at the back of its head. In the past Jackson's chameleon has been confused with Johnston's chameleon, *Chamaeleo johnstonii*, an egg layer from the mountains of Burundi, Rwanda, and Zaire.

Jackson's chameleon's habitat in the mountains of East Africa is moderately cool with high humidity. Although rainfall exceeds 50 inches (127 cm) per year, there are distinct wet and dry seasons. Daytime temperatures reach 80°F (27°C), and an average nighttime temperature is 59°F (15°C), although it can drop to as low as 42°F (5°C). Human population explosion in Kenya and Tanzania has led to the felling of considerable tracts of primary forest; as a result, Jackson's chameleon has adapted to living in secondary forest and plantations, the latter being particularly rich in insect life. In fact, plantations have the highest population density of Jackson's chameleons.

The horns on male Jackson's chameleons are not just ornamental. These two males have locked horns in a ritualized shoving contest over a mating territory.

Color and Defense

The basic coloration of Jackson's chameleon is a uniform green to yellow-brown with pale blotches forming a faint lateral line. This lichenlike coloring provides excellent

camouflage among foliage. Juveniles are dark green, almost black, with white triangular markings on either side of the dorsal crest that serve to break up their outline.

When a threat from a bird is perceived, the chameleon moves deeper into the foliage where the bird cannot follow. An alternative defense strategy is to relax its grip on the branch, fall to the ground, and move to the base of a shrub. If it detects a threat from other predators, such as venomous and nonvenomous tree snakes, it responds by flattening the sides of its body and darkening its color. The mouth gapes, and the chameleon swings its head around in an attempt to bite.

Breeding

While breeding males intensify their colors, those of receptive females become lighter. Following a successful mating, the female gains weight. She tends to stop feeding for several weeks prior to giving birth as the developing young take up more room. The gestation period varies depending on temperature. Gravid females spend more time basking, angling their body so that the underneath of the body gets the sun's rays. They give birth to live young, usually in the morning. The young emerge from the female's cloaca, each encased in a membrane that is deposited on a branch and from which the baby breaks free. They disperse among the foliage and catch their own food within hours.

Horns

The horns of Jackson's chameleons are not collected for use in traditional medicine, but many East Africans regard cutting off a chameleon's horn as an act of great courage.

Male Jackson's chameleons engage in combat, locking horns and using them to try to force their opponent off the branch. Larger horns are obviously an advantage, since an individual can inflict damage on its opponent without receiving any.

Hawaiian Habitat

In 1981 the Kenyan government halted the previous widescale export of Jackson's chameleons, but the owner of a pet store on the Hawaiian island of Oahu obtained a permit to import a few specimens. Because they arrived in poor condition, they were released into a garden to try to improve them. They found the climate similar to that of their native habitat in East Africa, and from this initial group scattered populations of Jackson's chameleon have become established on the islands of Oahu and Maui.

Common name King cobra (hamadryad)

Scientific name *Ophiophagus hannah*

Subfamily Elapinae

Family Elapidae

Suborder Serpentes

Order Squamata

Length Usually from 10 ft (3 m) to 16.6 ft (5 m) but can reach 18 ft (5.5 m)

Key features Large but fairly nondescript; body slender with smooth scales; yellowish- or greenish-brown in color; juveniles are more brightly colored, being dark with narrow yellow or cream chevrons across the back and a boldly barred head; spreads a long, narrow hood when aroused

Habits Active on the ground during the day or night

Breeding Egg layer with clutches of 20–50 eggs ; eggs hatch after about 60–70 days

Diet Other snakes; sometimes lizards

Habitat Primary forests

Distribution Eastern India, Indochina, and Southeast Asia

Status Locally common, but in a diminishing number of places

Similar species The Indian ratsnake, *Ptyas mucosus* (although it does not spread a hood); other hooded snakes in the region are all *Naja* species cobras (considerably smaller and with wider hoods)

Venom Highly venomous; life threatening within minutes, but bites are rare

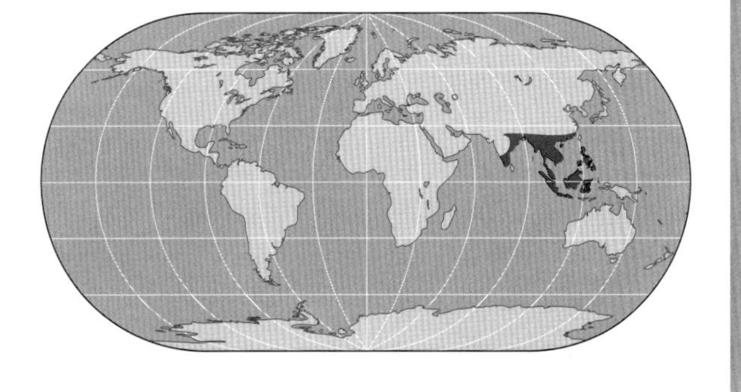

King Cobra

Ophiophagus hannah

At twice the length of an average human, the king cobra is by far the longest venomous snake in the world. Its closest rival in size is the black mamba.

King cobras are secretive and rather timid, preferring to slip away quietly rather than to confront danger head-on. That is probably just as well, considering that they often live in close proximity to humans in villages and on the edges of towns. If they were as aggressive as their reputation suggests, they would create havoc. There are, however, thought to be only about five king cobra fatalities a year.

Theoretically the king cobras can raise the front third of their length, which gives a large specimen a potential vantage point of up to 10 feet (3 m). When they put this tactic into practice, king cobras also let out a low-pitched hiss, likened by some witnesses to a growl.

There have been several occasions on which king cobras were mistaken for other snakes: One was unceremoniously dragged off a golf course in Singapore by its tail without retaliating, having been mistaken for a python. That said, a full bite from a king cobra is likely to prove fatal within half an hour or so because it injects large amounts of neurotoxic venom. Even elephants are known to have died from king cobra bites.

Loyal Partners

The breeding habits of king cobras are well known and unusual. The male and female remain together throughout the breeding season. The female lays 20 to 40 eggs (or exceptionally 50) in a large pile of dead leaves that she pulls together with her coils. Internally the nest is divided into two compartments, a lower one in which the eggs are laid, and an upper one where the female coils. She remains with the eggs to protect them from predators or disturbance, but she does not brood them in the same way as the pythons brood their eggs.

A king cobra eats a large snake. King cobras are capable of swallowing very large prey, since they themselves are enormous—longer than most crocodiles!

The male king cobra usually remains in the vicinity too. The eggs hatch after 60 to 70 days, and the young measure about 14 inches (35 cm) in length. The female is thought to leave the nest shortly before they hatch.

King cobras were placed in the genus *Naja* along with the other cobras for many years. However, they are now in a genus of their own, *Ophiophagus*, meaning "snake-eater." Although they may occasionally eat lizards, including monitor lizards, their preferred prey is other snakes.

Their alternative name of "hamadryad" is borrowed from the name of a mythical wood nymph that only lives as long as the tree of which she is the spirit. The myth could be tragically prophetic, because king cobra populations are badly affected by large-scale clearance of forests for lumber or agriculture. Once displaced, they may move into populated areas, and most of them will eventually be discovered and killed. Apart from fear and prejudice, king cobras are killed for the skin trade. Their flesh is also eaten, and their gall bladders are used in traditional Chinese medicine.

Snake Eats Snake

Snakes that eat other snakes are well distributed throughout the snake families. That is not surprising, since snakes are after all the ideal shape to fit inside other snakes. Although there are many species that eat other snakes alongside other prey, such as lizards, a number of species specialize in them.

In North America they include the larger subspecies of the ringneck snake, *Diadophis punctatus*, and the sharp-tailed snake, *Stilosoma extenuatum*, which preys largely on the rare crowned snake, *Tantilla relicta*, in Florida. In Asia the kraits, *Bungarus* species, and the Asian coral snakes, *Maticora*, are also keen snake-eaters, while in Africa most of the stiletto snakes, genus *Atractaspis*, and all the file snakes, genus *Mehelya*, are snake specialists. To complete the list, Australia has a snake-eater in the form of the black-headed python, *Aspidites melanocephalus*.

Komodo Dragon

Varanus komodoensis

Common name Komodo dragon

Scientific name *Varanus komodoensis*

Family Varanidae

Suborder Sauria

Order Squamata

Size Up to a maximum of 10.3 ft (3.1 m)

Key features Body very large; head relatively small; ear openings visible; teeth sharp and serrated; tail powerful; strong limbs and claws for digging; scales small, uniform, and rough; color varies from brown to brownish or grayish red; juveniles are green with yellow-and-black bands

Habits Spends much of the time foraging for food; digs burrows to which it retreats at night and during hot weather

Breeding Female lays clutch of up to 30 eggs (depending on size of female); eggs buried in earth and hatch after 7.5–8 months

Diet Insects, reptiles, eggs, small mammals, deer, goats, wild boar, pigs

Habitat Lowland areas ranging from arid forest to savanna, including dry riverbeds

Distribution Islands of Komodo, Rinca, Padar, and Western Flores in Indonesia

Status Vulnerable (IUCN); listed in CITES Appendix I; protected locally

Similar species None

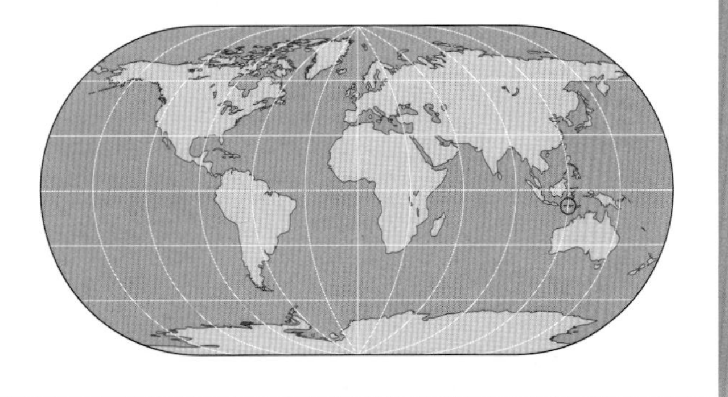

Growing to just over 10 feet (3 m) long, the Komodo dragon is the largest lizard in the world. Up to 5,000 Komodos are restricted to a few small islands in Indonesia, where they reign supreme.

THE KOMODO DRAGON IS KNOWN LOCALLY as the *ora* or *buaja daret*, meaning land crocodile, and is named after a mystical dragon famed for its size and ferocity. The total area occupied by these monitors is about 390 square miles (1,000 sq. km). It is found on the Lesser Sunda Islands of Rinca, Komodo, and Flores and the smaller islands of Gili, Montang, and Padar, although the latter does not have a permanent population. All the islands except Flores are now part of the Komodo National Park.

The native habitat of Komodo dragons consists of arid, volcanic islands with steep slopes. At certain times of the year water is limited, but during the monsoon season there may be some flooding of the area. Average daytime temperatures are 80°F (27°C). Komodos are abundant in the lower arid forests, savanna, and thick monsoon forest near watercourses. Adults prefer the more open areas of savanna with tall grasses and bushes. Hatchlings and juveniles are more arboreal and tend to stay in forested regions, where they spend much of the time on branches away from would-be predators.

Despite their enormous size and weight, Komodo dragons are quite fast runners and can swim well. Young Komodos are also adept at climbing.

Largest Lizard

Although the Crocodile tree monitor, *Varanus salvadorii* from New Guinea may be slightly longer (due to its long, slender tail) and weighs up to 200 pounds (90 kg), the Komodo dragon is classed as the world's largest living lizard. It weighs up to 330 pounds (150 kg). Its head is relatively small when compared with its large,

stout body. Unlike in many other monitor species, the nose is somewhat rounded and blunt. External ear openings are visible. The strong jaws, which are capable of crushing bones, contain 60 laterally compressed teeth with serrated edges (similar to those of flesh-eating sharks), which can be replaced many times. The highly flexible skull allows the monitor to swallow large pieces of food, and the muscular, powerful legs end with clawlike talons that are ideal for ripping open carcasses. A heavy, muscular tail makes an additional weapon to help overcome prey and is capable of delivering a crushing blow.

The body of a juvenile Komodo is more sinuous and, together with its sharp claws, enables it to pursue an arboreal lifestyle. Green background coloration with pale yellow-and-black bands provide effective camouflage against the foliage. Juveniles lead secretive lives,

Discovering the Dragons

Although there had been tales of "monsters" on Indonesian islands, it was only early in the 20th century that serious consideration was given to mounting an expedition to search for them. This was after a Dutch pilot had crashed in the area and taken back to the West information from tribes in the Lesser Sunda Islands about "huge monsters—land crocodiles more than 20 feet [6.1 m] in length." An expedition led by Major P. A. Ouwens, director of the Zoological Museum in Buitenzorg, Java, produced the first scientific description of the dragons; shortly after, in 1912, the Indonesian government closed the area. Hunting of the dragons was outlawed, and the numbers sent to zoos were restricted. However, myths of enormous 20-foot (6.1-m) dragons persisted.

After an exhaustive study of the creatures in the wild, in 1981 the American paleontologist Walter Auffenberg recorded the largest specimen encountered as 9.8 feet (3 m) long. To date the largest authenticated Komodo dragon is a 10.3-foot (3.1-m) male. It died in 1933 and is on display at the Tilden Regional Park in Berkeley.

Despite the existence of the dragons, dangerous felons were once exiled on Komodo Island. The survivors built Komodo village, which makes most of its money today from tourism and fishing.

spending most of their time foraging in trees, avoiding predation by adults. As they grow and become too heavy to live in trees, their diet alters, and they become more terrestrial.

Adult Komodos lose the juvenile sinuousness to become robust creatures. The rough scales give the skin a beaded appearance, and coloration changes to brown or reddish gray. Some individuals may have darker limbs and a patch of peach color around the eyes. Both adults and juveniles have a yellow tongue. Despite their size and weight, Komodos can move surprisingly quickly and are excellent swimmers. Some of the populations on the smaller islands are transient, swimming from island to island in search of food.

Top Predator

Much of the Komodo dragon's day is spent patrolling its territory. The core range containing burrows may cover an area of 1.2 square miles

Attacking and Scavenging

The Komodo dragon obtains food both by attacking and by scavenging. Using ambush techniques, it hides in the long grass along well-used mammal trails. To be successful, the monitor needs to be within 3 to 5 feet (1–1.5 m) of its prey. Rushing from its hiding place, it seizes a leg, and its sharp teeth sever the tendons to disable the prey. The dragon then kills its victim by a bite to the throat or by using its sharp claws to rip out its intestines. Should a victim escape after the first bite, it may still die. Initially, Komodo dragons were thought to be venomous. However, it is now known that their saliva (of which they produce copious amounts) contains 57 types of bacteria (seven of which are extremely infective) and an anticoagulant—all acquired as a result of eating carrion.

Eventually the prey succumbs either from shock, blood loss, or infection. The smell of its rotting body can easily be detected by other dragons from as far away as 5 miles (8 km). When several Komodos find carrion, a complex social structure is observed. Using their serrated teeth to rip off large chunks, large males eat their fill first, followed by smaller males and females. Any juveniles in the area wait until the larger dragons leave before descending from the trees to scavenge on any leftovers. A Komodo dragon can eat up to 80 percent of its body weight in one meal. It has been estimated that only 13 percent of a corpse is left by Komodos—the intestines, fur, and horns are the only parts not eaten.

A keen sense of smell enables Komodo dragons to seek out carrion from several miles away. Tearing at a carcass with their sharp teeth, the feast is soon over.

Komodo dragons are now the focus of a growing tourist industry. In some areas of Komodo Island the dragons are so well fed that they just lie around waiting for tourists to bring the next meal of goat or sheep. However, their future surivival is threatened by habitat destruction.

(2 sq. km), but feeding ranges, which may be shared, extend farther. It is not unusual for a dragon to cover 6.3 miles (10 km) in a day. Burrows are used to regulate body temperature. They enable the dragon to cool down during the hottest part of the day and serve as retreats for shelter and warmth at night, since they retain some of the daytime heat.

Komodo monitors are formidable predators at the top of the food chain. Juveniles feed on grasshoppers, beetles, small geckos, eggs, and birds, and move up to small mammals as they grow. Adults consume a variety of large prey, all of which has been introduced to their islands by humans, including goats, pigs, deer, wild boar, horses, and water buffalo. Smaller, weaker dragons make up about 10 percent of an adult's diet. Eye witness accounts tell of an adult Komodo eating a 90-pound (41-kg) pig in 20 minutes; on another occasion one ate a 66-pound (30-kg) boar in 17 minutes!

Courtship and Monogamy

Courtship rituals have been observed in most months of the year, but mating activity peaks in July and August. When they are ready to mate, female Komodos give off a scent in their feces that is detected by a male when patrolling his territory. He follows the scent until he locates the female and then sniffs all over her body. He rubs his chin on her head, scratches her back, and licks her body—tongue-licking gives him clues to her degree of receptivity. The female communicates that she is ready to mate by licking the male. Grasping her with jaws and claws, he lifts her tail with his, which allows him to mate. An unreceptive female hisses, inflates her neck, bites, and lashes with the tail to drive away the male.

Up to 30 eggs are laid either in a specially dug nest chamber and covered with earth and leaves or in a burrow. Female Komodos have been seen to use the nest mounds of the male brush turkey. By adding or removing material, the male bird keeps the mound in which his mate's eggs are laid at a constant temperature, making an ideal incubator for Komodo eggs. The hatchlings emerge about 8 months later and measure 15 inches (38 cm). Mortality rates are high, with many falling prey to larger Komodos, predatory birds, snakes, and feral dogs. As soon as possible, the hatchlings try to make for the trees and comparative safety.

It is interesting to note that monogamy (having only one mate) and courtship displays have been observed in many Komodo dragons. These large monitors are capable of inflicting fatal wounds and readily eat members of their own species. Therefore it would seem that pair bonding in this way enables them to recognize certain individuals and ensures the continuation of the species.

Dragons in Danger

There are estimated to be between 3,000 and 5,000 Komodo dragons in the wild, and males outnumber females 3 to 1. They have been placed on CITES Appendix I to control trade in them, but occasionally specimens are smuggled out illegally. The Indonesian government has also given them the highest level of protection, and they are regarded as "national treasures." They are classed as Vulnerable by IUCN. The threat to their survival comes from habitat destruction and poaching of their prey: Volcanic activity and natural fires can have a serious effect on their already restricted distribution, and the poaching of their prey by humans may also have serious consequences for the dragons.

Common name Kuhl's flying gecko (Kuhl's parachute gecko)

Scientific name *Ptychozoon kuhli*

Subfamily Gekkoninae

Family Gekkonidae

Suborder Sauria

Order Squamata

Size 8 in (20 cm) long

Key features A frilly gecko; a flap of skin is present along the flanks between the front and hind limbs, and there is a scalloped frill on the edges of the tail; feet are webbed, and toes have adhesive pads; head, body, and tail are brown, gray, or olive in color with broken bands and patches of darker coloration, making the gecko well camouflaged

Habits Arboreal and nocturnal

Breeding Female lays 2 hard-shelled eggs; eggs hatch after about 100 days

Diet Insects and spiders

Habitat Rain forests with high humidity; also enters houses

Distribution Southeastern Asia (India—Nicobar Islands—Myanmar, southern Thailand, Malaysian Peninsula, Borneo, Sumatra, and Java)

Status Common

Similar species Apart from 5 other species of flying geckos, the species is very distinctive

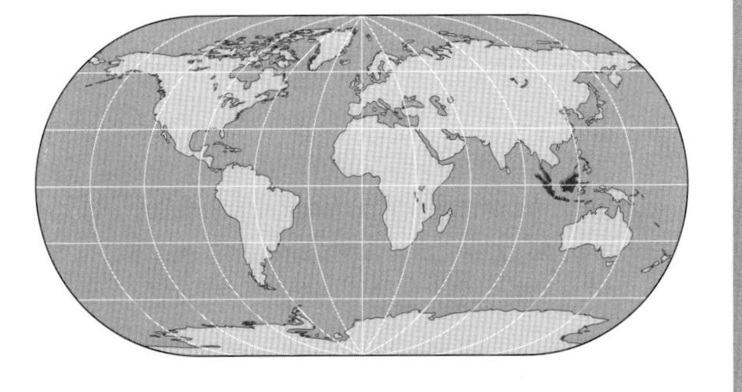

Kuhl's Flying Gecko

Ptychozoon kuhli

Southeast Asia is home to Kuhl's flying gecko and to a number of other "flying' reptiles," including snakes, frogs, and lizards.

FLYING, OR GLIDING, is not a widespread trait in reptiles. Only the flying geckos, *Ptychozoon*, the flying lizards, *Draco* species, and the flying snakes, *Chrysopelea*, have mastered it. All of them—39 species in total—come from the rain forests of Southeast Asia. One other lizard, the African blue-tailed tree lizard, *Holaspis guentheri* (sometimes regarded as two separate species), is thought to glide, but among the vast array of lizards in South and Central America, for instance, not one has taken to the air.

Some authorities think that the characteristics of the Asian rain forests, specifically the way in which the trees are spaced, may have contributed to the evolution of these "flying" forms. Tall, straight, hardwood trees (family Dipterocarpaceae) dominate many Asian forests, and the almost totally closed canopy shuts out light and prevents a shrubby understory from forming. By contrast, American and African rain-forest trees have more branches, and the forest takes on a more "crowded" character. In simple terms Asian rain forests are generally easy to walk through compared with South American ones.

Mastering "Flight"

No reptiles or amphibians are capable of powered flight. But by using flaps and flanges or by altering their body shape, some species can "glide" with a degree of control. The flying lizards, *Draco*, for example, always land head-up on the tree trunk they are aiming for by tucking their hind limbs under their body at the last minute so that they "stall." The flying snakes, *Chrysopelea*, can change direction by moving their bodies in a sinuous manner while they are gliding, and the flying geckos can even

This closeup of the front part of Kuhl's flying gecko shows the head detail and the heavily webbed front feet. The dull brown color of its head and body provide good camouflage against the bark of a tree.

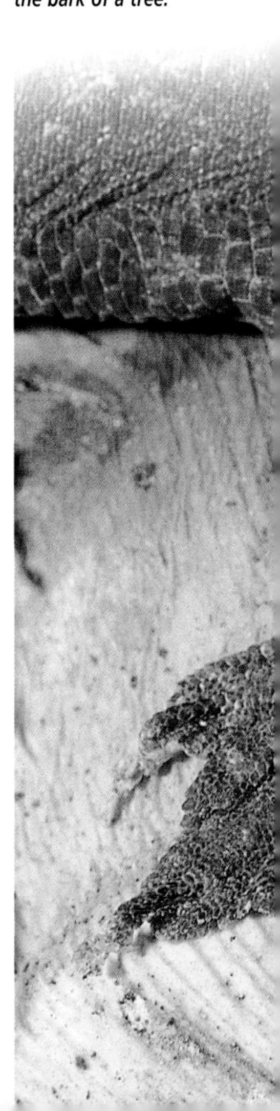

change direction, using their tails as rudders, and head back in the direction they came from.

Gliding in Kuhl's gecko is accomplished by means of heavily webbed feet and a fringe that runs around the edge of its tail. It also has a flap of skin along each of its flanks. During extended flights air pressure causes them to spread out as well. Unlike the flying lizards, the gecko has no control over the erection or folding of these flaps.

The flaps and frills almost certainly arose to enhance the gecko's camouflage and not for the purpose of flight. The gecko is colored and marked to match the lichenous branches and trunks on which it rests, and the flaps act as a cloak to eliminate shadows. When a Kuhl's gecko lands after a glide, it immediately "freezes" so that any predators that may be watching its movements lose sight of it. Again, this is unlike the flying lizard, which immediately runs up the tree trunk to regain the height lost through the glide. Like most geckos, Kuhl's flying gecko has adhesive pads at the tips of its toes that enable it to climb smooth surfaces, and also like many species, it sometimes occurs in houses and buildings.

Breeding

Male Kuhl's flying geckos are territorial and fight off intruders. Females apparently mate at any time of the year, although there may be

Who Was Kuhl?

Dr. Heinrich Kuhl was a German naturalist who lived from 1796 to 1821 and spent time in the East Indies. His short stay in the region was productive, since his name became associated with several fish, a frog, a lizard, a bird, and a mammal. As well as Kuhl's flying gecko, his name is celebrated in the kuhli loach, *Acanthophthalmus kuhli*, Kuhl's stingray, *Dasyatis kuhlii*, Kuhl's creek frog, *Limnonectes kuhlii*, Kuhl's lory, *Vini kuhlii*, and Kuhl's deer, *Axis kuhlii*. A family of Indo-Pacific marine fishes, the Kuhlidae, or flagtails, is also named after him. (The variation in the spelling of *kuhli* reflects the way it was spelled when each species was named.) *Ptychozoon* comes from two Greek words: *ptukhos,* meaning a fold, which refers to the fold of skin down the gecko's sides, and *-zoon*, meaning a living thing, or animal.

regional differences depending on climate. The eggs are usually laid in pairs, and a female can lay clutches every three or four weeks. They glue them to a suitable surface, often behind bark, and the eggs take about 100 days to hatch. The hatchlings are perfect replicas of the parents, complete with flaps and frills, and measure about 1.8 inches (4.5 cm) in total length. They grow quickly and can reach sexual maturity in less than one year.

Other *Ptychozoon* Species

There are six species in the genus altogether. One of them, *P. trinotaterra*, was only recently described and comes from Thailand and Vietnam, but the others have been known for at least 100 years. Four of them (*P. horsefieldii, P. intermedium, P. kuhli,* and *P. lionotum*) are quite widespread across the forests of Asia, and some overlap each other in places. *Ptychozoon rhacophorus*, however, is known only from Kina Balu, North Borneo, at an altitude of 2,100 feet (640 m). All these species look as though they could glide (and all are called parachute or flying geckos), but apart from *P. kuhli*, only *P. lionotum* has been seen to do so.

As it takes to the air, the flaps and frills on the body and tail of Kuhl's flying gecko spread out to allow it to glide between resting places on forest trees.

Taking to the Air

There are other types of gliding animals in Asian forests. Many frogs of the genus *Rhacophorus* use heavily webbed feet to "parachute" to lower levels. There are also flying squirrels, including 14 species in Borneo alone. Two species of flying "lemurs," *Cynocephalus variegatus* from Borneo and west Malaysia, and *C. volans* from the Philippines, are distinct from true lemurs (which live only on Madagascar) and are placed in an order of their own, the Dermoptera, meaning "skin-winged."

The Pteropodidae is the family to which flying foxes, or fruit bats, belong. About 166 species of these large, ungainly, fruit-eating bats are recognized. Apart from some species from Australia and a few others from Indian Ocean islands, they too are confined to South and Southeast Asia.

Common name Large sand racer

Scientific name *Psammodromus algirus*

Family Lacertidae

Suborder Sauria

Order Squamata

Size 9 in (23 cm) long

Key features Typical lacertid shape with a long tail and limbs and narrow head; scales heavily keeled, overlapping, and ending in a point; tail can be up to 3 times as long as head and body combined, and is stiff; color midbrown with 2 yellowish stripes down the back and another down each flank; males may have blue eyespots at the base of their front legs; hatchlings have reddish flanks

Habits Terrestrial, climbing occasionally; diurnal

Breeding Female lays clutches of 2–11 eggs that hatch after 5–6 weeks

Diet Mainly small insects and spiders

Habitat Dry, scrubby, or bushy places; often lives around the base of heather, gorse, and other dense bushes; rarely ventures into the open

Distribution Most of Spain and Portugal extending along the Mediterranean coast of France; also North Africa

Status Common but often overlooked

Similar species The Spanish sand racer, *P. hispanicus*, is smaller and has spots or dashes on its back rather than stripes

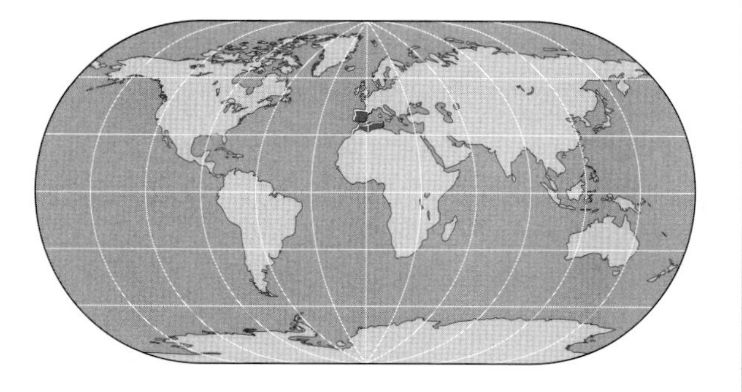

Large Sand Racer

Psammodromus algirus

The large sand racer is a handsome lizard that is probably more common than it would appear at first. That is because it has camouflage coloration and a retiring nature.

THE LARGE SAND RACER IS EASILY OVERLOOKED unless it is disturbed while basking, in which case it quickly runs away into the center of a large bush. It favors bushes that are armed with spines or prickles, such as gorse, bramble, or the introduced prickly pear, *Opuntia*, that has spread over much of southern Europe. It occasionally climbs up into bushes to bask or to search for food and also basks on top of logs and in brush piles. It is also found around human dwellings, in gardens, and in overgrown fields. Contrary to its name, it prefers hard-packed soils, such as clay, or stony ground rather than loose sand.

These lizards are very heat tolerant and are active in the middle of the day even in the North African summer. They stay active until dusk. If found in the open, they are often more easily caught than the wall lizards (Lacertidae) since they are slower and less agile. If handled roughly, they squeak and may also do so at other times, for example, when fighting or perhaps as a means of communication.

Their main predators are birds of prey and snakes, and their usual means of defense is to flee and hide in a hole or a crack in the ground. The shape of their scales, with drawn-out, backward-pointing tips, is thought to give the lizards some protection—it would make it hard for snakes, for example, to maneuver them in their mouths to line them up for swallowing.

Unwelcome Visitors

Like many lizards, the large sand racer is often troubled by small red "mites," which are actually mite larvae. In most lizards these pests

attach themselves to the spaces between the eyes and the surrounding scales or to the area around the eardrum, and cause irritation. They can also cause problems when the lizard tries to shed its skin. Some lizards, including the large sand racer, get around this problem by providing the mite larvae with a special patch of skin with no scales on it, called a "mite pocket," where the mites accumulate.

Breeding

The breeding season starts in about April in most places, and the male follows a female around, attempting to make contact. He bites her neck and twists his body under hers to mate. About 15 to 20 days after mating, the eggs are laid in moist soil in a shallow pit dug by the female. She lays between two and 11 eggs, and may lay a second clutch a few weeks later. The eggs hatch from July onward, and the hatchlings have reddish flanks.

***Psammodromus algirus** is the largest member of its genus. It lives in stony, sparsely vegetated areas and can be found sunning itself near human dwellings.*

Other Sand Racers

There are three other members in the genus *Psammodromus*. The Spanish sand racer, *P. hispanicus*, is smaller and lives in Spain and a small part of France. Two species, *P. blanci* and *P. microdactylus*, are similar to each other and occur in North Africa. The latter species is very rare and found only in the Atlas Mountains of Morocco.

The common fringe-toed lizard, *Acanthodactylus erythrurus,* is another lacertid that shares its habitat with the sand racers. It also lives in North Africa and Spain but prefers open ground, including dunes. This species and other members of the genus have a fringe of hairlike scales along the edges of their toes to help them run across loose surfaces. (*Acanthodactylus* means "spiny toes.") On hot surfaces, such as rock or gravel, it often straightens its forelegs when resting to raise the front part of its body clear of the ground and may even lift alternate limbs in turn, in a way similar to that of the shovel-snouted lizard, *Meroles anchietae* from Namibia.

Leaf-Nosed Vine Snake

Langaha madagascariensis

Madagascar's flora and fauna have evolved in isolation, producing species that are unlike any from other parts of the globe. Ninety percent of the island's reptiles and amphibians are found only there.

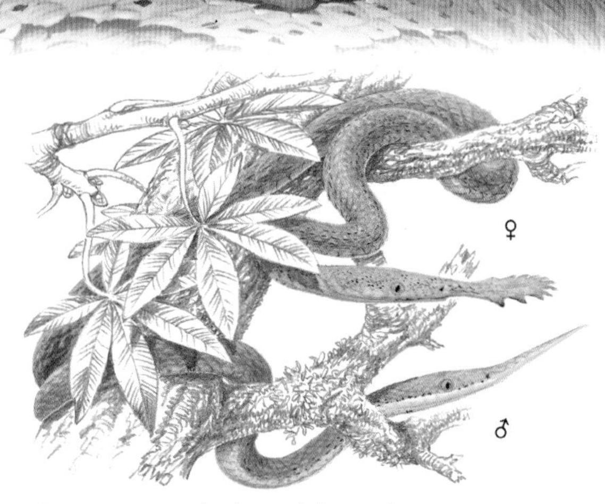

Common name Leaf-nosed vine snake

Scientific name *Langaha madagascariensis*

Subfamily Boodontinae

Family Colubridae

Suborder Serpentes

Order Squamata

Length 28 in (71 cm) to 35 in (89 cm)

Key features Body very slender; scales heavily keeled; tail long; eyes have vertical pupils; its most remarkable characteristic is the extension to the tip of its snout; color brown, but the male and female differ in their markings

Habits Arboreal and diurnal

Breeding Egg layer; small clutches of typically 3 eggs; eggs hatch after about 64–67 days

Diet Lizards; possibly other small vertebrates

Habitat Forests

Distribution Madagascar

Status Uncommon

Similar species 2 other members of the genus: the southern leaf-nosed vine snake, *L. alluaudi*, and the northern leaf-nosed vine snake, *L. pseudoalluaudi*; otherwise, it is impossible to confuse it with any other snake

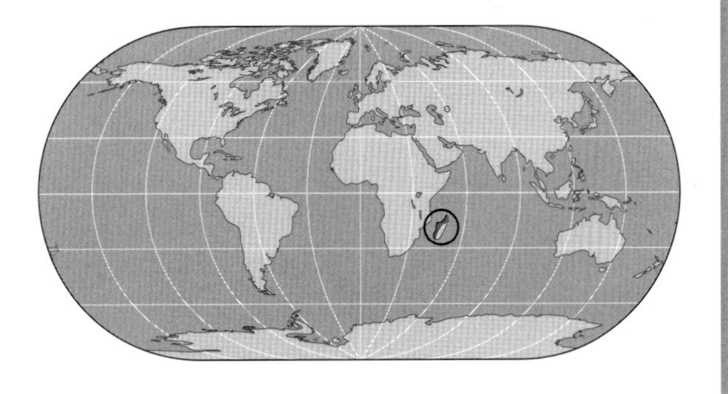

THE LEAF-NOSED VINE snake epitomizes the unique quality of reptilian and amphibian life on Madagascar. Originally described in 1790, it went largely unknown until the end of the 20th century, when interest in the island's biology grew as a result of habitat destruction and the near extinction of many unique forms.

Strange Snouts

The bizarre leaf-nosed vine snake has a thin, vinelike body with heavily keeled scales. Its pupils are vertical, even though it is diurnal. Its head ornamentation has no parallel in the snake world. Both sexes have long projections on the tip of the snout that almost double the length of the head. The projections are composed of numerous scales, but their shape varies according to the sex. Males have a long, tapering "horn" that is soft and pliable, and ends in a fine point. Females have a far more elaborate structure that has been likened to a leaf or opening bud. It is flattened from side to side and is edged with a series of jagged scales.

The function of these appendages is not known but might enhance the snake's camouflage by breaking up the outline of the head. Why the shape should be different between the sexes is a mystery, although some scientists suggest that males and females may occupy slightly different microhabitats in the wild. One of the problems in resolving puzzles of this kind is that leaf-nosed vine snakes, like other secretive species, are almost impossible to study under natural conditions. Their behavior in captivity is unnatural and masks the function of such structures.

The differences in color pattern and nasal structure between males and females are pronounced in* Langaha madagascariensis*, the Madagascan vine snake. The male (below) has a tapering pointed snout.

Apart from the difference in the shape of their horns, male and female leaf-nosed vine snakes differ in their colors and markings. Males are plain brown above and yellow below, with a thin white line marking the border where the two colors meet. Females are mottled grayish brown all over, with small irregular markings.

Stealthy Hunters

The snakes' colors and markings and presumably the shape of their nose appendages make them difficult to see when resting among vines and twigs. They usually rest completely motionless during the night and most of the day, and are active only during the middle of each day when the sun is hottest. They are stimulated by movement of the small lizards that make up the bulk of their prey, although they may also eat frogs and small mammals.

Sexual Dimorphism

Species in which males and females look different are said to be sexually dimorphic. Sexual dimorphism is common among birds: Males are usually more colorful than females because their plumage is used in courtship and territorial displays. Females are more subdued in color, making them difficult to see when they are sitting on eggs. A similar situation occurs in many lizard families, notably the iguanids, the agamids, and the chameleons, where males use their bright colors to advertise their territories to females and other males.

Snakes, however, do not take part in visual displays of this sort, and pronounced sexual dimorphism is rare. In snakes the main difference between the sexes is in size. There are other minor differences such as body proportion, head size, size of cloacal spurs in male boas, pythons, and some other primitive species, fang length in two or three cobras, *Naja* species, and tongue length in blunt-headed tree snakes, *Imantodes*.

The leaf-nosed vine snake and the other two species of *Langaha* are probably the most dimorphic snake species in the world. Other species that have a degree of dimorphism include several vipers, *Vipera* species, in which the males often have more contrasting markings than the females. In the banded rock rattlesnake, *Crotalus lepidus klauberi* from North America, males have a greenish hue, but the females are gray. In boomslangs, *Dispholidus typus*, the females are olive or brown, but the males are bright green, yellow, and black or black with a gray underside. As with *Langaha*, the reasons for their sexual dimorphism are not known, and we can only guess that the lifestyles of males and females are subtly different and require different color patterns.

Female Madagascan vine snakes have slightly more cryptic (disguising) markings, and their snout ends in a flattened leaflike structure.

Horns and Similar Structures

Snakes are conservative animals when it comes to ornamentation (compared with lizards, for example). The only area that has evolved ornaments is the head. A number of species from different families and different parts of the world have horns or other head appendages. They may consist of a single thornlike scale over each eye, as in the North African desert horned viper, *Cerastes cerastes*, and the South African horned adder, *Bitis caudalis*; or there may be a cluster of small horns, as in the Central American eyelash viper, *Bothriechis schlegelii*, and the many-horned adder, *Bitis cornuta* from South Africa.

Similarly, projections from the snout are seen in snakes from the Viperidae and the Colubridae. The nose-horned adder, *Vipera ammodytes* from eastern Europe, has a single fleshy horn on the tip of its snout, for example, while other adders, including Lataste's viper, *Vipera latasti* from Spain, have upturned snouts—seemingly halfway toward forming a horn. The African rhinoceros viper, *Bitis nasicornis*, has a group of enlarged and elaborate horny scales at the tip of its snout from which it get its common and scientific names. Among the colubrids the tentacled snake, *Erpeton tentaculatus*, has a pair of fleshly tentacles on its snout.

Two other groups of vine snakes, *Ahaetulla* from Southeast Asia, and *Oxybelis* from Central and South America, do not have nasal appendages as such; but when they feel threatened, they stick out their tongues and hold them rigid while remaining completely still. The tongues can often look like nasal appendages and, like them, probably help camouflage the snake's outline.

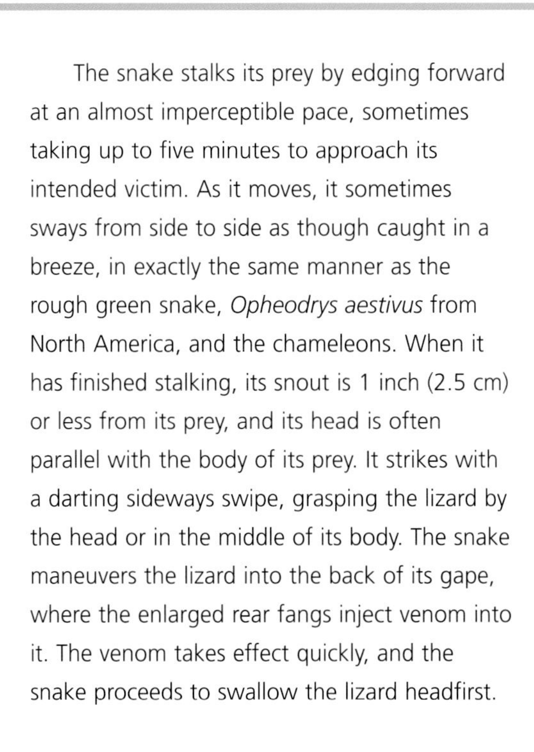

The snake stalks its prey by edging forward at an almost imperceptible pace, sometimes taking up to five minutes to approach its intended victim. As it moves, it sometimes sways from side to side as though caught in a breeze, in exactly the same manner as the rough green snake, *Opheodrys aestivus* from North America, and the chameleons. When it has finished stalking, its snout is 1 inch (2.5 cm) or less from its prey, and its head is often parallel with the body of its prey. It strikes with a darting sideways swipe, grasping the lizard by the head or in the middle of its body. The snake maneuvers the lizard into the back of its gape, where the enlarged rear fangs inject venom into it. The venom takes effect quickly, and the snake proceeds to swallow the lizard headfirst.

Breeding

From the little information available it seems that the leaf-nosed vine snake breeds in the spring. Two wild females at San Diego Zoo laid eggs at the beginning of November, with clutches of 10 and 11 eggs. The eggs were elongated, as in other slender species, and weighed just over 0.07 ounces (2 g) each. In captivity they laid their eggs in damp moss, but it is not known how they choose a nest site under natural conditions. They may have to come down to the ground to lay, which is unusual in such highly arboreal snakes. The eggs hatched after 64 and 67 days, and the young measured about 5 inches (13 cm) from snout to vent.

The unusual Madagascan vine snakes,* Langaha madagascariensis, *have a feelerlike structure on the tip of the snout. In males it is round and shaped like a spike, but in females (left) it resembles a fleshy leaf.

The Relatives

There are only three species in the genus *Langaha*. Apart from *L. madagascariensis*, there is *L. alluaudi* from the south of Madagascar. They also have similar nose horns, with females having a more elaborate design than males. The females have additional projections in the form of jutting scales above their eyes (supraocular scales). At the time of writing, the third species, *L. pseudoalluaudi*, is known from just a single specimen from the north of Madagascar. It was a female that subsequently laid three eggs, and it too had the leaflike nose horn. Until a male is described, we cannot be sure that it will differ from females, although it seems most likely that it will.

Leatherback Turtle

Dermochelys coriacea

The leatherback turtles are true giants. They are the largest of all marine turtles and the heaviest reptiles in the world. Their very distinctive shells have a leathery appearance.

Common name Leatherback turtle

Scientific name *Dermochelys coriacea*

Family Dermochelyidae

Suborder Cryptodira

Order Testudines

Size Carapace can be up to 8 ft (2.4 m) in length

Weight Up to 1,650 lb (750 kg)

Key features Carapace very distinctive with 7 ridges running down its length; surface of the carapace is effectively a rubbery skin rather than made up of scales; skin strengthened with very small bony plates; color dark with whitish markings; plastron bears about 5 ridges and varies in color from a whitish shade to black; flippers lack claws; front flippers extremely long; carapace of hatchlings has rows of white scales

Habits Often favors open sea, swimming widely through the world's oceans

Breeding Clutches consist of about 80 viable eggs; female typically produces 6–9 clutches per season; egg-laying interval typically 2–3 years; youngsters emerge after about 65 days

Diet Almost exclusively jellyfish

Habitat Temperate and tropical waters

Distribution Has the largest range of any marine turtle; found in all the world's oceans from Alaska to New Zealand

Status Critically Endangered (IUCN); listed on CITES Appendix I

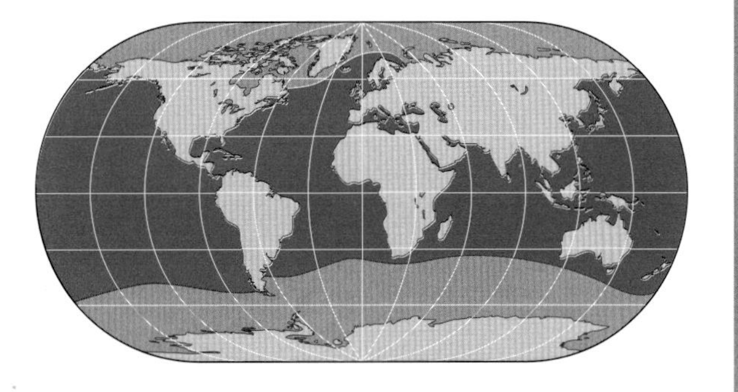

AT ABOUT 8 FEET (2.4 M) LONG and weighing up to 1,650 pounds (750 kg), the leatherback's bulk probably enables it to maintain a sufficiently high core body temperature that allows it to venture farther into temperate waters than any other species of marine turtle. Leatherbacks are apparently unaffected by sea temperatures even below 41°F (5°C), and they range as far north as the seas around Alaska. Their body is actually slightly warmer than that of their surroundings in these cold waters, which suggests that they have a basic mechanism to regulate their body temperature.

These turtles are also found in the oceans below the southern tip of Africa and off the Chilean coast as well as close to New Zealand. In fact, the largest leatherback recorded was not found in the tropics but was discovered stranded on a beach on the coast of Wales in the British Isles in 1988. It is possible that global warming and its effects on sea temperature are affecting the range of these turtles.

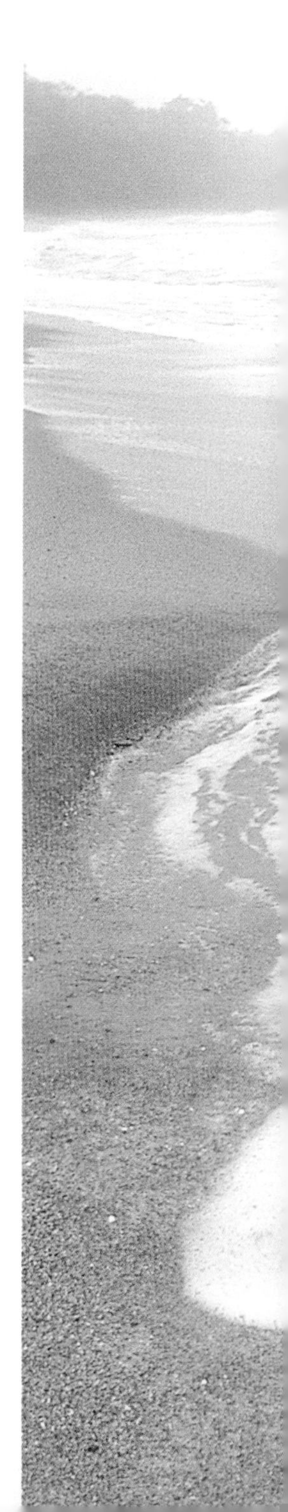

Remote Nesting Sites

Leatherbacks return to the tropics to breed. They often choose remote areas for this purpose, although there are about 50 nests recorded along the Florida coastline each year. They traditionally use beaches onto which they can haul themselves up without difficulty, and where they can come directly out of deep sea rather than swimming across reefs. This is possibly to protect their vulnerable underparts from injury and may explain why they tend to nest more commonly on mainland areas rather than islands. Unfortunately, these beaches can be badly eroded in storms, leaving the leatherback's developing eggs at greater risk of being lost than those of other marine turtles.

Egg stealing has been a threat in some areas in the past, but improved protective measures mean that it is less of a problem today. The oil in the leatherback's body was also used for the manufacture of many products, including cosmetics and medicines, but the introduction of synthetic substitutes has ended this trade. Leatherbacks are not hunted for their meat, which is regarded as unpalatable.

Although leatherbacks often lay eggs on their own, they sometimes nest in small groups. Their breeding range extends almost all the way around the world—from the Caribbean region across to the western coast of South Africa to India, Sri Lanka, Thailand, and Australia right across the Pacific to the shores of Mexico. Clutch sizes laid by leatherbacks in the eastern Pacific region tend to be smaller than those produced in other parts of their range.

What is suspected to be the largest breeding colony of leatherbacks in the world was only discovered as

Long Journeys

Tagging studies have revealed the remarkable distances that leatherbacks can cover in the world's oceans—one individual tagged on its nesting ground in Surinam, northern South America, was rediscovered on the eastern side of the Atlantic over 4,226 miles (6,800 km) from the original tagging site. Unfortunately, leatherbacks have tended to lose their tags more readily than other turtles, so fewer data are available, but it certainly appears that those encountered along the northern coast of South America regularly undertake journeys of over 3,125 miles (5,028 km). Switching the tag site on the leatherback's body from the front flipper to the inner side of the back flipper has helped, however, since the tags are exposed to less physical force in this area of the body. This should ensure that more information about their movements can be obtained.

Female leatherbacks, such as this one in Trinidad, come ashore to nest every two to three years on the warm sands of remote tropical beaches.

recently as 1976 thanks to the confiscation of a large number of leatherback eggs that were on their way to Mexico City. The trail led to an area known as Tierra Colorado on the Pacific coast.

Studies have since revealed that up to 500 leatherback females may come ashore to lay eggs there every night during the nesting period, mainly in December and January each year. It appears that, at least in this area, female leatherbacks return on their own with no males congregating offshore in search of mates.

An unusual phenomenon is the presence of small, apparently immature eggs found in the nests of leatherback turtles. Their presence may be linked in some way to the interval of time between the laying of the clutches, which is much shorter than in other marine species. It is often no more than seven to 10 days, and some eggs do not develop fully in this time. It takes about 65 days for the young leatherbacks to hatch and emerge at the surface, by which stage they are about 2.5 inches (6.3 cm) long. The hatchlings are unmistakable: The longitudinal ridges are well defined, and there are rows of white scales that appear as stripes along the length of the flippers.

It is quite straightforward to determine the sex of leatherback turtles, since males have much longer tails than females and, as in many other chelonians, a slightly concave plastron.

Death at the Hands of Humans

The leatherback's wide range means that it is very difficult to build up an accurate population estimate, but there are signs that the species is in trouble. This is not essentially because of hunting pressure but largely as a result of its feeding habits. Its rather slender jaws with their scissorlike action are used to capture jellyfish, which form the basis of its diet. Unfortunately, these turtles find it hard to distinguish between jellyfish and plastic detritus such as plastic bags and other similar waste floating on the surface of the sea. When seized and swallowed, these items are likely to get stuck in the turtle's gut, resulting in a slow and painful death. Controlling losses of leatherback turtles is exceptionally difficult, and there is no easy way of solving this problem.

There has been progress, however, in addressing some of the other threats facing leatherback populations. It was estimated that about 640 of these turtles were being accidentally captured in nets in U.S. waters annually. Many of them died through drowning or injuries sustained during their capture. Devices to keep turtles out of the nets were developed, and the law was changed to make their use mandatory in U.S. waters. Elsewhere, however (often in international waters), problems remain, with the turtles being caught in fishing nets or becoming entangled in ropes or lines. Even if the leatherback can free itself, the resulting injury can prove fatal. The leatherback's urge to swim, together with its specialized feeding habits, mean that nursing it back to health in captivity is often a difficult task too.

Imprinting Behavior

One strange phenomenon that has been repeatedly documented is the way in which, after she has completed the task of egg laying, a female leatherback circles the nest site, just as the young do once they hatch. It may be that this behavior somehow imprints onto the memories of the youngsters, aiding their return to the same place in due course. Current estimates suggest that there could be between 100,000 and 115,000 breeding female leatherbacks in the world's oceans today.

Predators

Leatherbacks tend to dive deeper than other turtles, which may give them some protection against being attacked. They are also well equipped to swim fast out of harm's way thanks to the propulsive power of their front flippers. They are longer than those of any other marine turtle and can extend to nearly 9 feet (2.7 m) in length.

Even once they are fully grown, however, these turtles still face a number of predators. Various sharks, including the notorious great

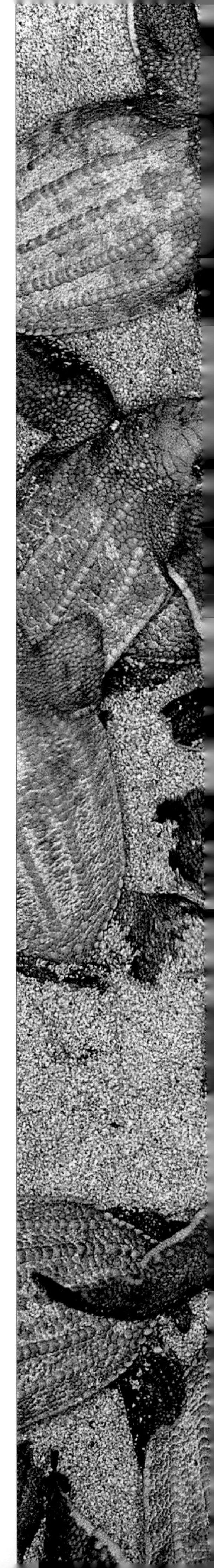

Ⓘ ***A leatherback turtle hatching on the Virgin Islands in the Caribbean. Hatchlings use a sharp tooth called an "egg tooth" to break through the eggshell.***

Ⓛ ***In French Guiana a group of young hatchlings have just emerged from their eggs. They must make their way to the ocean quickly to avoid predatory seabirds.***

white shark, *Carcharodon carcharias* from Australian waters, represent a hazard; killer whales, *Orcinus orca*, are also known to prey on leatherback turtles, the reptile's size being of little use against such fearsome predators.

Virtually nothing is known about the potential life span of these turtles, but for individuals that escape being hunted, it is thought to be measured in decades, as in the case of other sea turtles. While it is generally assumed that the leatherback turtle is solitary by nature, there have been accounts of sightings at sea of groups numbering as many as 100 individuals. Whether or not the groups are drawn together for mating purposes is unclear; it could simply be that they tend to congregate in areas where food is plentiful.

Common name Leopard gecko

Scientific name *Eublepharis macularius*

Subfamily Eublepharinae

Family Eublepharidae

Suborder Sauria

Order Squamata

Size 8 in (20 cm) to 10 in (25 cm) long

Key features This species has eyelids; head broad; body cylindrical; tail is thick and carrot shaped when animal is well fed; skin covered with small tubercles; toes lack adhesive pads; color yellow or tan with many small, dark-brown spots over the top of the head, back, and tail, sometimes with bluish background; spots on tail superimposed on wide black-and-white bands; juveniles are completely different with wide, saddle-shaped, dark-brown markings on a white or cream background

Habits Terrestrial and nocturnal; seeks refuge from the heat by day and from the cold in winter by living in underground burrows

Breeding Female lays 2 soft-shelled eggs; eggs hatch after 40–60 days

Diet Invertebrates, including insects, spiders, and scorpions; also other lizards

Habitat Desert and scrub regions in mountainous areas

Distribution South-central Asia (Pakistan, northwest India, Iraq, Iran, and Afghanistan)

Status Common

Similar species Other poorly known species of *Eublepharis* occur in the region and are similar to the leopard gecko

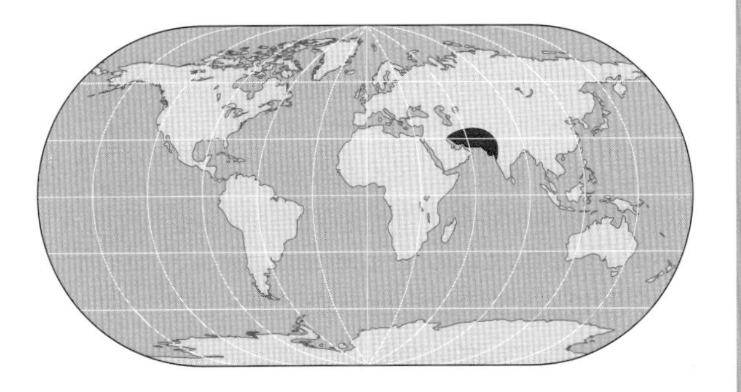

Leopard Gecko *Eublepharis macularius*

Leopard geckos were the first lizard species to be bred in large numbers for the pet trade. They still rank among the most popular because they are tough, attractive, and adapt well to an artificial environment.

THE LEOPARD GECKO IS A FAMILIAR REPTILE "pet" that is kept by the thousands by amateur enthusiasts and selectively bred to create a number of attractive color forms. The same qualities that make it a good pet also make it ideal as a laboratory animal. Most of the research on temperature-dependent sex determination (TDSD) was carried out on the leopard gecko, leading scientists to look more carefully at this process in other reptiles. The research has had important implications for conservation by means of captive-breeding programs in other species, especially tortoises and sea turtles.

The natural environment of leopard geckos is the arid plains and foothills in southern Asia and the Middle East, where they live among rock outcrops and desert scrub. They avoid loose, sandy surfaces. They survive this harsh environment by being strictly nocturnal and limiting their period of activity to times when temperatures are suitable. Because the desert cools rapidly, leopard geckos are most likely to be seen on the surface in the hours between sunset and midnight. After that they retreat into deep burrows to avoid the extremes of temperature. During winter they stop coming out of their tunnels altogether, and they live off the fat stored in their carrot-shaped tail. The tail will wither dramatically if the leopard gecko goes through a long period with no food.

Stiff Legs

Leopard geckos never climb and lack the adhesive pads on the tips of the toes that some other geckos have. Instead of pads they have small claws. Like most other members of the Eublepharidae (but in contrast to most other geckos), leopard geckos walk on stiff legs with their body held clear of the ground. They feed mainly on a variety of insects, spiders, and scorpions, and will also eat smaller lizards.

Leopard geckos appear to live in loose colonies centered around a rocky outcrop or a stony hillside. Several females and juveniles may live within a male's territory, but males do not tolerate each other. Leopard geckos lick each other's skin to determine sex, presumably picking up molecules that help with identification.

Breeding takes place in the spring, and females may lay several clutches of two soft-shelled eggs at intervals of a few weeks. They bury the eggs in moist sandy soil, carefully replacing the soil after laying and smoothing it over to disguise the eggs' whereabouts. The eggs usually hatch after between 40 and 60 days depending on the temperature. The sex of the resulting offspring also depends on the incubation temperature.

Leopard geckos get their name from the spots covering their body. Juveniles are banded and boldly colored, and appear quite different. The banding and color intensity gradually fade as the gecko matures.

Sex Determination and Temperature

The sex of the vast majority of lizards and other animals, including humans, is determined genetically by sex chromosomes (in humans XX produces females and XY produces males, for instance). The sex is fixed at the moment of fertilization and depends entirely on which type of spermatozoa (male gamete) reaches the ovum first.

However, a select number of reptiles, including the leopard gecko, use a method known as temperature-dependent sex determination (TDSD), in which the sex of the offspring is not fixed at the time of fertilization nor even when the egg is laid. It changes during the incubation period depending on the temperature. It appears that there is a "time window" during the early part of incubation when the sex can swing one way or another.

In leopard geckos lower temperatures of about 68 to 73°F (20–23°C) produce females and so do high temperatures in the range of 90 to 104°F (32–40°C). Temperatures of about 77 to 86°F (25–30°C) produce mostly males, and temperatures between each of these ranges produce a mix of both sexes.

Leopard Geckos as Pets

It is generally agreed among hobbyists that the leopard gecko is one of the best species of lizards for beginners to keep for several reasons. First, they are tough and undemanding in captivity and will thrive under a variety of conditions. Because they are nocturnal, they do not need ultraviolet light in which to bask. Second, they are calm and unhurried in their movements and less likely to escape than many other species, and they rarely bite. Most importantly, they can be bred in large numbers for distribution through the pet trade.

Many breeders have developed strains with particular color characteristics, some of which have fanciful names, such as "high yellow," "lavender," "striped," "jungle," and "albino." However, all these mutated varieties are equally easy to care for. Although females only lay clutches of two eggs, they will lay every two or three weeks over a long period in the spring and summer. The young are easily reared and reach maturity in one to two years.

A single adult male or a small group of juveniles or females can be kept in a medium-sized aquarium with a layer of paper, bark chippings, or small pebbles. They will always defecate in the same place, so they are easy to keep clean. Temperature is not critical but should be between 73 and 86°F (23–30°C) during the day. A drop of 9 to 18°F (5–10°C) at night will not cause any problems and may be beneficial. The best form of heating is an under-cage heat mat. Placing it at one end of the cage gives the gecko(s) a choice of temperatures. A hiding place should be provided at each end of the cage—the best design for this is an upturned flowerpot or plastic food container with a "door" cut out. The substrate inside the hide should be sprayed to provide a small area of high humidity that will help the gecko shed its skin properly when the time comes.

Leopard geckos can be fed crickets, mealworms, or wax worms from pet suppliers. Spiders and insects can be collected, but only if they come from places that have not been sprayed with chemicals. Too many waxworms can lead to the geckos becoming obese. They should be fed every day, preferably in the evening, although geckos in good condition (those with a plump tail) can easily go for two or three days without food. Large numbers of crickets should not be put in the cage if the carer is going away for a few days, since they can cause stress and damage to the geckos. If only cultured food, such as crickets and mealworms, is being used, each meal should be dusted with a calcium or vitamin-and-calcium preparation specially formulated for reptiles. They are available in specialized pet stores. Geckos fed on a variety of wild-collected food do not need any supplements.

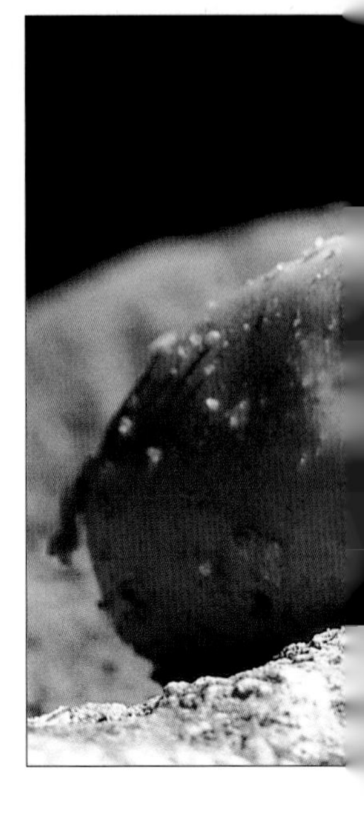

↑ Research on temperature-dependent sex determination in leopard geckos has helped halt the decline of some endangered species by skewing the sex ratio toward more breeding females.

Juvenile leopard geckos such as this one lack spots; instead, they are boldly marked with yellow, white, and dark-brown bands. Even when fully grown, leopard geckos only reach about 10 inches (25 cm) long, and their gentle temperament makes them ideal pets.

Some other lizard species and (as far as we know) all crocodilians have a system similar to that of the leopard geckos. But while other reptiles have TDSD, the details can differ. In many turtles, for example, low temperatures produce males, and high ones produce females. In some lizards the situation is the mirror image of this, with high temperatures producing males; while crocodilians and many other lizards follow a system similar to that of the leopard geckos.

Nobody knows how or why this system has evolved, although several theories have been put forward. But the implications for the future of some of these species are more obvious. If global warming continues, the sex ratios of many species could be affected, with a preponderance of males or females depending on which system they are locked into. On a smaller scale deforestation may remove shade from favored egg-laying sites, causing the overall temperature to rise, whereas other circumstances may cause temperatures to fall locally. We have no idea whether or not female geckos (or turtles or crocodilians) can assess the temperature of their egg-laying sites and act accordingly by repositioning their nests to counteract the changes. In all probability they cannot.

On the other hand, the implications for captive-breeding programs are more positive. By manipulating the temperature of the eggs, breeders can decide whether to produce mainly males or females. In some cases an abundance of females will be more useful than males. Before TDSD was properly understood, however, captive-breeding programs, including those in which wild turtle eggs were collected and incubated under controlled conditions, may have fallen foul of TDSD, producing offspring of only one sex.

A female leopard gecko sheds her skin and begins to eat it. The frequency with which these geckos shed their skin depends on many factors, including humidity levels.

Leopard Snake

Elaphe situla

Common name Leopard snake

Scientific name *Elaphe situla*

Subfamily Colubrinae

Family Colubridae

Suborder Serpentes

Order Squamata

Length From 30 in (76 cm) to 36 in (91 cm)

Key features Body slender; head narrow; scales smooth; background color yellowish or buff; a row of red (sometimes brown) blotches runs down its back with a row of smaller spots on the flanks; each spot is edged in black; in some parts of the range the blotches may be divided into 2 parallel rows of smaller spots, or the markings may consist of 2 black-edged red stripes down the back

Habits Mainly ground dwelling, but it may climb into dry walls or scree

Breeding Egg layer with clutches of 2–8 eggs; eggs hatch after about 60 days

Diet Small rodents; occasionally lizards

Habitat Dry rocky places such as scrub-covered hillsides, scrub, and fields

Distribution Southeastern Europe, extending into western Asia (Turkey and the Crimean Peninsula)

Status Common in suitable habitat

Similar species None in the area; rare individuals with brown blotches could be confused with young four-lined snakes, *Elaphe quatuorlineata,* or the transcaucasian ratsnake, *E. hohenackeri*, with whose range it overlaps in Turkey

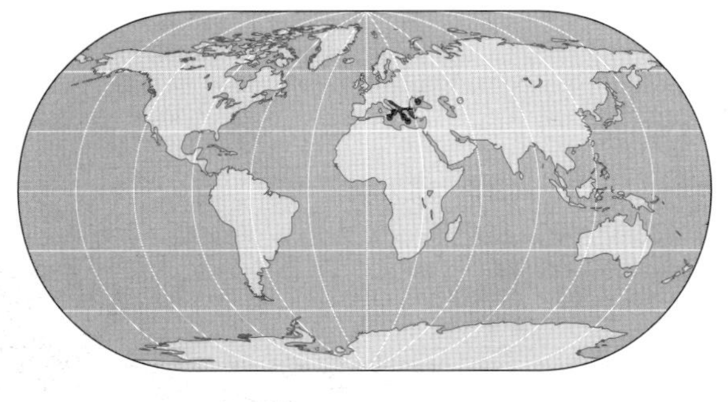

Arguably the most attractive European snake, the leopard snake is daintier than the North American ratsnakes to which it is closely related. Its colors stay bright from the day it hatches until it becomes an old adult.

The leopard snake's range includes the southern half of Italy, including the island of Sicily, Malta, the eastern Adriatic coast, right around the coastline of Greece and many Greek islands in the Aegean, as well as western Turkey and the Crimea. In these parts it is closely associated with the hot, stony fields and scrubby hillsides of the eastern Mediterranean.

In common with many snakes from this long-inhabited part of Europe, it is quite likely to turn up in overgrown gardens, barns, and even houses. The centuries-old drystone walls that surround small hay fields in this part of the world provide a source of food in the form of small rodents as well as plentiful basking and hiding places. Bales of hay stacked in fields in early summer also attract mice, and therefore leopard snakes, which sometimes shelter under them in the daytime.

Snakes are not popular around some Mediterranean villages, however, and are often killed on sight. It says a lot for the leopard snake's ability to go unnoticed that they are still quite abundant in places, although not easily found. They are active during the day, but they avoid the midday heat. At the height of summer they may hunt after sunset, or they may simply hide away during the hottest part of the year. They spend most of their time on the ground or in burrows, crevices in rocks, or piles of rubble, but they climb to the top of old walls to bask and also use low bushes for the same

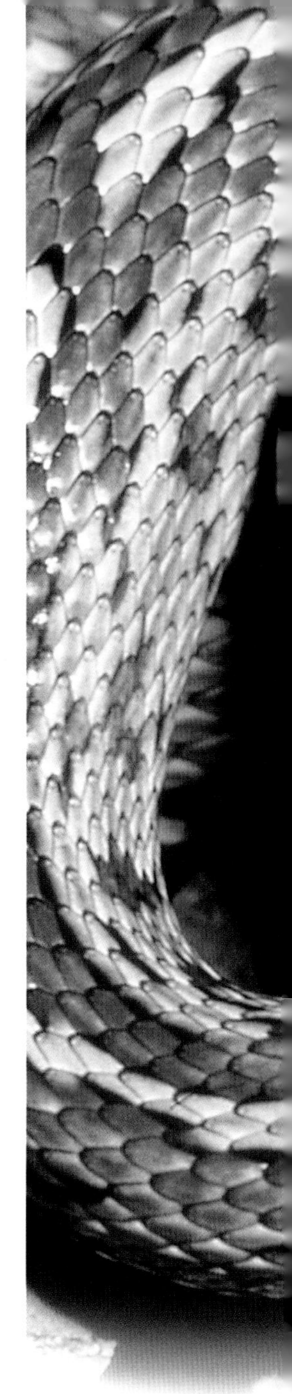

↑ ***A striped form of the leopard snake,*** **Elaphe situla,** ***photographed in the Cyclades, Greece. People in this part of Greece used to introduce the snake into their homesteads purposely to kill rodents—it was considered a good-luck charm.***

purpose. They hibernate in the winter, emerging in April and mating in May or June. Females lay small clutches of eggs usually numbering two to four, but occasionally up to eight. They hatch in about two months. This is the smallest clutch size among members of the genus *Elaphe*, although some of the closely related Asian species, such as the transcaucasian ratsnake, *E. hohenackeri*, may also have small clutches.

Egg Shapes

Most snakes lay oval-shaped eggs. A few lay large round eggs, and others, like the leopard snake, lay elongated eggs. Because the size of the young depends on the size of the eggs, there is a minimum size below which eggs cannot produce viable young. The exact size will depend on the diet of the young snake. Species that eat small frogs and tadpoles when they are young, for instance, can be smaller than those that eat food that is only available in large sizes, such as birds and mammals. Since young leopard snakes eat young rodents, they must be big enough to overpower and swallow them, and their eggs must therefore be relatively large compared to the adults. The only way a small female snake can produce and lay a large egg is by making it elongated. The tradeoff is that elongated eggs take up a lot of space in the oviduct, and so there is only room for a few of them—hence the leopard snake's small clutch size.

Other European Ratsnakes

There are four other European ratsnakes. *Elaphe quatuorlineata*, the four-lined snake, is a large species from southeastern Europe that extends into western Asia. *Elaphe longissima*, the Aesculapian snake, is a slender species from many parts of southern Europe. It is named for the Greek god of healing, Asklepios, and is the emblem of the medical profession in many parts of the world. *Elaphe lineata*, the Italian Aesculapian snake, is similar to *E. longissima* and is its counterpart in southern Italy. *Elaphe scalaris*, the ladder snake, comes from Spain, Portugal, and extreme southern France. Its scientific and common names refer to the ladderlike markings down the backs of juveniles (the Latin *scalarum* means steps, or ladder).

Leopard Tortoise

Geochelone pardalis

A large and attractively marked tortoise with a distinctive pattern and relatively high-domed shell, the leopard tortoise is widely distributed across sub-Saharan Africa.

Common name Leopard tortoise

Scientific name *Geochelone pardalis*

Family Testudinidae

Suborder Cryptodira

Order Testudines

Size Carapace to a maximum of 24 in (60 cm)

Weight Up to 70 lb (32 kg)

Key features Carapace domed and attractive with a variable pattern of dark markings on a light yellowish-horn background; skin is relatively light in color, but some individuals are darker overall than others; male has a longer tail than female, and its plastron is slightly concave; young hatchlings have egg tooth on snout

Habits Wanders through grassland and savanna; diurnal

Breeding Female lays individual clutches ranging from 5–30 eggs; eggs hatch after 8–18 months

Diet Eats virtually all plant matter, including dry grass when other food is in short supply

Habitat Open areas of country

Distribution Eastern and southern parts of Africa from Ethiopia south to the Cape

Status Relatively common but at risk from being hunted for food

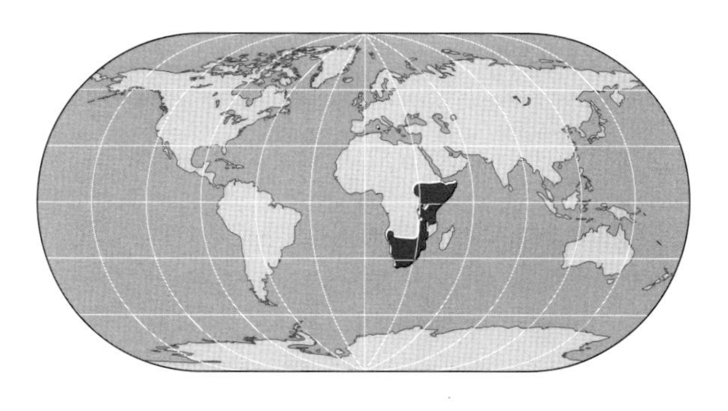

THE MARKINGS OF LEOPARD TORTOISES are surprisingly individual. Their coloration differs in some parts of their range, helping them blend in with the background of the particular locality. Their common name comes from the patterning on the carapace, which is similar to that of a leopard's spots.

Leopard tortoises rank among the most prolific of all species of tortoise in terms of breeding frequency. Females lay up to six clutches in succession with intervals of just a few weeks in between. Courtship is a violent encounter, however: The male simply rams the shell of his intended mate.

Males can be identified by having a longer tail and a slight concavity on the plastron, which helps facilitate mating. It is not uncommon for them to utter a wheezing, grunting sound with their mouth held open when they are on the female's back.

Underground Nursery

Each clutch of eggs is buried in a separate nest up to 12 inches (30 cm) below the surface. It can take a considerable period of time for the young tortoises to hatch, and they may spend as long as a year or more underground according to some reports. As in the case of other chelonians, they cut their way out of the shell. They use the temporary egg tooth that is present on their snout for this purpose and dig up to the surface.

Their carapace measures about 2 inches (5 cm) at this stage, and the hatchlings can weigh up to 1.75 ounces (50 g) each. They are likely to reach about 2.2 pounds (1 kg) by the

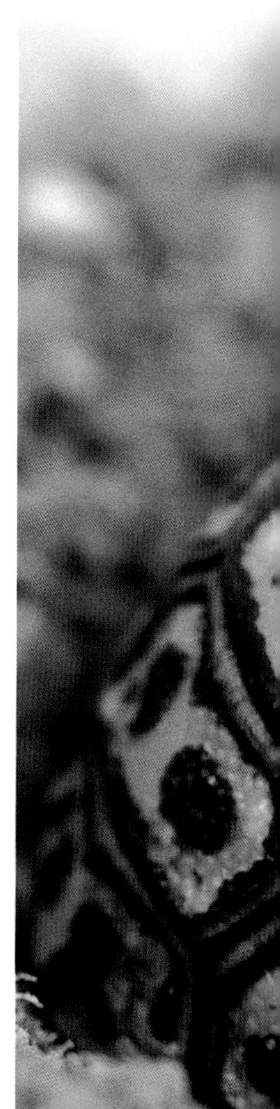

A newly hatched leopard tortoise emerges from its burrow in Transvaal, South Africa. Its highly patterned shell will act as a form of camouflage in open areas.

time they are eight years old, and they subsequently grow more quickly, doubling their weight every two years if food is plentiful. Both sexes are then able to breed by the time they are approximately 15 years of age.

A number of creatures prey on young leopard tortoises, ranging from birds such as large hornbills to various mammals. Grassland fires can represent a hazard too. Often, however, the tortoises retreat inside their shell, and the flames move over them very quickly, leaving them unharmed. They are more likely to die as the result of a fall, which can cause serious cracks to their shells.

Few animals disturb adult leopard tortoises, but they are hunted for their meat in various parts of their range. Their shells may subsequently be used to make musical instruments. As they grow older, the bright coloration tends to disappear, and their shell becomes grayish in color. Individual leopard tortoises are likely to live for over 50 years.

Homing Ability

Aside from laying sequential clutches of eggs at short intervals, these tortoises display another characteristic most commonly associated with marine turtles—they have a strong homing instinct. This was confirmed when a group of leopard tortoises were moved a distance of some 8 miles (13 km) from their regular home range. They subsequently returned within a period of two weeks. They even managed to scale a wire mesh fence that stood 4 feet (1.2 m) high. How these tortoises navigated their way back to familiar territory is unclear, but they may have relied on the earth's magnetic field at first. Then as they got closer to home, familiar landmarks such as trees may have provided guidance, since tortoises have excellent vision. There is an advantage for creatures such as tortoises in staying in a particular area—they will learn where food and, possibly more significantly, a supply of drinking water are likely to be found.

Subspecies

The appearance of the leopard tortoise can vary throughout its range, but only two distinct subspecies have been identified. The most widespread is *G. p. babcocki*, which extends from the Sudan to southeastern and eastern parts of Africa. In contrast, the nominate subspecies, *G. p. pardalis*, is confined to the southwestern corner of Africa, where its population has declined over recent years. Its carapace is not as tall overall, and it has a flat rather than a domed top to its shell.

LIZARDS

Recent figures put the number of lizard species at 4,713—just over 57 percent of all reptile species. Lizard classification at the family level is currently unsatisfactory, with a number of different systems in use. For the purpose of this volume they are divided into 20 families, but some authorities recognize more or fewer. In particular, the Iguanidae is sometimes divided into eight separate families instead of the one listed here.

Almost everywhere in the world (except the Arctic or Antarctic regions) lizards are likely to be the most conspicuous reptiles. The geckos that scurry across the walls and ceilings of restaurants in Southeast Asia, the spiny and side-blotched lizards that scamper from rock to rock in the American Southwest, and the colorful wall lizards that grace the hillsides, walls, and ruins of the Mediterranean region are proof that lizards can succeed in making a living in a variety of places and habitats.

Many lizards are highly visual in their communication with each other and can also be very colorful. In Africa brilliantly colored male agamas bask and display on prominent rocks, bobbing blue or red heads at each other. And in the Caribbean region small, colorful anole lizards live in and around gardens, hotel grounds, and even airports, flashing colorful dewlaps at each other like tiny semaphore flags. All of them are hard to miss.

Lizards are among the most well studied of all reptiles. An observer can sit quietly and watch the whole soap opera that is lizard life unfolding as individuals display, fight, mate, eat, and get eaten. The colonial existence of the side-blotched lizard, *Uta stansburiana*, for example, has been thoroughly explored, and many unsuspected and fascinating facts have come to light. However, other lizards lead more secretive lives and are poorly known. Many tropical species are hardly studied at all and may be known from just a handful of specimens.

Extreme Sizes

The largest lizard is generally accepted to be the Komodo dragon, *Varanus komodoensis* from a group of small islands in Indonesia. Males of the species average about 7.5 feet (2.3 m) in length and weigh about 130 pounds (59 kg). The largest specimen to be reliably measured was 10.3 feet (3.1 m) long and weighed 365 pounds (166 kg). It died in 1933 and is on display at the Tilden Regional Park in Berkeley, California. Female Komodo dragons are significantly smaller than males, averaging about 65 percent of their length and perhaps half their weight. A related lizard, Salvadori's monitor, *Varanus salvadorii*, is actually longer than the Komodo dragon and measures up to 15.6 feet (4.75 m) in length. It is a more slender species, however, and its tail accounts for nearly three-quarters of its length, so it is nowhere near as bulky.

At the other extreme many small lizards measure less than 3 inches (7.5 cm). The smallest species is the Jaragua dwarf gecko, *Sphaerodactylus ariasae*. It lives among leaf litter on an island near the Dominican Republic in the West Indies. This tiny creature is just 1.4 inches (3.6 cm) long, including its tail. It is not surprising that it remained unknown until 2001. Its length is one-hundredth that of the Komodo dragon, and the whole lizard is shorter than

From giant to midget of the lizard world. Above: At 7.5 to 10 feet (2.3–3 m) long the Komodo dragon, Varanus komodoensis *from Indonesia, is the largest lizard. At the other extreme the minuscule chameleon,* Brookesia peyrierasi *(right), is small enough to sit on a person's thumb.*

its name printed on this page. Several other geckos in the genus, all from West Indian islands, are only slightly longer. The tiny ground chameleon, *Brookesia minima,* and related stump-tailed *Brookesia* chameleons from Madagascar have a smaller overall length of 1.38 inches (3.5 cm), but because they have very short tails, their body length is slightly more than that of the dwarf gecko.

Legs and Locomotion

Lizards' shapes vary almost as much as their sizes. Typically they have four limbs, but they can be large and powerful, as in iguanas, or short and almost redundant. In normal locomotion the lizard moves one front leg and the opposite hind leg at the same time. Then the legs alternate, and the lizard goes along in a wriggling motion. In many species the hind legs are much longer than the front ones, and several run only on their hind

Who's Who among the Lizards?

Order Squamata (suborder Sauria)

Family Agamidae: 49 genera, about 430 species of agamas, dragon lizards, or chisel-toothed lizards, including the red-headed rock agama, *Agama agama*, and the thorny devil, *Moloch horridus*

Family Chamaeleonidae: 6 genera, about 131 species of chameleons and dwarf chameleons, including panther chameleon, *Furcifer pardalis*, African dwarf chameleons, *Bradypodion* sp.

Family Iguanidae: 44 genera, about 892 species of iguanas, basilisks, collared lizards, anoles, and lava lizards, including green iguana, *Iguana iguana*, plumed basilisk, *Basiliscus plumifrons*, American horned lizards, *Phrynosoma* sp.

Family Gekkonidae: about 75 genera, 910 species of "typical" geckos, including tokay, *Gekko gecko*, day geckos, *Phelsuma* sp., ashy gecko, *Sphaerodactylus elegans*

Family Diplodactylidae: 14 genera, 115 species of "southern" geckos, including Australian leaf-tailed geckos, *Phyllurus* species, golden-tailed gecko, *Diplodactylus taenicauda*

Family Pygopodidae: 7 genera, 35 species of flap-footed lizards, including Burton's snake lizard, *Lialis burtoni*

Family Eublepharidae: 6 genera, 22 species of eyelid geckos, including leopard gecko, *Eublepharis macularius*, and western banded gecko, *Coleonyx variegatus*

Family Teiidae: 9 genera, 120 species of tegus, whiptails, racerunners, and related lizards, including black tegu, *Tupinambis teguixin*, desert grassland whiptail, *Cnemidophorus uniparens*

Family Gymnophthalmidae: 34 genera, 179 species of spectacled lizards, including white spectacled lizard, *Gymnophthalmus leucomystax*

Family Lacertidae: 26 genera, 279 species of wall, green, jeweled, and related lizards, including viviparous lizard, *Lacerta vivipara*, common wall lizard, *Podarcis muralis*

Family Xantusiidae: 3 genera, 26 species of night lizards, including desert night lizard, *Xantusia vigilis*

Family Scincidae: about 115–124 genera, about 1,400 species of skinks, including monkey-tailed skink, *Corucia zebrata*, five-lined skink, *Eumeces fasciatus*

Family Gerrhosauridae: 6 genera, 32 species of plated lizards, including giant plated lizard, *Gerrhosaurus validus*

Family Cordylidae: 4 genera, 55 species of girdle-tailed lizards, including black-girdled lizard, *Cordylus niger*

Family Dibamidae: 2 genera, 10 species of blind lizards, including *Dibamus ingeri*

Family Xenosauridae: 2 genera, 6 species of knob-scaled lizards and crocodile lizards, including *Shinisaurus crocodilurus*

Family Anguidae: 12–14 genera, 113 species of alligator lizards, glass lizards, and slow worms, including southern alligator lizard, *Elgaria multicarinata*, and the slow worm, *Anguis fragilis*

Family Varanidae: 1 genus, about 57 species of monitor lizards, including Komodo dragon, *Varanus komodoensis*

Family Helodermatidae: 1 genus, 2 species of beaded lizards, including Gila monster, *Heloderma suspectum,* Mexican beaded lizard, *H. horridum*

Family Lanthanotidae: 1 species, Borneo earless monitor, *Lanthanotus borneensis*

Total: 20 families, about 418–429 genera, over 4,700 species

legs when they get up speed. This is known as bipedal locomotion and is most common in agamids, iguanids, and monitors. The South American basilisk, *Basiliscus basiliscus*, has extended the technique to include running across the surface of water and is known locally as the "Jesus Christ lizard." A second species, *B. plumifrons*, is also able to run across the surface. Some lizards use their hind legs to support themselves as they stand up to survey their surroundings or confront their enemies.

Toes can be long and spindly or short and stumpy depending on their function: Long toes are useful in climbing, while short ones are associated with digging. The toes of some species, notably many geckos, have broad pads immediately behind the claws that enable them to climb smooth surfaces. A few species of geckos even have webs between the toes. In the case of the so-called flying geckos, *Ptychozoon kuhli* from Southeast Asia, the webs enable them to glide from tree to tree. In the Namibian web-footed gecko, *Palmatogecko rangei*,

***Geckos' toes have sophisticated modifications that help them climb and stick to smooth surfaces. Seen here is the foot of the tokay gecko,* Gekko gecko.**

"Look, No Legs!"

An evolutionary tendency for limbs to disappear altogether or to be reduced to the point where their function is dubious has happened independently in several unrelated lizard families. The snake lizards, Pygopodidae, have lost their front limbs completely, and their hind limbs are reduced to barely noticeable scaly flaps. All 15 species of the blind lizard family, Dibamidae, are completely legless. Other families, such as the whiptails and racerunners, Teiidae, the girdled lizards, Cordylidae, and the microteiids, Gymnophthalmidae, have some species with well-developed limbs and others with reduced or absent limbs.

Within the Scincidae there is a very strong tendency toward limb reduction. All members of the subfamily Acontinae (with 18 species from southern Africa) lack limbs, as does the European limbless skink, *Ophiomorus punctatissimus*. Some Australian skinks in the genus *Lerista* have a pair of small hind legs and no front legs, and the genus *Chalcides* from the Mediterranean region shows the whole gradation from legged to legless skinks. Similarly, the alligator lizard family, the Anguidae, contains species with normal proportions as well as others with no limbs at all, such as the European slow worm, *Anguis fragilis*, the glass lizards, *Ophisaurus* from Europe and North America, and both species of the legless lizards, *Anniella* from California and Baja California.

Leglessness in a number of families seems to have evolved in response to a burrowing lifestyle and is a striking example of convergent evolution. Not all species are out-and-out burrowers, however—some live among thick vegetation where long limbs would be a hindrance. Species with small legs may use them when crawling slowly. However, when moving quickly or when burrowing, they lay them flat against their bodies.

The eastern glass lizard,* Ophisaurus ventralis *from North America, is legless and could easily be mistaken for a snake. However, unlike snakes, it has movable eyelids and external ear openings.

they act as snowshoes, allowing it to run swiftly over the loose, powdery sand on which it lives. Burrowing species tend to have small, reduced limbs. They move through loose sand by wriggling rapidly from side to side with their legs held against their bodies in a process known as "sand swimming." The lizard that is most adept at this is probably the North African sandfish, *Scincus scincus*.

Many lizards live in trees. Adaptations to an arboreal lifestyle may be slight, as in iguanids and agamids, whose long limbs and digits are equally useful in trees or on the ground. In extreme cases species are so well suited to climbing that they are almost helpless on the ground, for example, chameleons. Their digits are fused into two groups arranged opposite each other that act as pincers to grip branches and twigs. Their tails are prehensile to provide a fifth point of attachment, and their leaflike shape and legendary camouflage protect them against predators. Their elastic tongues enable them to mop up the insects that other lizards cannot reach. There are arboreal lizards in other families, including the geckos, anoles, and monitors, but none that are so wonderfully adapted to life high in the trees as the chameleons.

Scales

Lizards are covered in scales, but in no other group of reptiles is there so much variation in their size, shape, and arrangement. They can be large and overlapping or small, rounded, and close together (granular). Many species have a mixture of types. Sometimes, as in the chameleons, they have odd, studlike scales scattered throughout an otherwise uniform surface of small, granular scales. The scales of many species (for example, the girdle-tailed lizards, *Cordylus* species from Africa, and the alligator lizards, *Gerrhonotus* from North America) are thickened to form a kind of armor plating that protects them from drying out and from predators.

The scales of the Australian stump-tailed skink, *Trachydosaurus rugosus*, are among the largest and most knobby, making it look like a pine cone on legs. Other skinks have glossy, overlapping scales or heavily keeled scales arranged in rows. The Gila monster, *Heloderma suspectum*,

The jagged scales of the Texas spiny lizard,* Sceloporus olivaceus, *resemble the sharp edges of tree bark. They aid in camouflage as well as defense.

and the Mexican beaded lizard, *H. horridum*, have scales consisting of thick, circular beadlike studs arranged in regular rows and surrounded by thick skin. Many species, including the wall lizards, Lacertidae, and the night lizards, Xantusidae, have large, rectangular belly scales arranged in transverse rows.

Lizards of some families have small, platelike bones called osteoderms lying just below the surface. Notable examples are the skinks, Scincidae, the slow worm and alligator lizard family, Anguidae, and the African plated lizards, Gerrhosauridae.

Decorative Features

Many iguanids, chameleons, and agamids have flaps, frills, crests, or other ornamentation made up of (or covered by) scales. They are often larger and more prominent in males, and include the row of long, toothlike scales running down the center of the lizard's back and tail in iguanas and others. There are also the spiky scales found on the heads of some Asian agamas and on the necks and flanks of the American horned lizards, *Phrynosoma*. Other agamas from South Asia have horns on their snouts; but the most highly decorated horned lizards are the chameleons, especially Jackson's chameleon, *Chamaeleo jacksonii* from East Africa, in which the male has three long, tapering horns projecting from its head, resembling the prehistoric *Triceratops*.

The function of these accessories is often connected with display and communication to other members of the same species, but they also have a disruptive purpose, breaking up the lizard's outline when it is at rest and helping it avoid the attention of predators. Flaps and frills in species such as the leaf-tailed geckos, *Uroplatus* from Madagascar, and *Phyllurus* from Australia, have the same function. They act as a "cloak" to smooth out the lizard's shape and eliminate shadows.

The Senses

Many lizards rely heavily on sight to capture prey, avoid predators, and interact with their own species. However, as in many animals, they are better at detecting moving objects than stationary ones. Special types of eyes are found in several groups. Burrowing species often have only rudimentary eyes, sometimes covered with skin, as in the blind lizards, Dibamidae. Geckos, snake lizards, and night lizards do not have movable eyelids. Instead, as in snakes, the eyes are covered by a transparent scale (the brille) that owes its origins to fused upper and lower eyelids. Their eyes are therefore "closed" all the time, and the lizards see out of a window in the eyelid. They keep the brille clear of dirt and dust by licking it constantly with their tongue, and it is shed with the rest of the skin. In a few other species the eyelids are movable, but the lower one still has a transparent window. It probably protects the eye from windblown sand and dust and when the lizard is burrowing.

Nocturnal lizards, especially the geckos, have large eyes, sometimes with brightly colored irises. Like cats, they have vertical pupils that can be closed down to narrow slits in response to bright light, with just a few "pinholes" to see through. Chameleons have remarkable eyes. Each eyeball is surrounded by a turret of skin formed from the fused upper and lower eyelid, with just a small circular opening. By rotating the turret, the chameleon can look in any direction; and since the turrets move independently of each other, it can actually look in

The huge eyes with their brown-red irises and vertical pupils hint at the nocturnal habits of* Palmatogecko rangei, *the web-footed gecko from Namibia.

Like many lizards, *Lacerta vivipara*, the common or viviparous lizard from Europe, feeds mainly on invertebrates. This one is devouring a large cranefly.

two directions at the same time. When hunting, it can use one eye to look over its shoulder for predators while the other fixes on the prey.

Members of most lizard families have openings to their eardrums, but in some species they are covered with scales. Chameleons, however, are stone deaf. They have no ear opening and no middle-ear cavity. It is difficult to assess how important the sense of hearing is to lizards, but in geckos and snake lizards sound is a means of communication. Many of these species are nocturnal, and they use a variety of barks, grunts, and squeaks to keep in touch with each other. In fact, "gecko" is an alliteration of the sound made by some Asian species, as is "tokay," one of the more common species. Male barking geckos, *Ptenopus* species from southern Africa, call to each other from the entrances to their burrows with a loud, characteristic sound that lasts for several seconds and carries for hundreds of yards across the gravelly or sandy plains.

Food and Feeding

Between them the lizards eat a wide variety of food. Some species will eat more or less anything, while others are highly specialized. Examples of specialists include the thorny devil, *Moloch horridus*, which eats only ants; Burton's snake lizard, *Lialis burtoni*, which eats only other lizards; and the marine iguana, *Amblyrhynchus cristatus,* which lives exclusively on seaweed.

The majority of lizards are carnivorous, and the most common prey is insects and other invertebrates. They are abundant and of a size that is easily dealt with by most small- to medium-sized lizards. Larger prey, such as small vertebrates (including other reptiles and mammals), is only available to the larger, more powerful kinds, notably monitor lizards and the beaded lizards.

As a rule, hunting methods are unsophisticated: The lizard sees the prey, chases and catches it, and chews until it can swallow it. Many species rely on ambush tactics, but others search actively for prey. Most lizards probably use an intermediate strategy—they wait in one place, often basking; if they see prey within striking distance, they go after it, returning to their basking spot once the chase is over. Chameleons, of course, are unusual—they use their protrusible tongue to "zap" insect prey from a distance that can be greater than their own body length.

Enemies and Defense

Lizards have many enemies, not least other lizards. Defensive strategies have evolved along several lines. The first line of defense is to avoid being noticed. If noticed, the next best thing is not to be caught or to look as unappetizing as possible. As a result, many lizards are well camouflaged to match the surfaces on which they usually live (known as cryptic coloration): Sand dwellers are pale yellow; leaf dwellers are green; bark dwellers are gray, and so on. Plain camouflage is not common, however, and most cryptically colored lizards also have textures that match the surface. Foremost among them are the leaf-tailed geckos of Madagascar, *Uroplatus* species, and Australia, *Phyllurus* species, which can be impossible to see when they sit still on lichen-covered tree trunks. Chameleons come a close second in the camouflage stakes, but there are many more examples.

Many cryptic species sit very still even when approached, but most lizards run away when they sense they have been spotted. They may retreat into a burrow or into crevices in rocks or trees. Some species, such as the African flat lizards, *Platysaurus* species, can slip into narrow cracks. Others, such as the American chuckwallas, *Sauromalus* species, inflate their bodies to jam themselves in. Basilisks, *Basiliscus* species, are famous for their ability to run across the surface of water, and several other lizards, notably the green iguana, *Iguana iguana*, and the Chinese crocodile lizard, *Shinisaurus crocodilurus*, dive underwater to escape capture. A few lizards, such as the Asian flying geckos, *Ptychozoon kuhli*, and flying lizards, *Draco volans*, even take to the air, leaping from high branches and gliding down to safety.

As a last resort, lizards defend themselves by biting, bashing, or scratching their enemy. Many have long teeth and powerful jaws—a bite from a monitor, agama, or a tokay gecko can leave a surprised human bleeding. Only two species, the Gila monster, *Heloderma suspectum*, and the Mexican beaded lizard, *H. horridum*, are venomous. They have grooved teeth in their lower jaws along which venom flows (unlike venomous snakes that deliver venom through fangs in their upper jaws). Some lizards are good bluffers, squaring up to their aggressor and even jumping at it with an open mouth.

In Australia the central bearded dragon, *Pogona vitticeps*, and the frilled lizard, *Chlamydosaurus kingii*, raise their large beards or frills to intimidate their enemies. American horned lizards, *Phrynosoma* species, may squirt blood from their eyes, and juvenile bushveld lizards, *Heliobolus lugubris,* mimic beetles that squirt a noxious substance into the face of predators. Other examples of mimicry include some Australian snake lizards that resemble venomous snakes and a small gecko, *Teratolepis fasciata* from the Middle East, whose tail is supposedly shaped like the head of a viper.

Discarded Tails

Lizards from a number of families can discard part of their tail when under severe threat. The strategy, known as caudal autotomy, is especially well developed in the geckos, skinks, wall lizards, racerunners, whiptails, and the glass lizards, and is found in several other families. In fact, there are only five families in which caudal autotomy does not happen. The breakage occurs across a fracture plane found on several vertebrae in the lizard's tail. Associated with the fracture plain are bundles of muscles that cause the tail to come away from the rest of the lizard's body and to continue wriggling. While the muscles twitch, the tail holds the predator's attention, and the rest of the lizard can slip away. A replacement tail begins to grow immediately, but the new one will look different from the original. There is no limit to the number of times a regrown tail can autotomize. In some skinks the tail is brightly colored; as a result, predators are encouraged to attack the tail rather than the lizard's head or body.

⊖ **Uroplatus fimbriatus**, ***the leaf-tailed gecko, avoids detection by using camouflage colors that blend in with tree bark, moss, and lichens found in its native Madagascan habitat.***

⊖ ***If its camouflage colors fail to hide it from an enemy, the leaf-tailed gecko,* Uroplatus fimbriatus, *creates a startling warning display by opening its mouth wide to reveal a bright red tongue.***

Common name Loggerhead turtle

Scientific name *Caretta caretta*

Family Cheloniidae

Suborder Cryptodira

Order Testudines

Size Carapace 41 in (104 cm) long

Weight Up to 1,200 lb (544 kg)

Key features Very big head and powerful jaws; carapace heart shaped, lacking ridges in adults (but juvenile's carapace is ridged); carapace brown, often with light brown, reddish-brown, or black markings; plastron yellowish-brown in color; limbs paddlelike and have 2 claws on each

Habits Tends to breed farther from the equator than many turtles; relatively aggressive

Breeding Nesting interval typically 2–3 years but can range from 1–6 years; females come ashore to lay clutches of 100 eggs 4–7 times during the breeding season; eggs hatch after 54–68 days

Diet Mainly shellfish, including mussels, clams, and crabs; may eat some seaweed

Habitat Coastal areas, often in relatively shallow water; occurs in muddy waters as well as clear tropical seas

Distribution Wide range through the Pacific, Indian, and Atlantic Oceans, especially in southeastern United States; occurs as far north as Newfoundland and as far south as Argentina

Status Endangered (IUCN); listed on CITES Appendix I

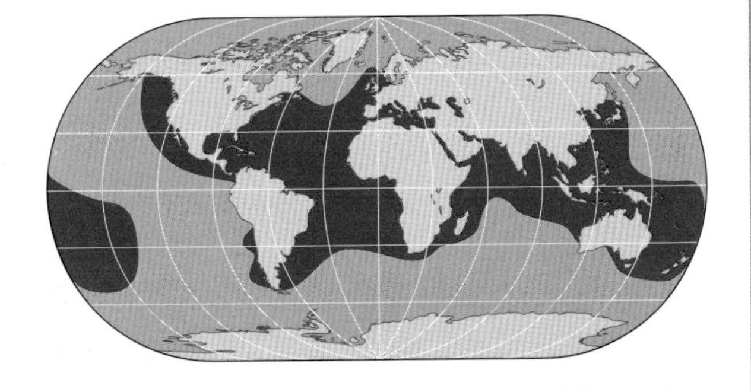

Loggerhead Turtle

Caretta caretta

The loggerhead turtle is one of the largest of the hard-shelled turtles. Despite a ban on international trade, the turtle is still considered to be vulnerable worldwide.

THE LOGGERHEAD IS THE LARGEST MEMBER of its family. Its skull length alone can be almost 12 inches (30 cm) and gives the turtle its common name. Its powerful jaws are used to crush the shells of invertebrates such as horseshoe crabs that feature prominently in its diet. The southeastern United States is one of its strongholds, with the beaches here being used by over one-third of the world's population. It is the commonest species seen by divers off the coast of Florida, where it frequents relatively shallow waters.

Nesting typically occurs between late August and the end of September in Florida, which is the most significant nesting area in the region. Small numbers of female loggerheads also lay their eggs on beaches in the Carolinas and Georgia. In both places they dig holes in which to lay their eggs, and the eggs take on average 54 days to hatch in Florida, extending to between 63 and 68 days in Georgia.

Cutting the Egg

When it is ready to hatch, the young turtle uses the sharp projection on the end of its nose, the egg tooth, to cut its way out. The plastron is curved when it hatches and straightens out later. Having nourished the young turtle during its development in the egg and for a while immediately after hatching, the yolk sac is soon absorbed into the body.

It usually takes about five days from the time that young loggerheads start breaking out of their shells until they appear at the surface of the sand. They often rest during this period, especially during the day when the sand above

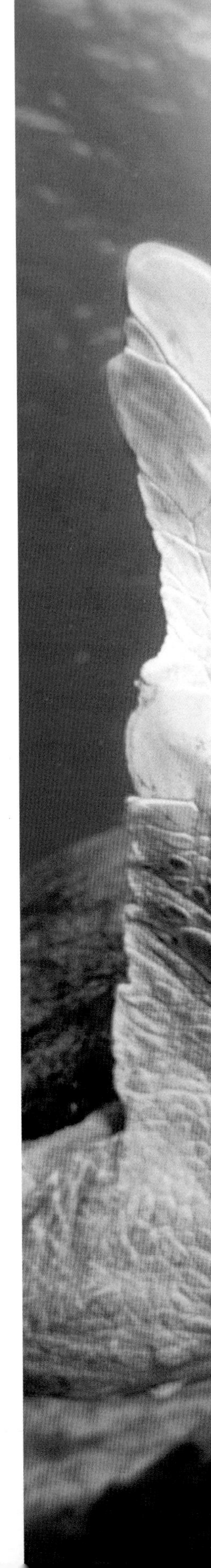

them is hotter. This instinctive reaction ensures that they only emerge under cover of darkness, when it is safer. The young loggerhead hatchlings measure just 2 inches (5 cm) and vary in color from light to dark brown on the carapace with yellowish underparts. The flippers are brown with very distinctive white edges.

Drawn to the Light

As in other sea turtles, young loggerheads orient themselves by light at first. They are instinctively drawn to the sea, where the light above is usually brighter than on land. Unfortunately, in areas where there has been marked beachfront development, the light from the land confuses young turtles and pulls them away from the sea. Weakened and disoriented, they are vulnerable to predators on the beach when dawn breaks.

Hatchling loggerheads that reach the water swim during the day and rest at night as they head to areas of sargassum in the ocean. It is thought that they drift in the fields of sargassum, traveling as far the Azores, which lie some 2,500 miles (4,023 km) from the beaches of Florida where they hatched. They remain there until they attain a shell length of about 20 in (50 cm) and probably return to the area around Florida for the first time 10 years later. Loggerheads are as likely to be found in muddy waters with poor visibility as they are to occur in clear tropical seas.

The turtles range along a vast area of the Atlantic coast and have been seen as far north as Newfoundland and as far south as Argentina. Juveniles in particular are found in large numbers well away from their traditional haunts. As many as 10,000 loggerheads are estimated to spend the summer months off the Chesapeake Bay region of the eastern United States, appearing from May onward. They overwinter in warmer waters farther south.

Young turtles that hatch in the Indian and Pacific Oceans are thought to head for the coast of Baja California. Huge numbers of juvenile loggerheads have been found there, and their movements have been confirmed by satellite tracking.

The Threat of Nets

Because they live mainly in relatively shallow coastal waters where they look for rich shellfish beds, loggerhead turtles are vulnerable to becoming entangled in shrimp nets and drowning as a result. Although they generally do not catch fish, they sometimes scavenge on dead fish used as bait; this habit can also lead to the turtles becoming caught in traps.

The head of the loggerhead turtle is relatively large compared with that of other turtles. In open sea loggerheads spend much of the time floating on the surface and feed on sponges, jellyfish, mussels, clams, oysters, shrimp, as well as a variety of fish.

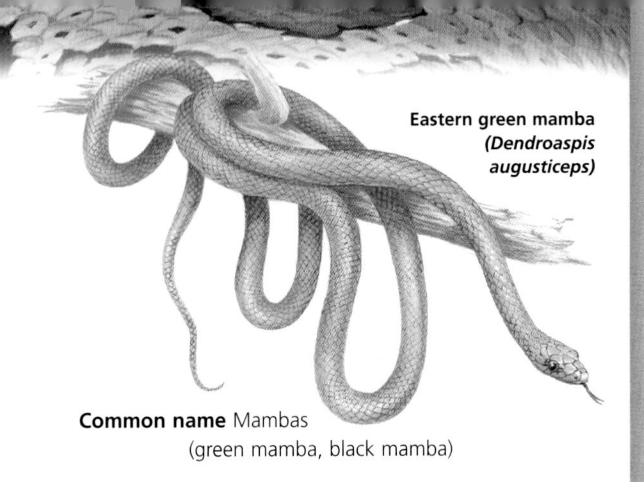

Eastern green mamba ***(Dendroaspis augusticeps)***

Common name Mambas (green mamba, black mamba)

Scientific name *Dendroaspis* sp.

Subfamily Elapinae

Family Elapidae

Suborder Serpentes

Order Squamata

Number of species 4

Length From 5 ft (1.5 m) to 14 ft (4.3 m)

Key features Body long and slender with smooth scales and an elongated head; 3 species are green; the 4th species is gray or brown; mambas are never black, except for interior of the mouth

Habits The 3 green species are tree dwelling; the black mamba lives on the ground among rocks or in low bushes and trees; active by day; very alert and very fast moving

Breeding All are egg layers, laying clutches of 6–17 eggs; eggs hatch after about 80 days

Diet Rodents, bats, birds, and their young

Habitat Forests (3 green species); savanna (black mamba)

Distribution Africa south of the Sahara from the west across to the Indian Ocean coast and south to eastern South Africa

Status Numerous in places, although forest species have suffered through habitat destruction; none are as common as in former times

Similar species Large adult black mambas are unmistakable, but juveniles of all species can be confused with many similar others

Venom Very potent; bites are often fatal, especially those of the black mamba

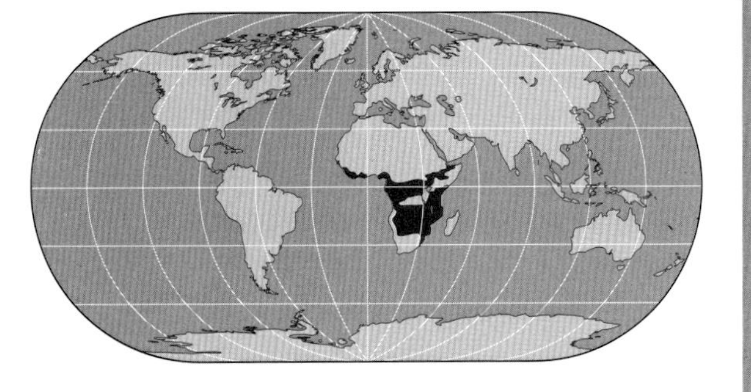

Mambas

Dendroaspis sp.

Mambas are the snakes most likely to strike fear into the hearts of Africans, even though several other species are responsible for far more fatalities.

MAMBAS, ESPECIALLY THE BLACK MAMBA, *Dendroaspis polylepis*, have a reputation for aggression, speed of attack, and fast-acting, potent venom. Although their venom is indeed powerful, and the black mamba will certainly stand its ground if necessary, mambas are mainly secretive snakes that prefer to avoid confrontation.

The three tree-dwelling species—*Dendroaspis angusticeps*, the eastern green mamba, *D. viridis*, the western green mamba, and *D. jamesoni*, Jameson's mamba—spend most of their time in the treetops, where they are rarely seen. The black mamba lives in rock piles, hollow tree trunks, termite mounds, and other similar places, especially in the savannas in East and southern Africa. However, it is usually more interested in reaching safety than in making unprovoked attacks.

Fatal Bites

Bites from mambas are extremely dangerous and can cause death unless immediate medical attention is provided. Until antivenom became available in the 1960s, nearly all black mamba bites were fatal. Today they still are fatal in rural areas where medical help is often too far away.

All mambas have neurotoxic venom that acts quickly, paralyzing the muscles of the victim. Symptoms include tightening of the chest and throat muscles, and gradual paralysis of the facial muscles, resulting in drooping eyelids, for instance. Venom yields can be as high as 100 to 400 milligrams in the largest black mambas; and since the lethal human dose is just 10 to 15 milligrams, each snake has more than enough to kill several people. Bad bites can become life threatening within minutes, and death occurs through respiratory failure if the victim is not treated.

The western green mamba, **Dendroaspis viridis,** *inhabits rainforest canopies in West Africa. Green species are shyer than black mambas. The large head scales are particularly noticeable on this green mamba.*

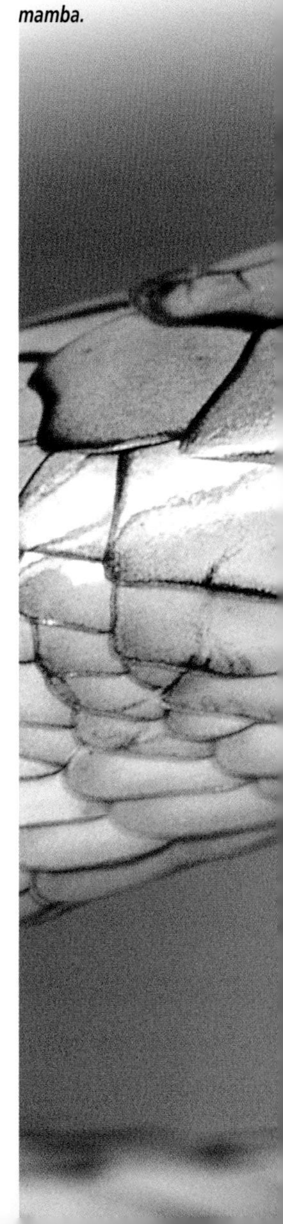

All three species of green mambas are less aggressive, and they produce less venom, so their bites are not necessarily fatal. In fact, green mambas are much less likely to bite, partly because they are tree dwelling and therefore less likely to come into contact with humans, and partly because they are shyer than black mambas.

Mambas are easily confused with harmless snakes. When they are young, the green species are almost indistinguishable from several harmless bush snakes, such as *Philothamnus* species, and adults are similar to the green form of the boomslang, *Dispholidus typus.* Black mambas are unmistakable when adult because they are so long, but juveniles are very similar to *Psammophis* species sand snakes, which are common in the same region.

Confusion over identification of mambas often works to the disadvantage of the lookalike snake species. People sometimes believe that every green or black snake in Africa is a green or black mamba (even though mambas are never black, despite their common name). The confusion certainly adds spice to tourist safaris but, unfortunately, often results in the death of a harmless snake.

The Meaning of Mamba

In Swahili—the most universal African language—"mamba" means crocodile. How the name came to be applied to mambas is a mystery but was presumably through a misunderstanding. The Latin name *Dendroaspis* means "tree asp." In ancient times the word asp meant any venomous snake and did not refer only to the viper of that name. The black mamba, which may be brown, gray, or olive, but never black, is named for the black interior to its mouth. It can be seen when the snake opens its mouth and gapes widely as part of its display of threatening behavior.

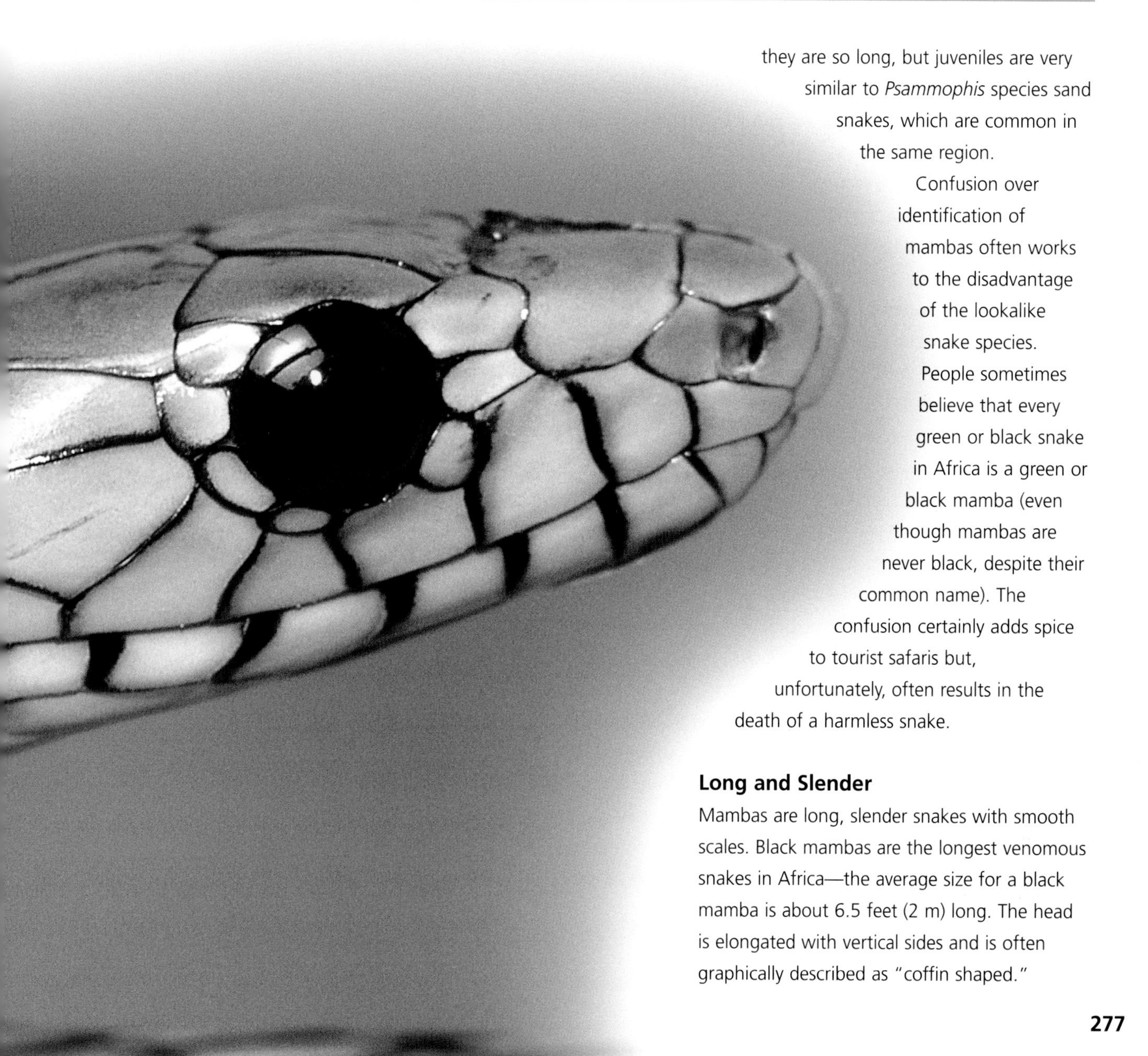

Long and Slender

Mambas are long, slender snakes with smooth scales. Black mambas are the longest venomous snakes in Africa—the average size for a black mamba is about 6.5 feet (2 m) long. The head is elongated with vertical sides and is often graphically described as "coffin shaped."

Their coloration is exactly as would be expected in camouflaged snakes and helps them blend into the background: The three tree-dwelling species are green, while the terrestrial one is brown or gray (never black).

The tree-dwelling species move through branches rapidly, but the black mamba is perhaps the fastest of all snakes. It has been timed at nearly 7 miles per hour (11 km per hour) over short distances. This is a fast jogging speed for humans. It is even more remarkable

The Snake Man and the Mambas

A famous snake hunter was C. J. P. Ionides, an Englishman of Greek descent who settled in Tanzania, East Africa, after retiring from a short career in the British army. He first became a big-game hunter, then a game ranger, and finally a snake collector. In the years immediately following World War II (1939 to 1945) snake parks, zoos, and laboratories were anxious to buy specimens of snakes from around the world for their collections. Ionides built up a successful business meeting the demand.

Although he caught and sold a wide variety of snakes, he is remembered for his skill in finding and catching venomous species, especially black mambas. On one expedition alone he captured 123 black mambas, and throughout his long career he caught several thousand of them, as well as over 3,000 eastern green mambas and many Jameson's mambas. He used a network of local farmers and villagers to locate snakes for him, but he insisted on making the captures himself so that the local people would not have to take unnecessary risks on his behalf.

In the beginning Ionides's equipment was crude, to say the least: He used a forked stick to pin the snake to the ground before grasping it behind the head. Using this technique on large black mambas can be especially dangerous, and Ionides narrowly escaped serious bites on several occasions. He developed better and safer equipment—first a leather noose on the end of a pole and eventually a grab stick similar to the device used by shopkeepers to lift goods off high shelves. On other occasions he would use caged birds as mamba bait. The snake would enter the cage through the bars, but having eaten the occupant, it would be unable to crawl out the way it had come in. "Iodine," as he was known, was bitten several times and had many close shaves. Despite his snake-catching exploits, he died of natural causes in 1968 at the age of 67.

since mambas can move quickly over rocky or scrubby ground, while humans often cannot.

Mambas often travel with their head and neck raised. If they feel threatened, they will rear up further, sometimes raising up to half their body length off the ground and spreading a narrow hood. At the same time, they open their mouth widely and hiss. Considering that black mambas can easily reach 9 feet (2.7 m) in length, an angry mamba can be eyeball to eyeball with an average human. Catching and handling a lively one is an experience that few herpetologists relish.

There is a suggestion that the black mamba is more aggressive during its breeding season, but this is not universally accepted. Males fight with each other, however, raising the front section of their bodies off the ground and trying to force each other back down to the ground. Male eastern green mambas, and probably the other two species as well, also engage in combat.

All mambas are egg layers, but breeding details are only known for the black mamba and the eastern green mamba. Female black mambas lay 12 to 17 eggs, often choosing termite mounds as a suitable location. Green mambas, which are smaller, lay fewer—up to 10 eggs—in hollow trees, logs, or leaf litter. The eggs of both species hatch in about 80 days, and the young measure 16 to 24 inches (40–60 cm). They are venomous as soon as they hatch. Black mambas are especially fast-growing snakes and can reach 6 feet (1.8 m) within a year.

Mambas under Pressure

Tree-dwelling mambas are restricted to canopy rain forests, especially in the coastal regions of Africa. The eastern green mamba occurs in East and South Africa, Jameson's green mamba comes from Central Africa, and the western green mamba is from West Africa. Many of the forests in these regions are disappearing rapidly because of excessive logging. Although this eventually eliminates the mamba populations, individual snakes may linger on for a while by adapting to alternative rural situations such as hedges in farms, plantations, parks, and even gardens. In the long term, however, they become isolated in colonies that eventually die out.

In view of their close proximity to human populations it is surprising that bites are not more common. In fact, the two species that are most affected by deforestation—the western green mamba and Jameson's green mamba—are so rarely seen that little is known of their biology. The eastern green mamba, which has been studied more extensively than the others, can occur in densities of two to three every 2.5 acres (1 ha) or 200 to 300 every half of a square mile (1 sq. km). Up to five of the snakes may live in the same tree.

A black mamba,* Dendroaspis polylepis, *in a threatening pose. It is a very long snake, and such raised poses could bring it to eye level with a human. Its black mouth gives it its common name.

Common name Marine iguana

Scientific name *Amblyrhynchus cristatus*

Subfamily Iguaninae

Family Iguanidae

Suborder Sauria

Order Squamata

Size From 30 in (75 cm) to 4.1 ft (1.3 m)

Key features Heavy bodied with muscular limbs and a powerful tail; a crest of elongated, toothlike scales runs along the center of the back and tail; top of the head covered with horny, conical scales of varying sizes; color usually gray, although some subspecies are more colorful with patches of red or turquoise

Habits Diurnal, basking by day on rocks and entering the sea to feed

Breeding Egg layer; female lays 1–6 eggs in tunnels on shore; eggs hatch after about 95 days

Diet Marine algae (seaweed)

Habitat Rocky seashores

Distribution Galápagos Islands

Status Protected under national legislation but possibly at risk in the long term from human pressures

Similar species The marine iguana is unmistakable; the only other iguanas on the islands are the Galápagos land iguanas, *Conolophus subcristatus*, and the much smaller lava lizards, *Microlophus* sp.

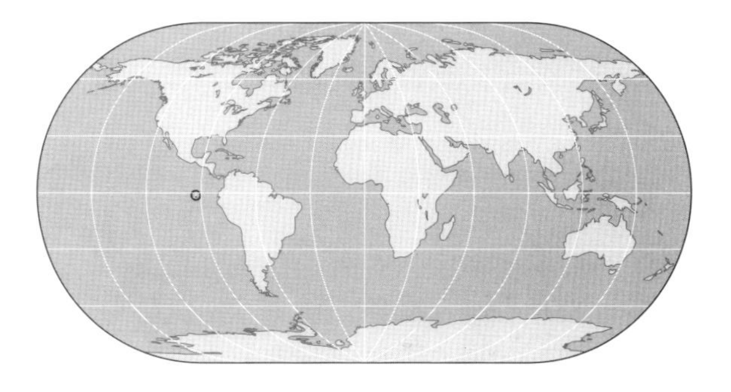

Marine Iguana

Amblyrhynchus cristatus

Marine iguanas are unique. They are the only seagoing lizards and the only lizards that feed on seaweed—a lifestyle that has resulted in many special adaptations.

AFTER VISITING THE GALÁPAGOS ISLANDS, the English naturalist Charles Darwin wrote, "The rocks on the coast are abounded with great black lizards between three and four feet long; it is a hideous-looking creature, of a dirty black colour, stupid and sluggish in its movements." We have no idea what they thought of him!

The Galápagos Islands straddle the equator and are about 600 miles (1,550 km) off the coast of Ecuador. The ocean there is very cold, but the Humboldt Current drives up the west coast of South America and brings with it the waters of the Antarctic, rich in nutrients and fish. Much of the land, on the other hand, is barren, formed by lava that resulted from the volcanic eruptions to which the Galápagos owe their existence.

Marine iguanas arrived on the Galápagos from South or Central America many thousands of generations ago and adapted to an environment in which food was readily available in the sea but not on the land. Their ancestors would have been one of the larger iguanas, possibly something like the spiny-tailed iguanas, *Ctenosaura*, that live on the mainland today. A few individuals (or a single gravid female) were swept out to sea on a raft of vegetation and drifted west until they washed up on one of the islands. Later they spread to all the larger islands as well as many smaller islets in the archipelago. Some populations, particularly on far-flung islands, evolved distinct sizes and coloration, and scientists recognize seven separate races, or subspecies.

Diving and Feeding

The iguanas' most remarkable talent is for diving, which is how they find the seaweed on which they feed. Although a typical dive lasts

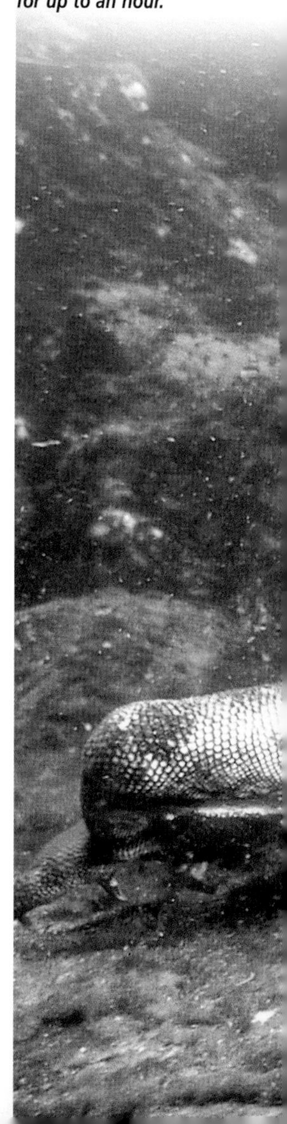

Marine iguanas are just as at home beneath the water as on land. Large adults can graze among rocks in shallows for up to an hour.

only for a few minutes, large iguanas can stay underwater for up to an hour and reach depths of 40 feet (12 m). The amount of time they can spend in the water is limited by temperature. The water is only a few degrees above freezing; being ectotherms, the iguanas lose mobility if their body temperature falls too low. Larger individuals (mostly males) can store heat in the core of their bodies, but females, young males, and juveniles get cold quickly. As a result, there are two different groups: divers and nondivers.

The feeding patterns of large males depend on temperature. They bask on the lava rocks during the morning until their core body temperature has reached the preferred level—about 100 to 104°F (38–40°C). Then they make their way down to the water regardless of the state of the tide and swim several hundred yards out to sea before diving down to the beds of sea lettuce that grow below the low-tide mark. Between dives they can crawl out onto emergent reefs or return to the shore to bask and top up their temperature. They swim by holding their limbs close to their body and swinging their tail from side to side to propel them through the waves and surf.

By late afternoon their body temperature is falling to the point where they can hardly

On the Galápagos Islands marine iguanas bask to raise their body temperature before plunging into the cold sea to feed.

Island Races

There is little or no gene exchange between the iguana populations on different islands because they are not able to swim large distances. Many of the more remote islands were probably only colonized once many thousands of years ago, perhaps by a handful of waifs after storms in the region. Other islands (those that are closer to their neighbors) may have had several "invasions." Scientists have recognized some of the differences and have divided the marine iguana into seven races, or subspecies. The first three subspecies, which appear to be closely related, come from the westernmost islands where the seas are richer. Consequently, they grow slightly larger than the other subspecies.

Amblyrhynchus cristatus cristatus—large, dark gray form from Fernandina Island
A. c. *albemarlensis*—largest form, also gray, from Isabela
A. c. *hassi*—large gray form from Santa Cruz
A. c. *mertensi*—medium form from Santiago and San Cristóbal
A. c. *nanus*—very small, black in color, from the remote northern island of Genovesa
A. c. *sielmanni*—medium-sized form from Pinta
A. c. *venustissimus*—a large, spectacular form from Española with deep red patches on its body

move, so they make their way back to shore and crawl slowly up the beach and rocks until they reach a place where they will spend the night. They prefer the company of hundreds of other large iguanas and form great heaps that help limit heat loss from their bodies.

Females and smaller males have a completely different routine. They do not dive but feed on the algae that is exposed at low tide. Females and subadult males may have to swim across short channels of water to reach it, but juveniles never take to the water. They lose heat too quickly and are vulnerable to predation in open water. An average-sized adult can survive on about 1 ounce (28 g) of algae each day partly because its metabolism is slow but also because red and green seaweeds are very nutritious. Brown algae, however, is poor in nutrients, and the iguanas refuse it unless they are desperate.

Head-butting matches between rival males can last for hours or even days. This pair of males belongs to the subspecies A. c. mertensi *from Santiago Island.*

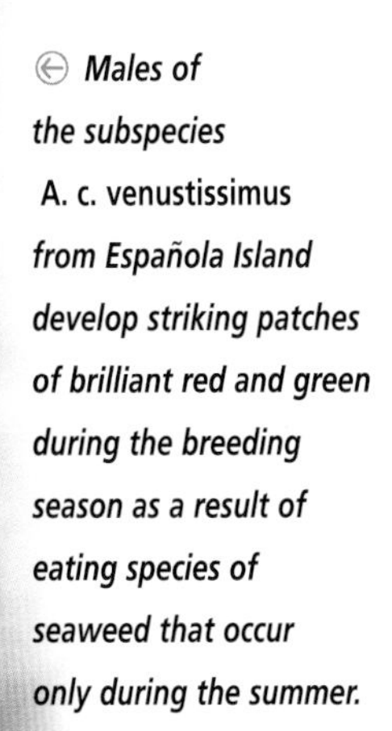

Males of the subspecies A. c. venustissimus *from Española Island develop striking patches of brilliant red and green during the breeding season as a result of eating species of seaweed that occur only during the summer.*

The iguanas' snouts are very short, and their mouths are bordered by tough scales so that they can crop the algae closely. Feeding iguanas are often pounded by large waves and seem as though they must be washed off the rocks and out to sea. However, they cling to the lava with very long claws like grappling hooks and remain in the same position until the surf subsides before continuing to graze.

Because they spend so much time in the sea and because they eat seaweed, marine iguanas accumulate a lot of excess salt in their system. To counteract this, they have salt glands just above their nostrils where the excess salt accumulates. While they bask, they eliminate it by snorting fine sprays of concentrated salt water from their nostrils at regular intervals.

Breeding

Mating takes place from November through to January or February but varies a little from one island to another. The large males defend territories containing many females; at this time the males become more colorful, often reddish or greenish depending on locality. They posture with much head bobbing and strutting. Rivals are chased away. If they persist, a fight breaks out. The head butting and shoving can last for several hours with short breaks between bouts. When two males are well matched, the bouts can continue on successive days with both males becoming bruised and bloody.

During mating the male mounts the female and grasps her neck in his jaws while he forces the rear part of his body under hers so that copulation can take place. Mating can last for several minutes. Small males, however, can "steal" matings lasting only a few seconds; to increase their chances of fertilizing the female, they transfer sperm into their hemipenal sacs before mating. Males probably take about 10 years to reach a dominant position in the colony, and each male controls a territory that can contain several dozen females.

A few weeks after mating, females look for suitable egg-laying sites. They are often several hundred yards back from the shore in sand or shingle. The eggs, numbering one to six, are laid at the end of a short tunnel and then covered over. Good sites are in short supply, and females often stay near their eggs to chase away other gravid females that may dig up the nest while excavating a tunnel of their own. During this time they look emaciated and in poor shape.

The eggs hatch after about 95 days, and the young iguanas dig their way out of the ground and make their way to the shore. They are very vulnerable to predation, especially by Galápagos hawks, snakes, and gulls. Hatchlings and juveniles live a secretive life hidden in crevices in the lava and only emerge for short periods to feed. They never go to sea; instead, they graze on algae in the spray zone. Nor do they bask, but close contact with the dark lava causes heat to transfer to their bodies.

Future Prospects

Marine iguanas have few predators once they are adult. Juveniles are eaten by the few snakes on the islands and by birds of prey. Gulls take a few of the hatchlings. On many islands humans have introduced feral populations of rats, cats, dogs, and pigs, which destroy nests and eat hatchlings. Up to 60 percent of juveniles may die during their first year.

Tourism may have some negative effects because of increased disturbance and pollution. On balance, however, it is probably beneficial, because it brings much-needed income to the islands and in particular to the Galápagos

National Park, which accounts for nearly all the area covered by the archipelago. Programs to eliminate feral goats and other pest species are funded by tourist dollars.

Marine iguanas are not found anywhere else in the world. They must be considered vulnerable to any serious environmental disaster along the lines of the oil tanker that sank off the islands a few years ago. Such events could wipe out whole populations. In addition, the human population of the Galápagos has swollen from 4,000 to 12,000 at the end of the 20th century, mainly through migrants from the mainland who are attracted by work in the tourism and fishing industries. Unless the influx of people is carefully managed, it is bound to put pressure on the environment.

Marine iguanas are protected by national legislation, but they are vulnerable to marine pollution and the effects of human development.

El Niño

The El Niño phenomenon has serious implications for the marine iguanas. During El Niño years the warm waters of the western Pacific push farther east than usual and prevent the cold, rich waters of the Humboldt Current from welling up to the surface. The result is more cloud and heavier rainfall in the Galápagos, killing much of the algae on which the iguanas feed. Up to 90 percent of the lizards may die during those years.

El Niño years used to occur roughly every seven years; but for reasons that are not fully understood, they now occur more often—sometimes every second year—and the frequency prevents populations from recovering. The hundreds of iguanas that used to be seen basking on Punta Espinosa on Fernandina, for instance, may soon be a thing of the past. Individual iguanas in good condition may be able to draw on stored food reserves, and there is evidence that they may also be able to shrink themselves by up to 2 percent to limit the damage.

The effects of El Niño are also significant for the other animals for which the Galápagos Islands are famous. The seabirds, including endemic gulls, boobies, albatrosses, and frigate birds, all rely on fish for their food, and the warmer water drives away fish stocks. Many seabirds fail to raise a family during El Niño years.

Matamata

Chelus fimbriatus

Common name Matamata

Scientific name *Chelus fimbriatus*

Family Chelidae

Suborder Pleurodira

Order Testudines

Size Carapace 18 in (46 cm) long

Weight 33 lb (15 kg)

Key features Head very broad, flat, and triangular when seen from above; conspicuous skin flaps on head for detecting prey; nostrils protrude at the end of narrow tubes on the snout; eyes very small and positioned on top of the head; head coloration chestnut-brown above; throat reddish in hatchlings but variable in color in adults; young matamatas have a chestnut-red carapace, which is darker (almost black) in some adults; red plastron becomes brownish with age; small feet show little webbing; scales separated by areas of rough skin are evident on the limbs; tail short and tapering to a point

Habits Sedentary; hunts by ambush; almost entirely aquatic but not a powerful swimmer

Breeding Female lays clutch of about 20 almost spherical eggs with hard, calcareous shells; eggs can take about 30 weeks to hatch

Diet Small fish; possibly also crustaceans

Habitat Prefers calm waters such as lakes and ponds

Distribution Northern South America; widely distributed in suitable habitat throughout the Amazon basin

Status Not presumed endangered nor subject to heavy hunting pressure

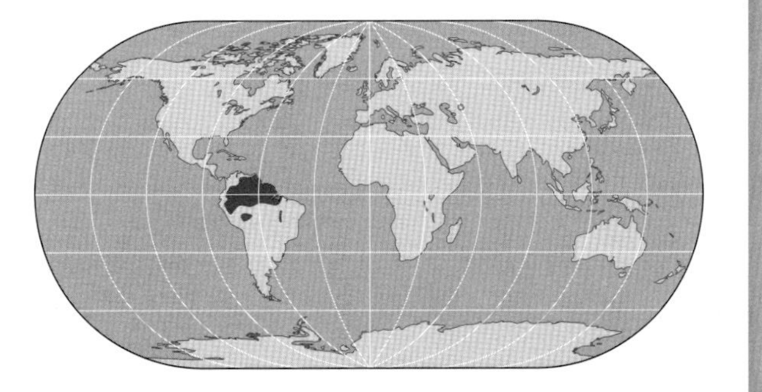

The highly distinctive profile of the matamata makes it one of the most instantly recognizable turtles in the world. Seen from above, young individuals look like fallen leaves.

THE BIZARRE APPEARANCE OF THE MATAMATA turtle does not serve only as camouflage—it is also significant in terms of the turtle's lifestyle. For example, the skin flaps on the matamata's head are important in helping it detect prey, since they are extremely responsive to the movement of fish and to vibrations in the water. In addition, the barbels in the throat area have a sensory function. Both the flaps and barbels themselves may even resemble edible items and therefore attract fish close to the turtle's head, where they can be snapped up.

Sensory Signals

Matamatas are sedentary predators. They wait for their prey to swim within reach rather than actively pursuing it, and therefore accurate sensory information is vital. Even when breathing, they do not surface—they simply extend their neck upward and rely on their snorkel-like nostrils to break the surface of the water in an inconspicuous fashion. The long nostrils can reach up more than 8 inches (20 cm) to the surface. Matamatas live in turbid water where visibility is poor, and so tactile senses are more significant than vision. This explains why their eyes are small. However, the tympanum on each side of the head is relatively large compared with other turtles. This has led to suggestions that sound transmitted in the water may also help the matamatas track the movement of possible prey.

As a target approaches, the matamata moves to extend its neck slowly and cautiously at first. It then allows the floor of its throat to expand very quickly, creating a current that literally sucks prey (such as fish) into its mouth. It then closes its mouth and forces out the

↑ Seen from above, the matamata's head is triangular with a long, slender snout. Its head and neck are covered in small skin flaps that move with the current, making the turtle look as if it were covered with weeds or algae.

water before swallowing the fish. This technique is made possible by the development of the hyoid bones and associated muscles in the throat. The jaws themselves are actually quite weak and cannot be used to restrain prey. They have no tough horny edges, reflecting the fact that matamatas swallow their food whole rather than breaking off pieces.

Regional Variations

Matamatas are common in western areas of South America such as Colombia, but there are parts of the Amazon region, particularly in Surinam, where the turtles are either extremely rare or simply not present. However, the species can be regularly seen both in neighboring Guiana and French Guiana.

Since they occur over such a wide area, a number of supposed regional differences have been recorded in matamata populations. The divisions are based not just on coloration but are related to shell structure, notably in terms of the shape and size of the intergular scute (the

Seagoing Turtles?

Since the distribution of the matamatas extends right down to the mouth of the Amazon, there is the possibility that these turtles may also venture into brackish water occasionally. Although this has yet to be confirmed for this region, there is a record of an old matamata that was found alive on the southwestern coast of Trinidad. It had large, dead barnacles measuring up to 0.6 inches (1.5 cm) growing on its carapace. This shows that it must have spent considerable time in brackish surroundings and may even have been at sea. An established population of matamatas is found on Trinidad, which lies off the northern coast of South America. In addition, it is not uncommon for individuals that have been carried out to sea from the Orinoco River to end up on Trinidad, notably in the vicinity of the Nariva Swamp, augmenting the population there. This may well explain the origins of the beached matamata.

There is also a record of another matamata being caught farther east on a beach in French Guiana at the mouth of the Sinnamary River, suggesting that they may be carried down flooded rivers. Although they are not strong swimmers, these turtles can clearly survive drifting in flood water and may sometimes end up hundreds of miles from their place of origin as a result.

scute that divides the gular scutes at the front of the plastron). However, identifying individuals from particular areas by the variation in scute patterns may not be a reliable method—as more matamatas from the same region have been studied, patterning has not been found to be consistent within a single population.

Color is probably a more reliable indicator of origins, particularly in juveniles. Those from the Orinoco region, for example, have bright pink areas on the throat, often with small spots forming faint lines. In contrast, young matamatas from the Amazon itself display three alternating red and blackish bands across this part of the body. Shell shape is also a guide: Older specimens from this area tend to have slightly concave sides to the carapace, while in matamatas from the Orinoco the carapace is more convex. Males may grow slightly larger than females in all cases and have longer tails.

Lengthy Incubation

Relatively little is known about the breeding habits of these turtles. The nesting period seems to vary across their range. Because of the long incubation period (which has been

Rounding up the Fish

Studies of captive individuals have indicated that matamatas are resourceful and adaptable hunters. A group of three kept at the Bronx Zoo in New York apparently learned to drive fish that were provided for them as food into the shallow end of their pool. They used their legs to keep the fish from slipping back past them into deeper water. Although the technique was initially used by just one of the turtles, its companions soon adopted this approach as well.

A slightly different technique was observed in a group of matamatas at the Beardsley Zoological Gardens in Connecticut: Each matamata angled its body to corral the fish so they were within reach, while keeping the rear end of the shell uppermost. Although similar behavior has not been documented from the wild, it is tempting to think that these sedentary predators will use whatever technique is best suited to capturing their quarry, either encouraging fish into the shallows or up against an impenetrable barrier of reeds, where they can strike effectively at close quarters.

The bright-pink throats of these two matamatas identify them as young individuals from the Orinoco region of South America.

Lying still on the bottom, the matamata waits for prey. Extending its long nostrils to the surface enables it to breathe without coming out of the water completely. Once the prey comes within range, the turtle uses its wide mouth and throat to suck it in.

recorded as nearly 30 weeks) it tends to start early during the dry season. Instead of laying on sandbars, females usually haul themselves up riverbanks to dig their nests. They are thought to produce just a single clutch of eggs annually, and young matamatas probably reach sexual maturity at between five and seven years old.

Matamatas are generally not hunted for food throughout their range, apparently partly as a result of their rather grotesque appearance. Their anatomy is also quite different from other turtles found in this part of the world, and their small, weak limbs provide little meat.

What's in a Name ?

The origins of the unusual name of the matamata are unclear, although they probably reside in Brazil. It has been suggested that the name may be derived from the Tupi Indian word for "staircase," which refers to the steplike patterning of the dorsal scutes. Alternatively, it could be from the Arruan word meaning "skin," with the repetition in the name reinforcing the turtle's "fleshy" appearance. In Venezuela it is known as *la fea*, meaning "the ugly one."

Mexican Beaded Lizard

Heloderma horridum

The Latin name for the Mexican beaded lizard means literally "horrible studded lizard." Its body is studded with beadlike scales, each containing a tiny piece of bone that gives it armor-plated skin.

Common name Mexican beaded lizard

Scientific name *Heloderma horridum*

Family Helodermatidae

Suborder Sauria

Order Squamata

Size Up to 35 in (89 cm)

Key features Head rounded; nose blunt; neck quite long; body heavily built; limbs short with long claws on ends of digits; tail long; scales beadlike; eyelids movable; color variable according to subspecies, usually some shade of brown with yellow or cream markings; adults of one subspecies totally black

Habits Diurnal and nocturnal depending on the weather; also climbs trees; during hot weather spends much of daytime in rocky crevices and self-dug or preexisting burrows

Breeding Female lays 1 clutch of 7–10 eggs that hatch after 6 months

Diet Eggs of reptiles and birds, nestlings, small mammals, occasionally lizards

Habitat Edges of desert, thorn scrub, deciduous woodland

Distribution Western Mexico and Guatemala

Status Vulnerable (IUCN); listed in CITES Appendix II; also protected locally

Similar species *Heloderma suspectum*

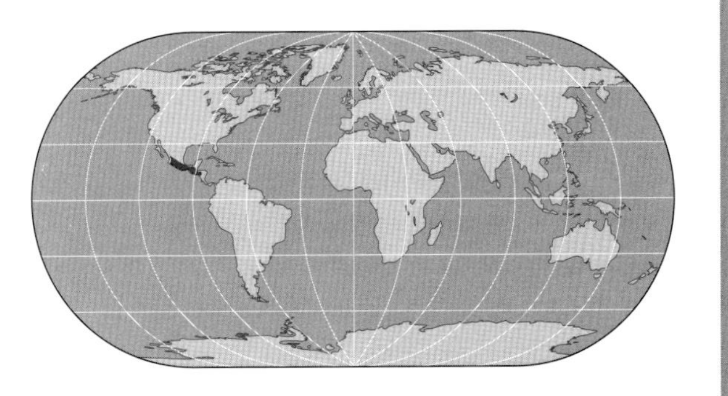

ALTHOUGH THE MEXICAN BEADED LIZARD belongs to the same genus as the Gila monster, *Heloderma suspectum*, there are some differences between the two species. The Mexican beaded lizard is about 13 inches (33 cm) longer and lacks the two elongated cloacal scales. Coloration is also more subdued in the Mexican beaded lizard, and the patch of light-colored scales on the head is absent. Also its tail is longer and more tapering than that of the Gila monster.

The Mexican beaded lizard is more likely than the Gila monster to climb trees to hunt for birds and raid nests for eggs and nestlings. Its tail, which is not prehensile, is used as a counterbalance during hunting. If the tail is lost to a predator, it will not regenerate.

The Mexican beaded lizard is typically found in dry, open forest areas with plenty of rocks and sparse vegetation. It burrows to avoid the midday heat.

Strong Jaws

The main enemies of the Mexican beaded lizard are coyotes, a few raptorial birds, and humans. Its jaws are strong enough to crush prey, most of which is slow moving, so it is thought that its

Names and Legends

It is often thought that the Mexican beaded lizard's scientific name, *Heloderma horridum*, was given partly based on its appearance and partly as a result of its reputation for being venomous. It has been called "one of the most repulsive lizards known to man" and "the terrible one with the studded skin." In Spanish the Mexican beaded lizard is known as El Escorpion. The name comes from a Mexican legend that tells of a beautiful but dangerous creature capable of inflicting its sting on the leg of a human.

venom is used primarily for defense. As with the Gila monster, it gives the impression of being slow moving but is capable of "turning and snapping with the agility of an angry dog" (Ditmars—American naturalist and author).

Humans are responsible for the destruction and fragmentation of considerable tracts of habitat for slash-and-burn agriculture. In the process many Mexican beaded lizards are suffocated in their burrows. Despite protection by CITES and the Mexican government, the creatures still suffer from overcollection for a lucrative, illegal trade. Their venom is no defense against these enemies.

Reproduction

In some parts of its range the Mexican beaded lizard may undergo a short hibernation period; in other parts it remains fairly active during the winter. Mating takes place in early spring. Two months later seven to 10 elongated eggs are laid in a burrow about 5 inches (13 cm) deep. Unlike those of the Gila monster, the embryos are more developed at the time of laying and so do not overwinter. About 6 months later the hatchlings emerge.

Identifying the Subspecies

Taxonomists claim that there are four subspecies of the Mexican beaded lizard, although not all are recognized as valid, since the characteristics used to identify them overlap considerably. In 2000, in an attempt to solve the dilemma, a program of genetic analysis using DNA of both captive and wild Mexican beaded lizards was started. The four subspecies have been identified as:

Heloderma horridum horridum—has a wide range in Mexico from Sonora through Oaxaca; coloration is lightish brown with pale yellow or cream markings; the head is darker brown to black

Heloderma horridum exasperatum—ranging widely through southern Sonora and northern Sinaloa in subhumid tropics as well as arid areas; specimens of this subspecies have more yellow than *H. h. horridum*

Heloderma horridum alvarezi—has a restricted range in northern Chiapas and was named after a Mexican botanist; it is smaller than the other subspecies, and although young specimens have the familiar yellow markings, adults lose them and acquire a totally black coloration

Heloderma horridum charlesbogerti—inhabits a relatively small area in the Rio Montagua drainage system; it has larger yellow markings that end at the armpits

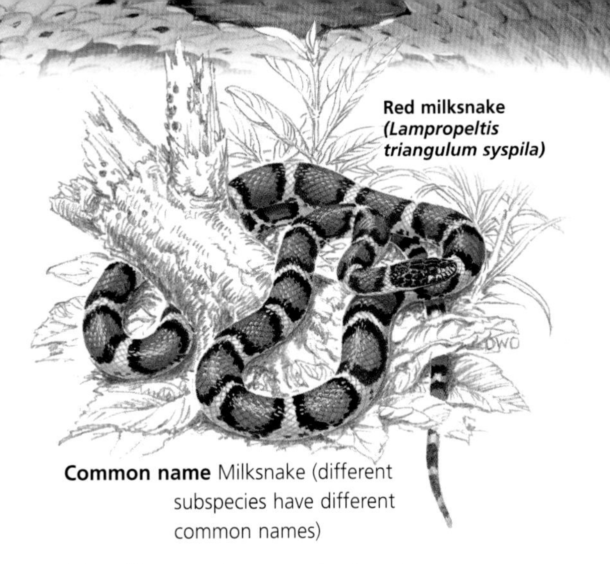

Common name Milksnake (different subspecies have different common names)

Scientific name *Lampropeltis triangulum*

Subfamily Colubrinae

Family Colubridae

Suborder Serpentes

Order Squamata

Length From 20 in (50 cm) to 6.5 ft (2 m)

Key features Cylindrical snake with glossy scales; head quite small with no distinct neck; eyes also small and pupils round; body usually marked with a combination of red, black, and white (or yellow), but some forms are brown and gray, while others become uniform black as adults

Habits Mostly nocturnal, although sometimes active in the late evening and early morning depending on locality; secretive

Breeding Egg layer with clutches of 4–15 eggs; eggs hatch after 40 to 60 days

Diet Small mammals, lizards, and other snakes

Habitat Extremely varied; found in almost every habitat within its wide range except the most arid deserts

Distribution From Canada in the north through Central America and south to Ecuador

Status Common in most places

Similar species At least 3 other members of the genus *Lampropeltis* with red, black, and white bands (triads) around their bodies, but they tend to be restricted to montane habitats

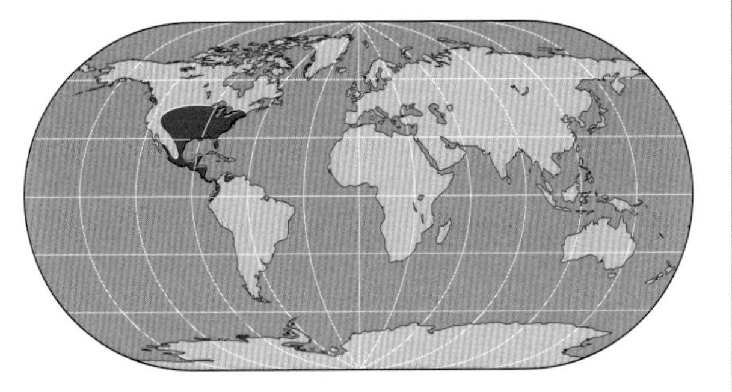

Milksnake

Lampropeltis triangulum

With many different forms or subspecies, the milksnake has one of the largest ranges of any terrestrial snake. It is found from Canada in the north to Ecuador, south of the equator.

In the United States milksnakes occur across all the central and eastern states as far west as Utah and Arizona. Farther south they are found along both coasts of Mexico but are missing from the highland areas in the north-central region. They occur throughout Central America and south into Colombia, Venezuela, and Ecuador. Within this huge north–south range, covering 3,600 miles (5,800 km), the milksnakes occupy a variety of habitats and are absent only from the driest deserts and the highest mountains. Only the European adder, *Vipera berus*, covers a larger area, while the American vine snake, *Oxybelis aeneus*, and the indigo snake, *Drymarchon corais*, are the milksnakes' nearest rivals in the New World. Neither of those species occurs as far north, but their range extends farther south into Argentina.

Adaptability

Milksnakes differ from the other wide-ranging species by the way in which they have adapted to a wide range of conditions and habitats. They live on beaches at the edge of the sea in Maine, in bog forests in the American Northeast, in semidesert regions in south Texas and northeastern Mexico, in tropical rain forests in Central America, and as high as 8,850 feet (2,700 m) in the Andes of Colombia.

They are equally adaptable in their choice of microhabitat. In the American Northeast they are common in woods, meadows, prairies, and around towns and timber buildings. (The name milksnake refers to the belief that they steal milk from cows because they are often found in barns searching for mice.) In the Southeast, however, they are strongly associated with pine

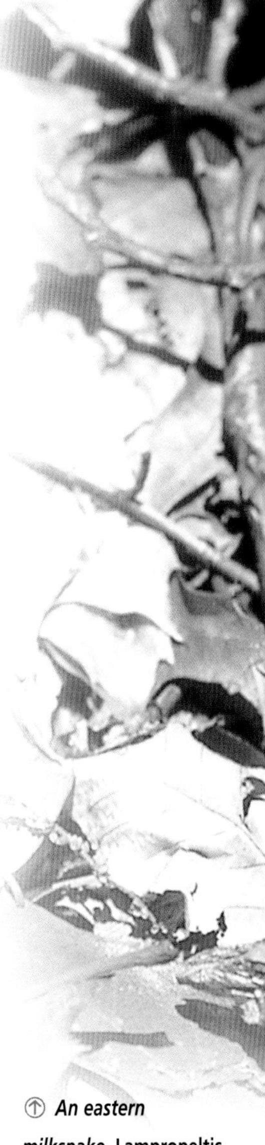

An eastern milksnake,* Lampropeltis triangulum triangulum, *rests on dead leaves with its head slightly raised. This subspecies relies on camouflage rather than warning coloration.

The Subspecies Problem

The concept of different subspecies within a species is not universally accepted among biologists. When it is used, it describes a geographically isolated population within a species' range that is distinct in size, shape, or color. The most obvious form of isolation in reptiles is when populations are confined to islands, mountain tops, or some other place surrounded by a "sea" of unsuitable habitat that prevents them from spreading. (Or, more importantly, it prevents gene flow from one population to another.) It is not the same as polymorphism, in which two or more distinct forms occur within a single population.

The 25 or so subspecies of milksnake are not isolated in this way, however. Their variation is gradual from one population to another, and it is difficult to see where one subspecies ends and another one starts. For this reason some experts argue that the milksnake should not be divided into subspecies at all.

However, if subspecies were not used, some other way of differentiating between the various forms would be needed. For example, the gray and brown eastern milksnake from the Northeast needs to be differentiated from the spectacularly colored Sinaloan milksnake from the Mexican West Coast. And at just 20 inches (50 cm) the diminutive scarlet king snake from Florida is quite different from the relatively gigantic Ecuadorian milksnake that measures over 6 feet (1.8 m).

King Snakes

Apart from the milksnake, there are seven other species in the genus *Lampropeltis*. They are known as king snakes. They all have smooth, glossy scales (the name *Lampropeltis* means "bright shields," referring to its scales). Several are tricolored snakes with alternating rings of red, black, and white.

Lampropeltis alterna, the gray-banded king snake, occurs in a multitude of patterns and lives in southern Texas, extending down into Central Mexico. Then there is *L. calligaster,* a species that consists of two different forms, the prairie king snake and the mole snake. They are gray or brown with darker blotches.

The third species is *L. getula,* the common king snake. It lives mainly in the United States and is a black-and-white or brown-and-cream species with several geographic races. *L. mexicana,* the Mexican king snake, is the fourth species. It is also a snake of many forms, some of which have red blotches on a gray, cream, or buff background. Others have rings of red or dark gray around the body.

Next, there is *L. pyromelana,* the Sonoran mountain king snake. It is a beautiful tricolored king snake with bands of red, white, and black and a white snout. It lives in the rocky outcrops of mountain ranges in Arizona, Nevada, Utah, and New Mexico, and in northern Mexico. The sixth species is *L. ruthveni,* the Querétaro king snake. It is another tricolor found in the Mexican states of Michoacán, Querétaro, and Jalisco. At one time it was classified as a milksnake.

Finally comes *L. zonata*, the California mountain king snake. It is similar to the Sonoran mountain king snake, but its snout is black. Isolated populations occur in California, Oregon, and Baja California, Mexico.

Sonoran mountain king snakes,* Lampropeltis pyromelana, *are mainly terrestrial but can be seen climbing trees and hunting in low bushes. They usually live in mountains or rock piles.

forests. They favor rotting stumps in which they hibernate and lay their eggs. Other populations hide under flat pieces of rock, and many have been collected so infrequently that it is not possible to build up a picture of their lifestyle.

Seasonal Habits

Milksnakes from tropical regions of Central and South America are probably active throughout the year. Even in the southern states they may hibernate halfheartedly, emerging for short spells whenever the weather is warm. Populations from the northern parts of the species' range, however, hibernate from October to April every year.

The eastern milksnake, *L. t. triangulum*, is thought to hibernate communally because several individuals—as many as 28—have been found in close proximity to each other in the fall. Similarly, gravid females (ones that are full of eggs) of this subspecies may congregate in certain places prior to laying their eggs.

Studies have also shown that eastern milksnakes may move seasonally to drier upland sites in the fall when looking for somewhere to hibernate and to moister lowland sites in the spring and summer during feeding and mating. Such migratory behavior has already been shown for a few other species, such as the northern adder, *Vipera berus*, in the United Kingdom and the carpet python, *Morelia spilota*, in Australia. There is every reason to suspect that this milksnake, which lives in similar situations, has similar adaptive behavior.

Living in a temperate climate, the eastern milksnake and another northeastern form often lay their eggs in piles of rotting sawdust, manure heaps, and decomposing vegetation. The heat produced by rotting material speeds up the development of the eggs, just as in the case of the European grass snake, *Natrix natrix*, which also lives in a cool climate.

Smaller in the North

The appearance of the milksnake also changes over the two continents on which it occurs. Some changes are minor and of interest only to taxonomists, but others are more dramatic. They concern the species' size, color, and pattern. Its total length can vary from about 20 inches (50 cm) in the southeastern states to over 6.5 feet (2 m) in South America. The difference in weight between the two forms is, of course, even greater. In many snakes that occur over a wide range of latitude (and in snakes in general), the largest individuals are found in the warmer tropical regions.

That is broadly true of the milksnakes: The largest forms live nearest the equator (*L. t. micropholis* in Ecuador and Venezuela), and they become smaller farther north toward Mexico and the United States.

↑ ***Pueblan milksnakes,* Lampropeltis triangulum campbelli *from Mexico, usually lay six to eight eggs in a clutch. The eggs take about 60 days to hatch.***

Carbon Copies?

Why does the milksnake vary so much from one part of its range to another? There are a number of different explanations as to why it differs in aspects such as size and color. One that is particularly attractive revolves around its mimicry of the venomous coral snakes of the region.

In order for mimicry to be effective, the mimicking species must be a good copy of the model. The coral snakes, however, tend to have small ranges. That means that in one place the milksnake shares its range with one species, while a few hundred miles away it may share it with a completely different coral snake. But coral snakes are not all the same; although they are mostly red, black, and white, their bands are often spaced differently. It is possible that natural selection has enabled each population of milksnake to copy the pattern of the particular coral snake with which it shares its range.

Ⓣ *A Mexican milksnake,* **L. t. annulata,** *hatches out. Except for the eastern milksnake, young of all subspecies start life as red snakes with black-and-white bands.*

Ⓔ *In Pennsylvania a juvenile milksnake basks on a leaf. The eastern milksnake,* **L. t. triangulum,** *is the most northerly subspecies.*

There is a blip in this trend, however. In the United States itself the largest form is found in the Northeast (the eastern milksnake), but the smallest is in Florida. (This is *L. t. elapsoides*, the scarlet king snake. Despite its common name, it is a form of milksnake.) The discrepancy can probably be explained by the burrowing habits of the scarlet king snake. It lives most of its life coiled inside decaying tree stumps, which would be difficult to do if it were a large snake.

Red, Black, and White Bands

Color variations can take many forms. The eastern milksnake is unusual, being gray with chestnut-colored saddles, or blotches, down its back. All the other milksnakes start life as red snakes with black-and-white bands, or triads, in the sequence red-black-white-black-red. However, many forms change color as they grow, and most get darker. The darkening occurs as black pigment infuses the tips of the white-and-red scales and gradually spreads. In most forms the red-and-white bands become darker but are still clearly visible.

In one form, the black milksnake from the Andes of Colombia, the black continues to spread until by the time the snake is fully grown, it is totally black save for a small dusky-white area on its chin. A common theory for this is that snakes in cooler climates (including higher altitudes) are darker because black pigment absorbs heat more efficiently. So why aren't they black from the start? Probably because tricolored milksnakes mimic venomous snakes from the area. While they are small, that is more important than their ability to absorb heat. When they grow up, they are larger than the venomous snakes they were mimicking, so changing color is no disadvantage.

Another color variation is found in forms such as the Mexican milksnake, *L. t. annulata*, the Pueblan milksnake, *L. t. campbelli*, and the Honduran milksnake, *L. t. hondurensis*, in which the pale bands are cream or even orange instead of white. This could simply be a case of chance mutation; and because it has neither advantages or disadvantages, the variation remains in the population by default.

How Many Bands?

Pattern variation concerns the number of bands, or triads, around the snake's body. The number can vary from 10 to 54, the largest score coming from the eastern milksnake and the smallest from a Mexican subspecies, the Sinaloan milksnake, *L. t. sinaloae*. The latter has wide red interspaces between its black-and-white bands, but other forms are more regularly banded. There is great variation within all the forms, and even snakes from the same litter can have different numbers of bands.

However, all these variations—in size, color, patterns, and numbers of bands—coupled with other factors such as scale counts, have been used to describe 25 different subspecies of milksnake in total. Although there is clearly a need to differentiate between the separate populations of such a highly variable species, some experts doubt the usefulness of describing such a large number of forms, and many even question the subspecies concept altogether.

Monitor Lizards

Monitor lizards in the families Varanidae and Lanthanotidae are quite closely related to the beaded lizards (Helodermatidae) and the alligator and glass lizards (Anguidae). They are also more closely related to snakes than any other lizard. They are all Old World lizards with a wide distribution in Africa, southern and Southeast Asia, Australia, and islands in the Pacific and Indian Oceans. Most of the species are concentrated in Indonesia and Australia, and in the latter they are referred to as goannas. The Komodo dragon, *Varanus komodoensis*, is the largest living monitor lizard, measuring 10.5 feet (3.2 m) and weighing over 330 pounds (150 kg). The short-tailed monitor, *V. brevicauda,* is the smallest with a total length of 10 inches (25 cm) and weighing just 0.3 ounces (10 g).

Within a relatively few years the number of species of monitor lizards has risen from 40 to 58. This is due to the discovery of new species and the elevation of subspecies to full species. On a number of Asian islands the water monitor, *V. salvator,* and the mangrove monitor, *V. indicus*, have developed a number of forms, all of which seem to be isolated from each other by either a physical or an ecological barrier, creating more confusion.

Varanids arose over 65 million years ago in Asia and spread to Africa and Australia. On the continent of Australia they divided into two forms: One evolved dwarfism, while the other remained large. The pygmy monitors (the dwarf form) make up the subgenus *Odatria*, while the largest belong to the genus *Varanus.*

Common name Monitor lizards **Order** Squamata

Family Varanidae 1 genus and 57 species

Genus *Varanus*—57 species in total: 27 from Australia, including the ridge-tailed monitor, *V. acanthurus*, the sand monitor, *V. gouldii*; 4 from Africa, including the Nile monitor, *V. niloticus*; 1 from Arabia; 25 from Asia, Southeast Asia, and Indonesia, including the Bengal or Indian monitor, *V. bengalensis*, and the Komodo dragon, *V. komodoensis*

Family Lanthanotidae 1 genus and 1 species—the Borneo earless monitor, *Lanthanotus borneensis*

Habitats

Monitors have adapted to a variety of habitats ranging from deserts to rain forest, mangrove swamps, and savanna. Their habitat can be divided into four types: aquatic, rocky outcrop, arboreal, and terrestrial.

Some species, such as the sand monitor, *V. gouldii* from Australia, have large ranges and tolerate a variety of habitats. Others exhibit special adaptations to a particular habitat. Merten's water monitor, *V. mertensi* from northern Australia, has a laterally compressed tail to help propel it through water and dorsally positioned nostrils with valves that enable it to lie in the water without choking. A number of species occurring in rocky outcrop habitats have adaptations to their tail scales. King's rock monitor, *V. kingorum* (also from northern Australia), has scales on its lower surface with small, spiny tips that support the lizard's body when at rest in a vertical position. The green tree monitor, *V. prasinus* from Australia and New Guinea, has adaptations for life in the rain-forest canopy. It has a prehensile tail, long claws, and a slender body that is mainly green.

The white-throated monitor, *V. albigularis* from sub-Saharan Africa, is a terrestrial species with powerful limbs for digging. The short-tailed monitor, *V. brevicauda*, spends most of its time in spinifex grass

Monitor lizards are voracious predators, able to tackle even dangerous prey. This savanna monitor,* Varanus exanthematicus *in Kenya, is attacking a spitting cobra,* Naja *species.

The Rise and Fall of the Varanids

True monitorlike lizards belonging to the genus *Paleosaniwa* lived more than 65 million years ago and are known from fossils found in Wyoming. About 15 million years later monitors of the genus *Saniwa* developed in the southern United States and northern Europe. While the European species thrived and spread, those in the United States died out. About 5 million years ago the first *Varanus* monitor appeared in India, at the same time as some species in Europe disappeared. Fossil remains in Africa of the semiarboreal *Varanus rusingensis* measuring 6.5 feet (2 m) long show considerable similarities to present-day African monitors such as the Nile monitor, *V. niloticus*.

Monitors reached the continent of Australia comparatively recently (2 million years ago). With a size of 23 feet (7 m) and weighing 1,320 pounds (600 kg) *Megalania*, which inhabited Queensland, New South Wales, and southern Australia, was one of the largest land-dwelling lizards that ever lived. It died out some 25,000 years ago after Australia was occupied by humans. The only remaining genus today in the family Varanidae is *Varanus*.

tussocks using its muscular tail for support. However, it is worth noting that these habitat specializations are not strictly adhered to. Water monitors, *V. salvator*, and Nile monitors, *V. niloticus*, climb trees, dig burrows, and swim.

Most monitors prefer a body temperature of 95 to 104°F (35–40°C). Species such as the white-throated monitor, *V. albigularis*, and the desert monitor, *V. griseus*, which both live in more temperate areas of Africa, are dormant during the winter.

Body Form

Although there is considerable variation in size, the basic body plan of monitors is consistent. Some species have shorter heads and snouts or enlarged scales on the neck, but there is not the dramatic variation in body form and scalation seen in other lizard families.

Monitors have a long neck and body with well-developed limbs, each of which has five toes and long, sharp claws that are curved and bend downward. The tail is long and muscular and can be laterally compressed or rounded and used to store fat. It has no fracture planes, so monitors cannot regenerate a missing part. The head is long with a pointed snout, fairly large ear openings, and eyes with large, round pupils. The teeth are sharp, and the tips point slightly backward. In some species

they are serrated. Body scales are usually small and dull. Species from hotter climates are lighter in color, and juveniles tend to be more colorful than adults.

Behavior, Senses, and Feeding

Monitors occupy large home ranges. When hunting or foraging, they have a swaying gait. They move their head from side to side and tongue-flick constantly. Like snakes, they use their long, protrusible, forked tongue to pick up chemicals in the air to detect food as well as mates.

Monitors are carnivorous predators. They tear flesh with their teeth and claws, and swallow large pieces. The ability to eat large lumps is due to three factors. First, the cranium is completely ossified (protecting the brain from the pressure of a large food item against the roof of the mouth). Second, the large head enables them to make a wide gape; and third, the hyoid bones in the throat are mobile, allowing the neck to be distended considerably. As monitors grow, they eat larger prey. Depending on species and habitat, large monitors eat small mammals, lizards, eggs, birds, fish, crabs, and snakes. Pygmy monitors feed mainly on insects and small lizards.

Most monitors have a clumsy walk. However, they can run fast over distance when chasing prey, since they have large lungs and muscles that store oxygen efficiently. They are opportunistic feeders, and some of them are the top predators, taking the place of large carnivorous mammals. In Australia there are no large predatory cats, and the Perentie monitor, *V. giganteus*, is at the top of the food chain. As its name suggests, it is a "giant," measuring 7 feet (2.1 m) long.

Defense

Habitat destruction by humans is the greatest threat to monitor lizards. They have no defense against it except to try to adapt to different conditions. In this

⊖ The crocodile tree monitor, Varanus salvadorii from New Guinea, uses its narrow snout to look for eggs and fledglings in birds' nests. This is a captive specimen.

Exotic Diets

The pygmy stripe-tailed monitor, *Varanus caudolineatus* from Australia, often tries to capture geckos, although they are too large for it to eat. However, partly as a result of panic and partly to distract the monitor, the geckos "throw away" their tails, leaving behind a tasty meal for the monitor.

White-throated monitors, *V. albigularis* from Africa, are proficient snake hunters and are thought to be immune to the venom in many African snakes. Gray's monitor, *V. olivaceus* from the Philippines, is unusual in that it is a fruit-eating species, although a considerable part of its diet is made up of snails.

The snout of the 10.6-foot (3.2-m) long crocodile tree monitor, *V. salvadorii* from Southeast Asia, is narrow and high, enabling it to forage under bark for vertebrates and to pick eggs and fledglings from nests. Despite its size, it is unable to eat larger prey items.

Dumeril's monitor, *V. dumerilii* from Southeast Asia, eats crustaceans and other hard-shelled items. As a result, its teeth tend to become blunt as the lizard ages.

↑ ***The white-throated monitor,* Varanus albigularis *from Namibia, opens its mouth to pant in an effort to cool down its body temperature.***

respect species that are both arboreal and terrestrial and whose range includes a diversity of habitats are more successful. Young monitors are preyed on by snakes, larger monitors, and raptors.

Monitors have a range of defensive measures. Even the tiny short-tailed monitor, *V. brevicauda*, shows the defensive behavior characteristic of monitors—hissing and swaying with its throat inflated and compressing its body to make itself appear larger.

The tail plays a large part in defense and is often used as a club. Other tactics to deter predators include the violent evacuation of cloacal contents and projectile vomiting, both in the direction of the perceived threat.

However, monitors have no defense against the spread of the cane toad, *Bufo marinus*, which is a particular threat in northern Australia. With the exception of the crocodile tree monitor, *V. salvadorii* from Southeast Asia, all monitors die after eating these toxic toads, resulting in the decline in numbers of some monitor populations.

Reproduction

Male monitors are highly territorial and do not hesitate to fight rivals. Standing on their hind legs, they grab each other with their forelegs, trying to push their rival to the ground. External sexual differences are difficult to detect. Sexual maturity is usually determined by size rather than age, and the female's readiness to breed is directly related to her body length.

All monitors are oviparous, laying eggs in burrows, tree hollows, and even termite mounds. Females of several species dig and open up termite mounds before laying their eggs in them. The insects repair and reseal the gap. Depending on species, about seven to 51 eggs are laid. Eggs of the dwarf monitors hatch after about 65 to 95 days, while the eggs of larger species can take up to 280 or 300 days.

Human Exploitation

For centuries indigenous peoples have hunted monitors for food, medicine, and their skins. In some parts of New Guinea skins of the mangrove monitor, *V. indicus*, are used to make drumhead covers for exportation. In lowland areas of New Guinea other groups prefer to use the skin of the crocodile tree monitor, *V. salvadorii*. The lizards are skinned alive, since they believe that the skin from a live specimen provides a better drum pitch when played. They also believe that the crocodile tree monitor is an evil spirit that is said to "climb trees, walk upright, kill men, and breathe fire."

Australian Aboriginals use the oil from the fat around the kidneys of the lace monitor, *V. varius*, as a remedy for several illnesses. The Aboriginals call it goanna oil. Nile monitors, *V. niloticus*, and Asiatic water monitors, *V. salvator*, are killed for their skins, which are exported for use in making bags, purses, belts, and shoes. Monitors are also collected for the pet trade.

Common name Nile crocodile

Scientific name *Crocodylus niloticus*

Subfamily Crocodylinae

Family Crocodylidae

Order Crocodylia

Size Largest official record is 19.5 ft (6 m) from snout to tail; large specimens today are usually no longer than 16 ft (4.9 m)

Weight Up to 2,300 lb (1,043 kg)

Key features Appearance variable, leading to the identification of numerous subspecies; body usually dark, sometimes blackish, with lighter underparts; young are olive-brown with blackish markings across the body; mouthparts broad and powerful

Habits Large individuals are aggressive and dangerous, seizing prey at the water's edge; uses speed and stealth for hunting; diurnal; sometimes bask on shore

Breeding Female usually lays 16–80 eggs in a clutch depending on age and subspecies; nest guarding known in both sexes; eggs hatch after about 2 months

Diet Young hatchlings feed largely on aquatic creatures, including invertebrates and amphibians; adults take much larger prey, including giraffes and even humans

Habitat Usually restricted to freshwater habitats; may be found on beaches and occasionally at sea

Distribution Africa, occurring over a very wide area south and east of the Sahara; also found on the island of Zanzibar

Status Some decline in local populations but relatively common overall; listed on CITES Appendices I and II

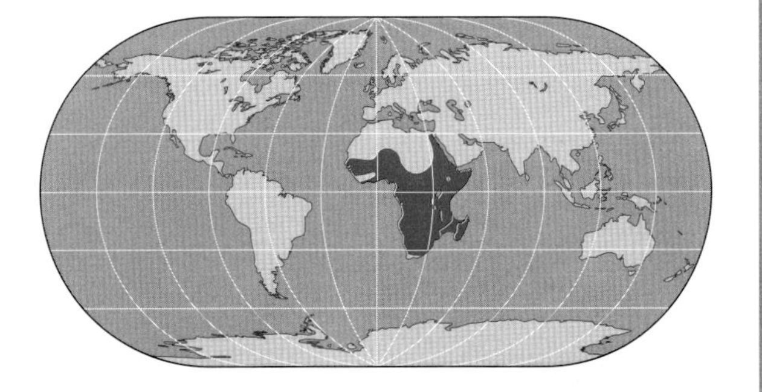

Nile Crocodile

Crocodylus niloticus

Nile crocodiles occur all over Africa, eating almost anything but rarely moving away from their favorite body of water. They are the largest African crocodiles, and their reputation as man-eaters is justified.

NILE CROCODILES ARE THOUGHT to kill about 300 people annually in Africa. The species has proved to be highly adaptable, and its population has withstood heavy hunting for over a century. The leather trade accounted for the deaths of over three million Nile crocodiles in just 30 years until 1980.

Crocodile Ranches

A ranching program at Lake Kariba, Zimbabwe, benefits from the high reproductive potential of these reptiles. Ranching entails taking a small percentage of the eggs laid by wild crocodiles under permit, hatching them artificially, and rearing the offspring in captivity. A crucial aspect of the program (which has become a model for others) is that some of the young crocodiles must be released back to the wild. When ranching began, Nile crocodiles were on the verge of extinction in Zimbabwe, but the current wild population has grown to over 50,000 individuals. The only significant drawback is that there have been an increased number of crocodile attacks on humans again.

The actual farming of crocodiles (in which breeding stock is retained on site and the eggs hatched there) began in Zimbabwe once the success of ranching became apparent. A unique, tamperproof tagging system has been developed to ensure that crocodile skins entering international trade can be monitored by the authorities, eliminating the risk of any illegal killing.

In the wild young hatchling crocodiles are shy. They remain in the shallows, feeding mainly on aquatic prey including amphibians, but they have a well-developed social structure. By two years old, when they are approaching 3.3 feet

Nile crocodiles may bask on shore before sliding into the water to protect themselves from the midday sun, opening their mouth in a wide yawn to allow heat to escape.

(1 m) in length, they start associating with each other in deeper water. After this stage they seek to join existing adult groups, but they are not always welcomed by the dominant bull (male). If challenged by a larger group member, a young Nile crocodile will raise its head out of the water in a submissive gesture. It will then retreat below the water again without any fighting taking place.

In spite of their reputation as Africa's most lethal aquatic predators, Nile crocodiles can occasionally come off worse in attacks, especially when elephants are involved. In one case the elephant responded by pulling the crocodile out of the water while it was still gripping onto the elephant's leg. Another member of the herd then trampled the unfortunate reptile to death, whereupon the elephant that had been attacked picked up the crocodile's body with its trunk and hurled it up into a tree. Similar behavior has been observed on more than one occasion.

Grooming Associates

One of the more remarkable aspects of the behavior of the Nile crocodiles is the way in which they allow certain birds (notably spur-winged plovers, *Vanellus spinosus*) to feed on scraps of food that attach to their teeth and also to remove parasites such as leeches from their bodies with impunity. This type of grooming by a species that would normally be considered to be prey is not restricted to crocodiles, however. It has also been documented in various communities of reef fish.

Evidence suggests that Nile crocodiles can become easily conditioned to eat humans. A particularly gruesome case occurred on the Zambezi River at a town called Sesheke. The local ruler, King Sepopo, disposed of his enemies by feeding them to the crocodiles; although the practice stopped with his murder in 1870, the crocodile population kept their reputation as man-eaters for decades after his death. This is perhaps not surprising, since the reptiles themselves can live for over 70 years.

Nile Monitor

Varanus niloticus

Nile monitors are most often seen basking on rocks and branches or in the water. Adults can easily outrun humans over short distances even over open ground and will invariably head for water when pursued.

Common name Nile monitor

Scientific name *Varanus niloticus*

Family Varanidae

Suborder Sauria

Order Squamata

Size Up to 6.5 ft (2 m)

Key features Body elongated with muscular limbs and sharp claws; skin tough with small, beadlike scales; tail laterally compressed; basic coloration olive-green to black with variable lighter markings

Habits Spends nonhunting daylight hours basking on rocks and branches or in water; uses burrows and old termite mounds at night; can swim well

Breeding Female lays up to 60 eggs in a clutch, often in termite mounds; eggs hatch after 150–200 days

Diet Insects, eggs, birds, small mammals, crustaceans, amphibians, snakes, lizards

Habitat Grassland, fringes of deserts, rain forests, even cultivated areas—almost anywhere providing there is a permanent body of water

Distribution Eastern and southern Africa from Egypt to South Africa

Status Common

Similar species None

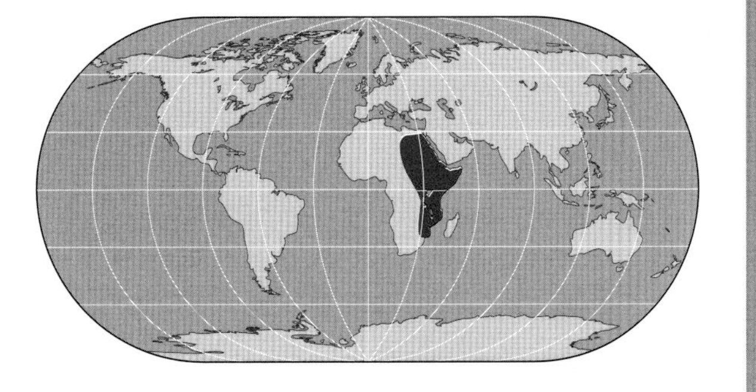

THE NILE MONITOR IS THE LARGEST LIZARD in Africa and probably the most widespread. Although absent from deserts, it is found wherever there are permanent bodies of water. It is common in all major river valleys and is often seen foraging for food in vegetation at the edge. In more temperate areas Nile monitors bask on rock outcrops, tree stumps, or branches overhanging water. They spend periods of the day partly submerged. In areas where water is present, Nile monitors remain active all year. Burrows, abandoned termite mounds, tree hollows, and caves within rock piles are used as nighttime retreats and during the dry season or in cooler months when activity is reduced.

There are two subspecies: *Varanus niloticus niloticus* (which has six to nine yellow dorsal bands and a bluish-black tongue) and *Varanus niloticus ornatus* (which has three to five yellow bands and a pink tongue). The latter prefers higher humidity and slightly cooler temperatures, and is restricted to the rain forests and coastal grasslands of central and western Africa. The general background coloration of both subspecies is olive-green to black with yellowish chevrons on the skull.

Agile Hunters

Nile monitors are quick, agile creatures with crushing jaws and powerful, claw-tipped limbs that make them efficient predators. Although they feed on carrion and forage in human garbage dumps, they get most of their food by hunting. To locate prey, they use vision and the tongue, the latter carrying scent particles to the Jacobson's organ in the mouth. Since prey varies with habitat and the season, Nile monitors tend to be omnivorous. They feed on insects, snakes,

lizards, fish, small mammals, birds, eggs, mollusks, and amphibians. They forage on and below ground, in trees, and in water.

As they age, changes occur in the structure of their teeth and skull. Young Nile monitors eat mainly active prey such as insects and lizards and have sharp, pointed teeth. Over time they are replaced with broader, blunter teeth. These changes occur as the monitors move to more sedentary prey with armor plating, such as crabs, mollusks, and turtles, which have to be crushed before they can be swallowed.

Some people believe that the Nile monitor is a more dangerous adversary than a crocodile of a similar size. With sharp claws and teeth and a powerful tail that it uses as a club, it can defend itself against most (but not all) predators. Large crocodiles, pythons, and cobras can overpower and kill a monitor. Monitors also come into conflict with humans when they raid poultry stocks, and they are hunted by humans. Their flesh is eaten, their organs and tissues are used in traditional medicine, and their skins are exported to make various leather-type goods.

Nile Monitors and Crocodiles

Nile monitors are important predators on crocodile nests, eggs, and young. Although a monitor usually digs up an unguarded nest, studies of their behavior in the wild show that monitors will cooperate with each other to raid a nest guarded by a female crocodile. While one monitor draws the crocodile away from the nest, its accomplice digs up the eggs. When an egg is located, the monitor grasps it between its jaws, raises its snout in the air, and cracks the egg, thus ensuring that both contents and shell slide down the throat with no waste. Large specimens eat eggs whole. Young crocodiles hiding at the edges of rivers are easy targets for Nile monitors; by preying on them, the monitors play an important role in controlling crocodile numbers.

Nile monitors are omnivores and often forage near water. Their strong teeth are adapted for crushing the shells of any mollusks or bivalves they encounter, such as these oysters in Tanzania.

Reproductive Behavior

Male Nile monitors are highly territorial and may rear up on their hind legs to engage in combat during the mating season. In other varanids the fight is over when the losing male retreats, but the victorious male Nile monitor inflicts painful bites on its defeated opponent. Mating and egg-laying times vary with location. In Senegal eggs are laid during the wet season in October to December. Gravid females have been found in Ghana in August and September, in Zanzibar in July, and in Tanganyika in November. Incubation periods vary depending on temperature, but on average the young hatch out over several days about 150–200 days after the eggs have been laid. They are sustained at first by the attached yolk sac. It is thought that the young that hatch from eggs laid in termite mounds eat termites for their first meal.

Nose-Horned Viper

Vipera ammodytes

Common name Nose-horned viper (sand viper and many other local names)

Scientific name *Vipera ammodytes*

Subfamily Viperinae

Family Viperidae

Suborder Serpentes

Order Squamata

Length From 24 in (60 cm) to 35 in (90 cm)

Key features Distinctive nose horn formed from 9 –20 scales; body color varies from silvery gray to reddish brown but usually has a well-defined, dark-edged zigzag down its back or a series of separate oval-shaped markings; underside of the tail may be yellow, orange, or red; males have more contrast in their markings than females

Habits Usually diurnal but active in the early morning and evening during very hot weather; terrestrial but quite a good climber, sometimes exploring shrubs in search of food

Breeding Live-bearer with litters of 4–20; gestation period about 4 to 5 months

Diet Mainly small mammals; also lizards and birds

Habitat Rocky places, especially hillsides

Distribution Europe from southern Austria (where it is rare) across the Balkan region and into Turkey, Armenia, Azerbaijan, and Georgia

Status Common in places

Similar species None within its range; several smaller European vipers have upturned snouts, but none has a horn like this species

Venom Highly toxic, causing swelling and pain; fatalities can occur if bites go untreated

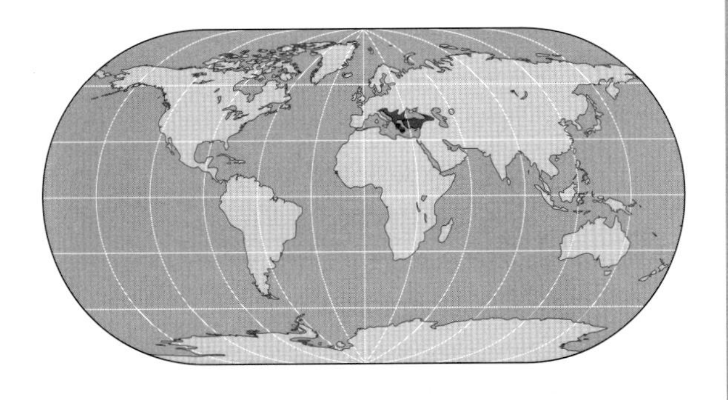

The nose-horned viper is the largest and most dangerous snake commonly found in Europe. It is the only one likely to cause fatalities.

THE NOSE-HORNED VIPER HAS potent venom, but bites are rare because it tends to live in places not frequented by humans. In addition, it is not aggressive, preferring to retreat quietly than to stand and face its enemies. If disturbed, it will hiss loudly; and if that does not have the desired effect, it will strike. However, events rarely reach this stage.

Nose-horned vipers live on hills and mountainsides, in open woods, scree slopes, abandoned quarries, and at the base of stone walls. They sometimes live in the dry walls separating the terraces of olive groves. Their main requirement seems to be dry rocky terrain, especially on south-facing slopes with scattered bushes and shrubs. The old common name of sand viper was misleading: In the European parts of their range they are most likely to be

Subspecies and Relatives

The nose-horned vipers are divided into several subspecies, some of which, such as *V. a. transcaucasiana*, are considered to be full species by some herpetologists. Others differ in only minor ways from the nominate form, *V. a. ammodytes*, which occurs over most of Europe. In the Balkan region it is replaced by the Mediterranean nose-horned viper, *V. a. meridionalis*, which also extends into Turkey.

One other European viper, Lataste's viper, *V. latasti*, has a "horn" on its snout. It is a much smaller species and lives in Spain, Portugal, and North Africa. The asp viper, *V. aspis*, usually has an upturned snout but lacks a real nose horn. It lives throughout much of Central Europe and Italy.

Altogether, some 27 species of *Vipera* are recognized, occurring in Europe, North Africa, and the Middle East.

Ⓘ ***A nose-horned viper feeds on a mouse. These snakes generally do no harm to humans and can even help by controlling populations of rodent pests in some areas.***

found at high altitudes—up to 6,000 feet (1,800 m) in places. Elsewhere, however, they live on lower, more level countryside. On the few Greek islands on which they occur, such as Naxos, they live as low as sea level but still among rocks. They are usually diurnal, lying in the sun or in open shade. In the hottest months they hide away at noon but may remain active well after sunset. There is some evidence that, in places, they move from exposed hillsides at that time of the year, preferring to shelter in lightly wooded valleys where it is cooler.

Ideal Camouflage

On the broken rocks and scree slopes where they live, their contrasting gray-and-black coloration and irregular zigzag markings make them surprisingly hard to see. In some places their background color has an orange or brownish wash, and throughout their range males are more brightly marked than females.

The nose-horned vipers may ambush their prey, but they also hunt actively, climbing across rock faces and poking their heads into cracks among jumbled boulders. They eat mostly small rodents, but lizards also figure in their diet, especially when the vipers are young. Juveniles may also eat invertebrates such as centipedes. They sometimes climb into low shrubs in search of nestling birds in the spring. Small items of prey are frequently eaten without the snake employing its venom, but larger ones are struck before being released, tracked down, and then eaten after they have died.

Nose-horned vipers hibernate from October to early April, using clefts in the rock to get below the frost level—even in the Mediterranean region many mountain ranges are covered with snow for several months. Mating takes place soon after the snakes have emerged, and about 15 young are born in August or September.

The venom of this species is the most potent of any European mainland viper, but two species with small ranges—the Milos viper, *Macrovipera schweizeri* from the Cyclades Islands of Greece, and the Ottoman viper, *Vipera xanthina* from Turkey and a few offshore islands—may be more dangerous. Since the fangs of all these species are quite long, up to 0.5 inches (1 cm) in adults, the consequences of a bite may be fatal to humans if not treated. Even so, humans are only likely to come into contact with these species where they inhabit drystone walls and field edges. Otherwise, they live on mountainsides among rocks and dry scrub, where shepherds, goatherds, and hikers may sometimes encounter them.

Painted Turtle

Chrysemys picta

Common name Painted turtle

Scientific name *Chrysemys picta*

Family Emydidae

Suborder Cryptodira

Order Testudines

Size Carapace up to 10 in (25 cm) long

Weight Approximately 2.2 lb (1 kg)

Key features Shell smooth with no keel or serrations along the rear of the carapace; in eastern painted turtle the central vertebral shields and adjacent side shields are aligned rather than overlapping; different subspecies identified easily by distinctive coloring and patterning; pattern of yellow stripes on head, becoming reddish on the sides of head and front legs; females grow larger than males; males have longer front claws than females

Habits Semiaquatic; often leaves the water to bask

Breeding Female may produce clutches of anything from 2 to 20 eggs; eggs hatch after 10–11 weeks on average

Diet Young tend to be carnivorous; mature painted turtles eat a higher percentage of aquatic vegetation

Habitat Relatively tranquil stretches of water, ranging from smaller streams to lakes and rivers; eastern form occasionally found in brackish water

Distribution Central and eastern parts of North America from Canada in the north to Mexico in the south

Status Relatively common

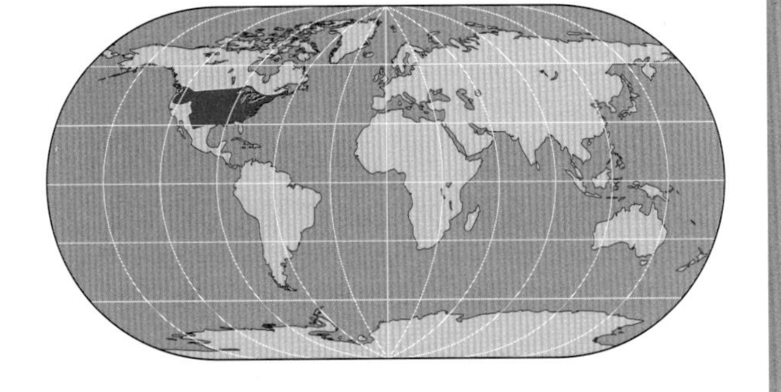

The painted turtle is the most widely distributed North American turtle and the only one whose range extends across the entire continent. Because of their size and attractive colors painted turtles are often kept as pets.

PAINTED TURTLES GET THEIR NAME from their bright coloration. Four distinct subspecies are recognized through their range. The eastern painted turtle, *Chrysemys picta picta*, occurs farthest east and is found from southeastern Canada down the Eastern Seaboard of the United States to northern Georgia and Alabama. A particular feature of this subspecies, which makes it virtually unique, is the way in which the central vertebral shields running down the back and the adjacent shields on either side are aligned rather than overlapping. This creates a distinctive pattern of lines running across the shell.

The eastern painted turtle can also be identified easily by its unmarked yellow plastron. This helps distinguish it from the midland painted turtle, *C. p. marginata*, which also has a blackish carapace but has a dark patch at the center of the plastron. This feature is apparent even in young hatchlings, although the exact pattern of markings on the underside of the shell differs according to the individual. As its name suggests, the range of this subspecies lies to the west of its near relative, extending across Canada from southern Quebec to Ontario and occurring as far south as Oklahoma and Alabama.

The southern painted turtle, *C. p. dorsalis*, is perhaps the most distinctive subspecies of all thanks to the yellow or sometimes reddish stripe that runs down the center of the carapace. The plastron is yellowish in color with no markings. As its name suggests, its range does not extend as far north as Canada; it is confined to the United States from southern Illinois southward to the Gulf of Mexico, ranging from Oklahoma to Alabama.

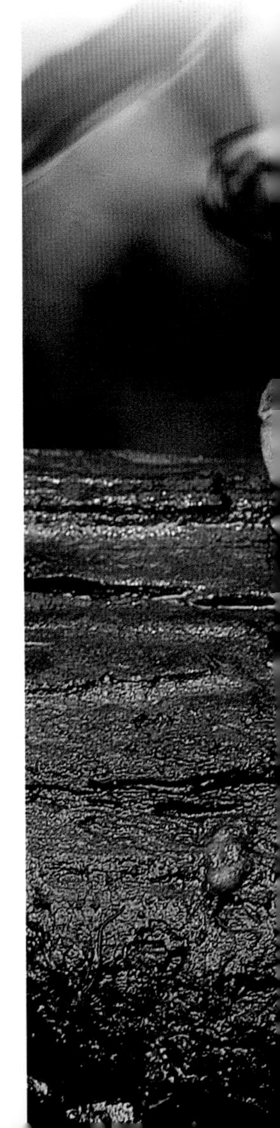

The eastern painted turtle,* Chrysemys picta picta, *has a greenish carapace with thick lines between the aligned scutes and red markings around the edge. The plastron is yellow.

The western painted turtle, *C. p. belli*, occurs farther west. It is found over a wider area than the other subspecies. It ranges farther north into British Columbia as well and is found in a number of localities in the southwestern United States—it even occurs in an area in Chihuahua, Mexico. It has distinctive lines over the carapace as well as elaborate patterning that extends to the edge of the plastron. It has potentially the largest size of all the subspecies. Where the different races of painted turtle overlap, however, they interbreed, giving rise to offspring with intermediate characteristics. This is known as intergradation.

Courtship and Breeding

Courtship begins in April. At this stage the reason for the male's longer front claws becomes apparent. He swims up in front of a female, approaching her headfirst. He gently uses his claws to fan water close to her face and then starts to touch her face. Assuming she is receptive, she responds by touching his face. The pair then swim to the bottom. The male grips onto the female, placing his legs at each corner of her shell, and mating takes place. A pair may remained joined in this way for up to 15 minutes.

The nesting period itself usually extends from May to July. Females typically emerge from the water to lay either soon after dawn or in the late afternoon. To make digging easier, they

Chrysemys picta belli, ***the western painted turtle, is the most widespread of the four subspecies and grows to a larger size than the others—up to 10 inches (25 cm).***

choose a site that has soft soil, often moving up onto a bank or even near a road, where the area is unlikely to become flooded. The nest itself is usually quite shallow, often less than 4 inches (10 cm) deep.

In northern parts female painted turtles tend to lay only once or twice during this period. Farther south they can lay throughout the season and produce up to four clutches. There are differences among the subspecies as well, with the relatively large western painted turtle producing more eggs per clutch than the southern race, which is the smallest. In any population large females invariably lay more eggs compared with smaller individuals. Breeding is unlikely to occur until the female is six years old; in some areas females may not lay every year if the temperature is too cold.

How Did the Subspecies Evolve?

There is a tendency to think that species are created as individual populations become isolated. However, this process can occur in reverse, which may be true in the case of the painted turtle. It is believed that at the time of the last Ice Age (about 18,000 years ago), today's painted turtles existed as three separate species. Their distribution was somewhat similar to their current ranges, except that they did not extend as far north because of the presence of the ice sheet.

The western painted turtle was present in the Southwest, and the southern painted turtle was confined in the vicinity of the lower Mississippi region. As the climate warmed, it is thought that these two distinct populations followed the retreating glaciers northward. They met around what is now known as the Missouri River, and thus began a process of hybridization that ultimately resulted in the development of the midland painted turtle.

In turn the midland painted turtles traveled northward as far as the eastern area of the Great Lakes. Sandwiched between their two ancestral forms, they occupied the area where they are found today. The movement of the eastern painted turtle up the Atlantic coast also resulted in contact with the southern painted turtle on the western side of its distribution. Instead of remaining isolated, therefore, the three original species came together, with the result that the painted turtle now enjoys the widest distribution of any aquatic chelonian on the North American continent.

The time taken for the eggs to hatch varies depending on local conditions, but incubation usually takes between 65 and 80 days. The gender of the offspring is determined by the temperature at which the eggs hatch—at higher temperatures of about 87°F (30.5°C) females are produced, but at temperatures below 77°F (25°C) males develop. It is unclear whether the female is influenced in this respect when choosing her nesting site. Young painted turtles have a rounded carapace. They are little more than 1 inch (2.5 cm) in length, but they grow fast at first, doubling their size within a year. This makes them less vulnerable to the many predators that they face when young.

↑ The different subspecies of painted turtles are distinctive. This is a southern painted turtle, **C. picta dorsalis,** ***distinguished by a reddish-orange stripe running down the center of the carapace.***

Basking in Groups

It is quite common to see painted turtles basking out of water, often with their hind legs stretched out behind them. This helps keep their shells healthy and raises their body temperature. In some areas it is not uncommon for up to 50 turtles to share a partially submerged rock or log when basking, often lying on top of each other. This may make them less vulnerable to predators—if one individual detects possible danger and dives back into the water, all the others will follow very rapidly.

Painted turtles display a regular daily routine, emerging to bask early in the morning. This means that they can raise their body temperature and therefore their level of activity at this stage, plunging back into the water to feed before basking again in the early afternoon. They then look for more food and finally burrow into the bottom of their stream or pond overnight.

Basking is particularly significant in temperate areas because, even if the water temperature is relatively low, the turtles can take advantage of any additional warmth provided by the sun. Painted turtles may enter a dormant period in northern parts of their range during the winter months; but although they become sluggish, they may not be completely inactive—they have been seen swimming even in ice-covered water. Their body temperature under these circumstances has been shown to be higher than that of their surroundings, suggesting that they have a primitive mechanism for heat regulation.

↓ Although they spend most of the time in water, painted turtles often sun themselves on a log, a rock, or the shore. They are often seen in large groups.

Common name Panther chameleon

Scientific name *Furcifer pardalis*

Family Chamaeleonidae

Suborder Sauria

Order Squamata

Size From 15 in (38 cm) to 22 in (56 cm) long depending on locality; female a little smaller

Key features Males and females differ in color, shape, and size; male has a prominent rostral ridge (less prominent in female); snout bears enlarged scales; males vary from blue to green to pink depending on location; females grayish fawn or light pinkish brown; both sexes have a lateral line of whitish or bluish oval blotches

Habits Lives on small shrubs or even weeds in deforested areas; on cool days it basks, on warm days it sits in partial shade

Breeding Female lays 2–4 clutches a year; clutch size 12–45 eggs; eggs hatch after 6–13 months

Diet Large insects, small mammals, small lizards and geckos; in some localities small frogs are eaten

Habitat Forest edge; agricultural and suburban areas

Distribution Eastern coast and coastal islands of northern and eastern Madagascar; small populations on Reunion Island and Mauritius

Status Common

Similar species None

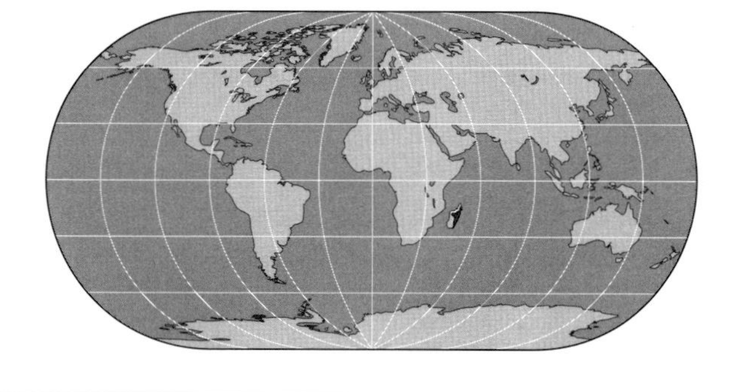

Panther Chameleon *Furcifer pardalis*

Numerous local myths surround the panther chameleon, probably due to the fact that it is not only common and conspicuous but also very colorful.

THE NATURAL RANGE OF PANTHER CHAMELEONS is the northern half of Madagascar. They occur from sea level to 3,900 feet (1,200 m), and the areas they inhabit are wet for eight months of the year. The northeastern part of their range receives 140 inches (355 cm) of rain a year, although there is progressively less rainfall the farther west and south one goes. Temperatures range from a daytime high of 96°F (35°C) to 50°F (10°C) at night during winter months, but frosts are restricted to the highest mountain peaks. Generally speaking, the climate is warm and humid with some seasonal fluctuations.

Panther chameleons dislike areas of deep shade. They prefer forest edge, agricultural fields, and suburban gardens. They are frequently seen in towns, crossing roads and moving from one garden to another. When using trees, they perch in the crown, but bushes up to 10 feet (3 m) high are their favorite perching places. Unlike many other species of chameleon, panther chameleons actually benefit from human disturbance: When an area of forest is cleared revealing shrubby undergrowth, they move in. They bask in the early morning sunlight to raise their body temperature; once they reach the optimum temperature, they move to perches in filtered sunlight.

Color Variants

More than any other species, male panther chameleons demonstrate geographical variation in color. Males from the island of Nosy Bé are light emerald-green or blue-green with yellow lips or uniformly turquoise-blue (sometimes referred to as the blue phase). Those from

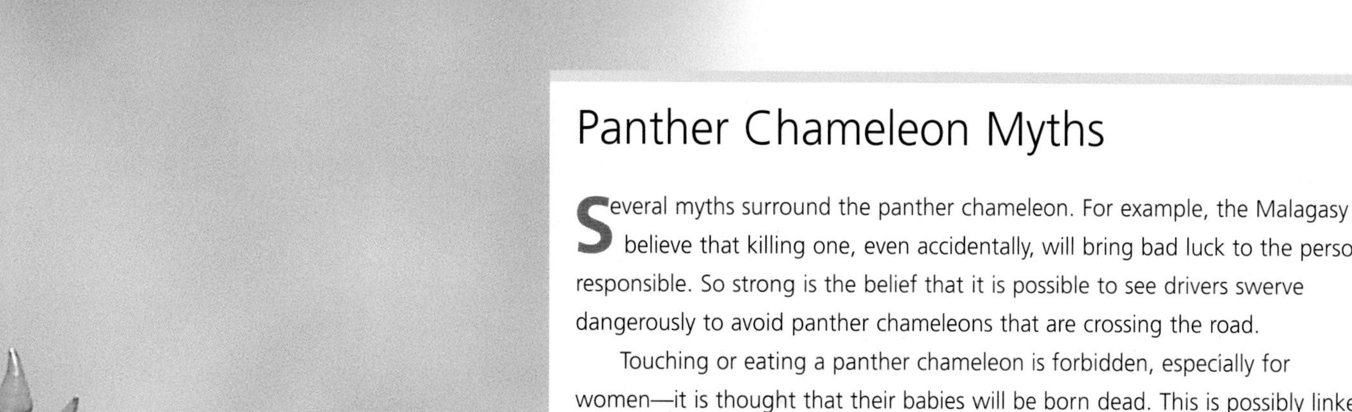

Panther Chameleon Myths

Several myths surround the panther chameleon. For example, the Malagasy believe that killing one, even accidentally, will bring bad luck to the person responsible. So strong is the belief that it is possible to see drivers swerve dangerously to avoid panther chameleons that are crossing the road.

Touching or eating a panther chameleon is forbidden, especially for women—it is thought that their babies will be born dead. This is possibly linked to the first myth, but the vivid coloration of the panther chameleons leads the Malagasy to believe they are poisonous.

Because they see these chameleons falling out of bushes or trees at certain times of the year, the Malagasy people believe that the panther chameleons are committing suicide. However, this phenomenon usually coincides with the dry season when food becomes scarce. The panthers are therefore forced into closer contact with each other, and in order to avoid fights, some fall either to the ground or onto lower branches.

Maroansetra and Tamatave regions are green with a light blue to gray dorsal crest and are capable of changing to a bright orange-red in a few seconds.

There are two color forms from Ambanja. Some males have a dark green body with vertical red bands. The other color form from the area is a blue panther, a stunning creature with a turquoise-blue body, head, and legs and with dark blue vertical bands. In contrast, its eye turrets are orange.

In the Diego Suarez region males are green with blue bands; when stressed, their body color changes from green to yellow, and the vertical bands turn red. Recently a pink form has been found at Ankaramy. Normal coloration of the Ankaramy form is grayish pink with pale turquoise vertical bands, legs, and head. The lateral line is yellow. When this creature becomes stressed, the background color turns pink, and the vertical bands, legs, and head become almost purple.

Geographical color variations among young female panthers have not been studied to the

No chameleon shows more geographical color variation than the panther chameleons. Note the opposed toes in this male, a useful aid for gripping.

same extent as in the males, but their colors are more subtle, varying from fawn to grayish fawn and light pinkish brown.

Social Displays

Panther chameleons are probably the most common and conspicuous chameleons in Madagascar. They are also among the most territorial. During male-to-male exchanges males undergo rapid color change, and their bodies are compressed from side to side. As a result, the ventral surface and gular pouch are extended. They curl their tail, and their eye turrets change color. One male moves slowly toward its rival; if it does not retreat, a bout of head pushing begins. Their strong jaws can tear flesh and break bones. If a panther is seriously injured, death is inevitable.

In male-to-female exchanges the males color up in the same way as when fighting a rival, but the color change is accompanied by vertical head bobbing. When nongravid females encounter each other, there is no change from their normal subdued coloration. Gravid (egg-bearing) females, however, are intolerant of all males and females, and they turn dark brown with contrasting salmon-pink blotches on the lateral line. They extend their gular pouch, which shows bright red markings.

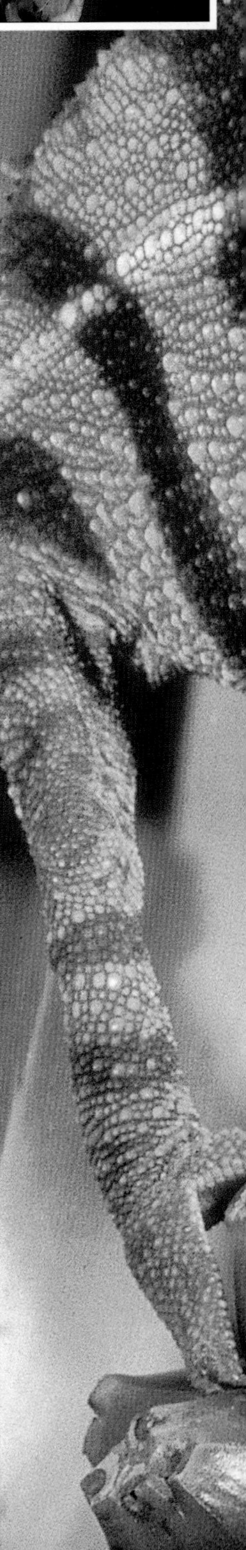

Color differences in panther chameleons reflect not only local variations but also the animal's response to danger or courtship advances. Above left: a blue-phase male from Nosy Bé. Above right: Young female showing coloration prior to becoming sexually active. Right: The face and legs of stressed males often turn red.

Incubation Times

Panther chameleons, as well as other species from Madagascar, lay eggs that are at a very early stage of development. In some parts of their range the eggs are laid just before a cooler, drier period—the winter. Development is suspended for one to three months (depending on the region), and the eggs undergo what is known as a diapause. When normal temperatures resume, development continues. This increases the overall incubation period to between 11 and 13 months. However, in areas with minimal seasonal temperature fluctuations the eggs do not undergo a diapause; instead, they hatch after six to eight months. The length of the diapause varies from region to region. The hatchlings seem quite tiny for such a large species, measuring just 1.5 inches (3.8 cm) long.

Breeding

The length of the breeding season varies from place to place. For example, in Nosy Bé females produce four clutches a year, but in Ambanja and Diego Suarez they only produce two. The reason is that the seasons become more pronounced the farther south one travels within the range of the panther chameleon. In Nosy Bé there is little temperature fluctuation, and humidity is relatively high for most of the year.

Males can be a little too zealous when approaching a female for mating. Unreceptive females often resort to biting to subdue an overamorous male. Female panthers become receptive as soon as they reach maturity, which can be as early as four months old. (During the breeding season they become receptive again about two to three weeks after laying a clutch.) The female indicates that she is receptive by turning a uniform pale salmon color. Females are receptive for four to five days. Gravid females usually choose to lay their eggs in the root system of shrubs and trees, since the ground there is less compact. The roots attract moisture, so the eggs will not dehydrate.

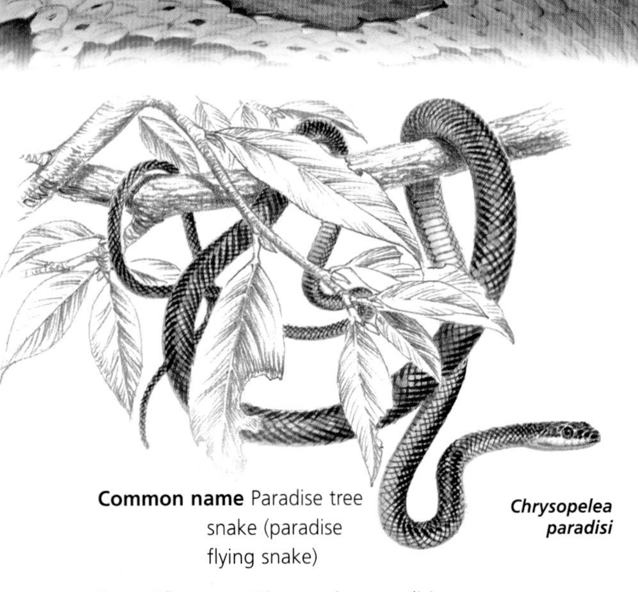

Chrysopelea paradisi

Common name Paradise tree snake (paradise flying snake)

Scientific name *Chrysopelea paradisi*

Subfamily Colubrinae

Family Colubridae

Suborder Serpentes

Order Squamata

Length 4.2 ft (1.3 m)

Key features Slender with a narrow neck, wide, flattened head, and a long tail; eyes large with round pupils; scales on body are black with a green center; green spots are largest on the lower scale rows; underside completely pale green; may also have a row of red spots down the center of its back

Habits Arboreal, rarely coming down to the ground

Breeding Egg layer with clutches of 5–8 eggs

Diet Mainly lizards—especially geckos—but also takes small mammals

Habitat Forests and plantations; also gardens and even houses

Distribution Southeast Asia (Andaman Islands, Myanmar, Thailand, peninsular Malaysia, Singapore, Sumatra, Java, Bali, Borneo, and the Philippines)

Status Common

Similar species Five other species in the genus have a similar body plan and coloration

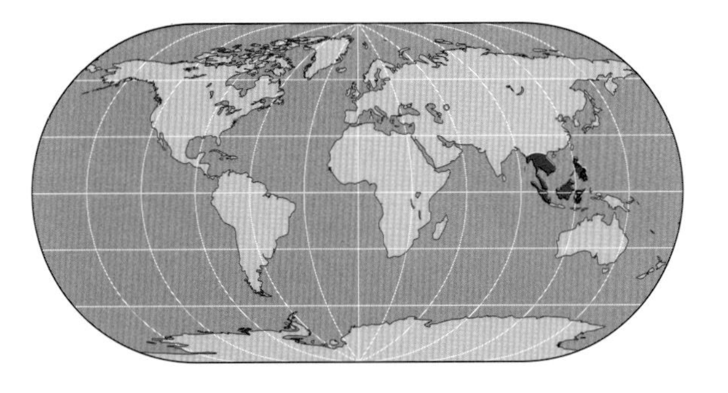

Paradise Tree Snake

Chrysopelea paradisi

A snake of the forest canopy, the paradise tree snake is one of the few snakes that has also made some progress toward mastering the air: The species is the celebrated "flying" snake of Southeast Asia.

THE PARADISE TREE SNAKE, and four other less-common members of the same genus, all from the same region, have the ability to glide from tall trees to lower branches or to the ground. Their ventral scales have a pair of ridges or keels near the point where their undersurface joins their sides. The ridges are useful when climbing vertical surfaces such as tree trunks, and they are also present in other proficient climbers such as some of the ratsnakes. In *Chrysopelea*, however, the part of the underside between the two ridges can be drawn up to form a concave surface when gliding.

The snake wraps its tail around a branch, draws up its body like a coiled spring, and suddenly launches itself into the air. The concave belly traps the air, and as long as the snake stays the right way up, it will come down in a controlled glide. By twisting its body slightly as it descends, it can even steer itself up to a point, so that it can land where it wants to.

Gliding between the Trees

It is not known why the paradise tree snake has evolved the ability to glide (flying is not an accurate description). It may simply be a method of getting from one place to another with the minimum of effort, but it is more likely to be a way of avoiding predators, such as birds of prey and climbing carnivorous mammals.

A few other snakes, such as the South American parrot snakes, *Leptophis*, jump out of trees and bushes if they are chased. Even some fast-moving ground dwellers, such as the African *Psammophis* species, leave the ground briefly as they race through grasses and other low vegetation, but only the *Chrysopelea* species have structural modifications that make controlled gliding possible.

Apart from its ability to glide, the paradise tree snake can scale vertical surfaces of trees with amazing ease, simply moving up the trunks using serpentine locomotion.

It is active during the day and often coils in the crowns of coconut trees in search of the skinks, geckos, and other climbing lizards on which it preys. It hunts by sight, grasping its prey with a sudden rush. The paradise tree snake is a rear-fanged species with a mild venom, and it holds its prey in its jaws until the venom begins to take effect. This can take quite a time if the prey is large. It is not dangerous to humans, although it is a nervous snake and may bite if cornered. It often comes into contact with people because, like a number of other tree snakes, it finds outbuildings suitable places in which to rest and hunt. It is even common in Singapore and other built-up places.

The paradise tree snake has one of the most intricately attractive patterns of all the snakes in Asia. Superimposed on the black-and-green markings there may be a variety of other patterns formed by scales containing larger and smaller green spots. This can lead to a barred pattern or a series of chevrons down its back. The red markings on its back are also variable but most commonly consist of four red scales arranged like the petals of a flower. Its head has bold black-and-green bars passing across the top.

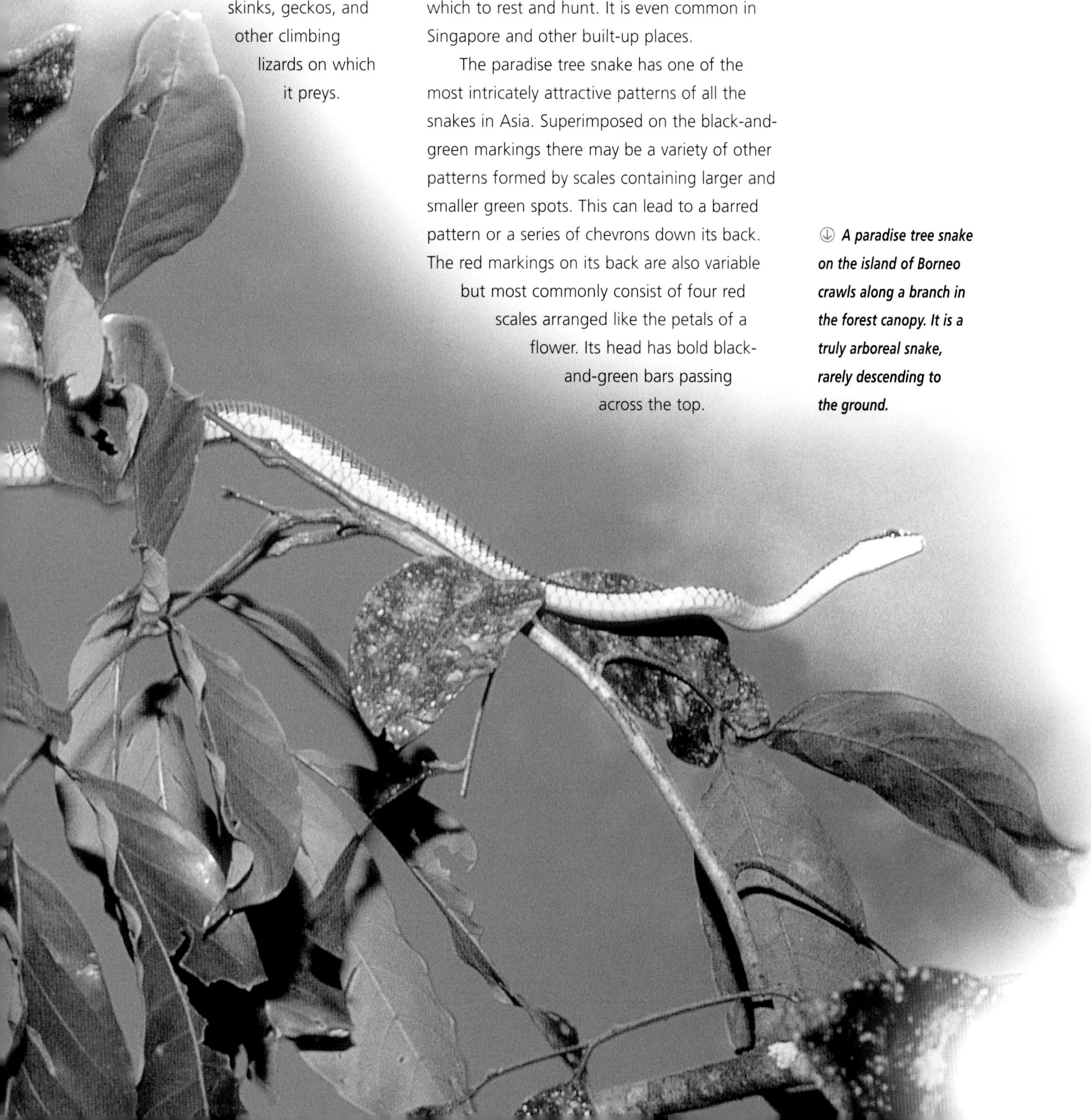

A paradise tree snake on the island of Borneo crawls along a branch in the forest canopy. It is a truly arboreal snake, rarely descending to the ground.

Plumed Basilisk

Basiliscus plumifrons

The colorful plumed basilisk lives in the forests of Central America. It is never far from water, which often provides it with an escape route when faced with danger.

Common name Plumed basilisk (Jesus Christ lizard)

Scientific name *Basiliscus plumifrons*

Subfamily Corytophaninae

Family Iguanidae

Suborder Sauria

Order Squamata

Size Males to 36 inches (91 cm); females to 20 inches (51 cm), of which the tail can account for three-quarters

Key features Adults green with black bars on the tail and lighter green or white spots on the flanks; both sexes have a crest on their head (the male's has two lobes); male also has separate crests on the back and tail; body and tail flattened from side to side; front legs and tail very long; juveniles are spidery in appearance with long, thin legs

Habits Arboreal or semiarboreal; diurnal

Breeding Egg layer that breeds throughout the year; female lays 4–17 eggs that hatch after about 65 days

Diet Small vertebrates, invertebrates, and plant material

Habitat Forests (usually in the vicinity of water)

Distribution Central America (eastern Honduras to southwestern Costa Rica)

Status Common in suitable habitat

Similar species There are other basilisks in the region, but *B. plumifrons* is the only bright green one; other green iguanids of similar size lack crests

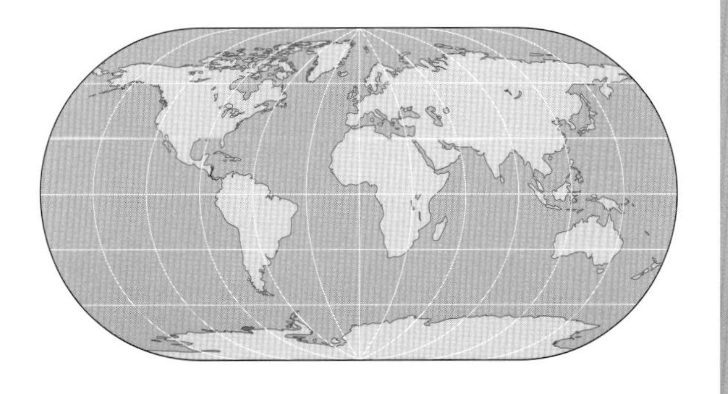

BASILISKS HAVE AN AMAZING ability to run across the surface of still water, a habit that has given them the alternative common name of Jesus Christ lizard. Their long toes have a fringe of rectangular, projecting scales that can rest on the surface film without breaking through, as long as the basilisk keeps moving. Once they get going, they lift the front of their body up, hold their front legs against their chest, and run on their long hinds legs, curving their tail up in an arc to act as a counterbalance. Small individuals do not even need to move very fast to remain on the surface and often seem to move in slow motion at first before accelerating across a patch of open water, leaving a twin trail of ripples. Sometimes a small group will break cover and run off together in formation. They are also good swimmers and can swim underwater for long distances.

Daytime Foragers

Basilisks usually occur in dense vegetation along streams and rivers, and often sleep on branches overhanging water. If they are disturbed by a predator, they are quick to drop into the water and may swim to the bottom. During the day they forage among the vegetation, eating more or less anything they find. Juveniles are almost entirely insectivorous; but as they grow, their diet becomes more varied, and they will also take leaves, fruit, and berries. Large adults sometimes tackle crustaceans such as shrimp and crabs, and have even been known to eat snakes and sleeping bats.

Like many tropical lizards, they breed throughout the year, although most activity takes place during the rainy season (May to September in Costa Rica, for instance). The

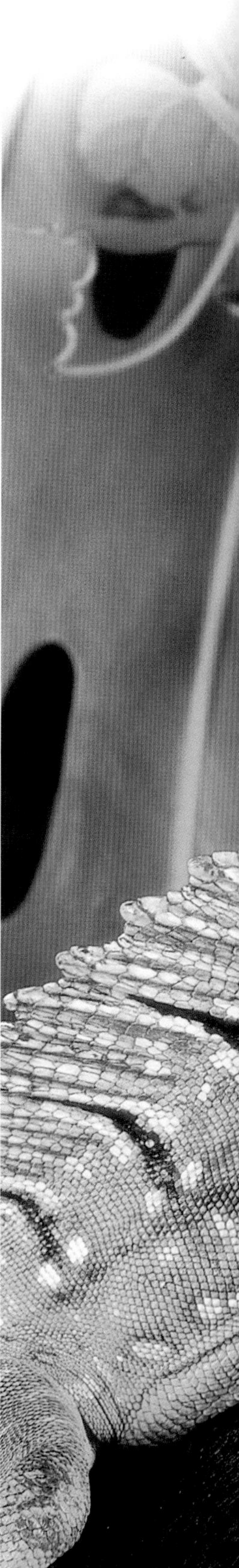

The plumed basilisk, Basiliscus plumifrons, *is one of the fastest lizards, able to run at speeds in excess of 6.5 feet (2 m) per second. It is also the most colorful species in the genus, the others being dull brown in color.*

female lays four to 17 eggs in a short nest chamber that the lizards excavate themselves. The eggs take about 65 days to hatch. The babies are dull in color and show no signs of the crests, which develop later as low ridges of skin that gradually increase in size. The crests on the head are fleshy and often become torn and somewhat ragged, while those on the back and tail of males are supported by elongated, bony spines projecting up from the vertebrae.

The Relatives

There are four species of basilisk altogether, ranging from Mexico to Ecuador and western Venezuela. All have crests, but only the plumed basilisk is bright green in color. They favor waterside habitats, except for the striped basilisk, *B. vittatus*, which can be found far from water in open places, including disturbed habitats such as pastures and coconut plantations. Like *B. plumifrons,* it is capable of walking on two legs (bipedal locomotion) across water if the need arises.

Mythological Creature

The basilisk is an ancient mythological beast with the legs and head of a rooster, the body, tail, and tongue of a snake, and the wings of a bat. The similarities to the basilisk lizard, therefore, are the crest (which is like the comb of a rooster), the long, snakelike tail, and the bright, staring eyes. The basilisk from mythology could kill everything it encountered with its stare, and the only defense against it was to hold up a mirror, whereupon its own reflection would frighten it to death.

The English playwright William Shakespeare often used herpetological metaphors, including one about a basilisk. In *Richard III* (Act I, Scene 2) Lady Anne responds to Richard's compliment about her eyes, "Would they were basilisk's, to strike thee dead!" It seems there is just no pleasing some folk!

Puff Adder

Bitis arietans

Common name Puff adder

Scientific name *Bitis arietans*

Subfamily Viperinae

Family Viperidae

Suborder Serpentes

Order Squamata

Length From 36 in (90 cm) to 6 ft (1.8 m)

Key features Body very stout with a large, broad head and rounded snout; scales heavily keeled; color varies slightly, but most have a dirty-yellow or brown body with large dark-gray, cream-edged chevrons or "u"-shaped markings down the back

Habits Terrestrial; nocturnal or diurnal depending on the weather

Breeding Live-bearer with large litters; the largest recorded was 156, a record for snakes; gestation period about 90–120 days

Diet Mammals, birds, and lizards

Habitat Very adaptable; absent from closed canopy forest and from the most arid deserts, but otherwise it can occur anywhere

Distribution Africa south of the Sahara with an isolated population in southern Morocco; absent from the Congo Basin; a small population also occurs in the Arabian Peninsula

Status Very common

Similar species None; all other large adders have colorful geometric patterns

Venom Relatively toxic, but the main danger comes from the large amount injected, easily enough to kill a human; venom is slow acting, and with treatment over 90 percent of victims recover

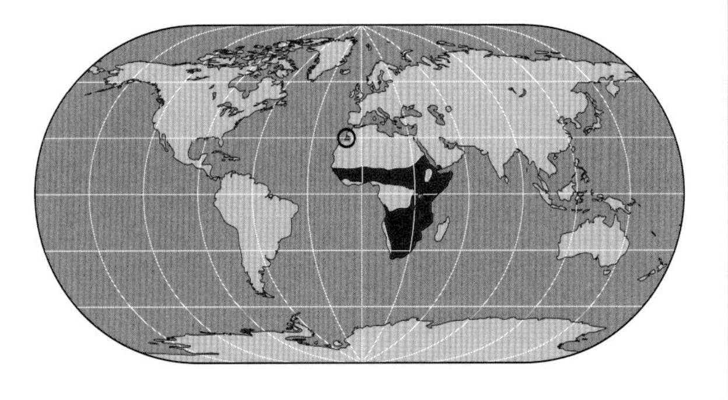

The puff adder is one of the most ubiquitous African snakes. Not only does it have a wide distribution, but it occurs in many different habitats and tends to be conspicuous because of its great size.

ADULT PUFF ADDERS ARE HUGE, sluggish snakes that often travel in a straight line like a caterpillar, leaving a distinctive shallow furrow behind them. The imprints of their ventral scales are often visible in sandy soil. Early in the year they are often seen crossing quiet roads where they sometimes stretch out motionless to absorb as much heat as possible, resulting in many casualties. They are so bulky that they cannot easily move from side to side in the typical serpentine manner. If they need to, however, they switch to this method of locomotion and can move surprisingly quickly. On loose ground this method of locomotion begins to change into a clumsy form of sidewinding.

Forceful in Defense

On the whole, puff adders are not easily frightened. If they are discovered in the open, they are likely to remain stationary for a time, then make off slowly. If they are prevented from moving away, they form a defensive "s"-shaped coil, raise their head slightly, inflate their body, and hiss loudly. Once heard, this sound is not easily forgotten.

Some individuals are bad tempered and will begin striking with little provocation, lunging rapidly and forcefully (the scientific name *arietans* is from the Latin *arieto*, meaning to butt or strike hard). They may almost leave the ground during the strike, but they immediately recoil, sometimes toppling over backward as a result of such force. This has led to a theory that puff adders can only strike backward, which is clearly inaccurate. Even while striking they often edge backward or sideways toward cover; and given the opportunity, they will turn and move away.

Most puff adder victims probably never see the snake that bites them until it is too late. Puff adders are pretty sluggish and have a habit of basking at the side of pathways. They are also very well camouflaged; and because they are often active at dusk, they are hard to spot. It is only too easy to step on or near one, causing it to make a reflex strike.

Being large, they can strike upward, above the height of most shoes or boots, so bites are often in the ankle or lower legs. Pain and swelling around the area develop quickly (although some bites are "dry," with no venom delivered). Swelling soon spreads through the affected limb and to the rest of the body, along with bruising and blood-filled blisters. However, the venom is relatively slow acting, taking two to four days to cause death, by which time it should be possible to get medical help.

Pofadder, South Africa

In northern South Africa, roughly halfway between the towns of Springbok and Kakamas, is the small pioneer town of Pofadder (the Dutch word for puff adder). This little town is not named after the snake, however, but for a local native chief who was given the name Klaas Pofadder by the occupying Afrikaners. Pofadder was an accomplished cattle rustler and a thorn in the side of local farmers, who eventually shot him. A Christian missionary working there in 1875 named the town in his honor.

A puff adder in the Sahara Desert in northern Africa. Its caterpillarlike locomotion leaves a straight, deep track in the sand. The snake relies on camouflage to escape detection and would rather freeze than move off.

Antivenom is available, and over 90 percent of victims recover. Complications are often caused by ill-advised first aid, for example, applying inappropriate tourniquets, slashing the wounds, or by overenthusiastic injections of antivenom. Despite the high recovery rate, puff adders are responsible for over half the serious snake bite cases in southern Africa and nearly all the fatalities. Death usually results from kidney failure.

Population Variants

Puff adders can make a living in almost any environment except for moist rain forests (where they are replaced by Gaboon vipers, *Bitis gabonica*)

Ⓣ *In the Kalahari Gemsbok Park in South Africa two male puff adders engage in ritual combat. They both rear up and attempt to pin their opponent to the ground.*

and for the most arid deserts. The Sahara forms a barrier to the north of their range, although there is a small relict population in southern Morocco that was left behind when the species retreated in the face of an expanding desert (in similar circumstances to the brown house snake, *Lamprophis fuliginosus*). The puff adders also live across the Red Sea in the southwestern corner of the Arabian Peninsula.

Snakes from the drier parts of the range are often dull and have a washed-out appearance to blend in with the sandy, dusty environment. Pale individuals from northeastern Africa are sometimes called *Bitis arietans somalica*, the Somali puff adder. This subspecies also has keeled scales under its tail (known as subcaudals). Puff adders from some highland areas have a brilliant yellow or orange background coloration, and populations from the Cape region of South Africa are similar. Male puff adders are smaller and more brightly colored than the females.

Variety of Prey

Puff adders hunt by ambush and eat most vertebrate prey such as rodents, birds, toads, and other snakes. They have even been known to eat tortoises, which they swallow whole, and African hedgehogs are often taken. They may catch and swallow small defenseless prey animals without injecting venom; but large prey is struck, released immediately, and tracked down later when the venom has taken effect.

Large individuals have huge gapes and tackle guinea fowl, sand grouse, hares, hyraxes, and even small antelope such as dik-dik fawns. They store large quantities of fat during the summer to tide them over an inactive period in winter, and this fat is taken as a cure for rheumatism by local tribespeople.

Large Litters

Males fight with each other during the breeding season and track down receptive females by following a pheromone trail. One female can have as many as seven males following her. The live young are born about three to four months after mating—at the end of the summer in southern Africa. In tropical parts the breeding season may be more geared to rainfall than temperature, with birth timed to coincide with the start of the rainy season.

Litter size depends on the female's size, as the following examples demonstrate: A female measuring 26 inches (65 cm) gave birth to 11 young, one of about 36 inches (90 cm) gave birth to 35 young, and a huge female had 147. This is the largest recorded litter for a wild puff adder, but the record for a captive is 156 young, produced in a Czech zoo.

With such enormous litters predation on the young is invariably high. Although the young puff adders are venomous from day one, they are eaten by monitor lizards, other snakes such as file snakes (*Mehelya*), storks, birds of prey, and even predatory fish—puff adders occasionally enter water.

Mountain Relative

The genus *Bitis* contains 16 species. The berg adder, *Bitis atropos*, is a dark-gray species with obscure markings that could easily be mistaken for a young puff adder. It is restricted to grassland habitats in mountains such as the Drakensberg in South Africa and along the South African coast, where its range overlaps that of the puff adder. The berg adder is a bad-tempered species that is quick to bite; but the

effects of its venom are mild, and there are no known fatalities. It is named after Atropos, one of the three fates in Greek mythology and daughter of Zeus and Themis. Atropos was the fate who cut the thread of life—an appropriate name for a venomous snake but perhaps overdramatic for this particular one, which is not one of the more dangerous species. One of the other two fates was called Lachesis, and it was she who measured the thread of life, determining how long it would be. Lachesis is the name given to the genus to which the bushmasters belong.

A Long-Lost Relative

Fossil snakes that can be positively linked to surviving close relatives are quite rare. *Bitis olduvaiensis* is an exception. It is notable for having been found (in the form of fossil vertebrae) at the famous Olduvai Gorge site in Tanzania, where early human remains believed to be 1.75 million years old were discovered by the anthropologist Dr. Louis Leakey and his wife Mary in the 1950s.

Closeup of a puff adder. The dark line through its eye and the patch of dark skin behind the eye that occurs in many venomous snakes are visible here.

Ultraviolet Protection

Many snakes have dark lines going through their eyes, designed to break up their outline and make it more difficult for them to be seen. Many venomous snakes have thicker lines or patches of dark skin behind the eyes, above their venom glands. In some cases black pigment (melanin) occurs beneath the surface of the skin. Some herpetologists think these areas protect the venom from being denatured by ultraviolet radiation (which causes chemical changes in some proteins). The black pigment absorbs UV rays and prevents them from penetrating. The skin of the puff adder contains small shiny cells, called iridophores, that reflect light rather than absorbing it.

Pythons

The python family contains four of the world's five largest snakes—the reticulated, scrub, African, and Indian pythons. Only the South American anaconda (family Boidae) is larger. There are 26 species of pythons (some authorities recognize up to seven more), and they all live in the Old World: three in Africa, five in Asia, and the rest in Australia and New Guinea. Although the family contains the longest snake (the reticulated python, *Python reticulatus*), it also contains several smaller kinds. The smallest species (the anthill python, *Antaresia perthensis* from Australia) rarely exceeds 16 inches (40 cm).

Regardless of size, all pythons are powerful constrictors. They kill their prey by throwing several coils around it and increasing the pressure. Death is by suffocation, not crushing, although bones may be broken in the process. Blood vessels may also be constricted, perhaps hastening the end for the victim.

Pythons are a relatively ancient group of snakes with several primitive features—a pelvic girdle, vestigial hind limbs in the form of small spurs on either side of the cloaca, and a functional left lung. However, the bones of the skull are only loosely attached, which is not a primitive characteristic. This allows pythons to tackle relatively large (even enormous) prey if necessary.

Heat Pits

Locating the prey is helped by the presence of heat-sensitive pits that are well developed on all pythons except two closely related Australian species, the black-headed python, *Aspidites melanocephalus*, and the woma, *Aspidites ramsayi*. Heat pits are also characteristics of the boas in the family Boidae, but the heat pits in pythons are situated within the labial and rostral scales, while those of the boas are positioned at the join between two adjacent scales. Although they obviously share a common ancestor, the two groups are distinct enough to be placed in separate families.

Another important difference is in their breeding habits: All boas (except one) give birth to live young, while all pythons lay eggs. Their distribution patterns are

Above a lake in Tanzania an African python,* Python sebae, *entwines its coils around a pied kingfisher.

Common name Pythons **Family** Pythonidae

Family Pythonidae—8 genera, 26–33 species of pythons, including Children's python, *Antaresia childreni*; black-headed python, *Aspidites melanocephalus*; amethystine python, *Liasis amethistinus*; carpet python, *Morelia spilota*; green tree python, *Morelia viridis*; blood python, *Python curtus*; royal python or ball python, *Python regius*; reticulated python, *Python reticulatus;* Bismarck ringed python, *Bothrochilus boa;* Papuan python, *Apodora papuana;* white-lipped (or D'Alberti's) python, *Leiopython albertisii*

also different. Ecologically, pythons fill many of the same niches that boas fill in America and Madagascar, but these parallels only go as far as the "typical" boas—there are no python counterparts to the sand boas and rosy boas of the subfamily Erycinae.

Types of Prey

As a rule, pythons are generalists. They can tolerate a wide range of climatic conditions, live in a variety of habitats, and eat many different kinds of prey (although populations in certain places may become specialized in eating certain prey species because they happen to be the most readily available). Three species of pythons are short and squat with short tails. Two of them (the royal python, *Python regius*, and the blood or short-tailed python, *Python curtus*) live in forests where they can easily conceal themselves. The third, the Angolan python, *Python anchietae*, lives among rock outcrops. These species rarely hunt actively. Instead, they ambush huge meals.

The giant pythons have more normal proportions than the short-tailed pythons, but their bodies become very heavy as they age. At this stage they often use their ventral scales to pull themselves forward in a straight line. They have carved a niche for themselves by tackling prey that other snakes can only dream of: gazelles, deer, pigs, dogs, and primates, including humans. When they are young and therefore more limited in their choice of prey, they are fairly active; but as they grow larger, they switch to a sit-and-wait strategy. Meals may not come along very often; but when they do, these pythons can stock up for several weeks or even months if necessary.

Ⓓ ***A woma python,* Aspidites ramsayi, *in desert near Uluru (Ayers Rock), Australia. Womas are endemic to Australia and are unusual in lacking the familiar-looking facial heat pits of most pythons.***

Other Pythons

A group of four Australian pythons with no collective common name is placed in the genus *Antaresia*. They are rather similar to each other and behave in much the same way as typical colubrid snakes such as ratsnakes and king snakes—apart from their ability to detect prey using their facial heat pits. Three pythons from the Pacific region are placed in genera of their own: the Bismarck ringed python, *Bothrochilus boa*, the Papuan python, *Apodora papuana*, and the white-lipped or D'Alberti's python, *Leiopython albertisii*.

The remaining pythons are placed in the genus *Liasis*, containing just three species, and the genus *Morelia*. Members of both genera are moderately large and occupy a variety of habitats in Asia and Australia. As its name suggests, the water python, *Liasis fuscus*, favors aquatic or semiaquatic places even though it feeds on birds and mammals. The green tree python, *Morelia viridis*, is also aptly named since it is a completely tree-dwelling species. Suprisingly, the carpet python, *Morelia spilota*, is not named for its favored habitat but for the intricate patterning on its body. Indonesian pythons include several species that also live in Australia (showing the geological affinity between fauna of the two regions) as well as several species that are unique to the Pacific region. They include four that are only found on New Guinea and small satellite islands, where they may coexist with the Pacific boas, *Candoia*.

Tropical Snakes

Pythons are mostly tropical in their distribution. If they occur outside the tropics, as in parts of Australia, they are restricted to warm environments such as deserts. Boas, on the other hand, especially the erycine species, have adapted to cooler places, usually by becoming smaller and by hibernating for several of the coldest months of the year. Pythons do not hibernate, although some of them may experience periods of relative inactivity during cold snaps or dry periods. Part of the reason for their failure to move into cooler places may be their breeding habits.

↑ *Maternal care of incubating eggs is an important way of ensuring the survival of the next generation. Here a female Burmese python,* Python molurus bivattatus, *coils around her egg clutch in Southeast Asia.*

Unlike other egg-laying snakes, pythons have evolved a system for ensuring that their eggs have the best possible chance of developing in safety. Female pythons of all species curl around their clutch, completely enclosing it within their coils. They remain there until the eggs begin to hatch, rarely moving more than a few feet away even to drink, eat, or sun themselves. Not only do they provide physical protection from predators, but they are able—up to a point—to control the temperature of the eggs. They do this by leaving the clutch to warm their body in the sun, then returning to the clutch to transfer the heat to it. In addition, by loosening or tightening their coils, they can help retain warmth or keep the eggs from

Egg Brooding—Who Cares?

Egg brooding is a form of parental care that is uncommon in reptiles as a whole and probably occurs in fewer than 3 percent of snakes. Parental care involves an investment of energy in the young or eggs over and above that used to produce them. There is nearly always a tradeoff between increasing the survival chances of the offspring and reducing the fitness of the caring adult—in this case the female. Because she is unable to leave the eggs to feed during the incubation, which may last as long as three months, she is unlikely to recover her body weight before the following breeding season. Under natural conditions, therefore, female pythons either lay smaller clutches or breed every other year at most. This behavior has probably prevented them as a family from spreading into regions that have a winter.

Having mated, a female python will spend six weeks with eggs developing inside her and a further eight to 12 weeks brooding them. In temperate climates that would account for much of her period of activity, leaving insufficient time for her to condition herself for a long period of hibernation. However, by brooding the eggs, the female may prevent them from being eaten—either simply by covering them or by directly defending them. To a certain extent, brooding offers some of the same advantages and disadvantages as bearing live young. It appears to have arisen in pythons as an alternative to the live-bearing breeding habits of snakes such as the boas.

overheating. They may also be able to regulate humidity in the same way. Some species—notably the Indian python, *Python molurus*—can even raise the temperature of their eggs by producing metabolic heat from their muscles, which they achieve by twitching repeatedly while the eggs are incubating. Although all pythons coil around their eggs, nobody knows for sure whether they all have the ability to produce heat internally.

Pythons in Folklore

Humans have been fascinated by pythons for centuries. Many of them live in places that are quite densely populated—in Southeast Asia, for example—and reactions to them vary from one culture to another. In Australia aboriginal folk stories describe a large snake, the rainbow serpent. It lives in deep waterholes during the dry season; but in the rainy season it leaves them and takes to the skies, appearing as a rainbow. The habit of many pythons of living around permanent waterholes in arid regions undoubtedly helped fuel such tales. Large snakes, probably pythons, also figure frequently in cave paintings by Aborigines and by the Bushmen, or San people, of southern Africa. Elsewhere in Africa pythons were worshipped, and African slaves carried their beliefs to the West Indies, notably to Haiti, where they took the form of voodoo cults. Today, large snakes are more often looked on with horror, and many are killed needlessly. Others are run over on roads, while in Southeast Asia pythons of at least two species are widely hunted for their skins, which are exported to western countries in the form of handbags, shoes, and other fashion items.

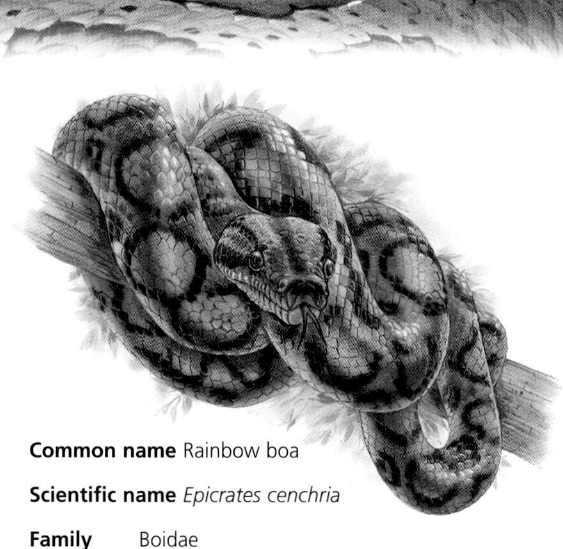

Common name Rainbow boa

Scientific name *Epicrates cenchria*

Family Boidae

Suborder Serpentes

Order Squamata

Length From 5 ft (1.5 m) to 6.5 ft (2 m)

Key features A powerful snake; body almost cylindrical with highly iridescent, glossy scales; markings varied according to 9 different subspecies

Habits Usually nocturnal; secretive and slow moving

Breeding Bears live young; litters contain 10–30 depending on subspecies—some have small litters of large young, others have large numbers of smaller young

Diet Mostly mammals

Habitat Rain forests, deciduous forests, open woodland, and pampas (grassland)

Distribution Central and South America from Panama to Argentina and Paraguay

Status Common

Similar species None, but variation in color and markings sometimes makes identification difficult

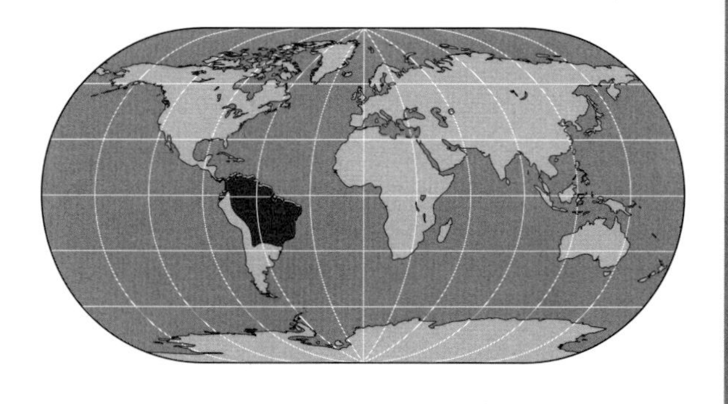

Rainbow Boa

Epicrates cenchria

Some forms of the rainbow boa are beautifully iridescent. In sunlight the surface of the scales shimmers with every color of the rainbow, giving the snake its common name. The iridescence is especially vivid just after shedding its skin.

THE RAINBOW BOA IN ALL its forms is a fairly slender but muscular snake with smooth, glossy scales. Its head is narrow, and there are nearly always five dark lines running from the snout and down its head: one down the center, a pair through (or just above) each eye, and two more below the eyes. Some stripes may be broken.

Color Variations

The markings on the snake's back consist of oval rings of black or dark brown running down the center and often broken along the midline. The centers of the ovals may be lighter in color than the areas outside them, as in the Argentine subspecies, *Epicrates cenchria alvarezi,* or they may be the same color, as in the Brazilian and Colombian subspecies, *E. c. cenchria* and *E. c. maurus*. There is an additional series of smaller markings, sometimes in the form of eyespots, on the flanks. They alternate with the dorsal markings.

The Brazilian subspecies and those from other countries neighboring the Amazon Basin are often larger and are considered to be the most colorful, often with rich orange or orange-red background colors. Those from cooler places are more likely to be predominantly brown or tan in color.

Some of the predominantly brown subspecies are thought to mimic pit vipers, such as those belonging to the genus *Bothrops,* which share the same habitat, and which have similar colors and markings.

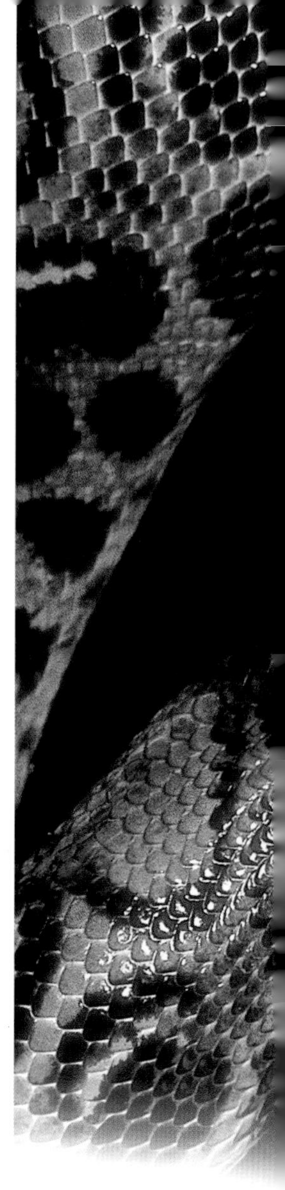

A rainbow boa in the Amazon rain forest in Ecuador. The surface of the snake's scales shimmers with color in the sunlight.

Rainbow boas have heat-sensitive facial pits, although they are not very deep and are present only on the upper jaw. They are generalists and can occur in a range of habitats. In much of South and Central America and on the island of Trinidad they live in rain forests, where they are quite secretive, spending the days in hiding and coming out at night in search of food. Large individuals crawl in a straight line and are not fast moving. They are able to climb and may rest up to several feet off the ground in sturdy trees and shrubs.

In regions where there are few or no trees, they are terrestrial. The Argentine subspecies, for example, rarely appears to climb, although its natural history is poorly known. Subspecies from cooler places are also usually smaller in size and differ in their behavior. While tropical subspecies are active throughout the year, those in cooler places become dormant during the cooler months. Their adaptability doubtless accounts for their wide range throughout much of Latin America and the Caribbean.

Fetal Membranes

Like all boas, rainbow boas bear live young, giving birth to litters containing between four and 25 young. The tropical subspecies have the largest litters, although each individual may be slightly smaller than those from less prolific subspecies. The young are born inside fetal membranes, or sacs, and usually break out within minutes of being born. Some observers have noticed females nudging and prodding the young while they are still encased in their membranes until they had worked themselves free. In other cases females that had just given birth inspected all the fetal sacs. If any proved to consist of infertile egg masses, the female ate them.

Where the Relatives Live

The rainbow boa is the only member of the genus *Epicrates* that occurs on the American mainland (and on the offshore island of Trinidad). All the other nine species live on various West Indian islands, including large islands such as Cuba and Jamaica and smaller ones such as the Turks and Caicos Islands. *Epicrates chrysogaster* and *E. exsul* live on the Bahamas. *Epicrates* species are among the most successful colonizers of Caribbean islands, which, although very rich in lizards, are relatively poor in species of snake.

Common name Reticulated python

Scientific name *Python reticulatus*

Family Pythonidae

Suborder Serpentes

Order Squamata

Length Typically from about 15 ft (4.5 m) to 18 ft (5.5 m); possibly up to 33 ft (10 m), but many reports are exaggerated

Key features Often huge with a large, wide head; dark stripe runs along center of head; another stripe joins each eye to the angle of the jaw; eyes orange; deep heat pits on snout and in lip scales; head covered with large scales; patterning complex, with angular patches of olive or brown with narrow borders of black and yellow; also a series of white markings, often triangular, along the flanks

Habits Mainly nocturnal; lethargic, spending most of its time doing nothing

Breeding Egg layer with very large clutches of up to 100 eggs

Diet Medium and large mammals and birds

Habitat Varied; forests, riversides, fields, plantations, and even cities

Distribution Southeast Asia (Indochina through the Malaysian Peninsula, Indonesia, Borneo, and the Philippines)

Status Common, but habitat destruction and slaughter for the skin trade are reducing numbers alarmingly in places

Similar species None in the region

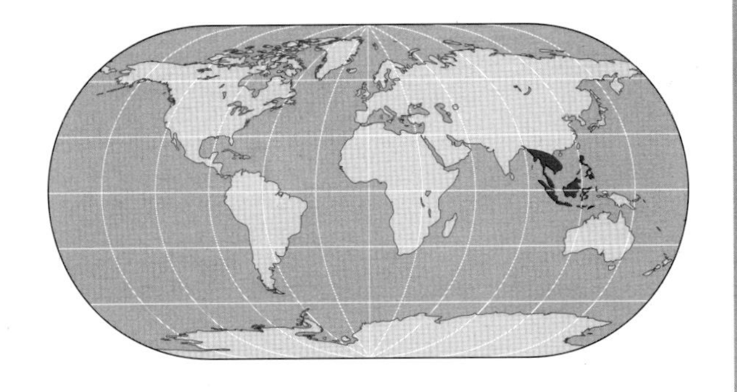

Reticulated Python

Python reticulatus

The reticulated python is famous, or infamous, because it occasionally eats humans. It is the most feared of the large snakes, even though the number of confirmed victims is relatively small.

THERE ARE SOME AUTHENTICATED accounts of reticulated pythons taking humans. Of all the giant snakes the reticulated python probably presents the greatest danger to humans, partly because of its size, of course; but there are other factors. Reticulated pythons often gravitate toward villages where there is an abundance of rats, poultry, and other domestic animals, and so they are frequently exposed to human scents and activities. People in Southeast Asia are often small and wear little clothing because of the warm climate; therefore their scent is stronger, and they appear less bulky. Significantly, most reports of python attacks are on children or women, who are relatively small.

The relationship with humans is not completely one-sided, however, because humans also eat reticulated pythons. The Dyak tribespeople of Borneo often eat pythons as well as any recent meals that might be in their stomachs. Because pythons that have recently fed are fairly immobile, it makes them easy prey. In other parts of Borneo, Chinese people traditionally keep pythons in the holds of their boats and in their warehouses to protect grain and other foodstuffs from mice and rats.

Exploitation

A more serious threat to the pythons is the skin trade. Most available figures are out of date but give some insight into the extent of the exploitation. Records of which species were involved, however, are hard to come by. Suffice it to say that reticulated pythons have always accounted for a good proportion of the snakeskin trade.

A hunter carries home his catch. Because of their size, pythons are a valuable food source for local people in many parts of the world.

In 1929 the United States imported about 5 million snake hides amounting to a value of nearly $3 million. By 1950 that had fallen to 4 million hides, while the United Kingdom imported 6 million (most of which were later re-exported, some to the United States). By 1992 the figure had dropped to 125,000 skins imported to the United Kingdom, about 60,000 being those of reticulated pythons. It is hard to say whether the fall in numbers is due to changing fashion, greater protection for snakes, or the increasing difficulty in finding snakes due to habitat destruction and overhunting. The latter is probably the most important factor.

Giant Specimens

Although the anaconda, *Eunectes murinus*, is usually regarded as the largest snake because of its great girth, the longest snake reliably recorded was a reticulated python. The specimen was killed by local people in the Celebes (nowadays called Sulawesi), Indonesia, in 1912. A civil engineering team working at a nearby mining camp accurately measured it at 32.8 feet (10 m).

Reticulated pythons often lie partially submerged in water waiting for thirsty prey animals to approach.

African Rock Pythons

The reticulated python's closest African relative is the African rock python, *Python sebae*, which is by far the largest snake in Africa. These snakes grow to 19.6 feet (6 m) in East Africa, although there is a dubious record of just over 32 feet (9.8 m) from West Africa.

Unlike the reticulated pythons, the rock pythons prefer open habitats, including savanna and acacia forests. They often frequent rocky places and river courses but avoid deserts and rain forests. Their range includes most of Africa south of the Sahara, as long as there is suitable habitat. They do not adapt to human settlement as well as other species, and they have been eliminated from large parts of South Africa where settlement and agriculture have occurred.

The African rock pythons look rather similar to the Indian pythons, *Python molurus*, but their markings are not as crisp or well defined. They are grayish-brown with darker brown markings down the back and on their sides. There is a wide, dark triangular mark on top of the head and a thin black line that passes from the snout through each eye and reaches the angle of the jaw.

African rock pythons eat a wide range of prey, including monkeys, gazelles, ducks, geese, pelicans, monitor lizards, and crocodiles. They may lie in wait near a water hole or submerged close to the water's edge for many days until a suitable victim approaches. They often kill and swallow their meals while still partly submerged. They also have a reputation for taking humans, and there are plausible accounts of them attacking children and small adults. After a large meal they are almost helpless and bask in the open to speed up digestion. At this time they are vulnerable to attack by predators such as jackals, hyenas, and humans.

In parts of southern Africa rock pythons are protected locally because they do a useful job in controling vermin such as cane rats. Although they are routinely killed for food (or often for no reason at all), there is no legal trade in the skins of African rock pythons.

The largest snake ever held in captivity was also a reticulated python, a female named Colossus. She was 28.5 feet (8.7 m) long and weighed 320 pounds (145 kg) at her heaviest. She lived at Highland Zoo Park, Pennsylvania, until her death in 1963.

In 1942 the National Zoo in Washington had one measuring 25 feet (7.6 m) and weighing 305 pounds (138 kg), and in 1959 the five largest snakes in American zoos were all reticulated pythons. Reticulated pythons are not among the most dynamic of display animals, however, since they spend the vast majority of their time doing absolutely nothing. Some zoos use the dubious technique of sending a keeper into the python's cage and dragging the snake to the front of the cage before the public arrives each morning!

Quick, Quick, Slow

Reticulated pythons illustrate well the way in which the rate of growth in snakes begins quickly and slows down in later years. Records show that one youngster grew from 24 inches (61 cm) to 6.5 feet (198 cm) in 18 months, while another, older snake took 11 years to grow from 11 feet (3 m) to 21 feet (6.4 m). Finally, a large individual from Thailand was 22 feet (6.7 m) when first imported to the United States and had grown to 27 feet (8.2 m) four years later, but was only 28.5 feet (8.7 m) long after another four years.

In other words, reticulated pythons grow rapidly (in length) during their first few years and gradually slow down as they approach maturity. Their growth rate dwindles to a barely measurable pace toward the end of their life, although like all snakes, they never (in theory) stop growing. Old reticulated pythons may become thin and fall into poor condition, however, and in the wild they would be killed by predators or starve to death.

Hiding and Hissing

Although we know something about the reticulated python's interaction with humans, little is known about their behavior in the wild.

Their best means of defense is camouflage. When resting coiled under low vegetation, they are hard to see because the irregular dorsal markings and the dark line that runs through the eyes disguises their outline. If discovered, the snakes often produce a loud, prolonged hiss designed to intimidate their enemies. Failing that, they may strike, sometimes bluffing but often in earnest, flinging their head forward with mouth agape.

There are mixed reports of the temperament of reticulated pythons. Some

↑ ***The presence of the photographer may have caused this reticulated python from Indonesia to open its jaws in such a threatening way.***

naturalists regard them as among the most aggressive of snakes, while others consider them docile and placid, only defending themselves in response to persistent tormenting. What is not disputed is their strength and the difficulty in handling them when roused. Some reports of encounters with these pythons relate how the snake managed to throw two or three coils around the snake collector, and in some cases it seems unclear who was capturing whom.

They are often cunning hunters. They may position themselves alongside game trails or near water holes for long periods of time in the knowledge that there is a good chance that suitable prey will eventually appear.

Large Clutches

The reticulated python is one of a very select group of snakes in which the females produce 100 offspring or more in one go. Of the other egg-laying species, two are also pythons (the African rock python, *P. sebae*, and the Indian python, *P. molurus*), and the other is the North American mud snake, *Farancia abacura*. Coincidentally, the mud snake is one of only a handful of colubrid snakes that coil around their eggs during incubation, as in the pythons, although it does not raise their temperature.

There are also a few live-bearing species that produce litters of 100 or more. They include two vipers (the puff adder, *Bitis arietans* from southern and East Africa, and the fer-de-lance, *Bothrops atrox* from Central and South America) and three colubrids (the African mole snake, *Pseudaspis cana*, the common garter snake, *Thamnophis sirtalis,* and the green water snake, *Nerodia cyclopion,* the latter two from North America). The record goes to a puff adder that produced 157 live young in a Czech zoo. These clutches and litters are not typical of the species but represent the limit of their reproductive capacity under ideal conditions.

The record for the largest clutch laid by a reticulated python is 103 eggs. Another clutch was found to contain 97 eggs, but the average appears to be between 40 and 50 eggs. The eggs are usually laid in disused animal burrows, in termite mounds, or inside caves.

Maternal Care

The female coils around the eggs throughout their incubation. She probably also provides them with additional heat by metabolic means if necessary, although this has not been well studied. Her presence may be mainly protective.

The hatchlings measure about 24 to 30 in (61–76 cm) when they emerge, and they grow very quickly at first. After they hatch, they may stay near the nest site for a week or more until they complete the initial skin shedding, after which they go their separate ways. They become sexually mature when they reach 9 to 10 feet (2.7–3 m) in length, which probably takes about five to six years.

The reticulated python has a complex geometric pattern. Along its back are irregular diamond shapes, usually flanked by smaller markings with light centers. This gives the snake a netlike (reticulated) pattern, hence its common and Latin names.

Giant Snakes of the Past

The Eocene epoch, which began roughly 60 million years ago, was the heyday of the boas and pythons. At that time the families were evolving into new species and spreading across the world. Fossils from this age include two large snakes from these families.

One was *Madtsoia bai*, which lived in what is now Patagonia (although the region had a much warmer climate in the Eocene). Based on fossil vertebrae, which is all they have to go on, scientists have estimated its length at more than 30 feet (9.1 m).

Another species, *Gigantophis garstini*, may have been even larger. It is known from fragments of jaw found in Egypt and could possibly have reached 60 feet (18.3 m). Scientists have calculated, however, that a snake over 50 feet (15 m) in length would be unable to support itself and move on land. If it was indeed longer than this, *Gigantophis* may have been aquatic or semiaquatic (just as large adult anacondas are today).

Either way, *Gigantophis* heralded the end of the dominance of boid snakes. It had died out by the end of the Eocene at exactly the same time as the smaller, perhaps more efficient colubrid snakes (belonging to the Colubridae) began to diversify and become widespread. The reticulated python and its giant relatives, therefore, are relicts of an age when boas and pythons were more common.

Ringneck Snake

Diadophis punctatus

Common name Ringneck snake

Scientific name *Diadophis punctatus*

Subfamily Uncertain, possibly Xenodontinae

Family Colubridae

Suborder Serpentes

Order Squamata

Length From 8 in (20 cm) to 30 in (76 cm)

Key features Body black, blue-black, dark- or olive-brown with a red or yellow underside and a collar of similar color just behind the head; scales smooth; head is small and hardly distinct from the neck; eyes also small

Habits Secretive, hides during the day; active at night or after rain in the early morning and evening

Breeding Egg layer with clutches of 2–10 eggs (usually 2–6) in June or July; eggs hatch after 28–42 days

Diet Invertebrates such as earthworms, small amphibians, lizards, and snakes

Habitat Damp places such as wood and forest edges, fields, farms, and gardens; in drier regions restricted to the areas around ponds and watercourses

Distribution Much of North America from coast to coast, including southern parts of Canada, extending to northern and north-central Mexico; possibly introduced to the Cayman Islands, West Indies

Status Very common to rare depending on locality

Similar species In Florida the red-bellied swamp snake, *Seminatrix pygaea,* or the worm snake, *Carphophis amoenus*, but they lack the red or yellow collar

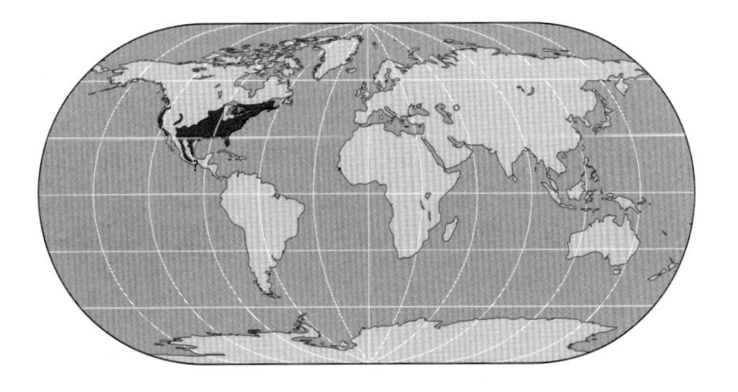

The secretive little ringneck snake is very fond of hiding and often goes unnoticed, even though it is one of the most numerous species in many parts of its range. It can occur in greater numbers than any other snake species.

THE RINGNECK SNAKE IS extremely prolific. In parts of Kansas ringneck snakes occur at the rate of 290 to 749 snakes per acre (719 to 1,849 per hectare). This is the greatest density of any known snake species.

Yet how can so many snakes go unnoticed? Apart from the fact that they are small, ringneck snakes spend most of the day hidden under flat rocks or in rotting logs and stumps. They also use human detritus in the form of discarded slates, tiles, and pieces of corrugated tin: 279 snakes were once found under just 24 pieces of tin. In an experiment 40 ringneck snakes were released into an area with 12 identical slates on the ground, but all the snakes grouped themselves under just four slates, leaving eight unoccupied. This suggests that they gain some advantage by clustering together, but nobody knows what that might be. However, it seems that they track each other down by smell. Ringneck snakes have a distinctive smell even to the poorly equipped human olfactory system.

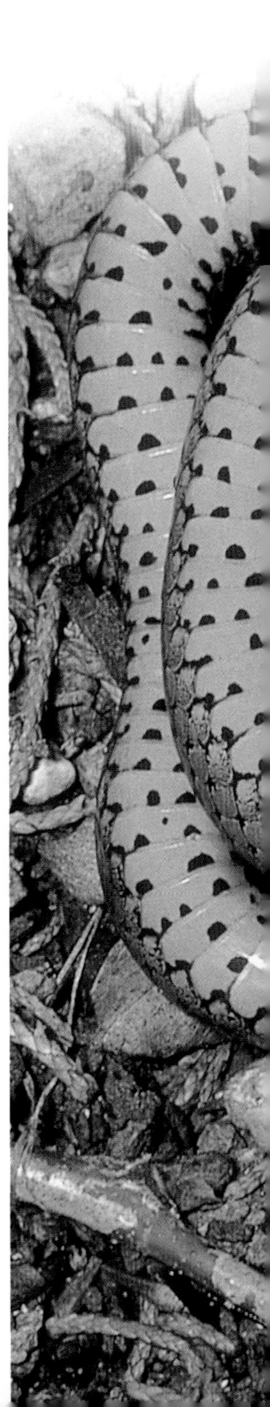

Damp Habitats

The most favorable conditions for ringneck snakes, and therefore the places where they are most common, are moist fields, woods, and gardens. Where they live in dry regions, they can only live in damp microhabitats, such as around the edges of pools in seasonal rivers and streams, or underground in rocky crevices and among moisture-retaining roots of cacti and shrubs. They emerge only at night, especially after rain. In the dry season they are not active on the surface at all. In parts of the southwestern United States they are absent over large areas because it is too dry, and many populations have become isolated from the rest of the species because of unsuitable conditions.

Foul-Smelling Substance

As well as using scent to communicate with each other, ringneck snakes use a strong-smelling secretion produced in their anal glands to deter predators, in common with several other species of snakes. They have a second line of defense, however, in the form of a dramatic display. If they are annoyed, they will coil up their tail like a corkscrew and show off their bright red, orange, or yellow underside. Apart from startling predators, the bright coloration is thought to deflect the predator's attention away from the snake's head, which remains hidden among its coils. If the predator attacks the snake's tail, it gets smeared with the foul-smelling anal secretion; and after attacking a few ringneck snakes, the predator may learn to leave them alone. If the tail display fails to produce the desired effect, the snakes may turn over onto their back completely, showing the bright underside of their body, and pretend to be dead.

Not all populations go through this sequence, however. Ringnecks from most of the eastern states have yellow bellies and do not display their tails, whereas those from the western states and Florida have red undersides and do display.

This regal ringneck snake from Arizona demonstrates its effective warning display by coiling its tail into a corkscrew and revealing its bright-red underside.

Although an accurate explanation for this is hard to find, scientists have noticed that scrub jays, *Amphelocoma coerulescens*, are common in places where ringnecks are most likely to display, but not in places where they do not. Scrub jays specialize in picking through leaf litter in search of food and could be expected to turn up ringneck snakes occasionally. It is possible that the tail display behavior has evolved in response to scrub jays or similar predators hunting in this way. In Kansas, where the snakes have been well studied, about two-thirds of juveniles coiled their tails when handled, compared with fewer than one-third of adults, so it seems that their behavior changes with age as well as with location.

Separate Identities

All subspecies of the ringneck snake intergrade with neighboring subspecies where their ranges come into contact, as is normal with subspecies. The regal ringneck, *D.p. regalis*, and the prairie ringneck, *D. p. arnyi*, interbreed over a wide area in Texas, for example, giving rise to a wide range of intermediate sizes (the prairie ringneck being significantly smaller than the regal ringneck).

In the Guadaloupe Mountains on the Texas–New Mexico border, however, they live side by side and do not appear to interbreed: Both subspecies maintain their identities, and the larger regal ringneck as well as the smaller prairie ringneck can be found in the same area with no intermediates. It seems that two populations were once separated by unsuitable habitat and evolved different characteristics, including size and, presumably, breeding behavior. When the populations came into contact again due to habitat change, they were different enough to prevent interbreeding. Interestingly, they did not evolve away from each other so much that they see each other as potential prey—the regal ringneck snake preys heavily on other snakes but does not appear to eat prairie ringnecks.

Unusual Biology

The ringneck is unique. There are no other species in the genus, and it does not seem to be very closely related to any other snakes. It is sometimes placed in the subfamily Xenodontinae ("strange-toothed snakes"), but scientists are not quite sure where it fits into the bigger picture.

Some aspects of its biology are quite unusual too. Despite its small size, it is very successful. Females take three years to reach breeding size. They lay small clutches of two to 10 eggs, but more usually two to six, and they only breed once each year. Even so, studies have shown that on average about 34 percent of newborn ringneck snakes survive their first year; and once they reach adult size, the survival rate goes up to 74 percent per year. That figure is very high compared to other snakes. Ringneck snakes probably live for 15 years, longer than many far larger snakes such as garter snakes, for example, which only live seven or eight years on average.

Female ringneck snakes lay their eggs in crevices, under flat rocks, in cracks in old buildings, and in abandoned small mammal and cicada holes. Several females may choose the same place to lay their eggs. Each egg is elongated like a sausage, and the shell is very thin and translucent. The embryo is quite well advanced when the female lays it, and it hatches after only four to six weeks.

The tiny young measure 3.5 to 5 inches (9–13 cm) long, and there are usually more males than females. Females are larger than males at hatching. This is unusual: In most snake species in which there is a difference in size between the sexes, both sexes start out the same size but grow at different rates. Male and female ringneck snakes, however, are different sizes from birth.

The young snakes probably eat small invertebrates. As they grow, their choice of food widens; and adults eat earthworms, salamanders, frogs, lizards, and other snakes. Their rear teeth do not have grooves but are enlarged. They produce venom that is not dangerous to humans. They subdue their prey by constricting it loosely while chewing on it, waiting for the venom to take effect. In this way they can overcome prey that is relatively large, especially if it is elongated, and ringnecks can eat snakes that are longer than themselves. They eat on average 27 meals in a season, and

the combined weight of their meals is about three times their own body weight. (Most other snakes eat more than this.)

In places where the summers are dry, such as Texas, ringneck snakes estivate (meaning that they remain inactive underground to avoid unfavorable conditions). They do this partly because the worms and amphibians that make up most of their diet are hard to find in dry weather. They spend the time hidden away in cracks in dried-out clay, under rocks, or in old rodent and insect burrows.

Variations

Generally speaking, ringneck snakes are aptly named. There are populations that have no collar, or only a very faint one, although they are still called ringnecks. One such population, the Key ringneck, *Diadophis punctatus acrirus*, lives on the Florida Keys, particularly on Big Pine Key and the Torch Keys, where its numbers have been reduced by commercial development. Fortunately, this area is also home to the endangered Key deer, *Odocoileus virginianus clavium*. A protected area has been set aside for the deer. By chance this may save the Florida ringneck snake as well.

Other variants are divided into about 13 or 14 subspecies. One that is very distinct is the regal ringneck snake, *Diadophis punctatus regalis* from some of the central and western states, including Arizona and Texas, and down into north-central Mexico.

This snake is much larger than the other subspecies, growing to 30 inches (76 cm) in places. It is steel gray in color and feeds almost entirely on other snakes. Its larger size may have evolved as a response to the dry environment in which it lives, since large snakes tend to retain moisture more efficiently than small ones (because they have a smaller surface-to-mass ratio). Its size may also enable it to eat larger prey since its preferred prey elsewhere (such as earthworms) is not found in dry places.

The ringneck snake employs partial constriction to subdue its prey. This regal ringneck from Arizona has enveloped a snake in its coils and is waiting for its venom to take effect.

Royal Python

Python regius

The royal, or ball, python is much better known in the pet trade than it is in the wild. Sadly, the species is collected in vast numbers for distribution to North America and Europe.

Common name Royal python (ball python)

Scientific name *Python regius*

Family Pythonidae

Suborder Serpentes

Order Squamata

Size From 36 in (91 cm) to 5 ft (1.5 m)

Key features Small and thickset with a small head and short tail; distinctive pattern of rounded brown or tan blotches on a black background; usually a row of blotches down the center of the back and a series of larger ones on its flanks, sometimes with one or more dark spots in them

Habits Nocturnal; terrestrial, resting on the forest floor among dead leaves or in burrows

Breeding Egg layer with small clutches of typically 3–6 large eggs; incubation period about 8 weeks

Diet Small mammals

Habitat Lowland and coastal forests and adjoining grasslands

Distribution West and Central Africa

Status Becoming rare in places due to continued exploitation for skins, meat, and the pet trade

Similar species None in the region; could possibly be confused with the Angolan python, *P. anchietae*, which is a similar shape, but lives farther south in southwestern Africa

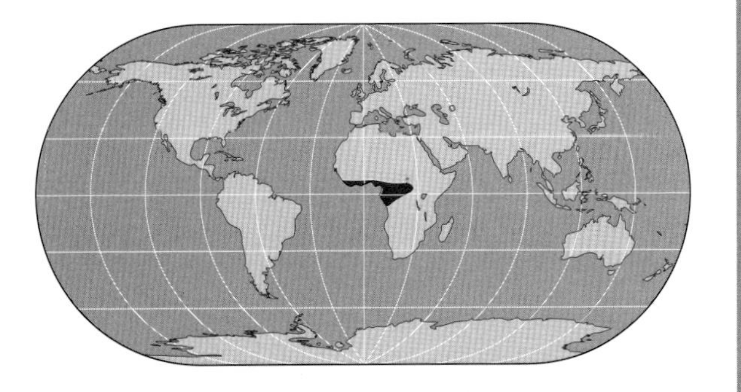

DESPITE DWINDLING NUMBERS in the wild, trade in royal pythons has been allowed to continue for reasons that are hard to understand. The area in which they live is subject to habitat destruction on a huge scale, and desert is encroaching on the northern part of their range. At present, the royal python is still considered to be fairly common in some places, but it is doubtful if the current rate of collecting can be sustained.

Although huge numbers of royal pythons are exported, virtually nothing is known of their lifestyle or natural history. They live in a region that has distinct wet and dry seasons. In the dry season they go underground to live in the burrows and tunnels of burrowing mammals. They are active in the wet season, using leaf litter and dense vegetation to hide themselves while they wait for prey to come within striking distance. They have large heat pits in the rostral scales of the snout and in the first four lip (labial) scales of the upper jaws. They eat mainly small rodents such as gerbils, mice, and rats.

Breeding takes place in the wet season, and females lay small clutches of large eggs, three to six being a typical clutch size. She coils around them in the usual python fashion and remains with them until they hatch about two months later. It has not been established whether or not she raises their temperature by shivering to produce endothermic heat, but it seems likely that she can do so if necessary.

The small shiny scales and beautiful colors and markings make royal pythons especially popular with the fashion industry. Unfortunately, their skins are exploited to make belts, shoes, and other fashion accessories.

"Balling" for Protection

The alternative common name of ball python alludes to the snakes' defensive habits. When threatened, they rarely attempt to bite. Instead, they coil themselves into a tight ball with their head in the center. They remain in this position

for as long as they feel in danger and only uncoil slowly after several minutes.

They are not the only snakes to use this technique: "Balling" is known in several species of wood snakes, *Tropidophis*, and in a number of much smaller species, including some of the blind and thread snakes.

The royal python also has enemies on a much smaller scale. Close examination of the skin of wild royal pythons often leads to the discovery of tiny parasites attached to the interstitial skin between the scales. They are ticks, blood-sucking relatives of spiders and scorpions (in the order Arachnida). Although many reptiles, including lizards and turtles, are parasitized by ticks, those that infest the royal python are unusual in that they are cunningly disguised to resemble one of the snake's own chestnut-brown scales.

Pits or No Pits?

The royal python has large and very conspicuous pits in the scales around its mouth. Other pythons (except for the black-headed python, *Aspidites maculosus*, and the woma, *A. ramsayi*) also have heat pits, but they are not always so obvious. There are also species among the boas that have pits, but they are not as numerous as the pythons. Neither the small erycine boas nor the members of the genera *Boa* and *Candoia* have pits. Recent research shows that some of the species that do not have obvious pits can nevertheless sense radiant heat through sense organs in the scales of the face. Where pits are present, they tend to point forward.

Many species, like the royal python, have the largest pits in their rostral scales at the tip of the snout. When viewed head-on, the pits on the labial scales of the royal python are clearly visible. They are located in parts of the face that are angled so as to have a clear "view" of the area in front of the head. A similar situation exists with the pit vipers, whose pits have evolved independently.

Common name Russell's viper

Scientific name *Daboia russeli*

Subfamily Viperinae

Family Viperidae

Suborder Serpentes

Order Squamata

Length From 36 in (90 cm) to 5 ft (1.5 m)

Key features Body has distinctive pattern of round brown spots, each edged in black, often surrounded by a thin white line; additional rows of similar oval spots present along each flank, alternating with the dorsal ones; background color is brown, gray, or pinkish; a broad dark stripe runs from the eye to the corner of the jaw

Habits Terrestrial; nocturnal

Breeding Live-bearer with litters of 20–63 (usually 30–45)

Diet Mammals and birds

Habitat Dry woodland, grasslands; also found around farms and villages

Distribution South and Southeast Asia from Pakistan through India and Sri Lanka across Indochina and into Thailand and Java; absent from the Malaysian Peninsula

Status Common

Similar species None

Venom Very dangerous

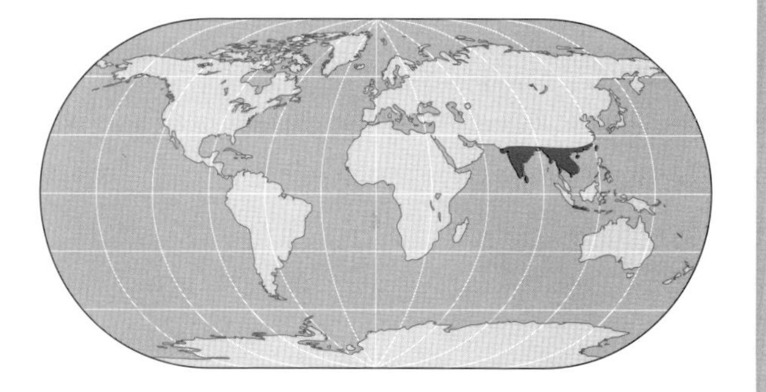

Russell's Viper

Daboia russeli

The beautiful colors and patterns of the Russell's viper belie the fact that this species is a major cause of human fatalities—more people are bitten by Russell's vipers than by cobras or kraits.

Russell's vipers live in a wide variety of habitats and are absent only from forests where the canopy shuts out the light. Elsewhere they often coil in the sun, well hidden among grasses and scrub, where their disruptive coloration provides good camouflage. They will ambush their prey from this position, but at night they become more active and set off to hunt. They often go into villages and even houses in search of rats.

The temperament of Russell's vipers is very like that of puff adders. They usually tolerate a certain amount of disturbance, often producing a long, low hiss if they feel under threat. The hiss is produced by forcing air out of its nostrils, which are unusually large. These vipers can strike with amazing speed if necessary, however, launching themselves in the blink of an eye from a seemingly peaceful resting position. They strike so hard that they may even leave the ground when hurling themselves at their enemy.

Young Cannibals

The diet of Russell's vipers consists of mammals such as rats, mice, squirrels, and kittens. They also take birds such as sparrows and occasionally agamid lizards and frogs. Young Russell's vipers are reported to be cannibalistic.

In experiments scientists have shown that Russell's vipers can sense a source of heat just as well as pit vipers despite the fact that they do not have the specialized facial heat pits of those species. Heat-sensitive nerve endings in the face detect the radiation instead. (Other vipers—perhaps even all vipers—are also able to

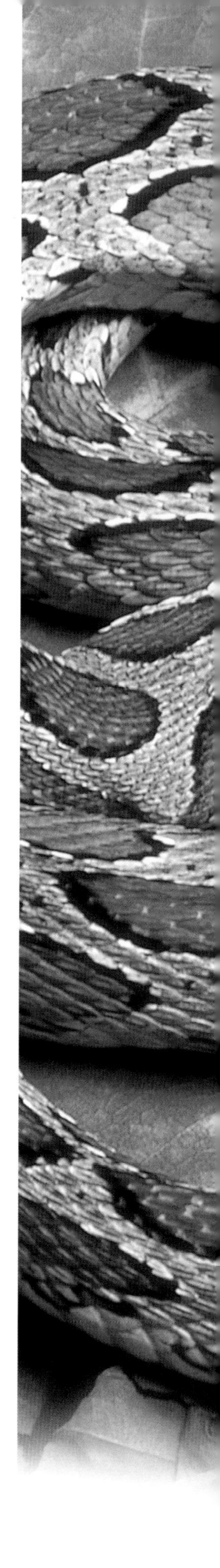

This closeup shows the beautiful markings of the Russell's viper. They make the snake very difficult to see when it is resting among vegetation.

do this). Without the pits, however, they may not be able to pinpoint the range and position of their prey as accurately, but no one has so far conducted experiments to investigate this.

Potent Venom

Russell's vipers deliver their venom via fangs that can measure well over 0.5 inches (1.5 cm) long in large specimens. An average-sized snake can deliver enough venom in one bite to kill two humans, although there are many examples of bites that resulted in a full recovery.

One victim that did not recover was the world's unluckiest dog. The dog was well and truly bitten by a Russell's viper and dragged the attached viper behind it for several yards as it tried to run away. Although it became rather ill, the dog had started to make a good recovery when it was bitten by another viper eight days later. This time it died in less than three hours.

The effect of the venom is that it rapidly coagulates the blood inside the blood vessels, leading to seizures and death from asphyxiation. Victims of bites suffer acute burning pain, swelling of the injured part, bleeding, discoloration, and sloughing (decay) of the tissue around the bite. Russell's viper venom has been introduced into hospital pharmacies, since the blood-clotting component can be isolated and used to help stem the flow of blood from wounds.

Spelling Mistake?

There is some confusion over the name of Russell's vipers. They are named for the early herpetologist Patrick Russell, author of *Indian Serpents* (1796). However, Russell's name was wrongly spelled "Russel" in the original description, so the snake's name was given that spelling. In scientific nomenclature once a spelling has been accepted as a valid name for a species, it can only be changed under exceptional circumstances (and this is not deemed to be one). This explains the seeming inconsistency whereby the common name is Russell's viper (two "l"s), while the scientific name is russeli (one "l").

Sahara Horned Viper

Cerastes cerastes

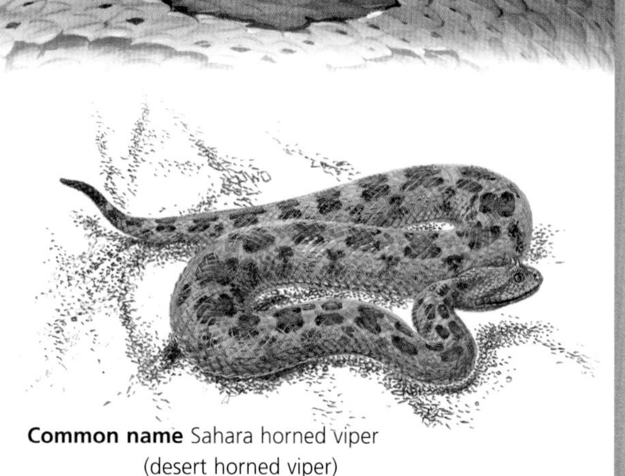

Common name Sahara horned viper (desert horned viper)

Scientific name *Cerastes cerastes*

Subfamily Viperinae

Family Viperidae

Suborder Serpentes

Order Squamata

Length From 24 in (61 cm) to 34 in (86 cm)

Key features Body short and stocky; heavily keeled scales give it a roughened, matt appearance; there is a long spinelike scale (horn) over each eye, but hornless individuals also occur; the back is buff or sandy colored with ill-defined darker blotches along the center and smaller dark patches on each flank

Habits Terrestrial; nocturnal

Breeding Egg layer with clutches of 10–23 eggs that hatch after 42–56 days

Diet Lizards, small mammals, and birds

Habitat Desert; found across the Sahara on loose sand and sandy gravel

Distribution North Africa from the Sinai Desert to the Atlantic coast

Status Common

Similar species Two other species of *Cerastes*: *C. gasperettii* is found in the Arabian Peninsula; *C. vipera* lives in the Sahara but is smaller than *C. cerastes* and never has horns

Venom Bites to humans are serious but rarely fatal

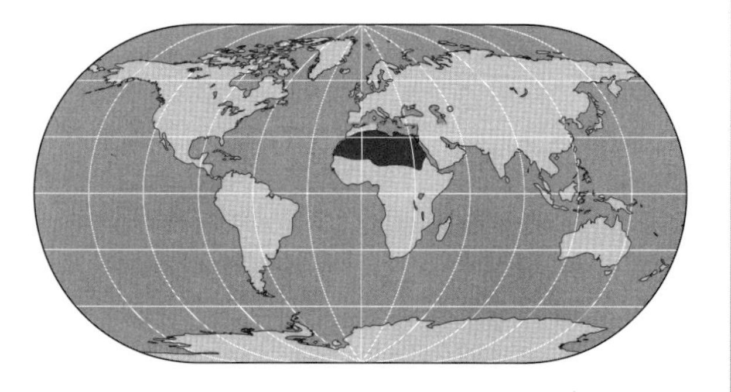

The horned viper is the Sahara's answer to the sidewinders of the American Southwest and the Namib Desert, but the "horns" of this species are more spinelike.

THE SAHARA HORNED VIPER and the American sidewinder, *Crotalus cerastes*, share part of their scientific name. *Cerastes* is simply the Latin word for horned viper and comes originally from the Greek *keras*, which means a horn.

The Sahara horned vipers live across North Africa from coast to coast, but not usually in the dunes of loose, windblown sand that make up the biggest part of the desert. Instead, they live in more permanent patches of sand, such as those that collect in hollows, on gravelly sand, and even on flat rock surfaces. They live among low shrubs and sparse grasses, but their numbers decrease as the vegetation increases.

The Sahara horned viper is well named, as this head-on view shows. The horns may help protect its eyes from injury or may simply enhance the snake's camouflage.

Avoiding Heat

Sahara horned vipers are also found near oases and palm groves. They are active mostly in the early morning and evening, especially in spring and fall. They move across the sand by sidewinding and leave the characteristic tracks for up to half a mile (1 km). In summer they become nocturnal, spending the heat of the day in the burrows of rodents or dab lizards, where the sun cannot penetrate.

As well as using their homes, they also eat small rodents such as gerbils, jirds and dab lizards, *Uromastyx acanthinurus*, including individuals up to 40 percent of their own body weight. They take small birds, such as wagtails and larks, but lizards including geckos, skinks, and lacertids are their most common prey. Young monitor lizards, *Varanus griseus*, have also been eaten, but this species usually turns the tables by preying on the snakes. Other predators of the Sahara horned vipers include foxes, desert hedgehogs, and short-toed eagles.

Faced with a predator, Sahara horned vipers go through a routine of passive warning in which they form a semicircular coil and rub the scales on their flanks together to produce a rasping sound. If that fails to have the desired effect, they inflate their body, hiss, and leap forward, often leaving the ground altogether while striking.

Related Species

Horned vipers from the Arabian Peninsula and parts of the Middle East used to be classified as subspecies of the Sahara horned viper (and the species was therefore known as the desert horned viper). The Middle Eastern population has been raised to a full species, however, and given the name *Cerastes gasperettii*. It is known commonly as the Arabian horned viper. As in *C. cerastes*, the species occurs with and without horns. The two species are almost identical except for minor differences in their scales, but they do not occur in the same places.

The third species in the genus is the Sahara sand viper, *C. vipera*. This species never has horns and is smaller than the other two. Loose shifting sands are its preferred habitat. Although its range and that of the Sahara horned viper are almost identical, they hardly ever occur together because of their subtle habitat differences. This species has a black tip to the tail, which it uses to lure lizards within striking range when it is almost completely buried in the sand.

Artificial Horns

Sahara horned vipers are favorites with snake charmers because the "horned serpent" has connotations of evil. In places where there are not enough horned vipers to go around, or where the showmen concerned are not too happy dealing with venomous snakes, harmless species have thorns implanted over their eyes to imitate those of the horned viper. Few tourists are in a position to know the difference.

Common name Saltwater crocodile (Indo-Pacific crocodile, estuarine crocodile)

Scientific name *Crocodylus porosus*

Subfamily Crocodylinae

Family Crocodylidae

Order Crocodylia

Size At least 23 feet (7 m) long; possibly up to 30 feet (9 m), but this is unconfirmed

Weight Possibly in excess of 2,418 lb (1,097 kg)

Key features Broad, powerful snout and strong jaws; a ridge runs from each eye toward the center of the snout; usually quite light in color when adult with overall gray or tan coloration; darker banding may be apparent on the tail with the smooth underparts creamy yellow; young are more colorful with clearer contrast between light and dark areas; has less bony protection on the head shield than any other living species of crocodile

Habits Highly aggressive predator; powerful swimmer; diurnal

Breeding Female lays 1 clutch of 60–80 eggs and guards them until they hatch after about 3 months

Diet Hatchlings prey on small aquatic animals; adults eat larger prey ranging from buffalo to sharks

Habitat Found in a variety of environments from rivers, lakes, and swamps to the open ocean

Distribution Eastern India, Southeast Asia, Papua New Guinea, Australia, and other Pacific islands

Status Relatively common; numbers have increased in some areas thanks to protective measures; IUCN Lower Risk; CITES Appendices I and II

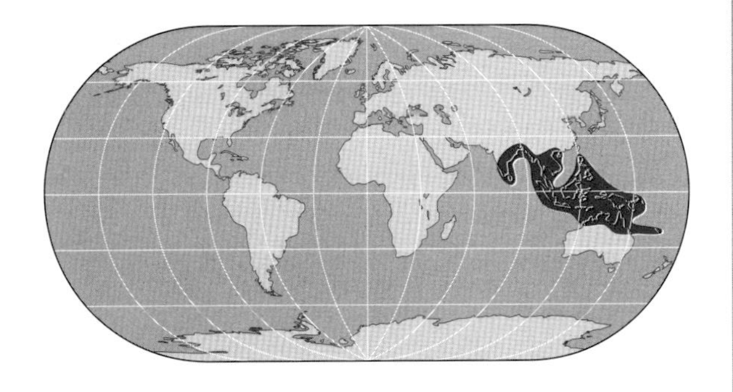

Saltwater Crocodile

Crocodylus porosus

The largest and most feared crocodile in the world, the saltwater crocodile is a confirmed man-eater, thought to kill as many as 1,000 people each year. Just like sharks, these crocodiles have acquired a gruesome reputation for congregating at the site of disasters, such as boating accidents, and preying on survivors.

THE SALTWATER CROCODILE is the most widely distributed of today's crocodilians. Its ability to travel long distances across the Pacific has enabled it to colonize islands far from the Asian mainland. Saltwater crocodiles have even been seen farther north in the vicinity of Japan. Recent research has also shown that the crocodiles on the Seychelles that went extinct in the 1800s belonged to this species rather than the Nile crocodile, *Crocodylus niloticus*, as previously thought. Some saltwater crocodiles that have been found even have well-established barnacles on their backs, indicating that they must have spent a considerable amount of time in the marine environment.

During the second half of the 20th century populations of saltwater crocodiles plummeted because large numbers were slaughtered for their skins, which are regarded as the most valuable of all crocodilian skins in terms of commercial trade. Wildlife management programs, especially in Australia, have since seen a resurgence in crocodile numbers, however. In addition, farming of these reptiles has resulted in a more sustainable trade.

Breeding

Courtship in saltwater crocodiles can be aggressive, and the males often fight among themselves in the breeding season. During

→ ***In the Adelaide River in South Australia a saltwater crocodile rears up out of the water as food is offered. Adults eat anything they can find, even large mammals.***

mating the male holds the female with his forelegs placed on either side of her neck. Males are sexually mature by the time they are about 16 years old, once they have grown to approximately 10.3 feet (3.2 m). Females start to lay at a smaller size, once they are about 7.2 feet (2.2 m) long and about 10 years old. They build a large nest mound for their eggs and guard the site diligently against potential predators such as monitor lizards, *Varanus* species, throughout the incubation period. They remain nearby for almost three months.

The female digs her eggs out of the nest once they are ready to hatch and carries the young hatchlings in her mouth to the safety of the water. At first, the young remain together in groups known as pods. Over half of them will die during their first year, a time when they are at their most vulnerable to predators because of their size. Young saltwater crocodiles can even fall prey to Australia's other crocodilian species, Johnston's crocodile, *C. johnstoni*, or be eaten by large snake-necked turtles, *Chelodina* species.

Attacks on Boats

Saltwater crocodiles sometimes exhibit strange behavior. Some large individuals regularly attack small boats, even persisting in damaging the craft when its occupants have been forced into the water. A particularly famous case of this kind involved a saltwater crocodile in Australia's Northern Territory, which became known as "Sweetheart." This individual is believed to have been responsible for a series of attacks that began in 1974 and only ceased after its capture in July 1979. This giant was found to measure 17 feet (5.2 m) long.

Although the propeller of the outboard motor was the usual target for the crocodile's attacks, Sweetheart actually bit directly through the aluminum hull of one boat. Luckily, although no one was hurt during these incidents, Sweetheart died soon after being caught, probably as a result of stress. It is now believed that the shape of the outboard motor seen from under the water and its resultant noise are confused by these reptiles with a rival crocodile entering their territory. Therefore they attack. This explains why it was only the boats (especially the motor, which is seen as the "head" of the invading crocodile) and not the people that were targeted by Sweetheart.

Common name Sand monitor (Gould's monitor, "racehorse goanna")

Scientific name *Varanus gouldii*

Family Varanidae

Suborder Sauria

Order Squamata

Size From 4 ft (1.2 m) to 5.3 ft (1.6 m)

Key features Body elongated but stout; head relatively small with pointed snout; tail rounded with dorsal crest on last part; front legs large and strong for digging; coloration variable depending on locality but usually tan to yellow with contrasting spots

Habits Terrestrial with a wide home range; spends much of the day foraging for food; rests in burrows during inactive periods

Breeding Average clutch size of 6–7 eggs laid in excavated nest in soil or in termite mounds; eggs hatch after about 3–4 months

Diet Lizards, small mammals, eggs, and birds

Habitat Subhumid to arid areas, deserts to jungle rivers; generally favors sandy soil; in the north of its range habitat includes tropical woodland

Distribution Most of mainland Australia (except the extreme south and Victoria); recently reported in New Guinea

Status Common

Similar species The Argus monitor, *Varanus panoptes*

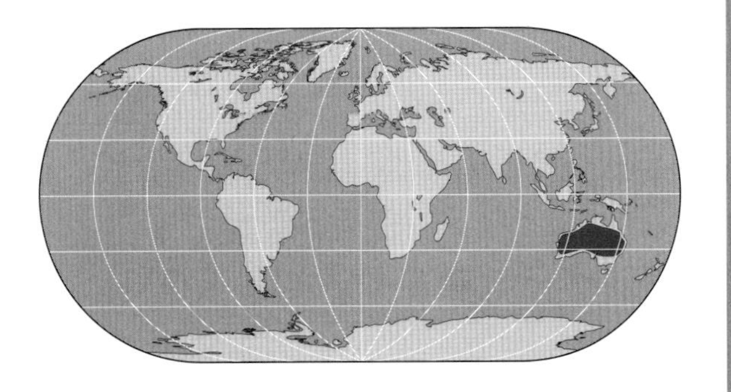

Sand Monitor *Varanus gouldii*

One of the largest Australian monitors, the sand monitor is sometimes known as the "racehorse goanna" because it is capable of extreme speed, often running on two legs over short distances.

DISTRIBUTED OVER MOST of the Australian mainland, the sand monitor is the most widespread goanna. Its habitats vary from the hottest desert areas to humid tropical forest. Sand monitors like to dig at the base of low vegetation, using their long, stout claws to excavate a sloping burrow with an expanded chamber at the end.

Although they swim reasonably well, they are unable to remain under water for more than a few minutes. They are mainly terrestrial and not very good at climbing. But if that is the only means of escape, they overcome their reluctance. Activity levels are seasonal. During the warm months (September to February) they are very active. From March to August activity levels fall dramatically, and they spend most of the time in burrows or tree hollows.

Sand monitors are capable of standing and running on their hind legs. This is known as bipedalism and appears to have evolved in a few monitor species partly to allow the creatures to look over tall vegetation or rocks. In the case of males they use this stance to grapple with rivals. While bipedal locomotion is fast, it cannot be sustained for very long.

Taxonomy

Two subspecies are recognized: *V. gouldii flavirufus*, which inhabits true sandy deserts and is the smaller of the two, and *V. gouldii gouldii*, which is found in areas with more rainfall and more vegetation. The latter is larger, since its habitats have higher densities of prey.

Sand monitors have a long, stout body with a small head and a pointed snout. The long, muscular front legs end with sharp claws and are ideal for digging in hard ground and into mammal tunnels. The tail is compressed from

side to side and helps in swimming. Color can be dark brown or black to light yellow and reddish. Large pale spots form rows between the neck and the base of the tail. The tail is paler with narrow bands, and the last part has no markings. A dark, white-edged streak runs from the eyes to the ears. These colors provide camouflage in desert areas and among grasses.

Sand monitors are opportunistic feeders, taking whatever they can whenever they can. Although they have been seen to eat carrion, they prefer to capture live prey, such as geckos, other lizards, some of the pygmy monitors, as well as young of their own species. They also eat reptile eggs, baby mammals, and young birds. In the northern part of their range they readily consume cane toads, *Bufo marinus*, but they are vulnerable to the toads' toxins and often die as a result.

Because they can grow up to 5.3 feet (1.6 m), adult sand monitors have few enemies. Juveniles are more vulnerable. When cornered, sand monitors inflate their throat, rise up on their hind legs, hiss, and lunge at their aggressor. They are frequently caught by Aboriginals who kill them and cook them.

Little is known about reproduction in the sand monitor. Mating has been observed during October to December. Eggs are laid in burrows dug in hard soil, inside tree stumps, or occasionally in termite mounds. Incubation time is thought to be about three to three and a half months, and the hatchlings emerge in January to March. Juveniles are more brightly colored than the adults.

Hunting Methods

Sand monitors hunt mainly by smell. They forage for over 1.2 miles (2 km) a day. In desert areas where food densities are low, *V. g. flavirufus* needs to travel farther in search of food. The monitors search for scent trails by swinging their long neck and head from side to side in a large arc, covering as much ground as possible, and constantly flicking out their long, forked tongue.

When they pick up a scent, they follow it to its source—usually a burrow. They use their front legs to dig up the prey, making sure they keep their head well down with their sharp teeth at the ready to snatch up the prey before it can escape.

Spreading its feet wide and raising itself to its full height, a sand monitor stands on its hind legs to look over tall grass or bushes. As it does so, it constantly flicks its tongue to test the air for scents.

Sea Snakes *Pelamis* and other genera

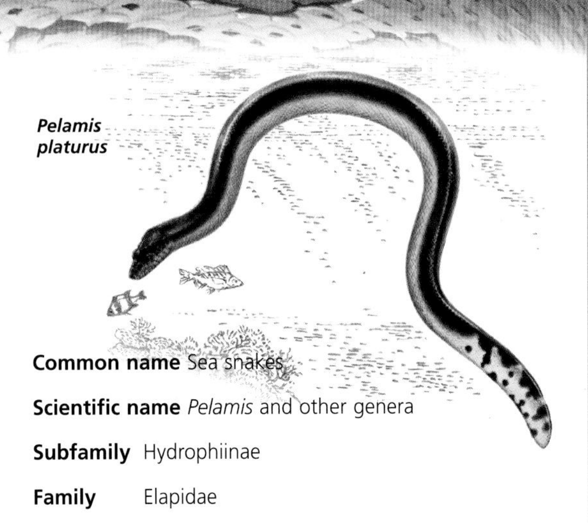

Common name Sea snakes

Scientific name *Pelamis* and other genera

Subfamily Hydrophiinae

Family Elapidae

Suborder Serpentes

Order Squamata

Number of species 47–50 in 14 genera

Length From 18 in (45 cm) to 9 ft (2.7 m)

Key features Body flattened from side to side toward the tail, which is oar shaped; eyes small and usually placed on the side of the head; nostrils valvular and can be closed when diving; ventral scales greatly reduced; some species have small heads and thick bodies; various colors but often gray, brown, or olive; some have black-and-white bands; 1 species is bright yellow and black

Habits Exclusively marine

Breeding Bear live young, giving birth at the water's surface

Diet Fish, crustaceans, and fish eggs

Habitat Open seas, estuaries, mud flats, mangrove forests, and tropical reefs

Distribution Northern Australia, the Indo-Pacific region extending to the Arabian Gulf; *Pelamis platurus* has a much wider distribution

Status Mostly common, sometimes numerous

Similar species Sea kraits have dark and light rings (as in some sea snakes) but less compressed bodies

Venom Depending on species, may be very potent; most species are reluctant to bite, but a few can be aggressive

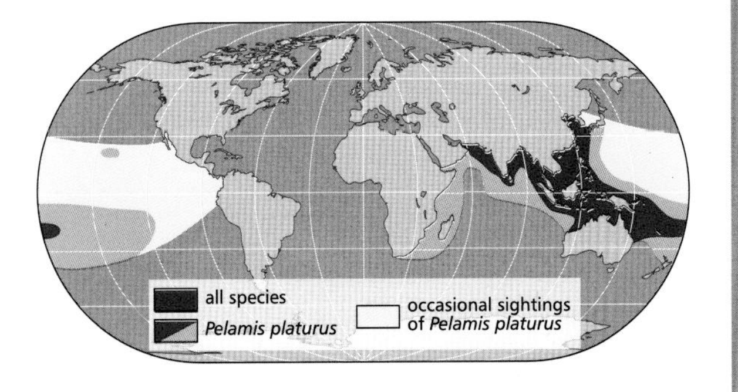

The "sea serpents" of ancient times were huge creatures that could rear up out of the ocean, coil themselves around ships, and pluck mariners from the decks. Real sea snakes have more modest proportions and are less ambitious in their feeding habits.

Snakes have colonized the oceans on several occasions. Fossils of snakes thought to be sea snakes are known from Tertiary deposits, but they had died out by the Eocene period more than 35 million years ago. The 50 or so species of living sea snakes are the result of at least three separate invasions from the land. A group of five species known as the sea kraits, *Laticauda*, is the least adapted to life in the sea.

The other species, the "true" sea snakes that bear live young, are more thoroughly adapted to the marine environment. They evolved from live-bearing terrestrial snakes, of which there are many in the Australian region. Even so, some are less well adapted than others: Three species live only in estuaries, among mangrove forests and on mud flats. They are *Ephalophis greyi*, *Hydrelaps darwiniensis*, and Merton's sea snake, *Parahydrophis mertoni*. They differ from other sea snakes by having wide ventral scales, cylindrical bodies, and only slightly flattened tails. In other respects, however, they resemble the rest of the subfamily, so they have characteristics of land snakes and sea snakes. A number of other species live around shallow reefs. They are the turtle-headed sea snakes, *Emydocephalus* species, and the snakes in the genus *Aipysurus*.

About 15 types of sea snake occur in the waters of the Great Barrier Reef. The olive sea snake,* Aipysurus laevis, *is the one most commonly encountered by divers.

Built for Swimming

All sea snakes are practically or totally helpless on land. Their unusual lifestyle involves some unique adaptations. Snakes in general are well suited to a life in water, and their basic body plan provided a good springboard for the challenges of their new marine environment.

Serpentine locomotion, for example, works well in the water and allows movement in three dimensions. Some species of sea snakes, notably the yellow-bellied sea snake, *Pelamis platurus*, can swim almost as well backward as they can forward simply by putting their sinuous movements into reverse.

Their elongated bodies contain large, elongated lungs with a large capacity. All sea snakes are venomous, having evolved from land elapids. Catching and restraining slippery fish underwater is helped enormously by the snakes' ability to subdue them quickly.

Special Adaptations

The sea snakes also evolved new characteristics to enhance their performance even more. In the locomotion department their bodies became flattened from side to side, especially toward the tail, which is oar shaped in most species. Species from open water even have flattened vertebrae to support the "oar" and give it more thrust. All sea snakes' nostrils are on top of the

Sea Snake Classification

47–50 species in 14 genera (excluding the 5 species of sea kraits, *Laticauda*):

Genus *Acalyptophis*—1 species from Australia, *A. peroni*
Genus *Aipysurus*—7 species from Australia and New Guinea
Genus *Astrotia*—1 species from the Indo-Malayan region to Australia, *A. stokesii*
Genus *Disteira*—3 species from the Persian Gulf to Australia
Genus *Emydocephalus*—2 species from Australia and South China Sea
Genus *Enhydrina*—1 species from Australia and New Guinea, *E. schistosa*
Genus *Ephalophis*—1 species found only in northeast Australia, *E. greyi*
Genus *Hydrelaps*—1 species from north Australia and south New Guinea, *H. darwiniensis*
Genus *Hydrophis*—22–25 species from South Asia to Australia
Genus *Kerilia*—1 species from Malaysia and Indonesia, *K. jedoni*
Genus *Lapemis*—4 species from the Persian Gulf to Australia
Genus *Parahydrophis*—1 species from north Australia and Aru Islands, *P. mertoni*
Genus *Pelamis*—1 species found throughout the tropics, *P. platurus*
Genus *Thalassophis*—1 species from Malaysia and Indonesia, *T. anomalus*

⇑ ***The turtle-headed sea snake,* Emydocephalus annulatus, *uses its hard, pointed snout, which resembles that of a turtle, to prod and break up clusters of fish eggs before devouring them.***

head and have valves that can be closed when diving. Extra waterproofing is achieved by means of a shaped rostral scale that seals the notch in the lower jaw (the lingual fossa) through which the snake pushes its tongue.

With the exception of the three estuarine species, the ventral scales of sea snakes are narrow, no wider than the dorsal scales. Stokes's sea snake, *Astrotia stokesii*, has two parallel rows of ridges along its ventral midline. Several of the open-water species have very thick bodies that taper sharply toward the head and neck, which are narrow. The turtle-headed sea snake, *Emydocephalus annulatus*, has a blunt, conical spine on its snout; and the head scales of both sexes of the eyelash sea snake, *Acalyptophis peroni*, have raised projections, which gives them a spiky appearance. No one knows the purpose of these unusual scales.

Seafood Diet

All sea snakes depend to some extent on fish for food. A few species, such as the olive sea snake, *Aipysurus laevis*, and Hardwicke's sea snake, *Lapemis hardwickii*, also eat prawns, crabs, squid, and cuttlefish.

However, most sea snakes are specialists, with distinct preferences for certain types of prey—nearly always slow-moving species that can be trapped in burrows or crevices. But the

yellow-bellied sea snake, *Pelamis platurus*, which lives in the open ocean, eats the small fish that accumulate around flotsam. This species effectively ambushes its prey by living among the debris that collects along "driftlines"—strips of calm water occurring where two ocean currents come together. Small fish are attracted to the debris because of the cover it affords; and by living permanently among it, the snakes have a constant supply of food. Because their vision is poor, it is thought that they use small changes in water pressure to detect and locate fish in the immediate area.

Eels are high on the list of preferred food, especially for *Hydrophis* species. The large genus contains some of the "thin-necked" sea snakes. Because of their shape they can force their heads and necks into eel burrows. They hunt by cruising over the seabed, probing for eels in every burrow they come across. There is an advantage to eating eels: Because they are long, the snakes can eat a much larger meal than they could with more heavily built fish.

The beaked sea snake, *Enhydrina schistosa*, feeds largely on catfish, which it finds by chance while swimming randomly just above the seabed in murky water. If it senses one nearby or makes contact with one, it strikes sideways and holds onto its prey with its jaws. The snake then turns into the current so that the fish is held against its jaws, making escape difficult. Once the snake's venom takes effect and the catfish stops struggling, the snake gets a better grip if necessary and starts to swallow it headfirst, now turning its head away from the current so that the flow of water helps keep the fish aligned.

Egg Specialists

The most specialized feeders, however, are three species of sea snakes that only eat fish eggs. They are the two turtle-headed sea snakes,

***Hydrophis* species have the flattened tail that is typical of sea snakes. Many sea snakes have black-and-white banded coloration.**

A Freshwater Sea Snake (A Contradiction in Terms)

Like the Solomons sea krait, *Laticauda crockeri*, one species of sea snake has become trapped in a freshwater lagoon long enough to have become different from its closest relatives and to be regarded as a separate species. The Lake Taal sea snake, *Hydrophis semperi*, lives in the lake of the same name on the island of Luzon in the Philippines. Scientists think that a massive prehistoric eruption blew the top off the Taal volcano, leaving a huge caldera (crater). Seawater rushed in, creating a lagoon 20 miles (32 km) in diameter, which sea snakes then entered. Over the centuries a smaller volcano closed the channel leading to the sea, isolating the sea snakes and allowing the seawater to become diluted with rainwater that drained from the surrounding slopes. Practically nothing is known about the natural history of these snakes.

Emydocephalus annulatus and *E. ijimae*, and *Aipysurus eydouxii*. These species apparently search in crevices and burrows for the buried eggs of gobies, blennies, and other bottom-dwelling fish.

Their teeth are extrmely reduced in size, and their venom is weak compared with that of other sea snakes, but they have specialized muscles in the head that probably help them suck up the eggs. The only other snakes that have comparable muscles are the blind snakes, some of which suck the contents out of invertebrates and their larvae.

Oceangoing Snakes

The highly specialized lifestyle of the yellow-bellied sea snake, *Pelamis platurus*, is unique even among sea snakes. The species has a black back and a yellow underside and lives in floating colonies. Individuals congregate at places where two ocean currents meet and create a narrow band of slack water, and where seaweed and other floating debris accumulates. They are the only sea snakes that "put to sea," drifting out over the deepest waters and going largely where the ocean currents take them.

Although other sea snakes are restricted to the South Pacific archipelagos and the coasts of southern Asia and northern Australia, the yellow-bellied sea snakes also occur across the Indian Ocean and down the East African coast as far as the Cape of Good Hope. The cold Benguela current prevents them from straying farther west, and to the east they extend as far as the southern Japanese islands. In midocean they occur as occasional strays, but enough snakes must have made the voyage east for them to have become established along the west coast of Central America, where they are relatively common. Their bright colors warn potential predators that they are venomous, and their skin and flesh also seem to be distasteful—in experiments birds and fish quickly regurgitated pieces of dead yellow-bellied sea snake that they were given.

The yellow-bellied sea snake is the most widely distributed of all sea snakes. Its yellow-and-black longitudinal striped pattern makes this species unmistakable.

Deep-Sea Diving

Although sea snakes normally forage in shallow water and come to the surface every half hour or so, some species can dive as deep as 500 feet (152 m) and stay under for up to two hours. In order to stay underwater for so long, their lungs need a greater capacity than those of land species. The part of the lung extending backward, known as the saccular lung, is especially large in sea snakes. It is nonfunctioning but forms a large air store. Unlike the saccular lungs of other snakes, its wall is thick and muscular, so the snake can force the stored air forward into the functional part of the lung (the bronchial lung), where the oxygen is extracted. The lungs may also act as buoyancy organs, assisting the snake in diving and surfacing.

Sea snakes are also unusual in the amount of gaseous exchange that takes place across the

A pair of olive sea snakes, **Aipysurus laevis**, *swim together during courtship in seas around the northern Great Barrier Reef.*

surface of their skin. Some species are able to absorb up to one-fifth of their total oxygen requirement despite their scaly covering.

Salt Balance

By living in seawater, which consists of a more concentrated solution of salts than their own body fluids, sea snakes have a tricky problem to overcome. There is a tendency for water to flow out of their body by osmosis, causing them to dehydrate. Unlike those of mammals, snakes' kidneys cannot excrete salt at higher concentrations than their blood, so a different method of excretion has to be found.

Their main line of defense is their skin, which is less permeable to water than that of other snakes. In addition, they have a special gland under their tongue, known as the sublingual gland, that accumulates excess salt (from their diet, for example). A duct leads from the gland to the sheath surrounding the tongue, so that each time they push out the tongue, a small amount of concentrated salty water is pushed out with it. Sea kraits and the marine file snakes (Acrochordidae) have similar glands.

Reproduction

All the true sea snakes give birth to live young. This is essential, since they cannot come ashore to lay eggs, and eggs laid in seawater would die. Mating usually takes place on the seabed, but yellow-bellied sea snakes, *P. platurus*, mate at the surface of the water, the male coiling around the female. Other species form huge aggregations or "slicks" in the breeding season, but their breeding behavior is rarely seen and impossible to study.

The little that is known about their habits leads us to believe that most species do not breed every year and that, on average, sea snakes have smaller broods than equivalent land species (in terms of total weight compared to mother's weight). Herpetologists think that is because the lungs in sea snakes take up a large proportion of their body cavity, and a heavy burden of developing young would keep them from swimming efficiently. Many species have only three or four relatively large young, but some have up to 18 smaller young.

In the Philippines sea snakes are collected in their thousands. Their flesh is eaten, and their skins turned into goods for the tourist trade, such as shoes and bags.

Sidewinder

Crotalus cerastes

The common name for the sidewinder refers to its method of locomotion. It moves quickly over loose sand with a characteristic sideways looping technique.

Common name Sidewinder (horned rattlesnake)

Scientific name *Crotalus cerastes*

Subfamily Crotalinae

Family Viperidae

Suborder Serpentes

Order Squamata

Length From 24 in (61 cm) to 30 in (76 cm)

Key features Body smaller and more slender than other rattlesnakes; head flat with raised scales over the eyes, giving the appearance of horns; pattern consists of large blotches of various colors interspersed with dark speckles; background coloration varies with the soil type and may be yellow, beige, pink, or gray

Habits Terrestrial; semiburrowing by shuffling its body down into loose sand; active at night; moves in a sideways looping motion

Breeding Live-bearer with litters of 5–18; gestation period 150 days or more

Diet Mainly lizards, especially whiptails; also small rodents

Habitat Deserts, especially where there are extensive dunes of loose, wind-blown sand

Distribution Southwestern North America

Status Common

Similar species The speckled rattlesnake, *Crotalus mitchelli*, has a larger rattle, but it lives mostly among rocks and does not sidewind

Venom Not very potent, causing pain and swelling around the site of the wound

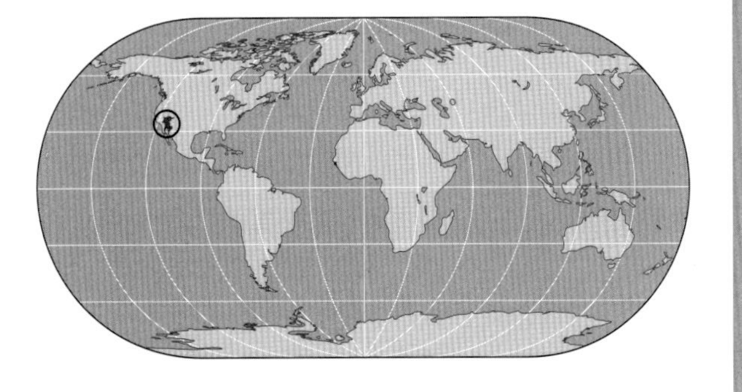

WHEN SNAKES ARE PLACED on a very smooth or a yielding surface such as sand on which it is difficult to gain a hold, many of them use a sidewinding locomotion in their attempts to move. This is also true of the sidewinder, which has developed it into an efficient and rapid way of covering loose ground.

Other desert species that sidewind include the carpet, or saw-scaled, vipers, *Echis* species from Africa and Asia, Peringuey's adder, *Bitis peringueyi* from southern Africa, the Sahara horned viper, *Cerastes cerastes* from North Africa, and the Patagonian lancehead, *Bothrops ammodytoides*. Another type of sidewinder is the bockadam or dog-faced water snake, *Cerberus rynchops*, from Southeast Asia and northern Australia, which sidewinds across the tidal mudflats on which it lives.

Desert Nomads

Sidewinders are unusual among snakes (especially rattlesnakes) in having no home range. They appear to wander randomly across the desert, covering several hundred yards each night in search of prey, probably looking for sleeping lizards in their burrows. In the morning the sidewinders' distinctive tracks often indicate the route they have taken. By daybreak they shuffle down into the sand to escape the heat, often choosing a patch at the base of a shrub where the sand is cooler. The next evening they emerge to continue their wanderings.

If the opportunity arises, they will ambush prey from their daytime resting position, striking up out of the thin layer of sand that covers them and holding on until their venom takes effect. This avoids the need to venture out in order to locate their dead prey later. If they had to leave their cover and move onto an open patch of sand, they would be very conspicuous.

The sidewinder,* Crotalus cerastes, *has horn-shaped scales bulging out above the eyes, earning it the nickname of "horned rattlesnake" in some places.

Diet of Lizards

They mainly eat desert lizards, especially whiptails, *Cnemidophorus*, which are almost unbelievably fast moving during the day when they are warm but easy to catch at night when they are cold and therefore not very energetic. Other prey species include side-blotched lizards, *Uta stansburiana*, and fringe-toed lizards, *Uma notata*. One captive sidewinder was seen eating a desert horned lizard, *Phrynosoma platyrhinos*, without any apparent ill effects; but on the other hand, a wild one was found dead with one of these spiky lizards lodged in its throat. Immature sidewinders apparently use their tails as lures for small lizards, moving them slowly across the ground so that the segments look like those of a crawling insect.

Apart from lizards sidewinders occasionally eat small desert rodents and birds, although opportunities to eat the latter must be very limited since they are hard to catch. Sidewinders have been found swallowing small rodents, apparently killed by traffic in which rigor mortis had set in. Vipers in general, including many rattlesnakes, will take advantage of food that they did not kill themselves.

Some Like It Hot—but Not the Sidewinder

Sidewinders come from the hottest parts of the American Southwest, including Death Valley, California, which has the highest-ever recorded temperature on earth. You might think that would make them more tolerant of higher temperatures than other snakes, but that is not so.

Sidewinders avoid the extremes of heat by being strictly nocturnal. If they are exposed to daytime temperatures, they soon die. Their somewhat flattened body shape enables them to gain and lose heat quickly from the soil or other surface because more of their body is in contact with the surface. As the temperature falls at night, they form a flattened coil on a warm surface (for example, on tarred roads) to absorb heat quickly and remain active. During the day they shuffle down into cool sand and lose the heat.

A sidewinder moves across the sand in the Anza-Borrego Desert, California. The sideways looping technique keeps much of its body off the ground, enabling it to move efficiently across shifting surfaces.

Skinks

Common name Skinks **Family** Scincidae

Family Scincidae 115–124 genera, about 1,400 species, but this classification is disputed and may be revised. Important genera include:

Genus *Acontias*—8 species of medium to large legless skinks from South Africa
Genus *Amphiglossus*—36 species of semiaquatic and aquatic skinks from Madagascar and other Indian Ocean islands
Genus *Carlia*—26 species of rainbow skinks from Australasia
Genus *Chalcides*—25 species of cylindrical or barrel skinks from Mediterranean countries and North Africa through to western Asia and India, including the three-toed skink, *C. chalcides*
Genus *Corucia*—1 species from the Solomon Islands, the monkey-tailed skink, *C. zebrata*
Genus *Cryptoblepharus*—29 species of wall or fence skinks from Australia, various Indo-Pacific islands, southeastern Africa, and Hawaii
Genus *Ctenotus*—95 species of striped skinks from Australia and 1 species from New Guinea
Genus *Dasia*—8 species of lizardlike skinks from Southeast Asia
Genus *Egernia*—31 species of smooth and spiny-tailed skinks from Australia, including the spiny-tailed skink, *E. stokesii*
Genus *Emoia*—74 species of slender skinks from coasts of Australia and various Indo-Pacific islands
Genus *Eumeces*—40 species of skinks from southern Asia, northern Africa, North and Central America, including the five-lined skink, *E. fasciatus*, and the Berber skink, *E. schneideri*
Genus *Feylinia*—5 species of snake skinks from Africa
Genus *Lerista*—79 species of ground-dwelling and terrestrial burrowing skinks from Australia
Genus *Mabuya*—124 species of terrestrial skinks from Southeast Asia, Africa, Central and northern South America, including the Cape skink, *M. capensis*
Genus *Neoseps*—1 species of burrowing skink from the United States
Genus *Nessia*—8 species of worm-eating skinks from Sri Lanka
Genus *Riopa*—4 species of forest-dwelling skinks from Africa and southern Asia, including the fire skink, *R. fernandi*
Genus *Ristella*—4 species of cat skinks with retractable claws from India
Genus *Scelotes*—23 species of terrestrial and burrowing skinks from Africa and Madagascar
Genus *Scincus*—3 species of sand skinks from North Africa and the Middle East, including the sandfish, *S. scincus*
Genus *Sphenomorphus*—112 species of forest skinks from India, Southeast Asia, Indonesia, and New Guinea
Genus *Tiliqua*—6 species of large, terrestrial skinks from Australia, New Guinea, and some Indonesian islands, including the blue-tongued skink, *T. scincoides*
Genus *Trachydosaurus*—1 species from Australia, the stump-tailed skink, *T. rugosus*
Genus *Tribolonotus*—8 species of casque-headed skinks from New Guinea, Solomon Islands, and New Caledonia, including the crocodile skink, *T. gracilis*
Genus *Tropidophorus*—25 species of semiaquatic skinks from Southeast Asia, Indochina, Malaysia, Borneo, and the Philippines
Genus *Typhlosaurus*—9 species of blind legless skinks from southern Africa

The family Scincidae has been described as the most species-rich lizard family. It has about 1,400 species in total, but there is some debate about the exact number of genera. Revisions are constantly being suggested, but the number is thought to be between 115 and 124.

The general description of a skink is a reptile with a flattened head, long body, tapering tail, overlapping scales, and short legs. However, there are many variations within the family. The head is often triangular shaped, and many species have an elongated or pointed snout. The top of the head is covered with large bony plates that are usually arranged symmetrically.

Although size and shape vary, most skinks have a cylindrical, elongated body with scales that are usually smooth, flat, and that overlap each other. In the New Guinea water skinks, *Tropidophorus* species, scales are keeled and roughened, while Stokes's skink, *Egernia stokesii* from Australia, has keeled, spiny scales.

The scales in skinks are supported by small bony plates called osteoderms. They are unusual in that each scale is made up of a set of smaller ones. (Other lizards have a single osteoderm in each scale.) This makes a strong but highly flexible structure.

Skinks' teeth are pleurodont, meaning that they are either fixed by calcified tissue to the inside of the jawbone, or they are located in shallow indentations in the jawbone. Teeth are lost and replaced throughout the skink's life, the new ones moving up from the tongue (lingual) side of the jaw to fill gaps left by any displaced

↑ The casque-headed skink, Tribolonotus novaeguineae *from New Guinea, shows the symmetrical bony plates on top of the head that are characteristic of the family.*

teeth. As the skink grows, the number of teeth gradually increases. The tongue is fairly short, broad, and fleshy. It is slightly notched at the end and covered with flattened, overlapping tubercles.

The tail varies among family members. In some genera (*Eumeces* and *Dasia*, for example) tails are long and slender, up to twice the body length, while that of the stump-tailed skink, *Trachydosaurus rugosus* from Australia, is short and fat with a bulbous end.

Lifestyle Adaptations

Skinks are common on all continents of the world except Antarctica. They thrive even on some of the more remote oceanic islands. In Australia they are the most successful reptile group with about 370 species. The abundance of skink species in Africa, India, Indonesia, and Australia has given rise to the theory that the family evolved first on the southern continents and spread north and west.

Some genera have a very wide distribution and are found in both the Old and the New World. Species of *Eumeces* can be found in North and Central America, Asia, and North Africa. Similarly, *Mabuya* species inhabit Central and South America, Southeast Asia, and Africa.

Although most skinks have adapted to a terrestrial, burrowing, or subterranean existence, there are exceptions. A few species show specializations for an arboreal life. The lizardlike skinks of the genus *Dasia* from Asia, Malaysia, Borneo, and the Philippines are medium sized with long, gradually tapering tails. Most of them also have an enlarged, flattened area under the toes to aid climbing. Living in forested hilly areas in India, the partially arboreal cat skinks, *Ristella*, are the only species that can retract their claws into a sheath formed by a large, compressed scale. This enables them to move swiftly over the ground. The green blood skink, *Prasinohaema virens* from New Guinea, spends much of the day foraging in trees and bushes aided by toes similar to those of anoles and geckos. Just one species—the large and bulky monkey-tailed skink, *Corucia zebrata* from the Solomon Islands—has developed a prehensile tail, a useful adaptation for its arboreal lifestyle.

The slightly flattened tail of the keeled, or water, skinks, *Tropidophorus* species, is a modification to a partly aquatic lifestyle. These skinks live on the banks of streams and feed on insects and crabs. The diving skink,

Amphiglossus astrolabe from eastern Madagascar, has an unusual lifestyle—it lives in flowing forest streams. In order to breathe in this watery habitat, its nostrils are directed upward and situated a little higher on the head than in other skinks. Once in the water, the diving skinks move against the current to hide beneath rocks. They swim, dive, and prey on anything they can catch.

Cryptoblepharus boutonii is a terrestrial skink living on Madagascar and various islands in the Indian Ocean. It has adapted to conditions near the sea and can be found moving from rocks into the intertidal area to feed on insects, crustaceans, and small fish. Its periods of activity are regulated by the movement of the tides.

In Australia the rainbow skinks, *Carlia* species, spend most of their time on exposed rock faces. Therefore their limbs and digits are flattened as well as being longer and more slender. Other Australian species, for example, *Egernia*, are more robust with spiny scales that allow them to wedge themselves securely into crevices in rocks and tree hollows.

It is among the ground-dwelling skinks that most adaptations can be seen from a terrestrial to a burrowing way of life. Skinks that have moved to a subterranean lifestyle have also tended to protect, reduce, or even close the openings of their sense organs. Most of them have sunken eardrums, the opening of which is either small or closed to keep soil or sand particles from damaging the membranes. In the Berber skink, *Eumeces schneideri*, and the sandfish, *Scincus scincus*, the ear openings are protected by three scales.

Digit and Limb Loss

Adaptations to the digits, limbs, and head have also evolved as a result of the move to a burrowing lifestyle. Many skinks of the genus *Mabuya* resemble wall lizards (family Lacertidae) in that their limbs are well developed, comparatively long, and have five digits. In other members of the family Scincidae digits and limbs have been lost or reduced many times over the generations.

The skink family shows a gradual transition from lizards with four long, strong legs to the legless, snakelike lizards. Skinks of the genus *Ablepharus* have retained reduced limbs, but depending on species, they have lost varying numbers of digits from both fore- and hind limbs. Interestingly, the forelimbs never have more digits than

In southern Australia the stump-tailed skink,* Trachydosaurus rugosus*, lives in desert grassland or sandy dunes. At night it finds shelter in hollow logs and ground debris.

the hind ones (except in the Australian genus *Anomalopus*, in which the skinks have more fingers than toes). Eventually forelegs disappear altogether, and hind legs are the last to vanish.

The various stages of digit reduction and limb loss are seen clearly in *Scelotes* species skinks. Bojer's skink, *S. bojeri* from Mauritius, has four quite well-developed limbs all with five digits, while the black-sided skink, *S. melanoplura* from Madagascar, has four shorter legs, also with five digits. The silvery dwarf burrowing skink, *S. bipes* from South Africa, has lost its forelimbs. Its hind limbs are very small with just two digits. Finally, Smith's dwarf burrowing skink, *S. inornatus* (also from South Africa), lacks all traces of external limbs.

Most skinks have movable, transparent lower eyelids that protect the eyes from dust and allow them to see when burrowing. This is a closeup of **Mabuya maculilabris*****, the speckle-lipped skink from Africa.***

Movable Eyes and Spectacles

In most skinks a transparent lid covers the area of the lower eyelid. It protects the animal's eyes against particles of soil or debris when burrowing and allows it to retain some vision. The transparent lid shows different stages of development according to species.

The tropical ground-dwelling skinks, *Mabuya,* have a lower eyelid that is covered with small scales, a few of which have a transparent disk, or window. They make up the center of the lid. In the brown ground skink, *Leiolopisma laterale* from the southwestern United States, the scales in the center of the lid have slightly larger transparent disks. In the Travancore skink, *Lygosoma travancorium* from Asia, the window takes up more than half of the lower lid. In the lidless skinks, *Ablepharus,* from Australia, East Indies, Africa, South America, and some South Pacific islands, the window covers the whole lower lid and stays in place all the time. In this final stage it is known as a spectacle.

The fusing of the eyelid in this way is unique to skinks. Small, active, diurnal skinks in arid regions can lose moisture from the surface of the eye, which would be detrimental to the animal's well-being. By capping the eye with a fixed, clear spectacle, the problem is overcome. In temperate and more humid regions many skinks have a movable lid that encloses the transparent window. In a number of subterranean skinks the eyes are reduced in size. The spectacle also acts like a filter, decreasing the amount of light entering the eye.

For species such as the Southeast Asian snake skinks, *Ophioscincus*, and the South African dart skinks, *Acontias*, that live in almost complete darkness underground, the resulting impairment in vision is not a handicap. These species have also lost the external ear that detects airborne vibrations. Instead, they rely on vibrations transmitted through the semisolid earth to locate their insect prey.

The Australian genus *Lerista* contains 79 species of ground-dwelling and terrestrial burrowing skinks. They thrive in dry conditions in habitats ranging from leaf litter to sand dunes. All have minute ear openings and either an eyelid fused to form a fixed spectacle or a movable eyelid enclosing a transparent disk, or window. The genus shows progressive loss of digits and limbs from those with well-developed limbs and five digits to two completely limbless species. *Lerista viduata* is a terrestrial ground dweller that shelters among leaf litter at the base of trees and shrubs. It has well-developed limbs with five digits and a movable eyelid. *Lerista apoda* lives in the sand beneath layers of leaf litter and is totally limbless. Its tiny eye lies beneath a spectacle. Its snout is flattened and protrudes, making an ideal tool for digging.

As limbs degenerate in skinks, the tail alters so that for the greater part it remains the same thickness as the body. Since the legs are normally used in locomotion, legless forms change from walking to a snakelike, slithering movement. Assisted by the thickened tail, which is more powerful than a long, thin one, the body is thrown in horizontal curves.

Loss of limbs means that the head has to be used for burrowing. A broad, blunt snout would be useless, so in many burrowing species, for example, the sandfish, *Scincus scincus*, the snout is drawn out into a wedge shape or pointed cone shape. This adaptation is sometimes accompanied by an upper jaw that protrudes over the lower one as in the South African dart skinks, *Acontias*, and the Florida sand skink, *Neoseps reynoldsi*. The pattern of the head scales is modified to form a sheath around the skull, and the fusion of scales makes the structure of the head a more rigid tool for digging.

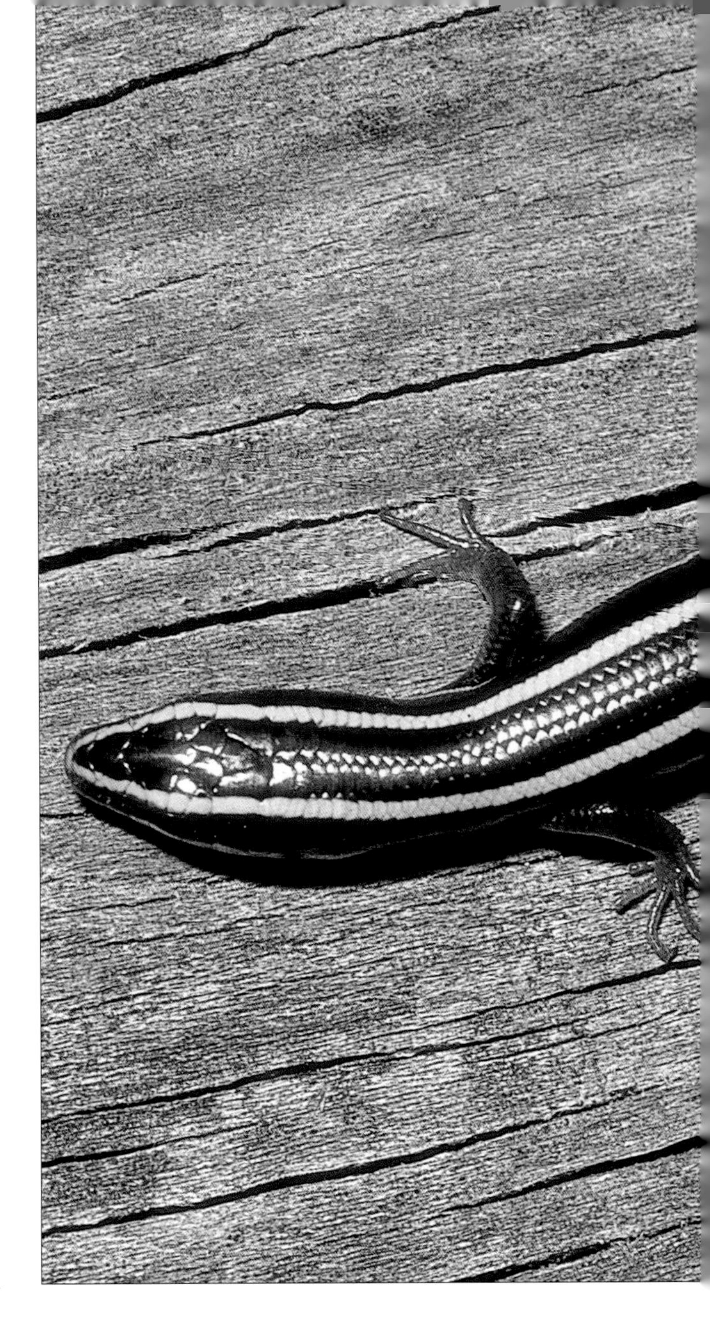

The striped skink,* Mabuya striata wahlbergi *from Namibia, is more brightly colored than some family members. Its brown snout contrasts with the turquoise underside.

Coloration

Although skinks tend to be various shades of brown, black, olive, and cream, many species have stripes, crossbars, spots, mottling, or blotches of differing sizes and thicknesses.

Skinks do not exhibit the rapid color change associated with chameleons, but some changes may occur as they mature and in the breeding season. For example, *Leiolopisma himalayana* males from South Asia develop an orangish-red band along the flanks in the breeding season. Members of the genus *Leiolopisma* also have colored tails—blue in *L. latrimaculata*, violet in *L. bilineata*, and red in *L. rhomboidalis*. The rest of the body is a dull brown. However, when these skinks bask in bright light, their scales give off an iridescence, and their

Tails—Use Them or Lose Them!

Skinks tend to rely on cryptic (disguise) coloration and the ability to disappear into the background to avoid danger. Many of them use an additional safety device, known as autotomy, whereby they can voluntarily detach part of their tail. While the predator's attention is distracted by the violently wriggling, disconnected piece of tail, the skink can escape.

Over a period of time the skink regenerates a new tail, although it is usually not quite as long as the original. Studies in the wild of the Great Plains skink, *Eumeces obsoletus*, have shown that the older the creature, the less likely it is to have an intact tail. Some species such as young Pecos skinks, *Eumeces taylori*, have conspicuously colored tails that add to their ability to distract predators. Not all species of skinks have a fracture plane in their tail (a point through one of the vertebrae where there is a special constriction to allow easy breaking); if a member of these species loses its tail, it will not regrow.

Having detached its bright blue tail, the five-lined skink, **Eumeces fasciatus,** ***scuttles away. The tail continues to twitch and creates a diversion. It will eventually regrow, but the new one will probably not be as long as the original.***

↑ ***A bright tail can be an asset in self-defense, as in the pink-tailed skink,*** **Eumeces lagunensis** ***from Baja California. Predators become distracted by the bright color and leave the skink alone.***

tail color stands out. Members of the genus often raise their tail and move it slowly from side to side, probably as a signal to warn off rivals and to attract a mate.

Diet and Feeding

Most skinks forage for insects, beetles, millipedes, small vertebrates, and plant and vegetable material. Collecting food is simply a matter of grabbing it in the jaws, crushing it with the teeth, and swallowing. The tongue is coated with sticky mucus that allows the skink to lap up ants and termites. Small species tend to have pointed teeth and are insectivorous. Larger species are more omnivorous. They supplement insect fare with plant and vegetable matter, including fungi. The Australian blue-tongued skinks, *Tiliqua*, are partly herbivorous, while, despite its size and stout build, the largest skink, *Corucia zebrata*, forages in trees for leaves and fruit. The teeth of herbivorous skinks are blunter than those of insect-eaters.

Any prey that struggles is held in the skink's jaws and banged or rubbed against the ground until it stops moving, after which it is eaten. Skinks of the genus *Nessia* from Sri Lanka have tiny, useless legs that cannot prevent their main prey item, earthworms, from sliding back out of their mouth. To overcome the problem and hold the prey in place, their teeth are pointed and curve backward.

Some species of skink are specialized feeders. Pink-tongued skinks, *Cyclodomorphus gerrardii* from Australia, feed mainly on slugs and snails. They have large, rounded teeth at the rear of the mouth that are used for cracking snail shells. Their specialized feeding technique involves approaching a mollusk while opening and closing their jaws and salivating. In this way their mouth becomes coated with a thick protective lining, allowing the skink to eat distasteful snails and slugs with no ill effects.

The mangrove skink, *Emoia atrocostata* from coasts around Singapore, swims and dives in seawater and feeds on crustaceans and small fish left in pools by the receding tide. A number of species of skinks will eat the contents of eggs found in the nests of ground-dwelling birds. On the Seychelles *Mabuya wrightii* uses its body to push the eggs of sea swallows from nests so that they fall onto the rocks below and smash. The skink scuttles down the cliff and laps up the contents.

Reproduction

Within the skink family all forms of reproduction occur. Some species produce eggs that develop in the female's body (ovoviviparous), others give birth to live young (viviparous), but about 60 percent of species are egg layers (oviparous). Live-bearing predominates in cooler areas or at higher altitudes where the incubation of eggs would be hindered by low temperatures. The advantages of warming the embryos within the female's body as she basks outweigh any mobility problems involved in carrying them.

In *Trachydosaurus rugosus,* the stump-tailed skink, one or two embryos develop in the female without eggshells and are supplied with nutrition from a placentalike organ. In the genus *Mabuya* some species are egg layers, but others are live-bearers. The Indian skink, *M. carinata*, lays up to 23 eggs, but *M. multifasciata,* the oriental brown-sided skink from Indonesia, gives birth to six live young.

Sexing is not as straightforward in skinks as in other lizards, since, unlike members of the Iguanidae, the Cordylidae, and the Agamidae, they lack femoral pores (enlarged pores along the inside of the thighs). In addition, males have no head ornamentation such as horns or high casques. Nor do they undergo dramatic color changes, unlike the chameleons. However, male skinks usually develop a broader head as they mature. In other lizard families, such as the Agamidae and the Iguanidae, males use signals—inflated dewlaps, darkened beards, or arm waving—to intimidate rivals or to indicate submission without resorting to fighting.

However, when two male skinks meet, especially in the breeding

Care of the Eggs

Some skinks in the genus *Eumeces* have developed a type of brood-care behavior in which the female curls around the eggs, cleaning and turning them. It is thought that turning the eggs keeps the same part of the egg from being in contact with the damp soil all the time and prevents them from rotting. There is no evidence to suggest that the female actually uses her body to warm the eggs. It has been said that although the female guards her clutch and keeps away small creatures that would nibble the eggs, she will flee at the first sign of any real danger.

season, they rush at each other using their strong jaws to bite and tear flesh or limbs. Occasionally the injuries inflicted result in death. Similarly, when finding a female, there is no courtship display. Males track females by scent. If a female appears receptive, the male licks her with his tongue and uses his jaws to grab her side. If she does not resist, he grabs the nape of her neck or her shoulder in a vicelike grip. He then places one hind leg over the base of her tail. Slipping his tail under her, he aligns their cloacae to allow mating to take place.

Egg clutches vary in size from one to 30 eggs depending on species. Usually they are deposited in small holes dug under rocks or logs, in crevices, in the roots of shrubs, or in tree hollows. The site must be protected and subject to fairly high humidity to keep the eggs from drying out. Incubation varies from five to 10 weeks.

Generally live-bearers produce fewer young than egg layers. Large species do not necessarily produce large numbers of young. The blue-tongued skink, *Tiliqua scincoides*, gives birth to 10 to 15 young (although it has been known to produce 25), while the similar-sized and related stump-tailed skink, *Trachydosaurus rugosus*, produces one or two youngsters, each nearly half the size of the adult female. The monkey-tailed skink, *Corucia zebrata*, the largest member of the family at 30 inches (76 cm) long, usually has just one offspring.

A female broad-headed skink,* Eumeces laticeps, *with her eggs in Florida. Females of this species usually lay between five and 20 eggs and remain with them until they hatch.

Slow Worm

Anguis fragilis

Common name Slow worm (blind worm)

Scientific name *Anguis fragilis*

Subfamily Anguinae

Family Anguidae

Suborder Sauria

Order Squamata

Size Up to 20 in (51 cm) but usually shorter

Key features Legless lizard; scales smooth and shiny; head no wider than its body; no distinct neck region; eyes small; tail longer than body when complete; color brown, sometimes coppery; females have a thin dark line down the center of the back and dark flanks; males usually uniform in color; juveniles look like females but are often more brightly colored

Habits Terrestrial; semiburrowing; active at night; occasionally basks in the day

Breeding Live-bearer; female gives birth to 6–12 young; gestation period 8–12 weeks

Diet Soft-bodied invertebrates, especially small slugs, snails, and worms; insects and small lizards

Habitat Damp places with plenty of vegetation, including woodland clearings, hedges, banks, gardens, parks, and railway embankments

Distribution Most of Europe except the southern half of Spain and the most northerly parts of Scandinavia; east Asia to west Siberia, the Caucasus, northern Turkey, northwest Iran

Status Common

Similar species Young European glass lizards, *Ophisaurus apodus*, but they are spotted; snakes are more supple, lack eyelids, and have a single row of wide scales down the belly

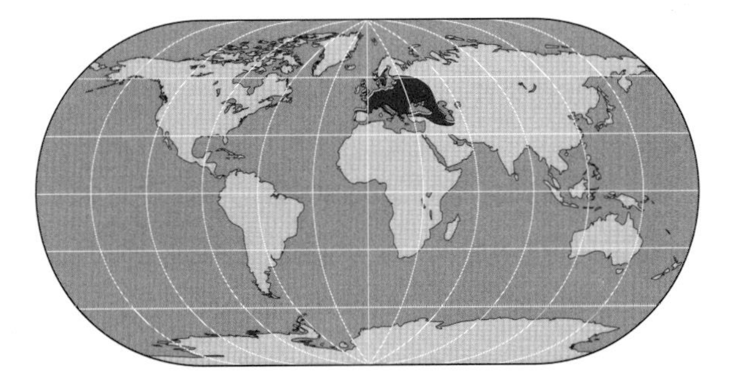

Sometimes mistaken for a snake, the slow worm is, in fact, a legless lizard. It is often found in gardens across Europe and is the gardener's friend, since it feeds on many invertebrate pests.

THE SLOW WORM IS OFTEN MISTAKEN for a snake, but even a superficial examination reveals that its movements are less agile, its eyelids are movable, and it has a short, notched tongue—all of which distinguish it from a snake. Also, as in other legless lizards, its tongue can only be extended when it opens its mouth slightly, while snakes protrude their tongue through a small notch in the lower jaw (the lingual fossa).

The slow worm's underside is covered by several rows of scales, with each scale being roughly the same shape as those on its back. (This also differentiates it from snakes, which have a single row of specialized scales along their underside.) When a slow worm is picked

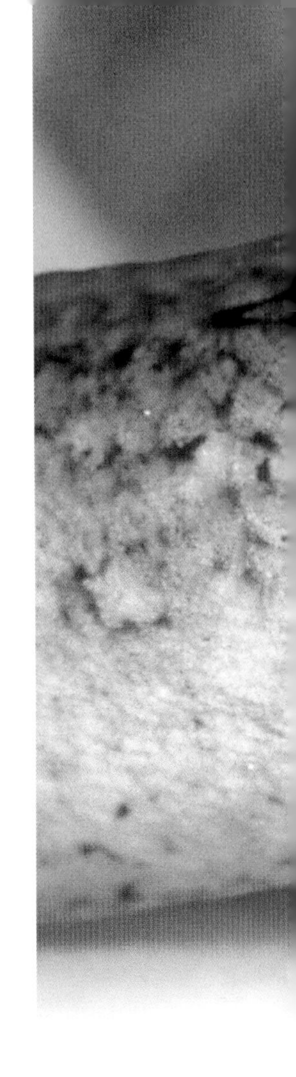

Spot the Difference

For a species with such a huge geographical range slow worms show surprisingly little variation. The population in the southern Peloponnese and the Ionian Islands, Greece, is more slender and has a proportionately longer tail, but it has now been reclassified as a separate species, *Anguis cephallonica*. In eastern Europe the subspecies *A. fragilis colchicus* has more rows of scales around its body on average, and males often have blue spots along the sides of their body, especially around the head and neck. Western individuals belonging to the nominate subspecies also have these spots sometimes, and the two forms are impossible to tell apart without information relating to locality. Other than these slight variations slow worms from northern Scotland look much the same as slow worms from Scandinavia, western Siberia, or the Balkans.

up, it may try to coil around your fingers but its body is neither as muscular nor as supple as a snake's, and it is unable to squeeze as tightly. Even so, many slow worms are killed needlessly in the belief that they are snakes.

Slow worms are secretive reptiles, rarely venturing out into the open except at night and then most commonly if it is damp, when their preferred prey—small gray slugs—are most active. They also eat snails, which they have to extract from their shells, and earthworms. These three items make up about 70 percent of their diet—the rest consists of insects and spiders. Slow worms also eat small vertebrates, including common and wall lizards, but they can only catch them on very rare occasions.

Slow worms are methodical feeders. They examine potential prey carefully by tongue-flicking. Then they raise the head slightly and make a short, deliberate, downward strike. If the prey moves, the process may be cut short, and the slow worm may react more quickly. After eating slugs or snails, it often wipes its jaws on moss or leaves to get rid of any slime.

Damp Habitats

Despite their wide range, slow worms' habitats tend to have several things in common. They prefer slightly damp habitats with plenty of ground cover in the form of grasses and low-growing vegetation. In agricultural regions they are often concentrated at the bottom of hedges and odd pieces of ground at the edges and corners of fields that get missed by the plow, although these places are becoming rarer. They also live on commons, "waste" land, and rough grazing, and are sometimes found in overgrown parks, gardens, and cemeteries. Railway cuttings and embankments are often rich in slow worms, which thrive on the sunny, south-facing slopes. These "dispersal corridors" often reach into the hearts of cities, and slow worms take advantage by spreading into adjacent gardens and vegetable plots.

Like many reptiles, slow worms use their notched tongue to find prey, flicking it in and out of the mouth to pick up scent particles in the air.

Slow worms rarely come onto the surface during the day. They prefer to stay under cover in disused rodent burrows or under objects lying on the surface. In suitable habitats one or more slow worms are likely to be found under flat rocks, slates, old sacking, carpets, or concrete slabs, basking in the heat that penetrates there. Because these places are also damp, they often act as magnets for slugs and snails as well, which makes them even more attractive to the slow worms. Slow worms will burrow through loose soil and leaf litter if necessary; but since they have no adaptations for burrowing, they nearly always use ready-made tunnels and underground chambers.

Fragile Tails

The slow worm's generic name *Anguis* means snake, showing that even the most famous zoologist of all, the "Father of Taxonomy," Carl Linnaeus, got it wrong some of the time. Its specific name, *fragilis*, is more accurate and stems from the lizard's tendency to lose its tail in response to an attack by a predator. In many populations nearly all adults have stumpy or regrown tails, often as a result of attacks by domestic cats. However, since they resemble worms, they have many enemies, including birds of prey, magpies and crows, domestic chickens, snakes, weasels, small boys, and other predatory mammals.

Although they have powerful jaws they rarely bite, and their best means of defense is to let the predator have just their tail. Because their tail is so long, this strategy works more often than it might in other lizards. Statistically, the slow worm has about a 65 percent chance that any attack will be directed at its tail. And unlike other lizards, the slow worm seems not to be unduly hampered by losing its tail (unless it is attacked again before a new one grows). In studies on female slow worms scientists found that those with damaged tails gave birth to roughly the same number of young as those with complete tails, indicating that the tail is not essential for accumulating food that can be used later in the reproductive effort.

Ⓣ ***Newborn slow worms are relatively small—2.8 to 4 inches (7–10 cm) long. The dorsal color is lighter than in the adults, usually a pale golden-brown or grayish silver, but their sides and underside are pitch black.***

Slow worms will bask in the open just after emerging from hibernation in the spring or, in the case of females, just before they give birth in the late summer. At these times they may aggregate, and several can sometimes be found in the same spot. Five pregnant females were once found basking on the flat top of a stump partially covered with a thin sheet of moss in a place where they had never been seen before (and where they have never been seen since). Because they tend to appear and disappear intermittently in this way throughout the year, it is difficult to know whether slow worms are common (or even present) in an area. There have been few thorough studies on them, but some ecologists estimate that in particularly favorable habitats there may be as many as 250 to 800 individuals per 1 acre (0.4 ha).

Winter is spent deep underground, often in rodent burrows, where the slow worms are

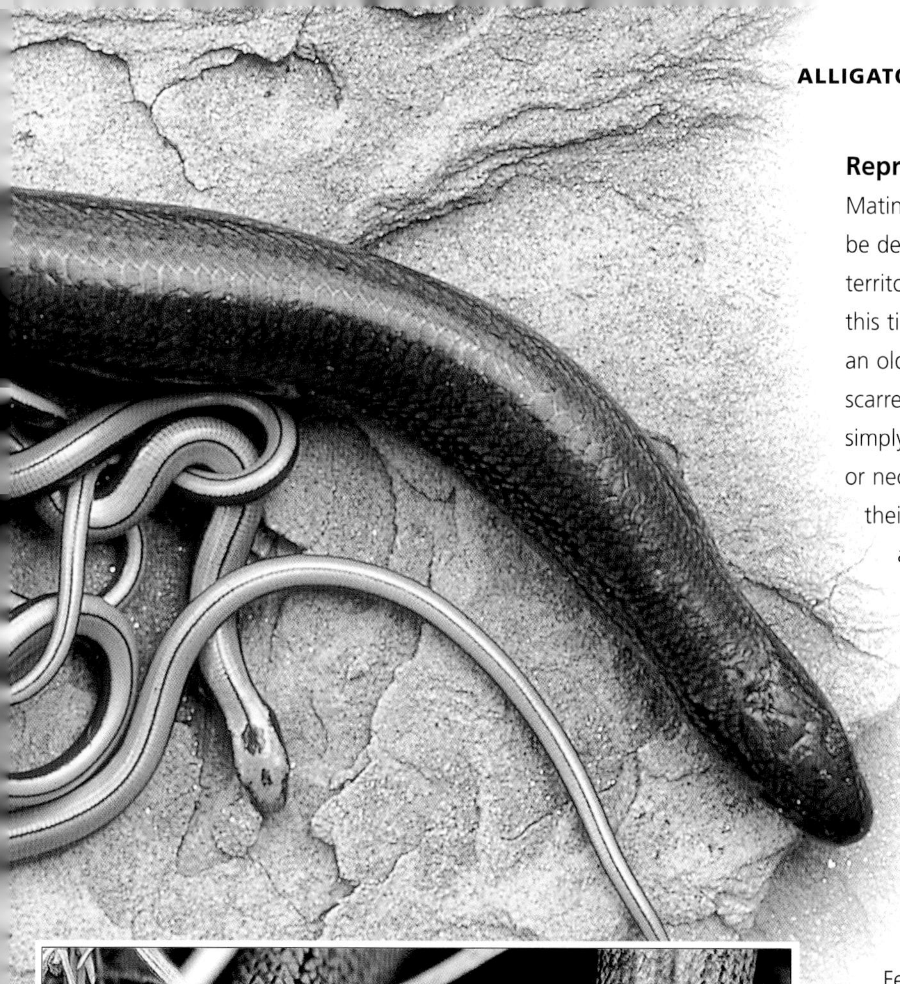

Ⓣ ***Courtship in slow worms is minimal, but mating takes up to 10 hours. This mating pair was photographed in the south of England.***

unharmed by frost. In northern Europe they do not emerge until March or even later, but in other parts of their range they may be active earlier than that depending on the climate.

They are most active near the surface between May and September, especially on warm days and during nights with light rain. Their preferred body temperature is 82°F (28°C), but they are active at temperatures as low as 58°F (14.5°C). If the temperature rises above 86°F (30°C), they retreat deeper underground until it becomes cooler.

Reproduction

Mating usually takes place in May, but it may be delayed until the end of June. Males are not territorial, but they will fight off other males at this time of the year by biting—it is rare to find an old male slow worm that is not battle scarred. Courtship is not very subtle: The male simply pins down the female by biting her head or neck, twists his tail around hers, and aligns their cloacae. Mating can last up to 10 hours, and females may mate with more than one male in a season.

Most females do not breed every year, however, but in alternate years. In the Netherlands about 70 percent of females "rest" in dry years, whereas up to 90 percent may become pregnant in years with high rainfall (when there is plenty of food). These figures probably vary across the species' extensive range.

Females continue to feed throughout their pregnancy, so they have a chance to breed two years in succession if conditions are good.

Females give birth from mid-August to October (or early November if gestation becomes drawn out during a cold summer). On rare occasions the female may not have given birth by the time she has to enter hibernation, and the young are born the following spring. Litter sizes vary greatly from three to 26, but most litters contain six to 12 young with an average of eight. As in other reptiles, larger females produce more young than smaller ones.

Young slow worms are born inside a membrane from which they escape by wriggling violently. They measure about 3 to 4 inches (7.6–10 cm) long and are a bright copper color. They double in length during their first year if all goes well. Males reach sexual maturity at about three years, females in five or six years, and they reach full adult size by the time they are about six to eight years old. Slow worms are very long-lived with an estimated life span of 10 to 12 years in the wild. In captivity, however, they can live much longer, with the record standing at 54 years.

SNAKES

There are nearly 3,000 species of snakes, and they account for about 42 percent of all reptiles. They form one of the subdivisions of the order Squamata, the other two being the lizards and the worm lizards. The snakes are divided into 18 families, although there is some dispute over this classification, and it will almost certainly change at some point in the future. In particular, a single large family, the Colubridae, contains over 1,500 species and is in serious need of revision.

Snakes are most obviously characterized by the things they lack! Even a superficial examination will reveal that they have no limbs, no eyelids, and no external ear openings. Despite this uniformity of design, snakes vary greatly in size, shape, color, and markings, so that most species are easily identifiable at a glance.

Size and Shape

The smallest snakes are certain blind snakes, which may be just 3 inches (8 cm) or less in length and are so thin that they would comfortably fit inside a wooden pencil from which the lead has been removed. At the other extreme the very largest boas and pythons can exceed 16 feet (5 m) in length and in exceptional circumstances may reach almost 23 feet (7 m). Most snakes, however, are between 8 inches (20 cm) and 6.5 feet (2m) long.

Snake shapes vary from long, extremely slender, long-tailed species to short, squat species with almost no tail at all. Their proportions largely reflect their lifestyles. For example, species that live in trees, such as slug- and snail-eating snakes, tend to be long and thin. This helps them bridge the gap between branches and also ensures that their weight is kept to a minumum. On the other hand, burrowing snakes like sand boas often have stout cylindrical bodies with no obvious neck and a short tail, a shape that helps them force their way through the soil.

Other snakes are even more specialized and include species that are adapted to life in the oceans and those that "swim" through loose sand, such as the shovel-nosed snakes. The vast majority of snakes lie between these extremes and are moderately slender with a tail that accounts for about one-fifth to one-third of their total length. Familiar species such as the North American ratsnakes, king snakes, and garter snakes, for example, have quite typical proportions.

Color and Markings

Most snakes are dull in color, blending in with natural surroundings such as dead leaves, sand, gravel, mud, or soil. Tree-dwelling species may be green. Markings of these camouflaged species are often fairly regular, consisting of blotches, bands, or stripes. A few species, such as the Gaboon viper, *Bitis gabonica*, are spectacularly marked with geometrical patches of contrasting colors that serve to break up their outline. Plain-colored snakes are relatively rare because plain-colored habitats are equally rare. A small proportion of species use warning

⊖ *The long-nosed vine snake,* Ahaetulla nasuta *from Southwest Asia. Its slender form and green color are adaptations to life in trees. Its horizontal pupils give a degree of binocular vision.*

Who's Who among the Snakes?

Order Squamata (suborder Serpentes)

Family Anomalepidae: 4 genera, 15 species of dawn blind snakes
Family Leptotyphlopidae: 2 genera, over 90 species of thread snakes
Family Typhlopidae: 6 genera, over 215 species of blind snakes
Family Anomochilidae: 1 genus, 2 species of dwarf pipe snakes
Family Uropeltidae: 9 genera, 47 species of shield-tailed snakes
Family Cylindrophiidae: 1 genus, 8 species of pipe snakes
Family Aniliidae: 1 species of South American pipe snake, *Anilius scytale*
Family Xenopeltidae: 1 genus, 2 species of Asian sunbeam snakes
Family Loxocemidae: 1 species of Central American sunbeam snake, *Loxocemus bicolor*
Family Acrochordidae: 1 genus, 3 species of file snakes
Family Boidae: 9 or 10 genera, about 41 species of boas
Family Bolyeriidae: 2 genera, 2 species of Round Island boas, including 1 probably extinct
Family Tropidophiidae: 5 genera, 25 species of dwarf boas or wood snakes
Family Pythonidae: 8 genera, 26–33 species of pythons
Family Colubridae: about 300 genera, 1,800 species of typical snakes
Family Atractaspididae: 8 genera, about 60 species of stiletto snakes and burrowing asps
Family Elapidae: 60 genera, about 305 species of cobras
Family Viperidae: 35 genera, about 250 species of vipers and pit vipers

Total: 18 families, about 455 genera, about 2,900 species

colors in much the same way as bees and wasps use them. These brightly colored snakes may be banded in black, white, yellow, or red and are often known as "coral snakes" regardless of where they live. All these species are members of the cobra family (Elapidae), and their coloration is intended to warn possible predators that they are dangerous and should be left alone. Sometimes nonvenomous snakes imitate their markings and gain an advantage by fooling predators into thinking that they too are venomous.

Skin and Scales

Snakes' colors and markings are the result of pigments in their scales. Every species is covered in scales, but their shape and texture vary. In addition, different parts of the snake are covered with scales that are also different in size and shape, and they often help identify species that are otherwise similar in appearance.

The scales covering the backs and flanks of snakes are known as dorsal scales, and they may be of several types. Many species, especially those that burrow and those that live in the sea, have smooth dorsal scales that fit closely together, reducing their resistance to soil, sand, or water. Others, like the vipers, have scales with a raised ridge, or keel, running along the center. Keeled scales are found on a variety of types, but are especially common on species that climb and on some semiaquatic species such as garter snakes and grass snakes. A few species (the saw-scaled vipers, horned vipers, and their mimics) use the keels on their scales to produce sound. File snakes in the family Acrochordidae are unusual in many respects and have unique granular scales, giving the snake a rough feel. The scales are very small and conical in shape, and are used to grip the fish on which the snake feeds.

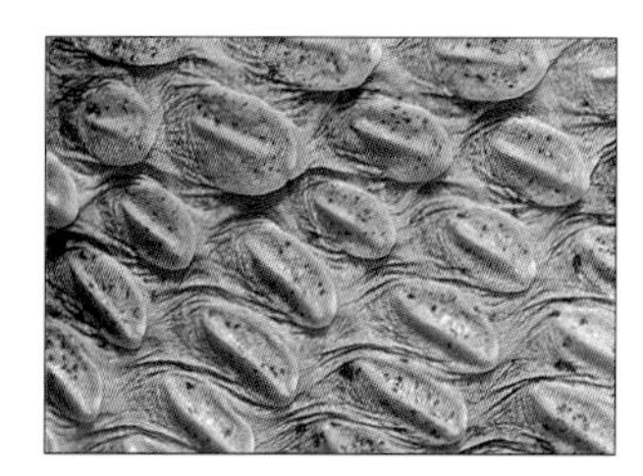

⊕ ***Left:*** **Cerastes cerastes***, the desert horned viper, has obliquely keeled scales with which it can produce sound. Right: The scales of the Sinaloan milksnake,* **Lampropeltis triangulum sinaloae***, are smooth and shiny. These examples also show extremes of color, from camouflage in the viper to conspicuous in the milksnake.*

The scales on the undersides of snakes are known as ventral scales. They are usually wider than dorsal scales and arranged in a single row beneath their bodies (a characteristic that immediately distinguishes them from lizards). These scales continue onto the underside of the tail as a single or double row (depending on the species), where they are known as subcaudal scales. Head scales may be large and platelike (as in nearly all colubrids and members of the cobra family) or small and granular (as in most boas and vipers). The scales bordering the mouth are nearly always large and are known as labial scales. Other head scales have specific names (for example, the "rostral" scale on the tip of the snout) but are not necessarily common to all species. A small number of snakes have some ornamentation in the form of horns on their snouts or over their eyes, and these may be hard and thornlike or soft and fleshy.

Internasal
Rostral
Nasal
Prefrontal
Frontal
Supraocular
Dorsal
Temporal
Parietal
Upper labial
1a

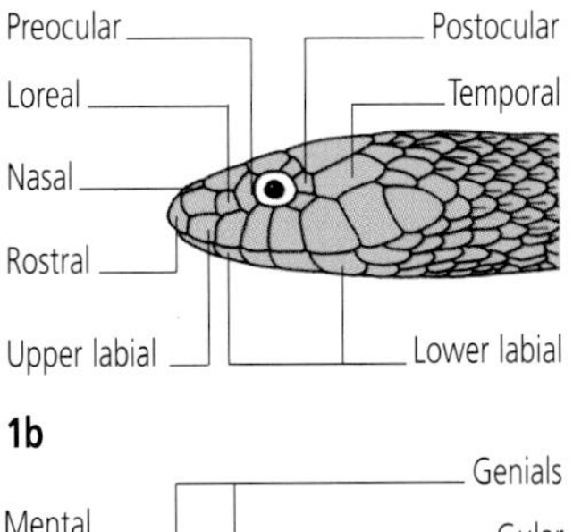

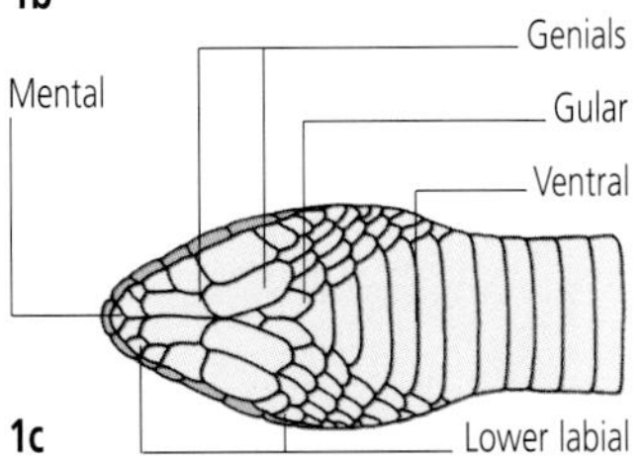

When a snake sheds its "skin," it only removes the outer layer of epidermis—the scales themselves and the pigment they contain remain on the snake. Shedding takes place at regular intervals as the snake grows and also serves to replace worn or infected skin, heal wounds, and remove parasites. Prior to shedding, the snake secretes an oily substance between the old and new layers of epidermis, making it appear dull and making its eyes opaque. This substance aids the shedding process.

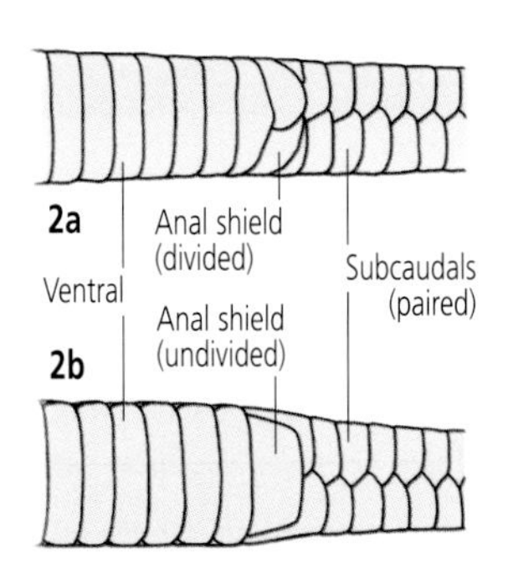

Ⓣ Ⓒ ***Above: the typical arrangement of scales on an advanced snake's head and upper body from above (1a), from the side (1b), and from below (1c); right: anal and tail scutes of an advanced snake (2a, 2b).***

A freshly shed skin is supple and tacky, although it quickly dries out and becomes brittle. A snake that has recently shed its skin has especially bright markings. Each segment of a rattlesnake's rattles is formed from the dead skin covering the end scale of its tail—they are retained when the snake sheds the rest of its skin; and as they accumulate, they are linked together loosely, forming a chain or rattle.

Locomotion

Despite their lack of limbs, snakes have a number of effective methods of getting around. The particular technique they use depends on two things: the size and shape of the snake concerned, and the material over (or through) which it is traveling. Lateral undulation and serpentine crawling are the terms applied when snakes move by wriggling their bodies from side to side. Most snakes use this method most of the time when they are moving across the ground or through vegetation. The snake flexes its body to form a series of "s"-shaped curves. The hind part of each curve pushes against small irregularities in the ground. As the snake moves forward, other parts push against the same point, resulting in a smooth, flowing type of locomotion. When snakes swim, they use the same basic technique except that their body pushes against the water.

Heavy-bodied snakes, especially large vipers, boas, and pythons, use caterpillar crawling (or rectilinear locomotion), in which they move forward in a straight line. The snake uses the edges of its ventral scales to hook over irregularities and pull itself forward. Different parts of

In the Anza-Borrego Desert, California, a sidewinder, Crotalus cerastes, *leaves a distinctive trail. Sidewinding locomotion is the most efficient way of moving across unstable shifting sand.*

the body pull at the same time as alternate sections stretch forward, so the snake's ventral scales move forward in a series of waves, and its body moves smoothly in a straight line. As well as large boas, pythons, and vipers, smaller snakes also use this technique when they are moving slowly, for example, in the final stage of stalking their prey.

Snakes moving along a narrow tunnel use a type of movement known as concertina locomotion. The snake jams its body against the sides of the tunnel by making a series of loops. It then uses the loops to anchor itself while the head and neck reach forward. Once they are also anchored, it pulls the rest of its body up before reaching forward again. A similar technique is used when a snake crawls along a branch or climbs up a rough tree trunk. In these situations, however, it first grips the tree with the ventral scales on the front part of its body while the back half is pulled up. Next, it uses the scales on the back half of its body and tail to hold on while the front half reaches forward.

Sidewinding is a much more specialized method of locomotion than the others. Only a few snakes, especially certain vipers, move in this way. It is most useful where loose sand would make other types of movement ineffective. The snake lifts its head and neck clear of the ground, and throws its head sideways, while the rest of its body stays where it is. Once the head touches the ground again, it acts as an anchor while the rest of the body follows. Before the tail comes to rest, however, the head is thrown sideways again to start the next cycle of movement. Because it happens so quickly, the snake appears to be looping across the ground at an angle of about 45 degrees to the direction it is facing. The North American sidewinder, *Crotalus cerastes*, which is a small rattlesnake, is the best known of the species that move like this, but a few other snakes in South America and Africa also sidewind, for example, the Sahara horned viper, *Cerastes cerastes*.

The Senses

Although snakes have sense organs similar to other animals, some do not work as well as in other species. Some work much better, however, and some are unique to snakes. Snakes sense the world in a very different way than humans and other animals. Except in a very small number of species, their sight is not very efficient. That is because at one point in their evolution they became burrowing animals with no need for eyes. The eyes therefore degenerated. Some snakes still live underground and have virtually no functional eyes. Others reemerged to live on the surface; but having lost the ability to see well, they had to reinvent the eye. Meanwhile, many of the eye's more advanced features were lost. For example, most snakes focus by moving the lens back and forth instead of the more effective method of changing its shape, as in other vertebrates. The result is that snakes can only see moving objects—stationary prey is often overlooked (although they may find it using other senses).

The size of the snake's eye and the shape of its pupil are often clues to its lifestyle. Nocturnal snakes may have vertical pupils (pupils that contract to vertical slits in the light), while daytime hunters may have large eyes with round pupils. A small number of species, including the long-nosed vine snake, *Ahaetulla nasuta*, and a few other tree snakes have horizontal "wrap-around" pupils, giving them a degree of binocular vision.

Snakes' hearing is also poor, since they lack external ears. However, they do pick up vibrations from the ground through their jawbones, and there is good evidence that they also hear airborne sounds of certain wavelengths.

What snakes lack in vision and hearing they more than compensate for in other ways. Their sense of smell is acute, and they use two organs—their tongue and the Jacobson's organ—to supplement it. Jacobson's organ consists of a pair of small depressions in the roof of the snake's mouth, matching the paired tips of the forked tongue and leading to the olfactory section of its brain (the part relating to the sense of smell). They flicker the tongue in response to stimuli such as sounds

Left: ***Vertical pupils and heat pits on its snout indicate that*** **Trimeresurus albolabris,** ***the white-lipped pit viper from Southeast Asia, is a nocturnal hunter. By contrast, the juvenile African boomslang,*** **Dispholidus typus** ***(above), has a large round pupil.***

Internal Anatomy

Under the skin snakes are highly modified due partly to their shape. They have no pectoral girdle or breastbone (sternum). Most have no pelvic girdle either, although some primitive snakes retain it, together with the vestiges of hind limbs. These vestigial limbs are visible externally as small thornlike claws or spurs. The skulls of the most advanced snakes are made up of light bones that are loosely tied together. That enables them to stretch their mouths to a huge extent and therefore to eat prey that is much larger than their own head. Primitive snakes, however, have more rigid skulls and eat small prey.

Snake's teeth are varied in number and form according to their family and their diet. Some species have very specialized teeth, or fangs. They are adapted for dealing with difficult prey—such as the hinged teeth in sunbeam snakes and the strongly recurved teeth of tree boas—or for injecting venom. Venom fangs may be hollow and positioned at the front of the mouth, as in vipers and cobras, or they may be grooved and positioned toward the rear, as in some colubrids. Vipers have very long fangs that are hinged so they can fold them up out of the way when they are not needed.

Many internal organs are elongated to fit into the stretched out body cavity. The paired organs, such as ovaries, testes, and kidneys, are staggered one in front of the other. Most snakes have only one lung, the right one, while the left lung is either absent altogether or greatly reduced in size. To compensate, the right lung is enlarged. An additional respiratory organ—the tracheal lung—may also be present. Snakes' stomachs are simple, just a muscular part of the gut taking up about one-third of its body length. The intestines are not as coiled up as they are in most other vertebrates.

The internal organs of most vertebrates are packed together in a relatively restricted abdominal cavity. In snakes, however, the internal organs are arranged in a linear way within the extended body cavity.

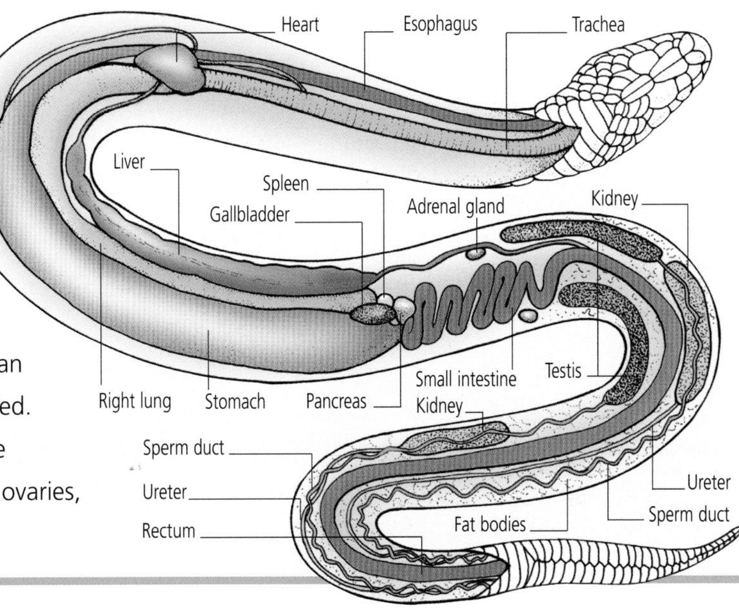

and vibrations. The tongue picks up scent molecules that are transferred to the Jacobson's organ when the tongue is withdrawn into the mouth.

Some snakes, including most boas and pythons and all pit vipers, have unique heat-sensing organs in their faces with which they can pick up heat radiated from warm objects or organisms. The organs, called "pits" or "heat pits," are lined with cells that contain numerous thermoreceptors, each connected to the brain. In the boas and pythons that have them, the pits are in or between the scales around their mouths, and they vary in number and size. In pit vipers, however, they are more highly developed and consist of a single pair of pits located just below an imaginary line joining the eye and the nostril. These organs are extremely sensitive and can detect temperature rises of just a fraction of a degree. As well as sensing the presence of a warm-blooded animal such as a mouse or rat, they can pinpoint its position exactly. They are so sensitive that they can allow the snake to strike at a particular part of its body, usually the thorax (where venom acts most quickly). All this can even be achieved in total darkness if necessary.

Lifestyle and Habitat

Snakes of some sort or another occupy most available habitats from the treetops to underground tunnel systems and from the most arid deserts to freshwater lakes and rivers, and even the oceans. The only habitat from which they are excluded is the polar one—their distribution is restricted to parts of the world where their body's operating temperatures can be reached for at least a few months every year. In this respect they are perhaps more limited than lizards, which have evolved a number of different ways of raising their body temperatures.

Lifestyle and habitat have to a large extent shaped and colored snakes. There are distinctions between specialists and generalists: Some species occupy a wide range of habitats, while others occupy narrow niches.

Distribution

There are snakes on every continent except Antarctica and some of the smallest oceanic islands, which they have obviously failed to reach. There are no snakes on New Zealand or Ireland, either. As with other reptiles, the greatest numbers of species and individuals are found in tropical and subtropical regions. Snakes from cooler climates may be active only in the summer, spending the winter in a state of inactivity or hibernation. The South and Southeast Asian regions are the richest in numbers of species for their size, with several unique families present. Central and South America are other "hot spots," while Australia—although having many species—has a relatively small number of families. The huge colubrid family, for instance, is represented in Australia by only a handful of species.

Feeding

All snakes are carnivores. Their prey species, however, range from ants to antelopes and pretty much everything in between. As with habitats, there are generalists and specialists, and the latter include species that will only eat birds' eggs, centipedes, earthworms, and worm lizards, to name just a few. Even generalists

Snakes on Islands

Snakes are present on many islands. They got there in one of two ways. Some islands became isolated from adjacent landmasses when sea levels rose or when pieces of land broke away. Madagascar is an example of one such large island, and there are instances of many small offshore islands, such as those in the Gulf of California. Islands of this kind may already have had snakes on them when they became isolated. Given enough time, some of those snakes will evolve into new and unique species. Others will be merely variants of the nearest mainland populations.

Other islands are volcanic, or coral atolls, and have never been part of the mainland. Examples are the Hawaiian Islands, which have no native snakes, and Galápagos Islands, which have four species, some restricted to a single island and others occurring on several. Snakes can only colonize islands such as these by rafting on floating vegetation, perhaps uprooted during a mainland storm. Getting to them is a chancy affair, and success is determined largely by their proximity to the mainland and the dominant ocean currents.

The Galápagos snake,* Dromicus biserialis, *is endemic to the Galápagos Islands and inhabits the southeastern islands. It is seen here eating a lava lizard.

may have preferences—for either warm- or cold-blooded prey, for instance. There are quite a few species, however, that will tackle both. Some species start out eating one type of prey as juveniles and switch to something completely different as they grow bigger.

There are many methods of subduing prey, all of them necessitated by the fact that snakes do not have any limbs. Depending on the type of prey, the snake may simply grab it in its mouth and swallow it live, coil around it and constrict it to death, or inject venom into it. The method chosen will often depend on the size of the prey and the likelihood of it doing the snake harm. In some cases the methods are determined by evolutionary developments, for example, certain behavior patterns or types of teeth, but individual snakes may also make decisions based on the prey they are facing at the time.

Because of their hugely flexible skull and the elasticity of their skin, members of the more advanced families are able to swallow relatively large prey animals. Swallowing may take less than a minute or more than an hour, depending on the size of the meal and how well the snake handles it. Some snakes seem particularly inept when it comes to getting their meals down.

Snakes use different methods to subdue their prey, including constriction. Here a spotted python,* Antaresia maculosus *from Australia, constricts a mouse and prepares to swallow it whole.

Venom

Members of four families—the colubrids (Colubridae), atractaspids (Atractaspididae), cobras (Elapidae), and the vipers (Viperidae)—produce and use venom. Venom evolved for subduing prey, but some snakes in these families have found a secondary use for it as a means of defense. Venoms are complex protein cocktails derived from digestive fluids and tailored to be most effective against the prey that the particular species eats or is most likely to encounter. The venom begins to digest the snake's victim before the snake swallows it.

Defense

The snake's best means of defense and the one on which it relies most heavily is to escape notice. To that end, many are well camouflaged, and the coloration within a single species may vary according to the color of the soil on which it lives. The snake's shape presents predators with an unusual problem: Since the snake can change its outline at will—from coiled to stretched out and anything else in between—the predator does not have the opportunity to build up a "search image." Additional "tricks," such as the geometric patterns that break up the snake's outline (known as disruptive coloration) and stripes that run through the snake's eye, add to the predator's confusion.

At the opposite extreme are the snakes that advertise their presence with bright colors and patterns. Many are venomous elapids and are called coral snakes regardless of whether they live in America, Africa, Asia, or Australia. (There are coral snakes on all these continents.) The theory is that most predators associate bold colors with danger and leave the snakes alone. Cashing in on this are a number of "false" coral snakes, harmless species that are brightly colored with similar patterns and are probably Batesian (harmless) mimics of venomous species.

When threatened, snakes behave in a variety of ways. The coral snakes often thrash about, presumably to flaunt their colors. Others try to escape or hide under objects. Some hide their head in their coils and display their tail instead, hoping to deflect the attack away from their head. Blunt-tailed burrowing species such as the rubber boa, *Charina bottae*, are well equipped for this, and many have scars on their tail as evidence that the system works. Other snakes seek to intimidate their enemies, turning defense into attack by hissing and striking, sometimes with the mouth closed but often in earnest. Both venomous and nonvenomous species may bite.

↑ ***The grass snake,* Natrix natrix *from Europe, feigns death. It can also adopt other forms of defense, including biting and exuding a foul-smelling secretion from its anal gland.***

← ***Representative species of two families of advanced snakes. Family Colubridae: the southern hognose snake,* Heterodon simus *(1), and the sand snake,* Psammophis condenarus *(2); family Elapidae: the common death adder,* Acanthophis antarcticus *(3), and the Indian cobra,* Naja naja, *in threat pose (4).***

Warning behavior is used to prevent the snake from being accidentally trampled. It may take the form of the buzzing produced by the tail in rattlesnakes or the rasping sound made by specialized body scales with oblique, serrated keels on saw-scaled vipers. Finally, a number of species, notably the European grass snake, *Natrix natrix*, the American hognose snakes, *Heterodon* species, and the African rinkhals, *Hemachatus haemachatus,* feign death by turning over on their back, gaping, and allowing their tongue to hang out.

Reproduction

Snake reproduction is not fundamentally different from that of their ancestral group, the lizards. The most important evolutionary decision they have had to make is whether to be bearers of live young or egg layers. In general, species from cool places are more likely to bear live young, while those from warm tropical places lay eggs. Superimposed over this pattern, however, is the tendency for members of some families to do one or the other regardless of where they live. So, for instance, all pythons are egg layers, and all boas (except the Calabar ground boa, *Charina reinhardtii)* bear live young.

Breeding seasons depend on climate, and many species are seasonal. They may mate in the spring after a period of hibernation, or they may have a wet season–dry season cycle. In the tropics snakes may breed throughout the year, with peaks of breeding activity corresponding to subtle variations in temperature or rainfall. Females of species living at high latitudes (such as the adder, *Vipera berus*, and the red-sided garter snake, *Thamnophis sirtalis)* or at high altitudes (such as several pit vipers) that spend up to eight months each year in hibernation will probably not be fit enough to breed in successive years. They may have a break for one, two, or more years.

Some species, such as the massasauga, *Sistrurus catenatus*, mate in the fall of one year and enter hibernation with a supply of sperm. The sperm fertilizes the female's eggs when she becomes active again the following year. Birth takes place several months later. The Arafura file snake, *Acrochordus arafurae* from northern Australia and neighboring parts, has such a slow metabolism that its breeding cycle is greatly extended, and each female may only breed once every 10 years.

Courtship behavior and social behavior in general are very poorly known for all but a handful of species. Observing snakes in the wild is extremely difficult because they spend a great proportion of their time out of sight. Recent developments using radiotelemetry to track snakes and obtain data from them without disturbance has added greatly to scientists' knowledge of snake behavior. However, with about 3,000 species to investigate, there are still many more questions than answers.

A black ratsnake,* Elaphe obsoleta obsoleta, *hatches from an egg. Hatchlings have a pale gray background color with black blotches. As the snakes mature, the color becomes darker.

Common name Spiny softshell

Scientific name *Apalone spinifera*

Family Trionychidae

Suborder Cryptodira

Order Testudines

Size Carapace length 6.5 in (16.5 cm) to 18 in (46 cm) in females; from 5 in (13 cm) to 9.3 in (23.5 cm) in males

Weight Approximately 2.2 lb (1 kg) to 3.3 lb (1.5 kg)

Key features Flattened, leathery shell lacks scutes and is circular; underlying color varies from olive to tan; patterning highly variable; spots (ocelli) on carapace have black edges; distinctive spiny tubercles on the front edge of shell are unique to this species; neck long with dark-edged lighter stripes; nostrils elongated and snorkel-like; limbs powerful and paddlelike

Habits Fast swimmer; predominantly aquatic but will emerge to bask on occasions; burrows into mud or sand beneath the water to hide with just the head exposed

Breeding Female lays 4–32 white, spherical eggs; eggs probably hatch after about 8–10 weeks

Diet Carnivorous; eats mainly invertebrates such as crayfish; larger individuals may take fish and amphibians

Habitat Still or slow-flowing waters that are often shallow with sandy or muddy bottom; may also occur in faster-flowing waters

Distribution North America from southern Canada south across the southern United States, including the Florida Peninsula, and around the Gulf Coast in Mexico

Status Generally quite common

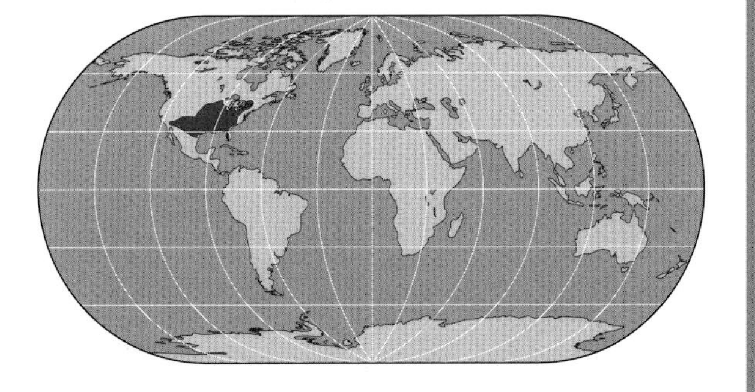

Spiny Softshell

Apalone spinifera

The highly aquatic spiny softshells seldom venture out of water except to bask. The Latin name spinifera *means "bearing thorns," a reference to the spinelike tubercles on the carapace edge just behind the head.*

Members of the family Trionychidae, to which the spiny softshell belongs, are widespread. They occur in parts of Africa and Asia as well as North America. The spiny softshell is one of the most widely distributed species. Its range extends from southern parts of Canada across much of the southern United States and south as far as Mexico.

Six distinctive subspecies have been identified. The eastern softshell, *Apalone spinifera spiniferus*, is relatively large with black borders around the spots (ocelli) on the carapace. In contrast, the western race, *A. s. hartwegi*, has much smaller ocelli as well as dots on its carapace. Unlike these two forms, the Gulf Coast spiny softshell, *A. s. asper*, has at least two lines at the rear of the carapace. The Texas race, *A. s emoryi*, has a pale rim to its carapace, which is much wider along the rear edge, and there are also white tubercles present on the rear third of the carapace. In the pallid subspecies, *A. s. pallidus* from west of the Mississippi and east of the Brazos River, the tubercles are more prominent, extending over the back half of the carapace. In the case of the Guadalupe spiny softshell, *A. s. guadalupensis*, the white tubercles cover most of the carapace and are encircled by black rings. In places where two different races overlap, however, they can be difficult to distinguish because they show characteristics of both populations.

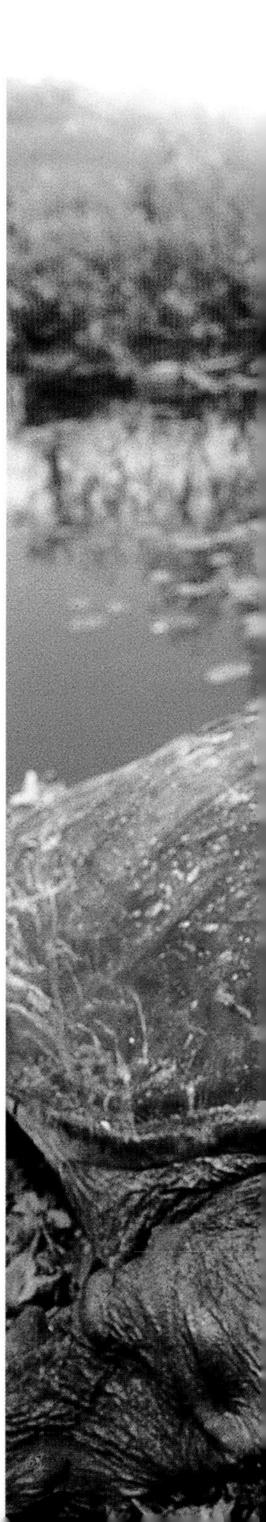

Pancake Turtles

Turtles in this family are sometimes known as pancake turtles due to their flat, circular shape. Spiny softshell turtles are adept at burrowing in mud or sand beneath the water, hiding away and leaving just their head exposed above the surface. They do not need to surface in order to breathe as frequently as other turtles, since they can absorb oxygen directly in the pharyngeal region of their throat as well as through their leathery shell. In extreme cases they are able to remain under the water for up to five hours. During the winter period in northern parts of their range they will hibernate in the water by burying themselves under several inches of mud. During this time they slow down their respiratory rate.

Mating occurs in the spring, and egg laying peaks during June and July. Females haul themselves onto land and dig a nest site quite rapidly using their hind feet. They sometimes complete this task in under 15 minutes. It is not uncommon for the nesting turtles to empty their bladder into the hole as they dig to make the soil particles stick together and help the excavation process. They invariably choose a sunny site close to water.

A female spiny softshell basks at the side of a river with its head raised, revealing the unusual long, piglike nose that is characteristic of the species.

Young spiny softshells start to emerge from the end of August through to October, but in the far north of their range the young may overwinter in the nest, emerging for the first time the following spring. They measure about 1.4 inches (3.6 cm). They appear to be relatively slow growing and only reach sexual maturity when they are about 10 years of age.

The species is known for its longevity—the largest females can be over 50 years old. They face relatively few threats except for pollution of the water, which can be a major hazard in some areas. Adults are sometimes caught for food, but generally these softshells are not subjected to heavy hunting pressure.

Their powerful feet mean that the spiny softshells are able to live in fast-flowing rivers but are equally at home in ditches. When seeking food, they often prefer to comb the bottom rather than swim actively in search of prey. Invertebrates such as crayfish are their main prey items, but larger individuals also prey on fish and amphibians, notably frogs.

Sunbeam Snake

Xenopeltis unicolor

The sunbeam snake is well named: When seen in sunlight, its large, glossy scales reflect every color in the rainbow, forming a kaleidoscope of colorful shapes that change as the snake moves.

Common name Sunbeam snake (iridescent earth snake)

Scientific name *Xenopeltis unicolor*

Family Xenopeltidae

Suborder Serpentes

Order Squamata

Length From 39 in (100 cm) to 4.3 ft (1.3 m)

Key features Dark brown with large, extremely glossy dark-brown or black iridescent scales; head very flat and spade shaped, covered with large platelike scales; no pelvic girdle or signs of vestigial hind limbs; juveniles have a wide white "collar" across the back of the neck

Habits Burrower, especially in damp soil or leaf litter; often turns up under rocks, logs, or garbage

Breeding Egg layer, with clutches of 6 or more large eggs

Diet Amphibians, lizards, snakes, and rodents

Habitat Lightly forested areas, plantations, parks, gardens, and other urban situations

Distribution Southeast Asia

Status Common

Similar species The Chinese sunbeam snake, *Xenopeltis hainanensis*, is almost identical; the 2 species were only recently separated

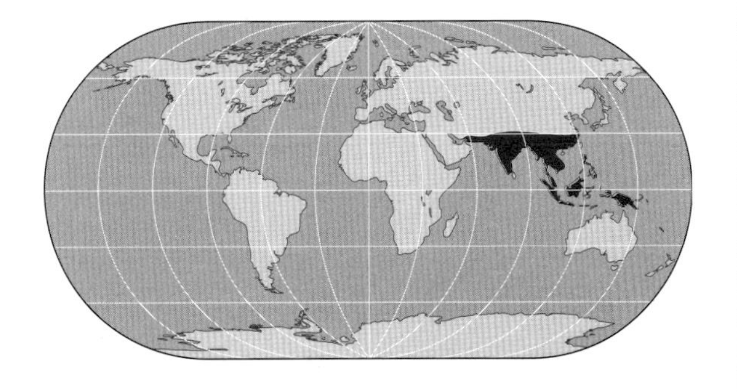

Although the sunbeam snakes look like typical snakes (they are of medium size with average proportions), they combine a number of primitive features with more advanced ones. Their ventral scales are significantly wider than the dorsal ones, unlike in many other species that have narrow ventral scales. They also lack all traces of a pelvic girdle and hind limbs. These are features of advanced snakes. On the other hand, they have teeth on the premaxillary bone, and the left lung is large and functional, both of which are primitive characteristics.

Sunbeam snakes are common and familiar snakes in Southeast Asia because they tolerate a wide range of habitats and conditions. They often live in gardens, parks, and wasteland on the outskirts of cities and towns. No doubt they are attracted by the high density of frogs, lizards, and small rodents that are in turn attracted by human garbage. Their natural habitat, however, is forested hillsides and lowlands, especially where the soil is moist and contains a high proportion of decomposing organic material such as leaf litter.

Although the head is flattened and shovel shaped, sunbeam snakes appear not to burrow through compacted soil at all. Instead, they prefer to live in burrows made by the small mammals that form part of their diet.

The skull bones of sunbeam snakes are more mobile than those of the more primitive species. Because of this, sunbeam snakes have a much wider choice of potential prey. Unlike the pipe snakes, for example, whose skulls are more rigid, they are not limited simply to eating elongated animals.

Sunbeam snakes are heavily collected and sold for the pet trade. Sadly, few survive beyond the first six months after capture. They are difficult to breed in captivity, so wild populations will remain under pressure while there is a demand for them as pets.

One-Way Passage

They also have unusual teeth that are all of equal size and are attached to the mouth by a piece of ligament that acts like a hinge. The teeth can be folded backward but not forward, and they act together like a ratchet. This means that food can move smoothly down the snake's throat but cannot easily come back out. A few other snakes—all unrelated—have the same arrangement, which is thought to help them eat smooth animals, especially skinks. They do not constrict their prey, but may hold it against a firm surface to subdue it before swallowing.

Breeding habits of the sunbeam snakes are unknown, but they usually lay clutches of six to nine eggs (occasionally a dozen or more) that hatch in 60 to 70 days. The young have a bold white bar across the nape of their neck, which disappears after they have shed their skin two or three times.

Iridescence

Snakes' colors are mainly created by pigments contained in cells called chromatophores that are embedded in their scales. The pigments are of various types and can produce a whole range of colors, of which the most common are black and brown. They are caused by melanophores (cells containing the pigment melanin). Sunbeam snakes and some other species, however, use iridescence to supplement their underlying pigmentary colors. The outer layer of their surface is thin and transparent. When light strikes the surface from an angle, it is split into its spectral components, and each wavelength produces a different color. Microscopic structures contained in the layer may also help scatter the light. This creates an effect known as iridescence, often made stronger by the dark coloration of the layers below. As the snake moves (or as the observer's position alters), the colors appear to change, just like a film of oil on a puddle of water.

Although many snakes have a degree of iridescence, the most striking examples are seen in the sunbeam snake, the American sunbeam snake, *Loxocemus bicolor*, and some forms of the rainbow boa, *Epicrates cenchria*.

The lustrous scales that cover the whole of the sunbeam snake's body sparkle with iridescence in sunlight. However, this beautiful gentle snake is secretive and usually ventures out at night.

Coastal taipan *(Oxyuranus scutellatus)*

Common names Taipan (coastal taipan), inland taipan (fierce snake)

Scientific names *Oxyuranus scutellatus* (coastal taipan) and *O. microlepidotus* (inland taipan)

Subfamily Hydrophiinae

Family Elapidae

Suborder Serpentes

Order Squamata

Length From 6.5 ft (2 m) to 12 ft (3.6 m)

Key features Body large and cylindrical; head narrow with vertical sides; usually some shade of brown; coastal taipan sometimes has a pale head, inland taipan may have a black head and neck; coastal taipans from New Guinea may be brown with a wide rust-colored stripe down the back, or they may be uniform black

Habits Terrestrial; diurnal; fast and alert

Breeding Egg layers with clutches of up to 22 eggs; eggs hatch after about 9–10 weeks

Diet Mammals

Habitat Forests, wooded grassland (coastal taipan); dry flood plains (inland taipan)

Distribution Northern Australia and southern New Guinea (coastal taipan); east-central Australia (inland taipan)

Status Common in places but shy and rarely seen

Similar species Adults distinctive because of their size; juveniles can be confused with other species, such as the eastern brown snake, *Pseudonaja textilis*, which is also dangerous

Venom Both have exceedingly powerful venom; that of the inland taipan is most destructive and is considered to be the most potent of any terrestrial snake

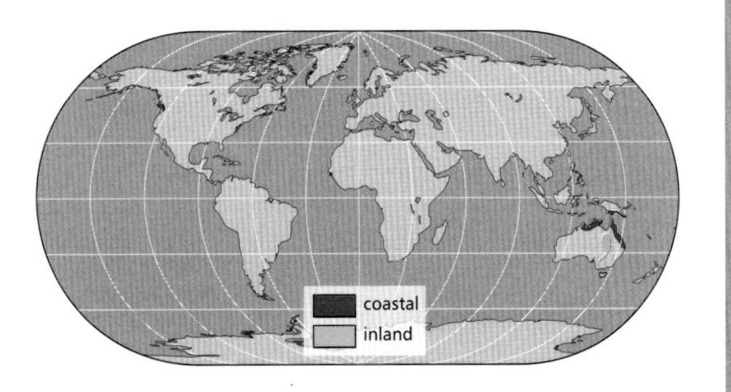

Taipans

Oxyuranus scutellatus and *O. microlepidotus*

Previously in two separate genera, the taipans were discovered to be more closely related to each other than to any other species. In 1981 they were placed together in one genus.

TAIPANS ARE GENERALLY CONSIDERED to be the most venomous land snakes. The inland taipan, *Oxyuranus microlepidotus*, can inject enough venom in a single bite to kill over 200,000 mice (more than twice the number that could be killed by a single bite from the coastal taipan, *O. scutellatus*, 50 times that of an Indian cobra, *Naja naja*, and 75 times that of a western diamondback rattlesnake, *Crotalus atrox*). The venom is a fast-acting neurotoxin, causing paralysis and respiratory failure. Taipans have the longest fangs of any Australian venomous snake, and a bite from either species could contain 60 mg of venom: more than enough to kill a small marsupial or rat, or several humans.

Ⓣ ***A coastal taipan,*** **Oxyuranus scutellatus,** ***eating a mouse. This aggressive, highly venomous snake is found along the coast of northern Australia.***

Ⓛ ***The inland taipan,*** **Oxyuranus microlepidotus,** ***produces the most toxic venom of any snake in the world. Its range is generally limited to a small area of western Queensland in Australia.***

Despite their alternative common name of fierce snakes, inland taipans are shy and have relatively placid natures, and there are no known human deaths from the bite of this species. (By comparison, Indian cobras are responsible for several thousand deaths each year.) Coastal taipans are also shy; but because they tend to come into contact with humans more often, bites are known, especially in New Guinea. In a sustained defensive attack taipans can bite repeatedly, inflicting several wounds before the victim has time to retreat.

Diet of Mammals

Taipans are diurnal ground-dwelling snakes that are most active in the morning and occasionally in the evening throughout the year, although they are most likely to be found in spring. Both species are unique among Australian elapids because they feed exclusively on mammals. Most other species eat a wide variety of cold- and warm-blooded prey.

They are active hunters, seeking their prey in crevices, burrows, and in the case of the inland taipan, in the deep cracks left in the drying mud at the end of the wet season. The inland taipan feeds largely on the plague rat, *Rattus villosissimus*, one of the few rodents in the arid Australian interior. When rat populations fall, which they do occasionally, the taipan resorts to other mammals. After biting their prey, taipans release it and wait a few seconds before setting out to find the carcass.

Numbers of taipans may be increasing due to changing farm practices, especially the increased growing of sugarcane, which is a favorite of plague rats and therefore their predators. The introduction of cane toads, *Bufo marinus*, may also have helped increase the numbers of tapians. The toad is highly toxic to most of the other snakes in the region, whose numbers have declined since the toad became widespread. This has eliminated some competition for the taipans, which do not eat toads and are therefore not affected by them.

Larger Males

Female taipans of both species are about the same size, but male coastal taipans grow larger than male inland taipans. In both cases, however, the males grow larger than females, even though they mature when they are smaller. Males fight with each other in the breeding season, rearing up and trying to force their opponent to the ground. Females of both species lay between seven and 22 eggs that hatch after about 66 days. The hatchlings are about 12 inches (31 cm) long and venomous from the start.

Common name Tegu (black-and-white tegu)

Scientific name *Tupinambis teguixin*

Family Teiidae

Suborder Sauria

Order Squamata

Size To 43 in (109 cm), of which just under half is the tail

Key features Large and powerful lizard with a cylindrical body; tail thick; head long; snout narrow; limbs are also powerful and end in long claws; the whole lizard is covered with small shiny scales that give it a glossy appearance; adults boldly marked in black and white or black and yellow, but juveniles are bright green

Habits Terrestrial and diurnal

Breeding Female lays 7–30 eggs that hatch after about 12 weeks

Diet Insects, spiders, other lizards, birds and their eggs, small mammals; also eats carrion

Habitat Rain forests in clearings and along river courses

Distribution South America (Brazil, Peru, Colombia, Venezuela, Ecuador, northern Argentina, Uruguay, Bolivia, Guiana, Surinam, French Guiana); Trinidad

Status Common in suitable habitat, although never in large numbers

Similar species Up to 6 other species in the genus occur throughout South America, all differing slightly in coloration, but the validity of some is in doubt; otherwise there are no similar species; juveniles could possibly be mistaken for jungle racers, *Ameiva* species

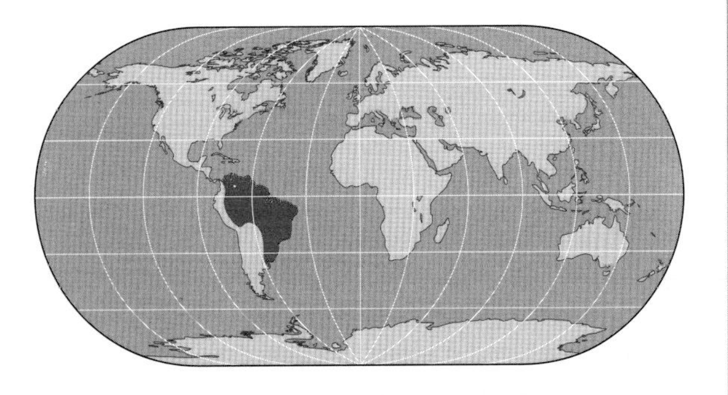

Tegu

Tupinambis teguixin

The tegu is among the world's largest lizards. In many of the places where it lives it fills the same ecological niche as small mammalian predators.

TEGUS ARE VERY POWERFUL LIZARDS and are able to catch large insects, other lizards, snakes, nestling birds, and small mammals. They are particularly fond of birds' eggs, and they often raid nests, including those of domestic fowl. A related species, the Argentine tegu, *Tupinambis merianae*, also eats fruit and has even been known to beg food from picnickers in public parks. The common tegu, however, is a confirmed carnivore.

Adult *T. teguixin* have quite well-defined home ranges centered on their burrows, where they can hunt most successfully without competition from neighbouring tegus. They use their long, sensitive tongue to detect potential prey by picking up scent particles from the atmosphere. Then they scratch or dig at the surface using their powerful limbs.

When they find a prey item, they grasp it in their jaws and, if it is large, bash it repeatedly against the ground in order to stun it. Once it stops struggling, they normally bolt it down in one piece, although they may tear large items into smaller pieces. Full-grown tegus feed mainly in the afternoon after they have raised their body temperature to about 86°F (30°C), but juveniles forage throughout the day.

Forest Dwellers

The natural habitat of tegus is dense primary rain forest. They are often found near the edges of wide rivers and lakes or in forest clearings, where they can use the open spaces to bask. They live in burrows that they either dig for themselves or take over from small mammals.

Tegus swim well; if their habitat becomes temporarily flooded, they make their way to

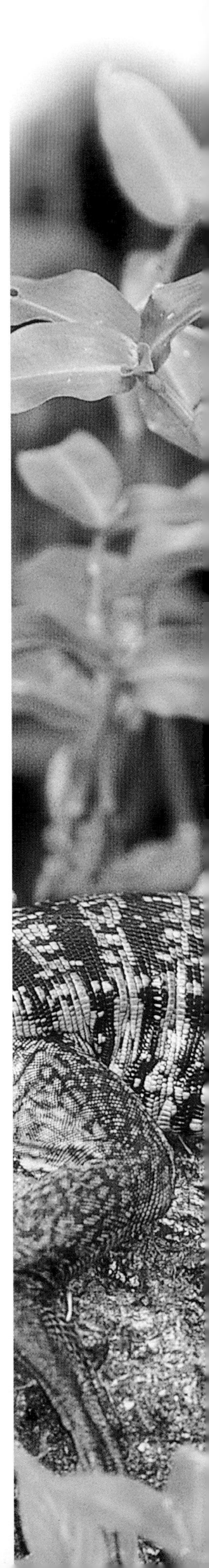

Lizard Parallels

The American tegus are the ecological counterparts of the monitor lizards, which are found only in Africa, Asia, and Australasia. Apart from having glossy scales, tegus could easily be mistaken for monitors—they have a raised, ambling gait, powerful limbs, a long and constantly flickering tongue, and natural curiosity. Like tegus, monitors often live in areas where there are few other predators and have been quick to fill this niche. And like the tegus, some monitor lizards associate humans with food and are especially common around villages and popular picnic sites. Finally, a number of monitor lizards also make use of termite nests to lay their eggs, taking advantage of the constant temperature and the security provided by the efforts of the insects.

higher ground to wait until the water level subsides. In addition, tegus often make their homes on the outskirts of villages and towns, where human activities create artificial clearings. They are also attracted by the prospect of finding easy prey and may even become pests by raiding poultry farms to steal the eggs. If they are unfortunate enough to get caught, they provide a welcome addition to the diet of local people.

Courtship and Egg Laying

The tegus breed at the start of the dry season so that their eggs are ready to hatch during the wet season, when there is plenty of food around. Courtship in tegus is dramatic. The male approaches the female sideways, puffing himself up and standing up on stiff legs to make himself look bigger. He makes regular snorting or "sneezing" sounds. Then he approaches her, grabs her neck in his jaws, and holds her body down with his hind legs. He twists his body under hers so that their cloacae are next to each other. Copulation can then take place.

When the female is ready to lay her eggs, she may dig a burrow in the soil or in leaf litter in a forest clearing; but she often chooses a termite mound. She breaks into the base of the mound using her powerful legs and claws, and makes a nest chamber.

All tegus are black with yellow or white bands across the back. The small glossy scales are arranged in regular rings around the body. This is* Tupinambis teguixin *in Trinidad.

She lays up to 30 eggs there, and the termites quickly repair the damage and seal in the eggs. The temperature and humidity in the mound are carefully controlled by the termites, so it makes an ideal incubator. The eggs hatch after about three months, and the young break out of the termite mound after it has been softened by rain. In contrast to the adults, they are bright green with black spots.

Warming Up

Like all reptiles, tegus rely on the sun to maintain a suitable body temperature, ideally about 86°F (30°C) in their case. They are not active at temperatures much below this. They raise their body temperature by occasional basking throughout the day. Large individuals take a long time to warm up and therefore do not emerge from their burrows until the middle of the day; they remain sluggish when the weather is cool.

In the colder parts of their range the tegus sometimes remain hiding in their burrows for several weeks during the worst of the cold weather, although they do not hibernate in the same way as the species from farther south.

Tegus' enemies include birds of prey, cats, and snakes. Juveniles are more vulnerable than adults, which have relatively few enemies by the time they reach full size. They defend themselves by using their tail as a whip or by biting and clawing their way out of trouble. In extreme situations they may discard their tail; although they can grow a new one, it is never quite as long as the original.

Hunting and the Skin Trade

Since tegus can have a habit of raiding poultry farms to steal the eggs, it is not surprising that they are often considered to be vermin. Over much of South America tegus are hunted and trapped by local people. Apparently they make good eating, and their skins are valuable for fashion goods. Lizard skin is a significant source of income for villagers, and the number of tegus captured for that purpose is estimated to be between one and three million each year. Populations living near towns and villages are the most vulnerable, while those in remote, forested areas are fairly safe—at least for the time being.

The Argentine tegu, *T. merianae*, however, lives only in the eastern part of Argentina. It inhabits open land often covered by tall clumps of pampas grass and is rarer than *T. teguixin.*

***Tegus are capable of eating large prey, which they can stun by beating it against the ground. Here* T. teguixin *is tackling a rattlesnake,* Crotalus durissus.**

Tegu Names and Species

The name "tegu" comes from an Amazonian Indian word that simply means lizard. Names within the genus *Tupinambis* are confusing. *Tupinambis teguixin* was previously used for the Argentine tegu, which is now called *T. merianae*. The black-and-white tegu discussed here used to be called *T. nigropunctatus*, a name that is no longer valid. Much information that was gathered in the past about these lizards is no longer useful because there is no way of knowing which species it related to.

The Argentine tegu is a hardy species that hibernates during the cold southern winter. It grows slightly larger than the black-and-white tegu. Most tegus in the pet trade are black-and-white tegus and are recognizable by their color and their small, shiny scales, which are similar to those of skinks. They are nervous and hard to tame. The Argentine tegu is rare in captivity, although it is bred in small numbers. It can be distinguished by small beadlike scales like those of a Gila monster.

Tupinambis duseni is a Brazilian species whose range overlaps that of *T. teguixin*. It is reddish in color, but its validity as a species is uncertain. A similar situation exists with *T. longilineus* (also from Brazil). *Tupinambis palustris*, which translates as "marsh tegu," was only described in 2002 from Rondonia, Brazil. It differs from *T. teguixin* in slight details of markings. *Tupinambis quadrilineatus* from central Brazil is also a recent addition to the genus, described in 1997. The seventh species is the red tegu, *T. rufescens*, which is restricted to Argentina and is a heavy-bodied species with reddish-brown coloration. Like the other Argentine species, *T. merianae*, it hibernates during the cooler months of the year.

Common name Thorny devil

Scientific name *Moloch horridus*

Family Agamidae

Suborder Sauria

Order Squamata

Size From 6 in (15 cm) to 7 in (18 cm)

Key features A weird-looking lizard; body squat, covered in large, thornlike spines; there is a very large spine over each eye, a raised, spiny hump on its neck, and 2 rows of spines along the top of its tail; dark reddish brown in color with wavy-edged, light tan stripes running over its head and down its body

Habits Diurnal; terrestrial; slow moving

Breeding Egg layer with a single clutch of 3–10 eggs; eggs hatch after 90–132 days

Diet Ants

Habitat Deserts

Distribution Western and central Australia

Status Probably common in suitable habitat but rarely seen

Similar species There is nothing remotely similar in the region

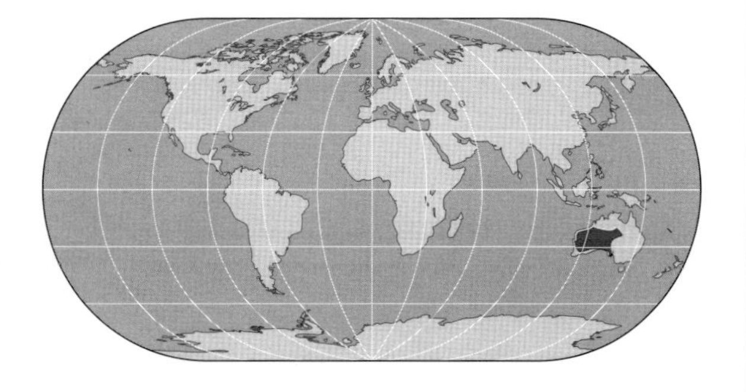

Thorny Devil

Moloch horridus

The thorny devil is Australia's answer to the American horned lizards, Phrynosoma, *with similar looks and lifestyle. Like the horned lizards, it is a dedicated ant-eater.*

THORNY DEVILS EAT NOTHING BUT ANTS, especially the very small species *Iridomyrmex flavipes* and others in the same genus. They lick them up at an estimated rate of 25 to 45 per minute. Their fecal pellets are round, glassy, and black. Unusually among lizards, each thorny devil uses its own individual place to deposit them, returning several days in a row until a small pile of distinctive droppings has accumulated.

Eating ants has its drawbacks: Many thorny devils are parasitized by nematodes (roundworms) that probably use the ants as an intermediate host.

Water Supplies

Drinking water is almost impossible to find in many of the dry, sandy places where thorny devils live, so the lizards have evolved an interesting system for capturing the dew that condenses on their body. A network of small channels covering their skin directs water by capillary action to the corners of the lizard's mouth. By gulping the water as it accumulates there, it can drink enough to compensate for the meager water content of its diet.

Thorny devils tolerate extremes of heat just as horned lizards do and remain feeding out in the open when other lizards have retreated into the shade. They have an average body temperature of 91°F (33°C). Even so, they are inactive during the hottest months of the year (January and February) and dig burrows into which they retreat to avoid the heat. They do not move around much in the coolest months (June and July) because they are too cold.

The lizards' activity patterns are easily studied by observing the distinctive tracks they make in their sandy habitat. In the summer they

Moloch horridus ***is an extraordinary lizard that inhabits the dry Australian interior. It is covered in warty, thornlike spines, and narrow channels between its scales draw precious droplets of dew or rain into its mouth.***

range less than 10 yards (9 m) from their burrows; within this "home range" there will be one or more ant trails, a few small bushes with a pile of dead leaves or a loose tussock of grass beneath them, and a defecation site.

The thorny devils spend each night and the hottest part of each day in a burrow under one of the shrubs. Each day they venture a short distance from their refuge to feed, defecate, and eat some ants. In the spring, however, they are much more active, and trails across the sand often extend for more than 100 yards (91 m). These more adventurous trips are probably the result of an urge to mate. Studies have shown that thorny devils home in on places where other members of their own species are, but nobody knows how they do it.

Desert Lookalikes

There are many good examples of convergent evolution among lizards, but the pairing of the Australian thorny devil, *Moloch horridus*, with the American horned lizards, *Phrynosoma* species, is one of the best. The horned lizards belong to the Iguanidae, so they are only distantly related to the thorny devil. Both are food specialists, feasting on large numbers of ants. (The thorny devil eats nothing else.) Both live out in the open and use cryptic (disguise) coloring to avoid detection by predators. And both are well endowed with spiny scales.

Apart from the novelty value of finding similar animals on opposite sides of the world, the similarities tell us something about the way in which evolution works. Each evolutionary line is constantly "fine tuning" itself in response to the prevailing conditions. In time, evolution often comes up with similar solutions to similar problems regardless of which part of the world is involved.

Spring Activity

The period of greatest activity is the southern spring and early summer (August to December), when mating and egg laying take place. Females lay a single clutch of three to 10 eggs each year. (They differ in this respect from the American horned lizards, which lay much larger clutches.) Females dig their nest tunnels on south-facing sides of sandy ridges, presumably because they are not subject to such extremes of temperature. The burrows have a right-angled bend in them, and the eggs are laid at the end in a chamber large enough to contain the clutch and a significant amount of air.

The female takes several days to dig the burrow. The temperature in the end chamber is a steady 88°F (31°C), and the eggs hatch after 90 to 132 days. The American herpetologist Professor Eric Pianka discovered that after the young had hatched and emerged from their nest, there were no traces of eggshells in the empty chamber. He concluded that the hatchlings had eaten them. If so, this is the only known case of reptiles eating their eggshells. The hatchlings are smaller versions of the adults, thorns and all.

Three-Toed Skink

Chalcides chalcides

At first glance the three-toed skink could be mistaken for a snake with its long, thin body and serpentine movement. However, its tiny limbs with three reduced toes on each one give the game away.

Common name Three-toed skink (seps, barrel skink, cylindrical skink)

Scientific name *Chalcides chalcides*

Family Scincidae

Order Squamata

Size Up to 17 in (43 cm)

Key features Body extremely elongated; color varies from olive-green to bronze, may be uniform or have dark, longitudinal lines; lower eyelids undivided with transparent disk; head conical; snout blunt; ear openings larger than nostrils; limbs and digits considerably reduced; 3 digits on each foot; tail 1–1.5 times the length of the body

Habits Diurnal or crepuscular depending on season; hibernates for 4–5 months; basks to raise body temperature before foraging for food

Breeding Live-bearer; female gives birth to 3–15 live young; gestation period 3–4 months

Diet Insects, spiders, caterpillars, centipedes

Habitat Moist, grassy areas to drier, stony regions

Distribution Italy, southern France, Iberia, North Africa

Status Common

Similar species *Chalcides mionecton*; *C. sepsoides*

THE THREE-TOED SKINK, or seps as it is sometimes known, belongs to the genus *Chalcides*, which contains 25 species. Within the genus the limbs of species vary from small to reduced with some being almost vestigial. As evolution in the genus has resulted in a reduction in limb size and number of digits, the body has become more elongated. As well as the change in body shape, the number of rows of scales around the middle of the trunk has decreased. The most primitive species is *C. ocellatus*, which has five digits on each foot and 26 to 34 rows of scales. *Chalcides chalcides* has three digits on each foot and 22 to 24 rows of scales. *Chalcides guentheri* has tiny vestigial limbs, no digits, and 20 to 22 rows of scales.

The three-toed skink is very agile and can catch flying insects. It usually moves like a snake with a winding movement, but it uses its reduced front legs to pick its way over hard surfaces.

Habitat and Behavior

The three-toed skink is found in a wide range of habitats in western Mediterranean countries and the Maghreb region of North Africa (Morocco, Algeria, and Tunisia into Libya). In fact, Morocco is believed to be an important evolutionary center for the whole genus.

In Italy the three-toed skink inhabits grass or clover meadows, and in northern Tunisia it is found in open cork-oak forests. In other parts of its range its habitat varies from herbaceous vegetation up to 16 inches (41 cm) high to reeds near lakes and salt marshes to shrubs on moist ground. It is also found in fallow fields, paddocks, and escarpments with dense grass cover. In the Atlas Mountains it inhabits drier, stony areas up to 7,590 feet (2,313 m).

Because of its shape people assume that the three-toed skink is a burrowing, desert-

Persecuted

The three-toed skink is long and slender. When at rest, it rolls up and resembles a snake. Because of its snakelike appearance and superstitions that are associated with snakes, it has been persecuted by local people. The ancient Latin name was *seps*, which comes from the Greek *sepein,* meaning "to rot." It was originally thought that the bite of these creatures was highly venomous and that it would cause the flesh of the victim to rot.

dwelling species. However, unlike other skinks with similar shape and limb reduction (such as the Florida sand skink, *Neoseps reynoldsi*), it is terrestrial. The reduced limb size is an adaptation not to burrowing but to moving on surfaces with dense vegetation, where progress by serpentine movement of the body is easier than walking or running. The skink is very agile and can move through damp grass and thickets at considerable speed while barely touching the ground. During slower movement on hard or rocky ground it uses its tiny front feet and presses its hind legs against its body.

Until May the three-toed skink is active all day. During June and July, however, activity takes place in the morning in order to avoid the greatest heat. From August onward the degree of activity reduces—in September or October (depending on temperatures) the skinks begin hibernation that lasts until February. Much of their daytime activity involves basking in the morning with either the front part or all of the body exposed. When they reach the right temperature, they shelter under stones.

Body Structure and Size

The three-toed skink has a very slender, elongated body that is oval in cross-section. It has four tiny limbs, each with three toes. The largest specimen recorded was a female that measured 17 inches (43 cm), 11 inches (28 cm) of which was the tail. Its foreleg measured just 0.3 inches (0.8 cm) and its hind leg 0.5 inches (1.3 cm) long.

There are five subspecies: *Chalcides chalcides chalcides*, the eastern form, has shorter limbs than *C. c. mertensi*, the western form. The other three subspecies are based on

Closeup of the front part of the body of the three-toed skink. Because it is nonburrowing, its ear openings are larger than its nostrils.

color variations. They vary from olive-green to bronze and can be uniform in color or can have several dark, longitudinal lines. There appear to be four dorsal pattern types that correspond to subspecies of the same name: *concolor* (uniform coloration), *vittatus* (four dark lines), *mertensi* (six dark lines), and *striatus* (11 lines). Specimens of all four pattern types can be found in Morocco, but *striatus* also occurs in Spain and the other three in Italy. The three-toed skink is occasionally confused with *Chalcides mauritanicus*, which lives in coastal sands, and with *Ophisaurus koellikeri*, the Moroccan glass lizard, which has no forelegs.

The snout of the three-toed skink is conical, and since the species does not burrow, the ear openings are prominent and larger than the nostrils. The lower eyelids have a transparent disk, or window. The body is covered with longitudinal rows of scales. Although not keeled, they are arched and underlaid with osteoderms (bony plates).

Insect Diet

Foods consumed by the three-toed skink include woodlice, centipedes, spiders, cockroaches, grasshoppers, earwigs, caterpillars, flies, and ants. Its feeding strategy is to observe

The most primitive member of the genus,* Chalcides ocellatus, *has limbs that are not as reduced as those of the three-toed skink and has five toes on each foot. It is a fast-moving, agile skink that lives on scrubland, grassy slopes, and dry, sandy areas.

Placental Development

The methods of reproduction in species in the genus *Chalcides* include both ovoviviparity (in which eggs develop and hatch inside the mother or hatch shortly after they are laid) and viviparity (in which females give birth to live young). The latter occurs in a number of other reptiles and involves the development of extra embryonic membranes next to the uterus. The three-toed skink, *Chalcides chalcides,* has the most advanced type of placenta. Not only does the embryo get water, nourishment, and oxygen from the placenta, its excretory waste is also taken away.

The same gene that is essential for placental development in mice has also been found in the three-toed skink, and its placenta has many aspects that are common to the placentas of mammals. Mature placentas of *C. chalcides* have been described as the most specialized found in any reptile, and they show a substantial transfer of nutrients from mother to fetus.

its surroundings with the front part of its body raised. Then quickly and deftly it shoots up from tufts of grass to catch the prey. When larger prey items are swallowed, it bends its head laterally in the same way as a snake. After lapping water, it holds its head horizontally.

Reproduction

Telling the sexes apart is quite difficult, but females tend to be larger than males. Research has shown that in many areas the skinks hibernate in groups at a depth of about 12 to 14 inches (30–36 cm). When they emerge after hibernation, males are particularly aggressive and will bite the necks and tails of their rivals. As a result of these confrontations, more than half of all adults and subadults have tails that have been regenerated.

Sexual maturity is reached in the third year. Mating begins from March onward, and females ovulate at the beginning of April. The gestation period lasts for three or occasionally four months, after which the females give birth to between three and 15 young. At birth the young measure 3 to 3.7 inches (8–9.5 cm). The number of live young depends on the size and age of the female, while their size at birth varies according to litter size—the more young produced, the smaller they are. Juveniles have voracious appetites and can double their size and weight in just a few weeks.

Predators and Defense

The three-toed skink shares its range with a number of predatory creatures such as the grass snake, the horseshoe racer, the false smooth snake, foxes, spotted genets, as well as egrets and aerial predators such as hawks and harriers. However, examination of the stomach contents of these predators reveals very few specimens of three-toed skinks, indicating that they do not seem to be particularly vulnerable. Their main form of defense is their agility and the fact that they are hard to catch even in low grass. Any specimens basking will evade their enemies by leaping through the air to land on another bush. If caught, many will twist and attempt to bite. The long, slender tail is very fragile; if necessary, the skink will cast it off to distract a predator and escape.

Common name Tokay gecko

Scientific name *Gekko gecko*

Subfamily Gekkoninae

Family Gekkonidae

Suborder Sauria

Order Squamata

Size From 8 in (20 cm) to 14 in (36 cm) long

Key features A large, heavy-bodied gecko with a massive head, prominent yellow or orange eyes, and a huge gape; conspicuous toe pads present; skin covered in small granular scales interspersed with raised tubercles; color bluish gray with evenly scattered spots of lighter blue and rust

Habits Naturally arboreal but also found on the walls of buildings; nocturnal

Breeding Female lays 2 (sometimes 3) spherical, hard-shelled eggs; eggs usually hatch after about 100 days

Diet Invertebrates of all sizes and small vertebrates, including other geckos

Habitat Forests, plantations, and buildings

Distribution Southeast Asia

Status Very common

Similar species The green-eyed gecko, *Gekko smithii*, is about the same size but gray in color with emerald-green eyes

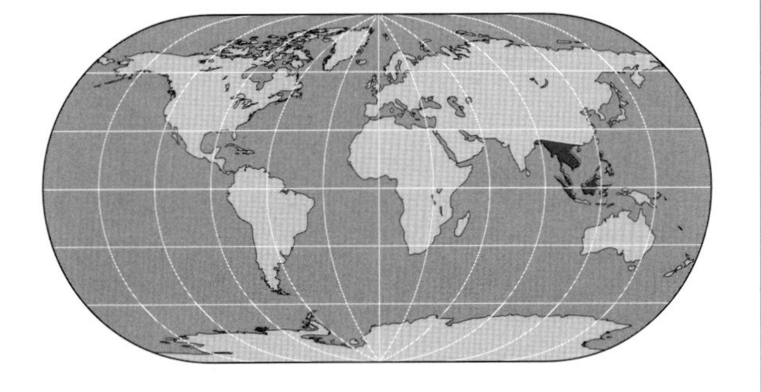

Tokay Gecko

Gekko gecko

The brightly colored tokay is the species that gives the whole family its common and scientific names. These large geckos make a raucous call that sounds, according to some listeners, like "gek-ko."

THE TOKAY GECKO IS ONE OF THE MOST VOCAL geckos. Its call can be heard for 100 yards (91 m) or more. While some people think the call sounds like "gek-ko," others interpret it as "toe-kay," hence its common name. Two tokays living near each other will have a "conversation" with calls echoing back and forth, not unlike the vocalizations of some frogs and toads.

This type of communication probably helps keep individual geckos spaced out and thereby ensures that they avoid competition. Calling is intensified during the main breeding season, when these geckos often call throughout the night. However, at other times of the year calling is less frequent, and they may be silent altogether at certain times. Their routine seems to vary from place to place, however, and some observers have noted that tokays call any time they are active.

Living With Humans

Tokays are forest species, living on vertical surfaces such as tree trunks and spending their days in cracks and behind loose bark. They have adapted very well to human activities, however, and are common around villages. Most homes will have a resident tokay, often living between the two layers of wooden boards around the outside of the house or behind pieces of furniture such as cupboards and closets. They are creatures of habit, and individuals will have favorite places to which they return regularly. They can even be relied on to emerge at roughly the same time every night.

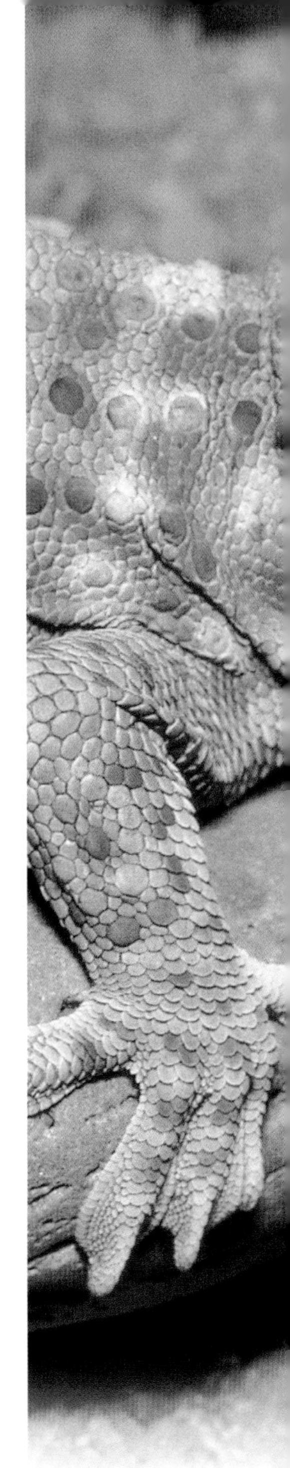

The jaws of the tokay gecko are strong, and its teeth are sharp, enabling it to eat quite large items of prey, such as this locust.

As soon as the sun begins to fade, the tokays will creep out from their daytime hiding places. They move cautiously at first before coming right out and heading for a favorite hunting position. If they are unsuccessful, they will move on after a while to try their luck somewhere else, but they have a well-defined "beat" that they patrol and that they defend against other tokays.

If they are in an area where there are electric lights, they often wait nearby, taking advantage of the moths, cockroaches, and other insects that are attracted to them. If necessary, they will alter their routine slightly to move to a window with a light in it. Tokays will also prey on the smaller geckos that live around buildings, although populations seem to vary in this respect, and in some places they live peacefully side by side.

Geckos on the House

Geckos have adapted perhaps better than most reptiles to human development. The species with expanded toe pads that live naturally on tree trunks or rock faces have moved easily into buildings. They are a regular feature in most accommodation—even modern hotels and restaurants—in Asia and elsewhere. There are plenty of hiding places in the eaves of roofs or in thatch, and the invention of electric lights has been of great benefit to geckos because of the flying insects that are attracted to them. House geckos have been quick to exploit this, and it is not unusual to see half a dozen or more lying in wait on or near the lights as soon as they come on in the evening. Others chase back and forth across the ceiling, chattering to each other and occasionally squabbling over a moth or a cockroach.

For this reason the common name of "house gecko" has been applied to a large number of species, including several *Gekko* species. However, the experts at house occupancy are the half-fingered geckos, *Hemidactylus* species, such as Brook's house gecko, *H. brookii*, the common house gecko, *H. frenatus,* and Moreau's house gecko, *H. mabouia*. These species have also been introduced around the world, traveling as stowaways among lumber, building materials, and food produce. In particular, the common house gecko is found throughout practically the whole of the tropical and subtropical world, including Florida, Hawaii, Mexico, and most other Central American countries as well as the whole of the Pacific region. It is probably the most widespread and numerous reptile in the world.

Attack and Defense

Tokays usually wait on a wall near a light with their head pointing down, ready to snap up any suitable prey that lands nearby. If they are really hungry, they will stalk their prey for a short distance before rushing at it from about 12 inches (30 cm) away.

They have a large gape, powerful jaw muscles, and sharp teeth, all of which are adapted for crushing their prey but are equally suitable for defending themselves. Receiving a bite from a large tokay is an experience best avoided because they can easily draw blood and cause painful lacerations. Once their jaws

The distinctive palm, or white-striped, gecko,* Gekko vittatus *from the Solomon Islands, has a single white stripe running along the center of its back.

Tokay Relatives

There are 30 geckos belonging to the genus *Gekko*. Most of them are smaller than the tokay, but a few approach it in size. The palm, or white-striped, gecko, *G. vittatus*, occurs from India through to the Philippines and the Solomon Islands. It is strikingly marked with a bold cream or white line running up the center of its back and then dividing into two lines that reach almost to the angle of each jaw. This species is more slender and less aggressive than the tokay. The green-eyed gecko, *G. smithii*, is found throughout Southeast Asia, including on many islands. It can grow slightly larger than the tokay and is mottled grayish brown in color with large emerald-green eyes. It also has a very loud call and is known locally as the "tok-tok" lizard. Other members of the genus are smaller, are not as widespread, and do not vocalize as loudly. They range from India throughout Southeast Asia, China, Japan, and several Pacific Island groups. A number of new species have been described from Vietnam since the country opened up to scientific exploration in the 1980s.

are fastened around something, they are very difficult to dislodge.

Adult tokays have few enemies, but they are sometimes eaten by snakes, especially climbing species such as the flying snakes, *Chrysopelea*, and a variety of ratsnakes. They put up a good fight and often escape, sometimes losing their tail in the process. One observer saw a fight that lasted for three hours, with the gecko and the snake both gripping the other in its jaws. The fight only ended when the pair was disturbed.

Humans are the tokays' biggest threat, especially in places such as China, where they are dried and sold in markets for food. They are also thought to have medicinal properties or to act as aphrodisiacs. As a result, tokay numbers are declining in some large cities.

Family Groups

Tokays appear to live in small groups with a single adult male and a number of females living within a territory.

The green-eyed gecko,* Gekko smithii, *is similar to the tokay gecko. It comes from Southeast Asia, where its loud call has earned it the name of "tok-tok" lizard.

The male mates with any of these females. Depending on the climate where they live, breeding probably takes place throughout most of the year.

Females lay two (or occasionally three) spherical, hard-shelled eggs. They attach them to a vertical surface, often between two layers of material; it can be under loose bark in the wild, in cracks between boards, or even behind furniture and pictures in a building. Each female may lay several clutches during the course of a breeding season with an interval of about one month between each clutch. Since they often return to the same place to lay their eggs and several females may use the same site, entire eggs and the remains of hatched ones are often found together.

The hatching time for the eggs is highly variable—from 60 to 200 days. Some of the variation is undoubtedly due to temperature differences, and the majority of eggs hatch in about 100 days. This species has temperature-dependent sex determination (TDSD), in which eggs incubated at higher temperatures—around 85°F (30°C) or more—hatch into males.

The hatchlings of both sexes are more boldly marked than the adults. They are dark gray with large white spots arranged in rows across their body and they have black-and-white banded tails. The head is proportionately larger than that of the adults.

Tortoises

Tortoises all belong to the family Testudinidae. They live exclusively on land, although they will sometimes wallow in temporary pools of water, usually as a means of cooling themselves. The giants of the group have some of the most restricted distributions of any chelonians, being confined to the Galápagos and Aldabran islands. The greatest diversity of tortoises exists in the African region, where species occur throughout the continent. The family is also represented farther north in Europe as well as in parts of Asia and the New World.

In the genus *Testudo* the Mediterranean spur-thighed tortoise, *T. graeca*, occurs in Europe and Africa. However, Hermann's tortoise, *T. hermanni*, is restricted to the northeastern Mediterranean and adjacent areas of Europe. It lacks the distinctive tubercle on each thigh that is present in *T. graeca* as well as having a longer and more pointed tail.

The biggest of the European species, however, is the marginated tortoise, *T. marginata*, which is found only in southern Greece. It has a distinctive flared appearance to the marginal scutes at the rear of the carapace. The carapace iteself can measure 12 inches (30 cm) long.

Common name Tortoises **Family** Testudinidae

Genus *Chersina*—1 species, the South African bowsprit tortoise, *C. angulata*
Genus *Geochelone*—15 species from Asia and the Americas, including the star tortoise, G. elegans; Galápagos giant tortoise, *G. nigra*; spurred tortoise, *G. sulcata*; plowshare tortoise, *G. yniphora*; leopard tortoise, *G. pardalis*
Genus *Gopherus*—4 species from North and Central America, the California gopher tortoise, *G. agassizii*; Texas gopher, *G. berlandieri*; Mexican gopher, *G. flavomarginatus*; Florida gopher, *G. polyphemus*
Genus *Homopus*—4 species from Africa, including the parrot-beaked tortoise, *H. areolatus*, and the speckled Cape tortoise, *H. signatus*
Genus *Kinixys*—3 species of hingeback tortoises from Africa, including Bell's hingeback, *K. belliana*
Genus *Malacochersus*—1 species from Africa, the pancake tortoise, *M. tornieri*
Genus *Psammobates*—3 species from Africa, including the serrated tortoise, *P. oculifera*
Genus *Pyxis*—2 Madagascan species, the Madagascar spider tortoise, *P. arachnoides*, and the Madagascar flat-tailed tortoise, *P. planicauda*
Genus *Testudo*—5 species from Europe, Africa, and Asia, including the Mediterranean spur-thighed tortoise, *T. graeca*; Hermann's tortoise, *T. hermanni;* marginated tortoise, *T. marginata*

African Species

In Africa populations of the Mediterranean spur-thighed tortoise are to be found living under semiarid conditions. They estivate during the hottest months of the year, when food and water are scarce. Farther south on the fringes of the Sahara Desert is the largest of the mainland tortoises, the spurred tortoise, *Geochelone sulcata*, which can reach a maximum carapace length of about 30 inches (76 cm). When weather conditions are bad, it retreats into underground burrows that it digs using its strong legs.

The distribution pattern of the hingeback tortoises that form the genus *Kinixys* reflects the diversity of habitat in which African tortoises can be found—from tropical rain forest to grassland, scrub, and semidesert. Members of the genus include the widely ranging Bell's hingeback, *K. belliana*, which differs very markedly in coloration in different locations—some individuals have plain brown carapaces, while others have very evident patterning on their shells.

The difference in markings is thought to be linked to the type of environment in which the tortoises occur. For example, the eroded hingeback, *K. erosa*, is a species that occurs in tropical forest and can even swim quite well if necessary. It is mahogany-brown in color, which helps conceal its presence on the forest floor. (The patterning of the leopard tortoise, *Geochelone pardalis*, also found in Africa, serves a similar purpose.) The third member of the genus, Home's hingeback, *K. homeana*, has a much more vertical rear to its shell and tends to be a lighter brown color. In common with the other species it is able to seal

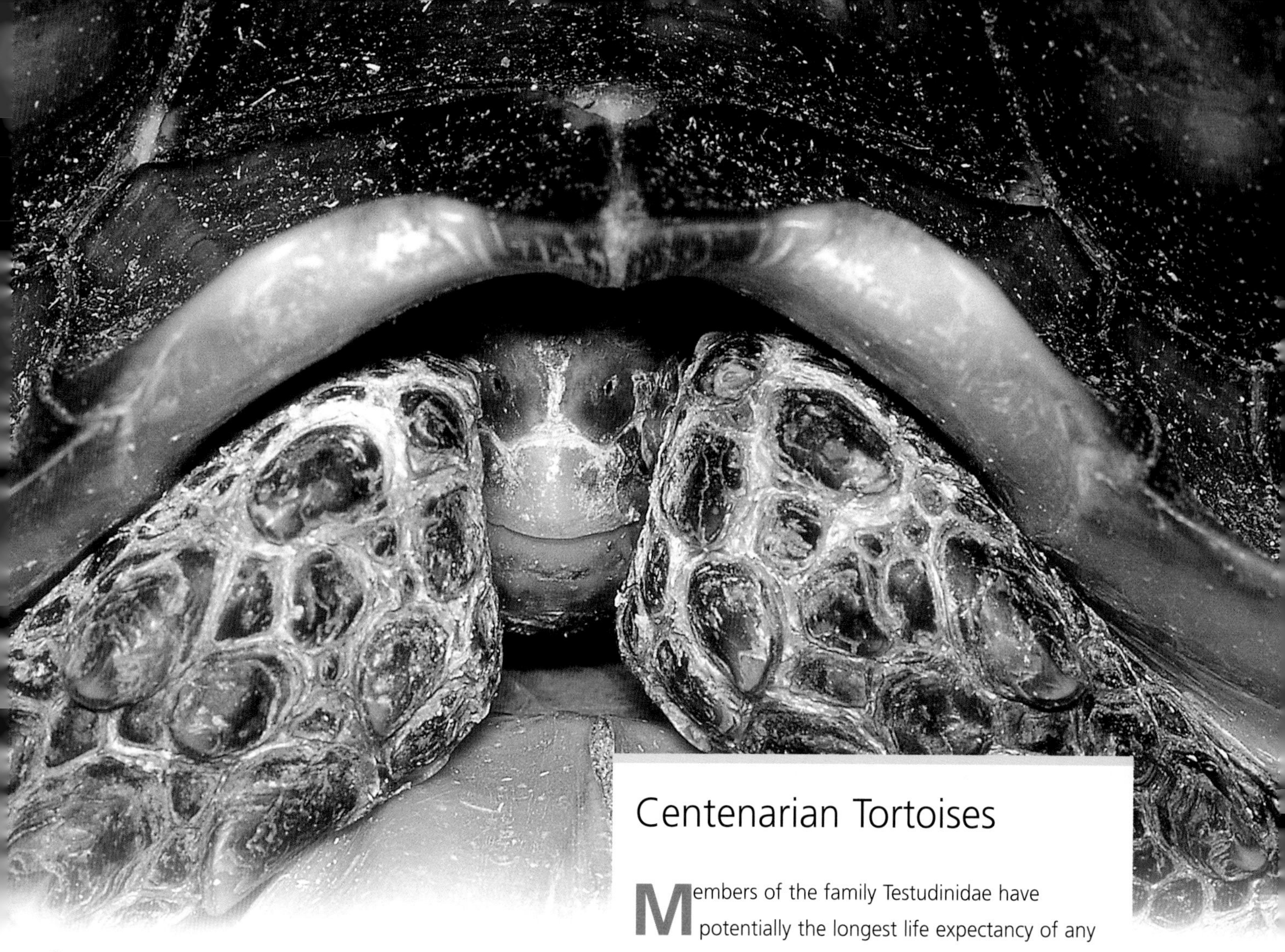

↑ ***Home's hingeback tortoise,* Kinixys homeana *from Africa, hides in its shell. African hingeback tortoises are characterized by a single hinge located about three-quarters of the way along the back of the shell.***

itself in its shell thanks to its hinged plastron. Although it is a forest-dwelling species, it tends to occur in slightly drier areas than the eroded hingeback.

Africa is also home to the world's smallest tortoise, *Homopus signatus*, the speckled Cape tortoise, whose range extends from southern Namibia down to the western Cape. It may only reach a carapace length of 2.4 inches (6 cm) when adult. These tiny tortoises live among rocks, where they can hide away in relative safety. They generally emerge from their retreats only in search of food in the early morning.

Another even more unusual African tortoise found in rocky terrain is *Malacochersus tornieri*, the pancake tortoise, so-called because of its flattened appearance. Occurring in East Africa, this species has sacrificed the rigidity of its shell for the ability to move very quickly and

Centenarian Tortoises

Members of the family Testudinidae have potentially the longest life expectancy of any vertebrates. As a result of the popularity of certain species as pets, it has been possible to discover their potential life span. In the case of the Mediterranean spur-thighed tortoise, *Testudo graeca*, the oldest individual for which reliable information is available lived at Powderham Castle in Devon, England, from 1914 until its death in 2004. It had previously been kept by a family nearby; when it died, this particular tortoise was about 160 years old.

Giant tortoises in the genus *Geochelone* may live even longer, however, as revealed by four individuals that were moved in 1776 from the Seychelles to the nearby island of Mauritius. The last member of the group died after becoming accidentally trapped in a gun emplacement in 1918. Contemporary accounts of its size when it arrived on the island 142 years earlier suggest it was already an adult and so it must have been close to 200 years old when it died.

↑ ***Madagascan members of the family are under threat from the pressure of hunting and loss of habitat.*** **Geochelone radiata,** ***the radiated tortoise, is listed as Vulnerable (IUCN).***

to retreat from danger under rocks. It uses its body and claws to anchor itself in place, making it very difficult to dislodge from its chosen hiding place.

The islands off eastern Africa are home to various tortoises. However, on Madagascar all four species face a very uncertain future largely because of hunting pressures and habitat disturbance. The status of *Geochelone yniphora*, the plowshare tortoise, notable for the very distinctive prong at the front of its plastron that is used by the male in mating, is especially critical on the island. Wild pigs have proved an additional threat to this species by digging up and destroying its nests. The radiated tortoise, *G. radiata*, with its attractive pattern of light streaks radiating out from a dark background remains more numerous but is considered Vulnerable (IUCN). Its carapace is markedly domed, which emphasizes its patterning. However, young hatchlings have relatively flat shells, albeit still brightly marked.

Asiatic Tortoises

Patterning similar to that of the radiated tortoise is seen in the star tortoise, *Geochelone elegans*, which originates from parts of India, as well as in its Burmese relative, *G. platynota*. Both species are likely to have descended from the same ancestral line as the radiated tortoise. The division probably occurred when the ancient southern continent of Gondawaland split up—India drifted northward and joined with Asia, while the adjacent region of Madagascar remained in the Indian Ocean.

There are seven species of tortoise currently found in Asia, and in general the others are much more subdued in coloration. An unusual feature of the elongated tortoise, *Geochelone elongata*, however (apart from its relatively long, narrow shell), is the skin in the area around the nostrils, which turns pinkish as the time for nesting approaches. This is caused by increased blood flow to that part of the body, suggesting that the tortoises rely heavily on their ability to detect scents in order to locate a partner. The largest of the Asiatic species is the Burmese brown, *G. emys*, achieving a carapace length of at least 18 inches (46 cm). It too has interesting breeding habits—it is thought that the female remains near the nest to guard the eggs for at least a few days after laying.

Tortoises of the New World

The four species of gopher tortoise in the genus *Gopherus* are restricted to North America, where they inhabit fairly dry terrain. Farther south the Amazon Basin is home to two tropical forest species in the genus *Geochelone* that are similar in appearance to the gophers, leading to the belief that they diverged from a common ancestor about 7 million years ago. The yellow-footed tortoise, *Geochelone denticulata*, is the larger of the two, with a maximum carapace length of 29 inches (74 cm), whereas the more colorful red-footed tortoises, *G. carbonaria*, rarely grow larger than 18 inches (46 cm). Both feed to a significant degree on fruit as well as greenstuff. Interestingly, in spite of their close relationship and the fact that their distribution overlaps in some parts of their range, they do not interbreed. They avoid this by recognizing specific courtship movements made by the males, which differ in each species. Clues relating to smell may also be used.

The most southerly representative of the tortoises in the New World is the Chaco tortoise, *Geochelone chilensis*, which lives in grassland areas in Chile and Argentina. It occurs quite close to the tip of South America, but it avoids extreme temperatures by retreating to underground burrows during cold or dry periods.

→ ***As its name suggests, the yellow-footed tortoise,*** **Geochelone denticulata,** ***has patches of yellow on its limbs as well as on its shell and tail. This is a young individual from South America.***

The Harshest Environment

Horsfield's tortoise, *Testudo horsfieldii*, has the most northerly distribution of any tortoise—the plains of Kazakhstan, part of the former USSR. The tortoises emerge from hibernation in spring and are active for only about three months before retreating to their underground burrows by July. This is a reflection of the harsh climate in which they occur, where the winters are very cold and the summers are hot and dry. Suitable vegetation is only available during this short spring period. They must also use this time to mate and, in the case of females, lay their eggs before returning to their burrows.

Land Tortoises in Australia

There are no land tortoises occurring in Australia today, but in the past there used to be a population of bizarre horned tortoises, known by their family name of meiolaniids. Their appearance was quite unlike that of chelonians that we would recognize today. These unusual tortoises were characterized by a pair of hornlike projections on their head. The "horns" could be spaced up to 24 inches (60 cm) apart. It is thought that, together with hard plates that were present on the upper surface of the tail, they helped provide these large tortoises with "armor" to protect them against would-be predators.

The ancestors of this group arose during the Cretaceous Period around 100 million years ago in the region that is now South America. Later they reached Australia, occurring on the southeast of the continent and also on nearby islands, notably Lord Howe Island off the coast of New South Wales in eastern Australia, where they survived until as recently as 120,000 years ago. They appear to have died out in South America much earlier, however, around 70 million years ago.

The way in which the meiolaniids arrived in Australia from South America has been a source of puzzlement, although members of the family Chelidae (snake-necked turtles) are still found on both continents today. It is believed that the ancestors of both families were actually present in Antarctica (which served as a land bridge to connect these continents) at a time when it lay much closer to the equator and was therefore significantly warmer. The discovery of the fossilized remains of chelonians in Antarctica dating back to the Miocene Epoch about 20 million years ago appears to confirm this theory.

Common name Tuataras

Scientific names *Sphenodon punctatus, S. guntheri*

Family Sphenodontidae

Order Rhynchocephalia

Size From 20 in (50 cm) to 31 in (80 cm); males larger than females

Key features Body squat; color of adults olive-green, gray, or black with a speckling of gray, yellow, or white; newly hatched animals are brown or gray with pink tinges; head has a pink shield and striped throat; sometimes distinctive light patches occur on the body and tail

Habits Nocturnal burrowers

Breeding Egg layers; average clutch size from 6 to 10; incubation period about 11–15 months

Diet Invertebrates such as beetles, crickets, and spiders; also the eggs and chicks of seabirds; frogs, lizards, and young tuataras also eaten occasionally

Habitat Low forest and scrub usually associated with colonies of burrowing seabirds

Distribution About 33 small islands and rock stacks off the coast of New Zealand

Status *S. guntheri* is listed as Vulnerable (IUCN); *S. punctatus* has a very limited range

Similar species None

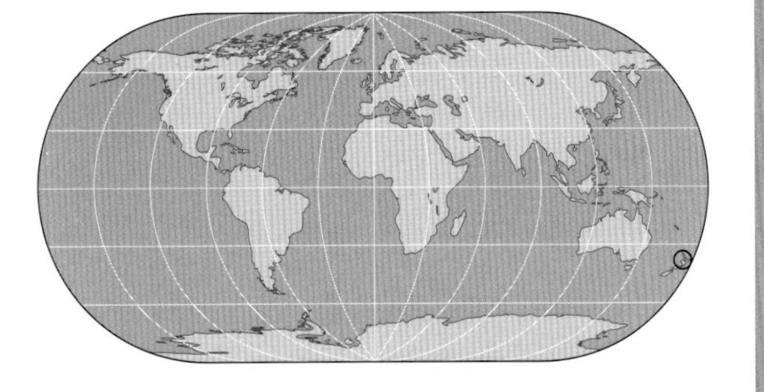

Tuataras

Sphenodon punctatus and *S. guntheri*

The tuataras are the sole survivors of a group of reptiles that have existed for about 220 million years. They are found only on some small islands around the coast of New Zealand.

THE TWO SPECIES OF TUATARAS, *Sphenodon guntheri* and *S. punctatus,* are found on offshore islets around the New Zealand coast. They are the only members of the order Rhynchocephalia (recently renamed the Sphenodontia in some publications). They belong to a branch of reptiles called the Lepidosauria (scaly reptiles), which also contains the squamates (lizards, snakes, and worm lizards).

The other reptilian branches are the Archosauria, containing the crocodilians and a number of extinct lines such as the dinosaurs, and the Chelonia (turtles and tortoises). Strangely, the Archosauria also includes the birds, making them more closely related to reptiles than to mammals. Similarities between the skulls of tuataras and crocodilians led to early classification systems placing them close together in evolutionary terms, but the similarities are now believed to be coincidental.

Early Reptiles

The rhynchocephalians were among the first reptiles to appear, at around the same time as the crocodiles (and the dinosaurs), even though they have no direct connection to them. The tuataras, therefore, have an ancestry that goes back to the Triassic Period about 220 million years ago at the beginning of the "Age of the Dinosaurs." Tuataras are sometimes referred to as "living fossils," but it should be stressed that it was their ancestors that lived at the same time as the dinosaurs—the tuataras as we know them today arrived later.

Fossil rhynchocephalians have been found in many parts of the world, including Africa, Madagascar, Europe, and South America. At the time they evolved, the southern landmasses

In the wild tuataras are mostly nocturnal. They make their homes in burrows and emerge at night to prey on insects of various types and occasionally the eggs of seabirds.

were all joined into the supercontinent of Gondwanaland, and they were able to spread out and diversify across a wide region. Twenty-four genera have been identified from fossils, and no doubt there are plenty more waiting to be found. Others were never lucky enough to become fossilized. As the landmass broke up into the continents, rhynchocephalians became extinct throughout the world until by 60 million years ago they had completely disappeared, except on the relatively small landmass that became New Zealand. Their survival there was due to New Zealand's early separation from the rest of the landmasses and consequently its lack of large predators. So they lived

an untroubled life in splendid isolation until the arrival of humans about 1,000 years ago. That event heralded their disappearance from the mainland and several of the larger offshore islands, mainly due to introduced animals (although tuatara remains have been found in kitchen middens of the earliest inhabitants, suggesting that they were eaten). Today they live only on about 30 small islands, mostly less than 25 acres (10 ha) in area. Five of them are in the Cook Straits between New Zealand's North and South Islands, and the rest are in the Gulf of Hauraki and the Bay of Plenty off North Island's northeast coast.

The tuatara was first described in 1831 and placed in the Agamidae. Historically, tuataras were thought to be lizards and were given a variety of names until 1867, when a scientist named Dr. Albert Gunther, working at the British Museum, recognized their link with the rhynchocephalians (which were thought to have been extinct for a long time). The tuatara was named *Sphenodon punctatus*. (*Sphenodon* means "wedge toothed.") Then in 1989 Dr. Charles Daugherty, working at Victoria Museum in Wellington, concluded that there were two species involved. The new one, from North Brothers Island, was named *S. guntheri* after Dr. Gunther. The Maoris, who were the original colonizers of New Zealand, gave them the common name tuatara, which means "peaks on the back," a reference to the jagged crest on the back and tail of both species.

What Makes Them Different?

Apart from their skulls, tuataras have several characteristics that set them apart from other reptiles. Unlike any lizards, they have a single row of teeth in the lower jaw that fits between two rows of teeth in the upper jaw. The arrangement enables the tuataras to grind and shear tough foodstuffs, such as the outer skeletons of some of their insect prey. The teeth are fused to the upper edge of the jawbone (and are known as acrodont). As a result, tuataras have only a limited ability to replace worn-out teeth, so those of older animals are often worn completely level with the jaw. Most lizards, on the other hand, have pleurodont teeth, a condition in which they are loosely attached to the inside surface of the jawbones, although members of the Agamidae also have acrodont teeth. A second difference is that male tuataras do not have an external copulatory organ. Instead, sperm is transferred from the male's cloaca directly to that of the female. Tuataras lack external openings to their ears, while the majority of lizards have them.

Tuataras are brown, olive, or reddish brown in color with no distinct markings other than a light speckling of white or yellow. They have a crest consisting of a raised ridge topped with a row of toothlike scales, which is higher and larger in males than in females. They grow to 20 to 30 inches in length (50–76 cm) with a maximum weight of about 2.2 pounds (1kg). Males are larger than females, which are significantly lighter and have narrower heads.

Tuataras and Birds

Tuataras live in coastal forests at densities between 3,000 and 5,000 in every acre (1,300–2000 per ha). They live in burrows that they can dig themselves, but on some islands they also make use of burrows belonging to a variety of species of seabirds that use the islands as nesting sites. The main species are the common diving petrel, *Pelecanoides*

A Third Eye

When tuataras hatch, they have a translucent scale on top of their head. It covers a well-developed structure resembling an "eye" that is situated between two large bones on top of the skull (the parietal bones). A similar structure is found in some lizards, but it is most obvious in young tuataras.

The function of the "eye" is poorly known; but it has a lens and a retina, and it is connected to the brain. It is thought to be sensitive to light but not capable of producing an image, and many scientists think that it helps the animals regulate their exposure to the sun and control their daily activity rhythms.

urinatrix, known locally as mutton birds, and fairy prions, *Pachyptila turtur*.

Tuataras are nocturnal. They search for their prey by sight, mostly in the area immediately around their burrows. The presence of the seabirds and their guano encourages a high density of scavenging invertebrates, especially the giant crickets known locally as wetas, which the tuataras eat. They also eat the eggs and the young of the nesting birds occasionally, along with lizards and smaller members of their own species.

Life in the Slow Lane

Because they live in a cool climate, tuataras have evolved a slow metabolic rate. They maintain a lower body temperature than most other reptiles and are active when their body temperature reaches 45°F (7°C) or more. However, they sometimes bask in the sun at the entrance to their burrows, when their body temperature can reach 80°F (27°C). Their preferred body temperature seems to be about 60°F (15°C).

The effect of the low metabolic rate is to slow down their development and growth and to prolong their lives: Tuataras only become sexually mature after about 10 years old, and they continue growing until they are 20 to 35 years old. They may breed until they are at least 60 years old, possibly for much longer, although more research will be needed to confirm this. So far, captives have lived for 77 years, but a life span of 100 years is not unlikely.

Every other aspect of their lives moves at a slower pace too. For example, females only

⊕ ***A female tuatara watches the approach of a male* Sphenodon punctatus *in a tree on Stephen's Island, New Zealand.***

produce eggs every two to five years. The eggs take up to three years to develop in the female's ovaries, and it takes seven months more for the shell to form around them. Their incubation period is the longest of any reptile.

Breeding

The mating season coincides with the southern summer (January to March). At this time the males become more territorial and display to each other and to females by raising their crests. After mating, the females ovulate but do not lay their eggs until the following spring when the shell has been formed. Then they migrate to open areas where they can dig tunnels in which to lay their eggs. The tunnels are about 8 inches (20 cm) in length, and the females defend them for a few days after laying to keep other females from digging in the same place and disturbing their eggs.

Tuataras lay 10 to 15 soft-shelled, elongated eggs. Incubation takes from 11 to 15 months, but that includes a period during the winter when the embryo's development is arrested. Eggs laid in captivity and incubated at a constant temperature develop more quickly. As in turtles, crocodilians, and some lizards, for example, the leopard gecko, *Eublepharus macularius,* the hatchlings' sex depends on the temperature at which the eggs develop: Higher temperatures produce males, while lower temperatures produce females.

Just before they hatch, the eggs absorb water. They swell, and their shells become taut. The embryo develops a small "egg tooth" on its snout, which it uses to slit the shell. The egg tooth falls off about two weeks after hatching.

Defense

Adult tuataras have little need to defend themselves because they have no predators. But juveniles are vulnerable to predation by larger tuataras as well as by owls, hawks, and kingfishers, so they live in small burrows that they dig for themselves or under rocks or logs.

Tuataras can shed part of their tail, a process known as autotomy. The tail breaks across a fracture plane in one of the vertebrae rather than at the junction of two, and the bone, nerves, and blood vessels all break cleanly. The discarded part of the tail continues to twitch to attract the attention of the predator while the tuatara escapes. The stump heals quickly: Four pairs of muscle bundles emerge from the stump immediately after the break, and they gradually bend inward, sealing off the nerve ends and blood vessels to prevent the loss of blood and further damage. A conical scar forms, skin grows over it, and the stump slowly elongates over a period of several weeks. The vertebrae of the original tail are replaced by rods of cartilage, and new blood vessels and nerves extend into the new section. Scientists assume that most of the widespread tail damage is due to fighting—attacks by predators could probably not account for the high proportion of tuataras with regrown tails.

Conservation Program

Tuataras have few natural enemies on their isolated islands. The main threats are from introduced predators such as dogs, cats, and rats, and from habitat destruction. There are no islands where rats and tuataras coexist—islands on which there were once good populations of tuataras that have since disappeared all have introduced rats on them. On some islands all the tuatara populations consist only of adults, which is worrying because it suggests that none of the young are surviving. On the other hand, tuataras coexist well with seabirds and are most successful where there is a healthy population of petrels, prions, and shearwaters.

Tuataras have been totally protected by law since 1953. All the islands on which they currently live are small and have steep cliffs that make landing by boat difficult. In addition, landing is restricted to scientists and conservationists who have special permission.

Recently a program to breed tuataras in captivity has been developed and seems to be producing good results. Eggs are collected from healthy populations and hatched under controlled conditions. The hatchlings are then

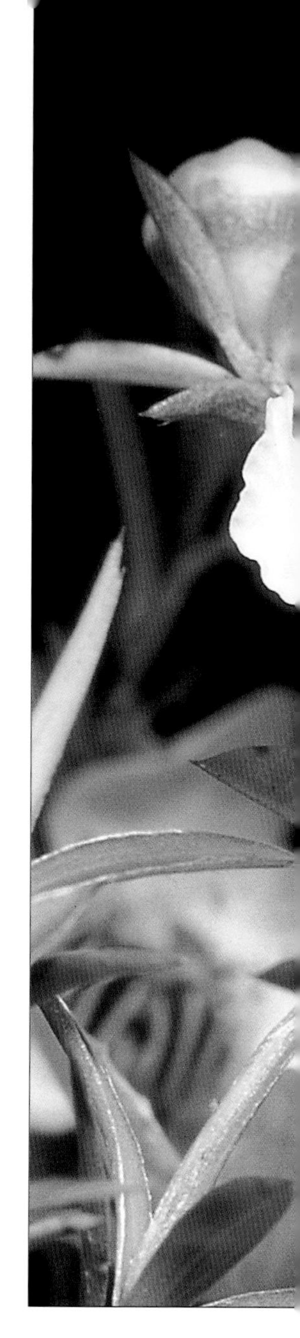

The common green gecko, Naultinus elegans, is endemic to New Zealand. As its name suggests, it is the most familiar green gecko on the North Island. It can be green, yellow, or patterned with both colors.

Reptile Neighbors

Very few reptiles managed to reach New Zealand before it became separated from the main southern landmass. Before humans arrived, the tuatara would have been the dominant reptile species throughout the country. Nowadays the two main islands (North Island and South Island) have representatives of two lizard families but no snakes, turtles, or crocodilians at all.

The 40 or so species of lizards belong to the Gekkonidae and the Scincidae in roughly equal numbers. All 18 of the geckos are in the genera *Hoplodactylus* and *Naultinus*. They are endemic to New Zealand, and unusually, they are all live-bearers (there are only a handful of other live-bearing geckos in the world). The skinks are slightly more numerous, with 22 species all in the genus *Oligosoma.* They are also live-bearers, with the sole exception of *O. suteri* from Great Barrier Island, North Island, which lays eggs.

released on islands where former populations were wiped out. Since 1995 three islands have been cleared of invasive rats, and juveniles have been reintroduced. The plan will be judged a success when the introduced animals produce a second generation of tuataras.

Altogether it is believed that there are over 60,000 common tuataras distributed among 29 islands, so their future is relatively secure. The Brothers Island tuatara, *Sphenodon guntheri*, however, only occurs on the North Brothers Island in Marlborough Sound. The total population is about 400 animals. Although they are quite safe at present, they are vulnerable to accidental introduction of rats and to habitat destruction, especially by fire.

TURTLES

The order Testudines encompasses all the reptiles variously known as turtles, terrapins, and tortoises. The order was formerly known as Chelonia; as a result, this group of reptiles is often referred to collectively as "chelonians." There is no strict rule about the use of the popular names, although the word "turtle" is frequently applied entirely for marine species. In North America there is a tendency to use this description for all aquatic forms, including those found in fresh water, although the term "terrapin" may be applied in some cases. Land forms are often described as tortoises and are identifiable by their relatively domed appearance.

The most obvious feature common to turtles as a group is the presence of a body casing in the form of a shell. It enables the reptile to withdraw its head, limbs, and tail to a variable extent, giving protection against predators. The shell itself is made up of two parts, and the individual segments of the shell are known as scutes.

The scutes can be colored to provide camouflage (known as crytic coloration). They are also distinctive enough so that, like fingerprints, they enable individuals to be identified. The scutes correspond to the epidermis, or outer layer of skin, in soft-skinned animals, and they lie on top of the bony layer. Just as in a brick wall, the links in the bony layer do not correspond exactly to the pattern of scutes directly above, which adds to the strength of the shell structure. The upper part of the shell is described as the carapace, and the underside is known as the plastron. The two areas are connected by so-called bridges that extend down each side of the body and are also made of the same material as the scutes.

The First Chelonians

The emergence of turtles as a group dates back to the late Carboniferous Period about 280 million years ago. The first examples are believed to have been rather like armadillos, with a semiflexible body casing that gave them some protection as well as allowing them reasonable freedom of movement. The first clearly recognizable forerunners of today's chelonians have been unearthed in rocks in Germany dating back to the Triassic Period approximately 220 million years ago. They were similar to the turtles of today, but unlike today's turtles, they had teeth both in their jaws and extending across the roof of the mouth. The distinctive pattern of scutes had already developed across the carapace. There were typically four, but occasionally five, vertebral scutes running down the center of the body, protecting the spinal column beneath. There were five corresponding scutes over the ribs (costal scutes) on each side of the vertebral scutes, and there were many smaller scutes around the edge of the shell, corresponding to the marginal scutes seen in chelonians today. These early

Turtles are at their most graceful in open water. This young loggerhead turtle, Caretta caretta, *uses its powerful front limbs to "fly" through the ocean off the Azores.*

Who's Who among the Turtles?

Order Testudines

Suborder Cryptodira (hidden-necked turtles)
- **Family** Carettochelyidae: 1 genus, 1 species, the pig-nosed turtle, *Carettochelys insculpta*
- **Family** Cheloniidae: 4 genera, 6 species of marine turtles
- **Family** Chelydridae: 2 genera, 2 species of snapping turtles
- **Family** Dermatemydidae: 1 genus, 1 species, the Central American river turtle, *Dermatemys mawii*
- **Family** Dermochelyidae: 1 genus, 1 species, the leatherback turtle, *Dermochelys coriacea*
- **Family** Emydidae: 34 genera, 86 species of basking turtles
- **Family** Kinosternidae: 4 genera, 21 species of American mud and musk turtles
- **Family** Platysternidae: 1 genus, 1 species, the big-headed turtle, *Platysternon megacephalum*
- **Family** Testudinidae: 9 genera, 38 species of tortoises
- **Family** Trionychidae: 5 genera, 23 species of softshell turtles

Suborder Pleurodira (side-necked turtles)
- **Family** Chelidae: 9 genera, 38 species of snake-necked turtles
- **Family** Pelomedusidae: 5 genera, 20 species of Afro-American side-necked turtles

Total: 12 families, 76 genera, 238 species

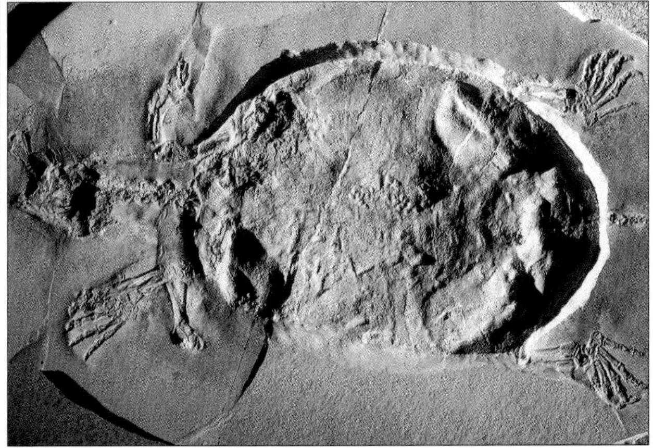

Fossil turtles dating back to the Triassic Period have been found in limestone deposits in Germany. They share many characteristics with today's turtles, including the development of scutes on the carapace.

forms were aquatic. Only much later in their evolutionary history did some turtles start to colonize land. They were apparently similar in size to freshwater turtles seen today, with a maximum shell length of about 24 inches (60 cm).

Turtles appear to have spread quite widely during the Jurassic Period (which began 200 million years ago) and continued their anatomical development. The teeth disappeared, and the shell began to fuse more completely. However, the early turtles were unable to take full advantage of its protection, because they could not withdraw their head or their limbs under the bony casing. It is believed that turtles increased significantly in size during the Cretaceous Period, which ended about 65 million years ago with the demise of the dinosaurs. The increase in size may have been triggered partly in response to the massive predators that had developed in the seas of that era. The largest member of the order ever recorded, known as *Archelon*, had a shell that measured at least 12 feet (3.7 m) long. This is a third longer than that of the leatherback, *Dermochelys coriacea*, which is the biggest turtle in the world's oceans today.

Retracting the Neck

As turtles evolved, two separate mechanisms for retracting the neck into the front opening between the carapace and the plastron developed. Today turtles are described as belonging to one or the other of two groups: the side-necked turtles (suborder Pleurodira), which retract their necks horizontally, or the hidden-necked turtles (suborder Cryptodira), which retract their necks vertically.

All turtles have eight cervical (neck) vertebrae, but those of members of the suborder Pleurodira are more developed. The dorsal spines running down the top of the neck bones are taller, and the transverse processes on the sides of the vertebrae are also enlarged, allowing for the attachment of muscles that retract the head to the side, leaving the neck and head a little exposed at the front of the shell. This group includes the snake-necked turtles, some of which have necks that are longer than the shell.

The vertebrae of members of the other group (the suborder Cryptodira) are reduced, making the neck more flexible and allowing the turtle to withdraw it into the shell by folding it back on itself in a tight vertical "s" shape. Most of them can withdraw their head completely inside the shell. All but two of today's families of chelonians belong to this latter group, including all those in North America, Europe, and mainland Asia.

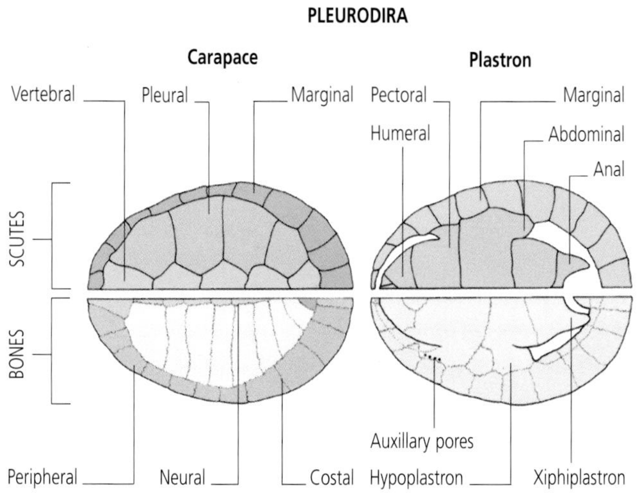

Ⓣ Carapace structure. Side-necked turtles (suborder Pleurodira) have 13 scutes and 9–11 bones in the plastron, and the pelvis is fused to the shell. Hidden-necked turtles (suborder Cryptodira) have 11–12 plastral scutes and 8–9 plastral bones.

Specialized Body Parts

The skeletal structure of turtles has become highly specialized. The cervical vertebrae are flexible. The next section of the vertebral column, extending right down to the tail, is fused within the carapace. But the tail itself is very mobile. Since the ribs are fused within the shell, they cannot be used for respiratory purposes. Instead, turtles rely on respiratory muscles located around the leg pockets. They force air out of the lungs, which are themselves quite rigid structures. Breathing in occurs as a result of the lower pressure in the turtle's body. In the case of aquatic chelonians the surrounding water creates extra pressure on the turtle's body. As a result, breathing in requires more muscular activity than breathing out.

Aquatic chelonians breathe when they surface; but some, such as the softshells in the family Trionychidae, are also able to extract oxygen from water in their cloaca, so they do not need to surface as regularly. These particular turtles, along with the matamata, *Chelus fimbriatus*, also have elongated nostrils that act like snorkels, allowing them to breathe without having to break the water surface. Some species in the Trionychidae, for example, *Trionyx*, can also extract oxygen directly from water in the pharyngeal region at the back of the throat. This area is highly vascularized (supplied with blood vessels) for this purpose, and water is exhaled through the nostrils.

Turtles' limbs are protected by a covering of scales, and their feet end in claws. In the case of tortoises the claws tend to be relatively broad and blunt, since they wear them down as they walk. In aquatic chelonians they are often much narrower and sharper. This is partly because some of them use their claws to tear food apart while holding it in their jaws before swallowing it. In the case of the sliders, *Pseudemys* species (family Emydidae), the claws are a means of distinguishing the sexes, since the front claws are much longer in males. They use these claws as part of their courtship ritual to fan water gently in the face of a female and touch her chin.

Many turtles, for example, the loggerhead musk turtle, *Sternotherus minor*, have small fleshy swellings known as barbels in the chin area. They are well supplied with nerves and play an important part in scent detection. They also have a tactile function, providing information to the turtle as it moves over the substrate. This is especially valuable in turbid waters, where visibility is poor.

The Senses

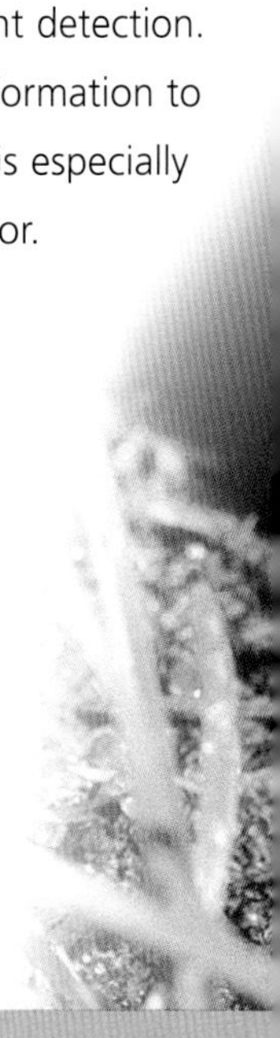

All chelonians rely on their sense of smell to figure out what is edible and to find a mate. This function is linked with Jacobson's organ, a structure in the roof of the mouth connected to the brain. Turtles can detect scents under water. They can also see well and have color vision, which

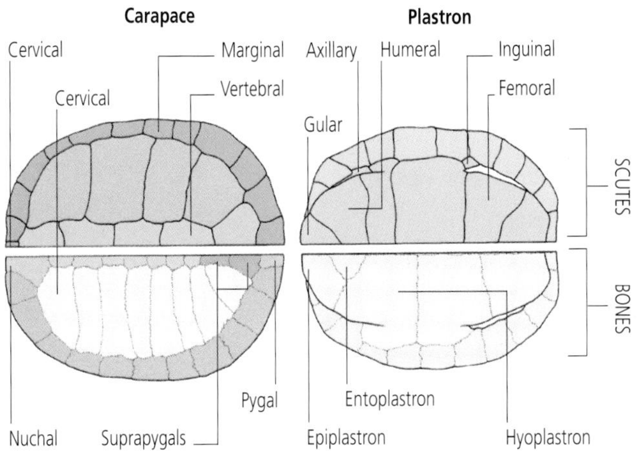

may help tortoises in particular locate fruits and plants some distance away. Hearing is not a significant sense in turtles generally. They have no external ear openings, but the tympanic membrane is evident on each side of the head behind the eyes. This may explain why they rarely vocalize except when mating.

All turtles reproduce by means of eggs, but clutch size varies dramatically according to species, ranging from a single egg to relatively large clutches numbering over 100 in some cases.

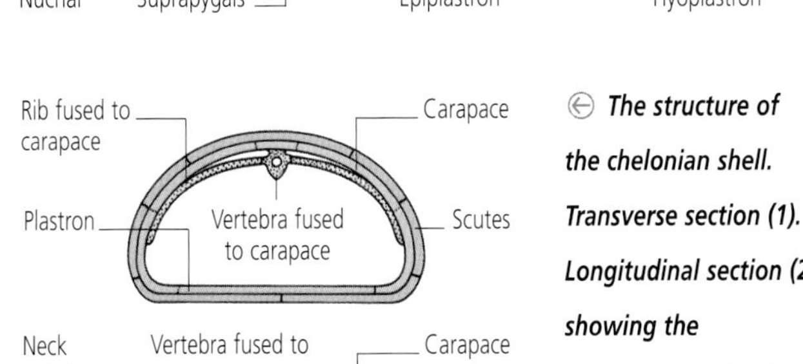

The structure of the chelonian shell. Transverse section (1). Longitudinal section (2), showing the arrangement of pelvic and pectoral girdles.

The eastern box turtle,** Terrapene carolina **from the southeastern United States, is one of the chelonians that have colonized land. It can be found in forests or meadows wherever there is enough moisture.

VENOMOUS SNAKES

Although some members of the Colubridae (the "typical" snakes) and the Atractaspididae (the African burrowing snakes) produce venom, in no snakes is venom production and delivery more advanced than in the Elapidae and the Viperidae. All the members of these two major families are, theoretically at least, venomous. They are the cobras, mambas, coral snakes, sea snakes, and their relatives (Elapidae), and the vipers and pit vipers (Viperidae). Both are large families, containing about 300 and 250 species respectively, and are widely distributed. There are no elapids in Europe, however, and few in North America. Nor are there any vipers or pit vipers in Australasia.

Members of the two families differ from each other most significantly in their venom fangs. In elapids the fangs are usually fixed, while in the vipers they are hinged. A few elapids, such as mambas, have a certain amount of mobility in their teeth, but not to the same degree

Snakebites—Perception Versus Reality

Snakebites are responsible for thousands of human deaths each year. Exact figures are difficult to arrive at because medical records in some places are either nonexistent or hard to get. Estimates give the number of victims as between 25,000 and 100,000 every year, mostly in the Far East and Africa. The carpet vipers, especially the wide-ranging saw-scaled viper, *Echis carinatus*, are usually thought to account for most of the deaths. They live in Africa and the Middle East, and are common, well camouflaged, and prone to lying along paths where they are frequently stepped on. The Indian cobra, *Naja naja*, and Russell's viper, *Daboia russeli*, are responsible for large numbers of bites in Asia. In Central and South America the terciopelo, *Bothrops asper*, and the common lancehead, *Bothrops atrox*, are also common and well camouflaged, and cause many deaths. In North America there are only a handful of fatalities each year, mostly caused by bites from eastern and western diamondback rattlesnakes, *Crotalus adamanteus* and *Crotalus atrox*. Other dangerous species include the puff adder, *Bitis arietans* in Africa, and the habu, *Trimeresurus flavoviridis* from Japan.

It is significant that the list of top killers does not contain many notorious species such as rattlesnakes, bushmasters (*Lachesis* species), the king cobra (*Ophiophagus hannah*), mambas (*Dendroaspis* species), and taipans (*Oxyuranus* species). They are spectacular and widely feared but cause very few deaths. That is partly because they are rare, at least in the places where humans are concentrated, and partly because they are shy and inoffensive. To be really dangerous, a snake needs to be common in places where humans live and work, and it has to be ready to bite. In the developed world snakebites are extremely rare. Australia, which has more venomous than nonvenomous snakes, averages about 10 fatalities a year. North America and Europe each report about 15 to 20 deaths annually. These relatively low figures would be even smaller were it not for casual snake handling by snake keepers and foolhardy members of the public.

The deadly fangs of a western diamondback rattlesnake,* Crotalus atrox *from North America. This common rattlesnake probably bites more people than any other species in North America, although the overall number of fatalities is small.

Who's Who among the Venomous Snakes?

Family Elapidae: 60 genera, about 305 species of cobras, mambas, coral snakes, and sea snakes, including:
Malaysian blue coral snake, *Maticora bivirgata*; black mamba, *Dendroaspis polylepis*; rinkhals, *Hemachatus haemachatus*; Indian cobra, *Naja naja*; Cape cobra, *Naja nivea*; king cobra, *Ophiophagus hannah*; yellow-lipped sea krait, *Laticauda colubrina*; coastal taipan, *Oxyuranus scutellatus*

Family Viperidae: 35 genera, about 250 species of vipers and pit vipers, including:
hairy bush viper, *Atheris hispida*; copperhead, *Agkistrodon contortrix*; Gaboon viper, *Bitis gabonica*; horned adder, *Bitis caudalis*; puff adder, *Bitis arietans*; terciopelo, *Bothrops asper*; Sahara horned viper, *Cerastes cerastes*; eyelash viper, *Bothriechis schlegelii*; western diamondback rattlesnake, *Crotalus atrox*; Russell's viper, *Daboia russeli*; Malayan pit viper, *Calloselasma rhodostoma*; sidewinder, *Crotalus cerastes*; massasauga, *Sistrurus catenatus*; adder, *Vipera berus*

Wagler's pit viper,* Tropidolaemus wagleri, *is from Southeast Asia and is perhaps the most well known of the green tree-dwelling pit vipers in the region.

as the vipers. In vipers and most elapids the venom-delivering fangs are hollow with a small opening at the tip through which the venom emerges. Because vipers' fangs can be folded down when not in use, they can be longer than those of elapids, allowing them to penetrate their victim more deeply. Elapids, on the other hand, often deliver a faster-acting venom, and the six or so snakes with the most potent venom are all elapids. It is worth pointing out that stiletto snakes (*Atractaspis* species in the family Atractaspididae) also have folding fangs, but they evolved independently from those of vipers.

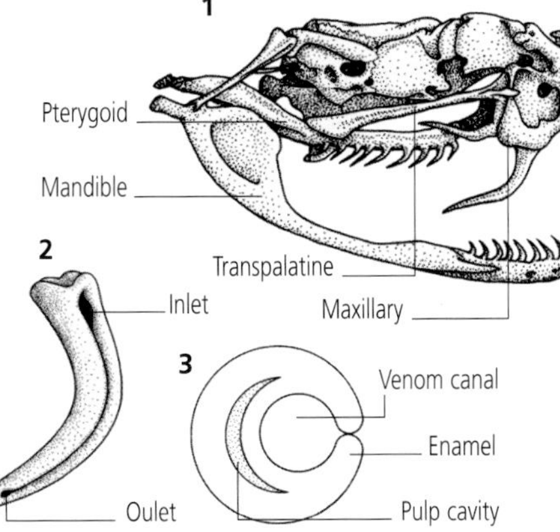

Vipers have specialized teeth in the upper jaw, as in this rattlesnake skull (1); fang showing the venom inlet and its outlet near the tip (2); cross-section of a cobra's fang, showing the canal through which the venom is delivered (3).

Although it is dangerous to generalize, members of the two families tend to have different lifestyles. Members of the Elapidae are either secretive species living and hunting in leaf litter or burrows (the American coral snakes, *Micrurus* species, for example) or fast, lively snakes that are active during the day, have good vision, and chase down their prey. Mambas, *Dendroaspis* species from Africa, the king cobra, *Ophiophagus hannah* from Asia, and the taipans, *Oxyuranus* species from Australia, are all examples of species fitting that profile.

Vipers and pit vipers, on the other hand, tend to be heavy-bodied snakes, often superbly camouflaged, that sit and wait for food to come to them. Most are nocturnal; and although there are groups of tree-dwelling species, few of them burrow, and none is completely aquatic.

In Australasia, where there are no vipers, three (or possibly four) species of elapids have moved into the niche usually occupied by vipers. They fit the generalized description above in every respect and have acquired the common name of death adders, even though they are not related to adders. They give birth to live young, another characteristic of vipers, although some vipers lay eggs. Most elapids, on the other hand, lay eggs—the sea snakes and some related land forms are notable exceptions to this rule.

Venomous snakes use their venom primarily for killing their prey. Its secondary use is as a means of defense. Many species, while capable of giving painful or fatal bites to humans, rarely attempt to do so. Others, like the rattlesnakes and the coral snakes, use warning behavior or coloration before resorting to attack—venom is costly to produce in terms of metabolic effort, and the snake prefers not to waste it by biting animals that cannot be eaten, such as humans. Only one group of species, the spitting cobras of Africa and Asia, seem to have evolved a venom-delivery system that is used exclusively for defense, although they bite in the usual way when hunting prey.

Venom is stored in the venom glands situated behind the snake's eyes. Some species—usually long, slender species with narrow heads—have greatly extended venom glands that reach well down their bodies. The glands are surrounded by muscles known as the masseter muscles. The muscles squeeze on the glands and force the venom through the venom ducts into the base of the venom canal that runs up the fangs. The snake therefore has control over whether or not it injects venom and how much it injects. To avoid waste, venomous snakes sometimes make "dry" bites with no venom. The proportion of dry bites probably varies according to species and the snake's condition. Some scientists estimate it at about 20 percent, while others put the figure at 50 percent.

Venom Composition

Snake venom is derived from the saliva and other digestive juices. In snakes venom serves the dual purpose of starting the digestive process and helping subdue the prey. It may also have a defensive function. The chemical composition of the venom varies from species to species and sometimes even between populations of the same

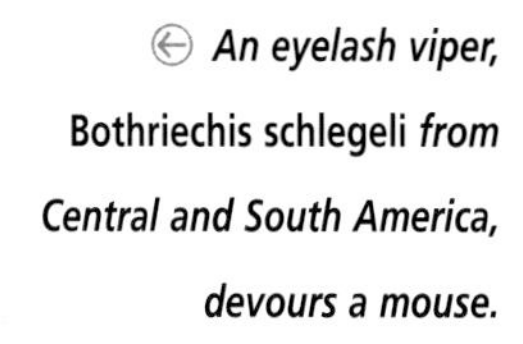

An eyelash viper, Bothriechis schlegeli *from Central and South America, devours a mouse.*

In Guyana the common lancehead, Bothrops atrox, *is "milked" by forcing its fangs through a membrane stretched over a glass vessel and applying gentle pressure to its venom glands.*

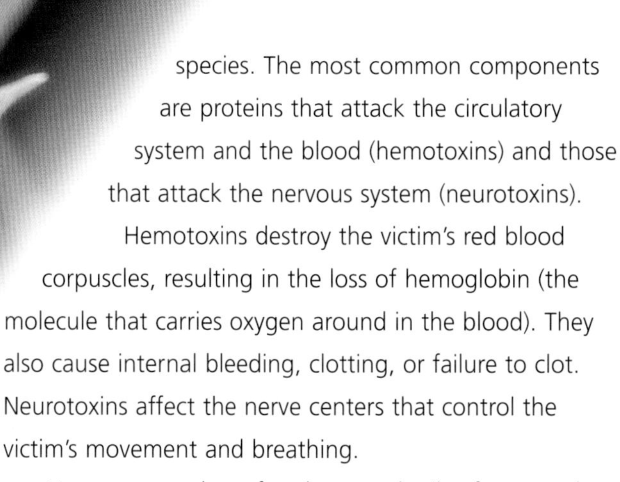

species. The most common components are proteins that attack the circulatory system and the blood (hemotoxins) and those that attack the nervous system (neurotoxins). Hemotoxins destroy the victim's red blood corpuscles, resulting in the loss of hemoglobin (the molecule that carries oxygen around in the blood). They also cause internal bleeding, clotting, or failure to clot. Neurotoxins affect the nerve centers that control the victim's movement and breathing.

Venomous snakes often have cocktails of venom that include hemotoxins and neurotoxins. Death of the victim may result from one type of venom or from a combination of the two. Vipers often have a high proportion of hemotoxins, while elapid venom usually contains more neurotoxins. There are, of course, always exceptions to the rule.

Generally speaking, neurotoxins kill animals more quickly than hemotoxins, but hemotoxins are often produced in greater quantities, so the end result tends to be the same. Snakes "target" their most common prey species by evolving the type of venom that will kill it most effectively. Some species produce one type of venom when they are young but switch to a slightly different mix as they grow and as their diet changes. Viper venom may also contain cytotoxins (cell-destroying toxins) that damage or kill the animal's tissues.

Venom and Medicine

Snake venoms have been used in medicine for hundreds of years. Research is under way to look at the effect of snake venoms on heart attack and stroke victims. Snake venoms contain factors that can prevent or encourage blood clotting. Strokes and heart attacks are caused by clots forming in blood vessels that have already been made narrow by the buildup of fatty tissue. If the anticlotting component of venom can be extracted, it might be used to prevent this from happening. Russell's viper (*Daboia russeli*) venom is widely used to test for certain abnormalities in blood-clotting mechanisms.

Venoms are extracted from snakes by "milking," usually in a laboratory or at a commercial snake farm, and then dried before being used as a medical preparation or for raising antivenom. Because only small quantities of venom can be extracted from each snake, the venom is extremely valuable and can fetch $1,000 to $1,500 for a fraction of an ounce (1 g) or $1 million for just over 2 pounds (1 kg) in its dry crystalline form. Antivenoms can be specific to a particular species of snake (monovalent) or effective against a number of closely related species (polyvalent). For example, an antivenom that acts on the venom of all rattlesnakes is known as polyvalent Crotalinae; one that works against the venom of all the snakes from a region is known as polyvalent Australian, polyvalent South African, and so on.

Vipers and Pit Vipers

Members of the family Viperidae are all venomous and have relatively long, hollow fangs. The fangs are attached to an upper tooth-bearing bone, the maxilla, which is shorter than in most other snakes and mobile. When not in use, the fangs lie folded under the roof of the mouth and are covered with a fleshy sheath. However, when the snake strikes, the maxilla and therefore the fangs are rotated forward due to the pushing action of two long bones in the skull—the pterygoid and transpalatine bones. By the time the fangs reach their target, the snake's mouth is agape, and the fangs point forward so that they stab rather than bite.

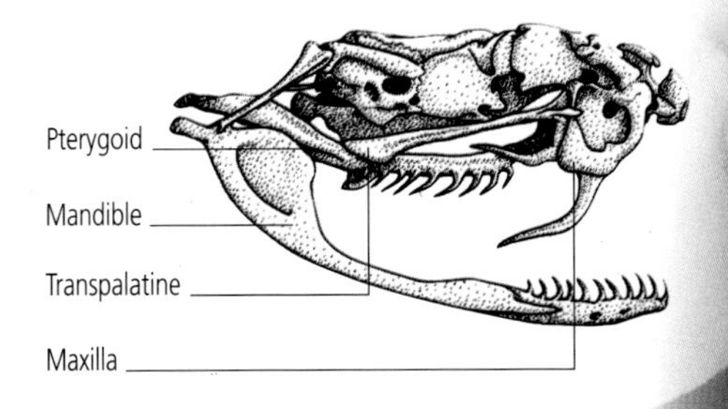

Cross-section of the jaw of a rattlesnake, showing the pterygoid and transpalatine bones that push the maxilla bone (holding the fangs) forward when the snake is ready to strike.

The family Viperidae includes the adders, night adders, vipers, bush vipers, rattlesnakes, and other pit vipers, as well as species with distinctive and sometimes evocative local names, such as the 100-pace snake, *Deinagkistrodon acutus*, the bushmaster, *Lachesis muta*, and the fer-de-lance, *Bothrops lanceolatus*. As a group they are among the most successful snakes, and vipers of one sort or another are found farther north and farther south than any other snake species. For example, the adder, *Vipera berus*, is the northernmost snake, and the Patagonian lancehead, *Bothrops ammodytoides*, is the snake that reaches farthest south.

Common name Vipers and Pit Vipers **Family** Viperidae

Family Viperidae 4 subfamilies

Subfamily Azemiopinae—1 species, Fea's viper, *Azemiops feae*

Subfamily Causinae—1 genus, 6 species of night adders, *Causus* spp.

Subfamily Viperinae—13 genera, 77 species of typical vipers, including the hairy bush viper, *Atheris hispida*; puff adder, *Bitis arietans*; horned adder, *Bitis caudalis*; many-horned adder, *Bitis cornuta*; Gaboon viper, *Bitis gabonica*; Sahara horned viper, *Cerastes cerastes*; Russell's viper, *Daboia russeli*; carpet or saw-scaled vipers, *Echis* spp.; nose-horned viper, *Vipera ammodytes*; adder, *Vipera berus*

Subfamily Crotalinae—21 genera, 166 species of pit vipers, including the copperhead, *Agkistrodon contortrix*; eyelash viper, *Bothriechis schlegeli*; terciopelo, *Bothrops asper*; Malaysian pit viper, *Calloselasma rhodostoma*; western diamondback rattlesnake, *Crotalus atrox*; sidewinder, *Crotalus cerastes*; bushmasters, *Lachesis* spp.; massasauga, *Sistrurus catenatus*; Pope's pit viper, *Trimeresurus popeiorum*

Members of the Viperidae are also found at higher elevations than other snakes—the Himalayan pit viper, *Agkistrodon himalayanus*, occurs up to 16,000 feet (4,900 m), and several others come close. Vipers have not reached Madagascar or Australasia, however. All vipers in the Americas are pit vipers, but the Old World is home to all four subfamilies.

The most unusual member of the family is Fea's viper, *Azemiops feae*. This rare species is from the remote foothills of the Himalayas, where it lives in cool cloud forests. It has smooth scales on its body and large scales covering the top of its head. Few people have seen a Fea's viper, and its relationship with other vipers is unclear. It is placed in a subfamily of its own, the Azemiopinae.

The six species of night adders in the subfamily Causinae—all from Africa—also have smooth scales and large scales on the top of their heads. The night adders tend to specialize in eating toads and are not considered dangerous to humans.

↑ *A sedge viper,* Atheris nitchei *from Central Africa, gets ready to strike with its mouth in a characteristic open pose.*

Typical Vipers

The "typical" vipers, or Old World adders, in the subfamily Viperinae are short and stocky. They have a wide, triangular, spade-shaped or pear-shaped head. The scales are heavily keeled, giving the snake a rough texture, and the head is often covered with many small scales. However, several species in the genus *Vipera* have at least a few large head scales as well.

They live in a variety of habitats, including rocky hillsides and mountains, meadows, scrub, and deserts. Species from sandy deserts, notably Peringuey's viper, *Bitis peringueyi* from the Namib Desert of southwestern Africa, and *Cerastes cerastes*, the Sahara horned viper from North Africa, move by sidewinding—making them in some way counterparts to the American sidewinder, *Crotalus cerastes*. The bush vipers from Central Africa are arboreal (tree dwellers), and other species sometimes climb into low vegetation in search of prey. Many adders are camouflaged, and the coloration of a single species may vary from place to place according to the soil type on which it lives. African *Bitis* species have especially intricate markings, and the Gaboon viper, *Bitis gabonica,* is often singled out as a classic example of disruptive coloration (meaning that the snake's pattern breaks up the outline of its body). This massive snake and its close relative the puff adder have huge fangs. They are among the most impressive and most feared snakes in Africa.

A number of adders have "horns," or protuberances, on various parts of the head. They sometimes consist of a single thornlike scale over each eye, as in the Sahara horned viper, or a cluster of scales on the snout, as in the rhinoceros viper, *Bitis nasicornis* from West Africa. A third variation is a cluster of scales over the eyes, as in the tree-dwelling Usambara bush viper, *Atheris ceratophorus* from Tanzania. The function of the horns is uncertain, but they may protect the snake's eyes from sand blown by the wind or from vegetation. However, if that were the case, snakes from other families should be similarly endowed, but they are not. The nose-horned viper, *Vipera ammodytes* from Europe, and some closely related species all have a single fleshy horn on the snout made up of several scales. Since vipers are all sit-and-wait predators, it is assumed that these different devices serve simply to break up the outline of their head for defense purposes.

The bush vipers have scales that are heavily keeled (with a ridge running down their center), especially the so-called hairy or spiny bush viper, *Atheris hispida* from East Africa, in which the scales taper to a raised point. Keeled scales are central to the defense strategy of carpet and saw-scaled vipers, *Echis* species. These snakes make a loud rasping sound by drawing two opposing sections of their body against each other, engaging small teeth on the keels of the scales. Carpet vipers are small, very common, and very ready to defend themselves, making them perhaps the world's most

Living with Danger

On a rural road in the Yucatán Peninsula in Mexico shortly after dusk small groups of workers can often be seen walking home from their work in the fields. Most wear open sandals, while some wear no shoes at all. Forest growth comes right up to the road, and most people walk within a foot or two of the edge. One particular night two terciopelos, *Bothrops asper*, were found on the road, as well as a coral snake. Another huge terciopelo crossed the road and disappeared into the vegetation. It was a sobering demonstration of the realities of living in a region where seriously dangerous snakes are commonplace, and the wonder is that snakebite accidents are not more numerous than they are.

dangerous snakes. Old World adders and vipers eat a wide range of prey, typically small mammals, nestling birds, and lizards. Orsini's viper, *Vipera ursinii* from Europe, eats invertebrates such as centipedes and grasshoppers. So do the carpet vipers, *Echis* species (which is surprising for snakes that have evolved some of the most powerful venoms in the world). Most members of this subfamily bear live young, an adaptation that has undoubtedly helped them make a living in cold environments. Some species from warmer regions, however, lay eggs.

Pit Vipers

The pit vipers are placed in the subfamily Crotalinae and are easily distinguished from other vipers by the presence of heat-sensitive facial pits. Although they fulfill the same function as the heat pits in the lips of boas and pythons, they evolved independently and are more sophisticated.

Heat pits consist of two compartments divided by a diaphragm. The inner chamber is connected by a narrow duct to a small pore in front of the snake's eye and probably serves to balance air pressure on either side of the diaphragm. The outer chamber is directed forward and detects infrared radiation emitted by warm objects. It detects mostly warm-blooded prey but also lizards that have been basking and whose body temperature is greater than their surroundings. The pits are extremely sensitive: Some pit vipers can detect temperature differences of as little as 0.001°C, allowing them to locate prey and strike accurately even in total darkness. In much of Latin America large pit vipers are known collectively as *cuatro narices*, which is Spanish for "four nostrils," a reference to the facial pits.

There are nearly 160 species of pit vipers. They occur in Asia and in North and South America. Like the Old World vipers, their heads are covered with small scales. There are a few exceptions, such as the copperhead, *Agkistrodon contortrix* from the southeastern United States and northeastern Mexico, which has large plates on the top of the head.

Pit vipers have colonized a number of different habitats from swamps to deserts, although there are no totally aquatic species nor any burrowing ones, perhaps because the pits would be more of a hindrance than a help in such environments. There are plenty of arboreal (tree-dwelling) snakes, however, such as the genera

The common lancehead,* Bothrops atrox *from South America, has a triangular (lance-shaped) head. It has a reputation for being aggressive, and its bite is highly venomous.

Coping with the Cold

The ability of vipers to live in colder places than most other snakes is due to a combination of factors. The family's predominant mode of reproduction is viviparity, meaning that the females carry their eggs inside their body until they are ready to hatch and give birth to live young. Some species also provide nourishment via their own bloodstream. This gives them the option of basking and therefore regulating the temperature of the developing young. Basking is also an important part of the daily behavioral pattern of many vipers, and they are often well camouflaged so that they can afford to place themselves in the open to make the most of the heat of the sun when it is available. Most vipers are dark in color, small, and stocky, allowing them to absorb and retain heat efficiently. A number of populations of different species produce individuals that are melanistic (they have high levels of black pigment in their skin). When they occur at high latitudes or altitudes (which they often do), their dark coloration may help them warm up quickly and remain active for longer. This offsets any disadvantage they incur through being less well camouflaged than others. Some populations consist entirely of melanistic individuals.

A melanistic (all-black) adder,* Vipera berus, *photographed in Britain. Melanistic individuals have better heat-retaining properties than other snakes.

⇧ ***A white-lipped pit viper,* Trimeresurus albolabris, *swallows a frog. Pit vipers usually prey on warm-blooded creatures but also eat reptiles, amphibians, and insects.***

Trimeresurus in Asia and *Bothriechis* in Central and South America. The world's longest viper is the bushmaster, *Lachesis muta* from Central and South American rain forests and plantations. It can grow to 6.3 feet (2 m) or more in length, but it is a relatively slender species.

Rattlesnakes

The rattlesnakes, of which there are just over 30 species, are among the most easily identified snakes because of the distinctive tail end. However, two or three rare species from small islands in the Gulf of California have "lost" their rattle through the evolutionary process. The rattle is formed from shed outer layers of the very last (terminal) scale, which is shaped so that the segments do not fall off when the snake sheds it skin. The snakes use their rattle as a warning instrument, vibrating their tail and causing the segments to click together rapidly to create a buzzing or fast ticking sound. All rattlesnakes are from the Americas; and most live in either arid habitats, rocky mountainsides, or open grassland, including montane meadows. One species, *Crotalus durissus*, lives in South America and prefers open clearings to forest floors. Another species, the massasauga, *Sistrurus catenatus*, lives in swamps in the northern parts of its range in the northeastern United States and Canada. The sidewinder, *C. cerastes*, lives in sand deserts of the American Southwest and owes its common name to its method of locomotion. There are no arboreal rattlesnakes, but some species, including the rattleless kinds, are thought to climb into bushes to stalk roosting birds.

Pit vipers feed mainly on warm-blooded prey such as mammals and birds, using their heat pits to hunt effectively at night. Large rattlesnakes can handle prey up to the size of hares and ground squirrels. The cottonmouths, *Agkistrodon piscivorus*, have perhaps the least specialized diet of all snakes. They eat insects, fish, frogs, turtles, baby alligators, and birds' eggs. Some populations lurk near seabird colonies to pick up fish that are dropped and will even peel dried fish carcasses from rocks where they have fallen.

Larger Males

As is often the case in snakes, the growth rate of female vipers slows down after they reach maturity. That is because they begin to divert a high proportion of their food into producing eggs. Males, however, do not bear such high reproduction costs and are able to feed and grow larger, but this only occurs where large males have an advantage over small ones. If there is no benefit to being larger, they may simply spend less time searching for food and remain smaller. When combat is involved, larger males nearly always have an advantage over smaller ones, so males do grow larger. It is important to note, however, that the males are not necessarily larger than females, as is often stated—only larger than they would need to be if they did not fight. Male rattlesnakes are invariably the larger of the two sexes—the sidewinder, *Crotalus cerastes*, being an exception—but in Eurasian vipers, *Vipera,* females are nearly always the larger sex.

A snake cult devotee holds a Bible in one hand while grasping four live snakes during a convention in North Carolina in 1948. Police later raided the convention and confiscated the snakes.

Snake Cults

In the New Testament the Gospel according to Mark says, "They shall take up serpents; and if they drink any deadly thing, it shall not hurt them; they shall lay hands on the sick and they shall recover."

Naturally, someone just had to put the Gospel's words to the test. The first person to do so was a man named George Went Hensley of Grasshopper Valley, Tennessee. Hensley caught a large rattlesnake and found he could handle it without being bitten. Flushed with his success, he demonstrated his "gift" to the congregation at the local church, which decided to get in on the act.

Pretty soon church services throughout the Appalachians took on the appearance of an early version of a television game show, with participants swaying, stamping their feet, and generally working themselves up into a frenzy. Then they plunged their hands into a box full of live copperheads and rattlesnakes, pulled them out, and draped them over their bodies. It says a great deal about the tolerance of the pit vipers that the cult was able to flourish for a while.

The cult spread south, amid growing publicity, but came to an abrupt end when George Hensley's luck ran out. He was killed at the age of 70 by a snake bite during a church service in Florida in 1955. Before his death he claimed to have survived 400 other bites. There are no copperheads or timber rattlesnakes in Florida, and the cult was using local species, including the deadly eastern diamondback rattlesnake, *Crotalus adamanteus*. This species has a less understanding nature and a more powerful venom.

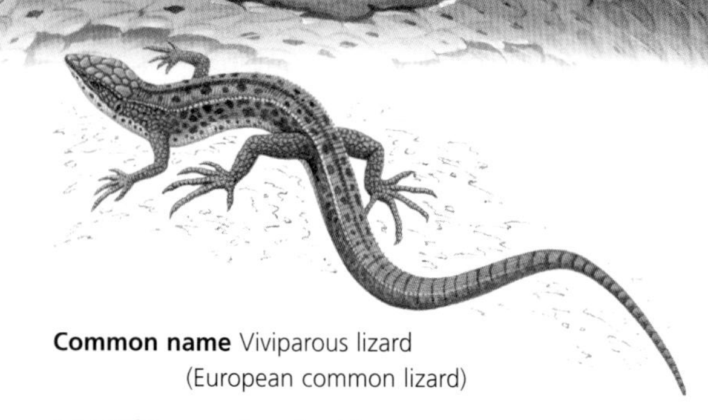

Common name Viviparous lizard (European common lizard)

Scientific name *Lacerta vivipara*

Family Lacertidae

Suborder Sauria

Order Squamata

Size 6 in (15 cm) long

Key features Small but robustly built with a cylindrical body, short legs, and a long tail; head short and deep with a rounded snout; neck thick; color and markings highly variable; most are some shade of brown but can also be olive or gray; females often have a plain back with a single dark stripe down the center and dark markings on the flanks; males are darker overall and have small markings consisting of light spots with black edges (ocelli); underside can be white, yellow, or orange

Habits Diurnal; mainly terrestrial

Breeding Most give birth to live young with litters of 3–11 born after a gestation of 8–13 weeks; some populations lay eggs

Diet Small invertebrates; also ants' eggs and larvae

Habitat Very adaptable but absent from forests, cultivated fields, and grazed meadows

Distribution Most of Europe except the Mediterranean region but including northern Spain east to north Asia as far as the Pacific coast and Sakhalin Island, Russia, and Hokkaido Island, Japan

Status Common in suitable habitat

Similar species All small lacertids are difficult to identify in the field, but the common lizard is usually darker than most; the others tend to have slightly flattened body shapes

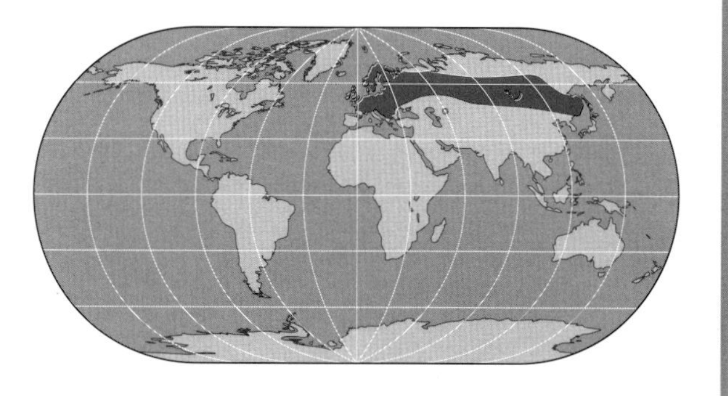

Viviparous Lizard *Lacerta vivipara*

Despite its insignificant appearance, the viviparous lizard is robust and a survivor. It lives farther north than any other reptile and has one of the largest continuous ranges of any lizard.

THE VIVIPAROUS LIZARD OCCURS in Lapland and along the shore of the Barents Sea in Arctic Russia. Its distribution reaches 70°N in Norway—over 200 miles (322 km) inside the Arctic Circle and farther north than any other reptile. Its range is huge—from the Atlantic coast of Ireland and France across Europe and the vast steppes of Central Asia to the Pacific Ocean in China and Russia and even to the islands of Sakhalin (Russia) and Hokkaido (Japan). It has probably the largest continuous range of any lizard and possibly of any nonmigratory vertebrate.

Varied Habitat

Within this range it is very common in parts but scarce or absent in others. It has certain habitat requirements and struggles to live where they are not available. It needs a fairly humid environment and often lives among grass and other lush vegetation that retains moisture. In the north of its range it lives in lowlands and occurs on damp heaths, bogs, dunes, sea cliffs, and road and railway embankments. It needs open spaces to bask in and is most often seen at the edges of paths, clearings, or among piles of logs or rocks. Small colonies quickly disappear if their habitat becomes overgrown to the point where they cannot bask. The lizard also occurs on mountains up to 8,200 feet (2,500 m), where it is restricted to damp meadows, moors, marshes, and ditches.

Despite a long list of potential habitats, it is absent from intensively cultivated land, dense forests, and short turf except at the edges of these habitats, where there is more of a patchwork of vegetation types and open spaces. In the north and in mountains it is most likely to be seen on south-facing banks and

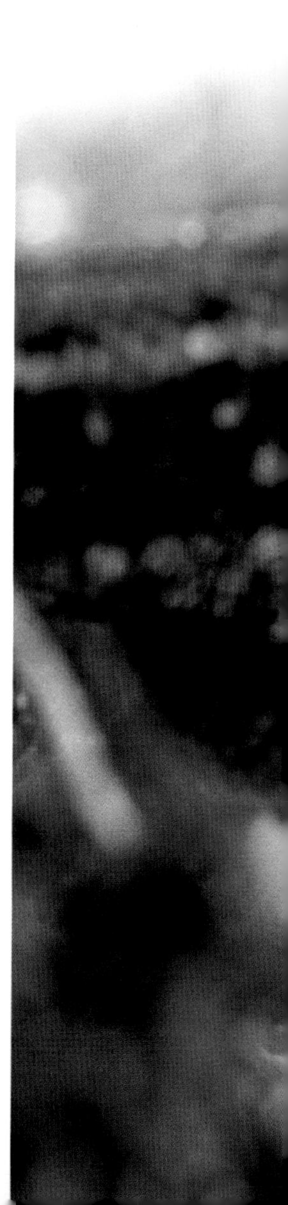

embankments. Railway cuttings have provided it with a corridor into towns and cites, and it can also be found in cemeteries, where it is sometimes seen basking on stonework.

In particularly favorable places to the north of its range with ample basking sites and plenty of food, it can occur in very high densities, sometimes as many as 40 to 400 lizards per acre (100–1,000 per ha). At these densities lizards appear to be everywhere, basking in every available clear space and clambering through vegetation in search of food.

Viviparous lizards are good swimmers and sometimes enter water voluntarily to move from one tussock of grass to another (in flooded marshes, for example) or to catch insects floating on the surface. They will also dive to the bottom to escape from predators and can stay under water for several minutes.

Viviparous lizards hunt small invertebrates such as insects, spiders, snails, and earthworms. They stun their prey by shaking it before swallowing it whole.

Prey and Predators

Viviparous lizards feed on small invertebrates, especially spiders, caterpillars, flies, and other species. They will also eat ants' eggs and larvae (but not adult ants, apparently). Large prey items are grasped in the mouth and shaken or bashed to death before being swallowed. Caterpillars are sometimes chewed from end to end to extract the insides while the skin is left, but distasteful caterpillars, as well as wasps and bees, are rejected.

The lizards' main natural predators are birds, including kestrels, buzzards, and members of the crow family; small mammals such as weasels and hedgehogs; and reptiles such as larger lizards and adders, *Vipera* species, smooth snakes, *Coronella austriaca*, and whip snakes, *Coluber* species, where they occur. In places viviparous lizards are the staple diet of the adder, *V. berus*. In towns and villages viviparous lizards are hunted by domestic chickens, cats, garden birds, and children. Their main line of defense is to run away and hide in dense vegetation or, if caught, to shed their tail. Urban and suburban populations of viviparous lizards often have a higher incidence of broken and regrown tails than those living in more natural habitats.

Young viviparous lizards, which are very small, are also eaten by spiders, frogs,

toads, slow worms, and a wide range of other opportunistic predators. The number of viviparous lizards that survive to the age of one year may be as low as 10 percent. However, once they reach adult size after two or three years, they have fewer predators and can live for as long as 12 years.

Temperature Regulation

Viviparous lizards are classic "shuttling thermoregulators." In other words, they bask in the sun until they have raised their body temperature to their preferred level, then they go off and do other things. They may hunt for food or search for a mate and return to bask if they need to top up their body temperature. They aim to raise their temperature to about 86°F (30°C), which is optimum for most populations, although those living in mountainous districts regularly operate at temperatures as much as 9°F (5°C) below this.

Obviously, the periods of time between basking sessions will vary according to the ambient temperature, but on days during spring and fall they usually return to their basking spot within two to five minutes. A patient observer can easily watch this activity pattern unfold. During midsummer they may bask off and on for an hour or two in the morning and again in the evening, but they can maintain their preferred body temperature in the middle of the day without exposing themselves.

In midsummer the first lizards emerge from their nighttime retreats at around 6 A.M. and can still be active at 7 P.M. Occasionally a disturbed individual will disappear from one basking spot only to reappear a few feet away. However, these lizards usually return to the same place, and a favored rock or patch of bare sand will be occupied by the same individual throughout the active season.

Ⓣ Although it is adaptable in terms of habitat, the viviparous lizard prefers open, damp areas such as marshes, moors, sand dunes, hedgerows, bogs, and ditches.

Lizard Colonies

Viviparous lizards live in loose colonies often centered on a good basking area, such as a south-facing bank with hiding places and somewhere to spend the winter. Their active season begins as early as February in warmer parts of their range, and they have been seen running around while there is still snow on the ground if the weather is warm. Equally, they may be delayed until April or even May in colder regions. Males and juveniles emerge from hibernation before females.

Mating takes place soon after the lizards emerge from hibernation, usually from late March to early June in England, for instance. Unusually among lizards, males are not strongly

Ⓡ In the European sand lizard, Lacerta agilis, *the male's green coloration is most vivid in the breeding season. The female mates with several males, but lays a single clutch of eggs.*

territorial. Most males remain in the same small area, however, and mate with all the females living there. Other males move around the edges of the colony; and although fights between them and the territory holder may break out, they are usually successful in obtaining some matings. Females often mate with more than one male.

Mating is not an elaborate or tender affair. The male simply grasps the female in his jaws and attempts to bring their cloacae together by twisting his body under hers. If she is not receptive, she bites back, and the male usually releases her right away.

Pregnant females bask more often and for longer periods than males or nonpregnant females. On cool days they are reluctant to take cover even when closely approached. They flatten their body to increase the area over which they can absorb the sun's heat and are often quite conspicuous.

The young are born in mid- or late summer. The female finds a secluded place to give birth, and they are usually all born during the course of a day, often within an hour. The babies are enclosed in a thin, transparent membrane at first, but struggle free within minutes and go

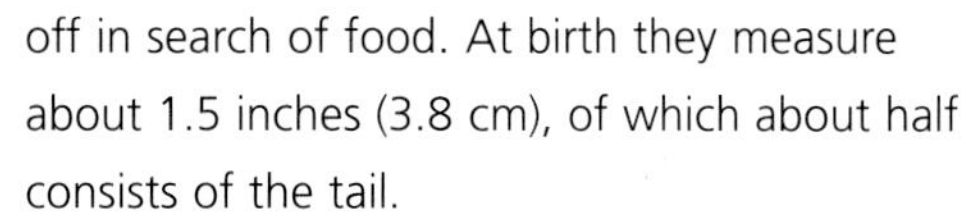

off in search of food. At birth they measure about 1.5 inches (3.8 cm), of which about half consists of the tail.

The young lizards are very dark in color—almost black—with no markings, and they can be mistaken for insects. By the end of the summer when they enter hibernation, they will have grown to about 3 inches (7.6 cm) long. Their tail will be proportionately longer by then, and their coloration will have become lighter. Signs of the adult's markings will have begun to appear. By the end of their second year they

More Than One Father

In lizards and other animals males are often territorial as a means of guarding their mate—by preventing other males from entering the territory where "their" females live, the territorial males have exclusive access to them. Things are not always (or ever) as simple as that, however. Males that move from one territory to another can often "steal" matings before they are driven off by the territory holder. They are sometimes called "sneaky" males, and in some circumstances there will always be a certain proportion of them. Females accept sneaky males because, by producing litters that have more than one father, they stand a better chance of getting at least some vigorous young—they are not putting all their eggs in one basket, so to speak.

Viviparous lizards are not especially territorial, even though most males tend to stay in the same place and will half-heartedly chase off strange "floating" males that come too near. This should mean that females are often mated by more than one male and therefore that litters of young have mixed paternity. Experiments with wild colonies in France have borne this out. By DNA-testing all the young in each litter (by taking a small sample of tissue from the tip of the tail), researchers found that 30 litters out of a total of 44 tested had more than one father. Of these 28 had at least two fathers, and the other two had at least three fathers. The researchers compared the results with others from different localities in France and with semicaptive colonies. They found that, broadly speaking, their results were typical.

Does this mean that male viviparous lizards are making a mistake by not defending their territories more vigorously? Actually, no. In the European sand lizard, *Lacerta agilis*, which is related to the viviparous lizard but is much more territorial, they found that four out of five clutches (80 percent) contained DNA from more than one male. Other experiments with lizards from other families in different parts of the world showed that the incidence of multiple paternity varied from 25 to 82 percent. Compared with these results, viviparous lizards appear to be no worse off than the more territorial species.

will have reached about 4 inches (10 cm) in total. Males will be sexually mature the following spring when they emerge from hibernation. Females take an extra year to reach breeding age.

Egg Laying or Live-Bearing?

The viviparous lizard lives up to its name over nearly all of its range, but in a few places it lays eggs. (Viviparous means giving birth to live young.) Females from populations in the Pyrenees, northern Spain, and parts of Slovenia lay one clutch of between one and 13 eggs that hatch in four to five weeks. The eggs are sometimes laid in communal sites—as many as 60 have been found together in various stages of development. The most likely explanation is that these localities are warmer than the places where the species occurs farther north.

Live-bearing is an adaptation to a cold environment, because by basking, the female can help the embryos develop more quickly. This helps explain why the species is so successful in cooler environments. Live-bearing places an extra burden on females, however, because they have to carry their offspring for longer and may make themselves more vulnerable to predation in the process.

Although it is very unusual for a single species to lay eggs in some parts of its range and give birth to live young in others, it is not such a complicated business as it might seem. The viviparous lizard is, in fact, ovoviviparous. This means that instead of laying its eggs, it simply keeps them in its oviduct until they are ready to hatch and then "gives birth" to fully formed young. The eggshell is reduced to a thin, transparent membrane. If the female

Males in the genus **Lacerta** ***often develop bright coloration during the breeding season, as in this male Schreiber's lizard,*** **L. schreiberi** ***from Spain.***

A female viviparous lizard gives birth. The young are enclosed in a transparent membrane and are born after a gestation period of about two to three months.

deposits them early, as she does in some southern locations, she becomes an egg layer.

Other lizards are genuinely viviparous—the developing embryo derives nourishment from its mother through a placenta. American skinks belonging to the genus *Mabuya* have the most highly developed placentas, with female *M. heathi* from Brazil providing 99 percent of their litters' body mass in this way. Scientists think that viviparity (including ovoviviparity) has arisen independently many times—up to 45 times according to some—in lizards.

The Relatives

The genus *Lacerta* contains 38 species in all, including a number of newly described ones from Turkey and the Middle East. Another three new species, *L. aranica, L. aurelioi*, and *L. bonnali*, are from very small areas in the Pyrenees on the borders of France and Spain; they were formerly considered to be forms of the Iberian rock lizard, *L. monticola*, a colorful montane species from several mountain ranges in the Iberian Peninsula. Other small species are found in the Balkan region.

Another section of the genus includes a series of larger species, measuring from 8 to 16 inches (20–40 cm) and heavily built. They are commonly referred to as "green lizards," even though they also include the European sand lizard, *L. agilis*, which is often light brown or reddish brown in color (the males, however, develop green flanks in the breeding season). The green lizards are found in central and southern Europe, the Middle East, and parts of North Africa. The remaining species, the eyed lizard, *L. lepida*, is by far the largest of the genus, being many times the size of the viviparous lizard.

Western Diamondback Rattlesnake

Crotalus atrox

Common name Western diamondback rattlesnake

Scientific name *Crotalus atrox*

Subfamily Crotalinae

Family Viperidae

Suborder Serpentes

Order Squamata

Length From 30 in (76 cm) to 7 ft (2.1 m)

Key features Body has large diamond-shaped markings along the back, each outlined with lighter scales; background color gray to brown, may be reddish brown or pink; head large and rounded with a wide dark stripe from the eye to the angle of the jaws; tail conspicuously banded in black and white, ending in a large rattle (series of horny segments made of keratin)

Habits Nocturnal in summer but active in late afternoon and early morning in the spring and fall; terrestrial

Breeding Live-bearer with litters of 4–25

Diet Small mammals up to the size of young prairie dogs and rabbits

Habitat Desert, semidesert, arid scrub, and dry grassland

Distribution North America almost from coast to coast; ranges from southeastern California and the Gulf of California to the Gulf Coast of Texas and south into Mexico

Status Very common in places

Similar species The Mojave rattlesnake, *C. scutulatus*, is very similar, and its range overlaps in places

Venom A potent hemotoxin leading to severe symptoms and possibly death unless treated

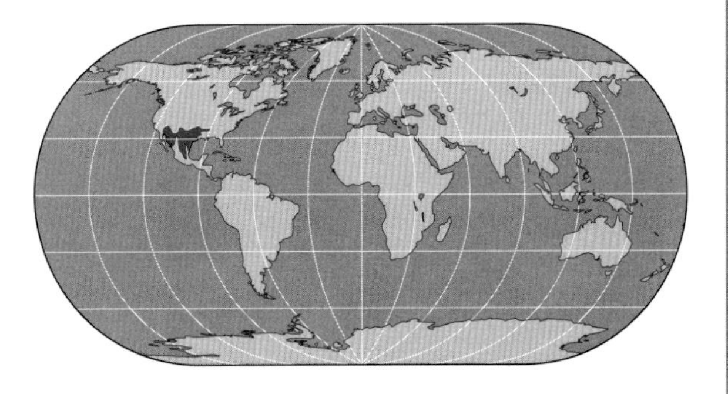

The western diamondback rattlesnake is at the top of the food chain. It is also an important element in the ecosystem of the desert and semidesert regions of the American Southwest.

THE WESTERN DIAMONDBACK rattlesnake is a large snake that preys on mammals up to the size of weaned jackrabbits, young prairie dogs, and adult ground squirrels.

While most snakes are fairly inconspicuous and have little effect on the ecosystem, this is not true of the western diamondback. It is common to see one crossing a quiet desert road at night in summer. In fact, it is not unusual to see six or more during a 100-mile (160-km) drive across southern Arizona or western Texas. There will be at least the same number of roadkill diamondbacks. Many are deliberately run over by misguided individuals who think they are acting in the best interests of humans. If numbers of this and other large rattlesnakes drop significantly, rodent and rabbit populations may increase with fewer predators to control them. This will have an effect on vegetation, for example, and may have repercussions for the whole ecosystem.

A western diamondback, Crotalus atrox, *strikes a defensive posture with its head raised, its body in an "s"-shaped coil, and its rattle in the upright position.*

Sensitive to Change

Population densities are thought to be about 75 snakes per square mile (2.5 sq. km) over much of their huge range. The snakes prefer dry habitats but are not restricted to deserts. They also live among scrub, in grassland, and on rocky hillsides. Although they are in many respects generalists, subtle alterations in their environment can cause dramatic population changes. Grazing by livestock, for example, often leads to a reduction in the number of plant species in an area and creates open patches of soil. That in turn may adversely affect the numbers of small mammals (the main food of diamondbacks). The same changes may favor lizards, and that allows Mojave rattlesnakes, *C. scutulatus*, which eat lizards, to move into the area. These changes can be quite abrupt. Within a few miles diamondbacks can go from being numerous to nonexistent and replaced by Mojave rattlesnakes.

Western diamondbacks also occur on rocky and gravelly hillsides, but only if there is sufficient vegetation to support a healthy population of rodents, rabbits, or prairie dogs. Where the soil is thin and the vegetation sparse, other species (notably the tiger rattlesnake, *Crotalus tigris*, and the speckled rattlesnake, *C. mitchelli*) replace western diamondbacks. They are more adept than the western diamondback at hunting among rocks and probably include a high proportion of lizards in their diet.

In the higher montane habitats that dot the landscape in this part of the world there are yet more species, such as the rock rattlesnake,

Hidden among bark in the Huachuca Mountains in Arizona, a banded rock rattlesnake,* Crotalus lepidus, *basks after eating a large lizard.

Rattlesnake Roundups

Rattlesnake roundups date back to the 17th century. They began as organized hunts to clear good grazing land of venomous snakes that might endanger settlers, their children, and their livestock. Because rattlesnakes tend to be concentrated around their dens early in the year, killing large numbers in a single day was relatively easy. At a single hunt in Iowa in 1849, for instance, two men killed 90 rattlesnakes in one and a half hours. By the 19th century the roundups had begun to develop a carnival atmosphere, with individuals competing with each other to see who could catch and kill the most snakes. People took picnics, and charities and civic authorities often sponsored the events to raise money. Snake handlers demonstrated their expertise; and the snakes were later killed, and their skins and flesh were sold.

In recent years, when the risk of being bitten by a wild rattlesnake is almost nonexistent, roundups are still held but purely for entertainment and profit. The event in Sweetwater, Texas, for example, was responsible for the slaughter of 70,773 rattlesnakes—mostly western diamondbacks—over a 16-year period. At the Morris snake hunt, Pennsylvania, 751 timber rattlesnakes (an increasingly rare species) were killed in nine years, while the Keystone reptile hunt in the same state accounted for 3,205 deaths in 17 years.

In the beginning, roundups were timed to catch the snakes in the open, after they had emerged from their dens. Nowadays the snakes are targeted while still underground, using dynamite and gasoline fumes to drive them out. Many snakes die in the dens, along with other wildlife, including harmless snakes, tortoises, and small mammals.

A Jaycee (member of the Junior Chamber of Commerce) displays a diamondback rattlesnake at the Sweetwater Rattlesnake Roundup, Texas, in 1995.

C. lepidus, and a host of small, specialized rattlers. The western diamondback is, therefore, a generalist species found over a wide geographical range and that adapts to a variety of different conditions. However, it is ousted in certain smaller patches of the environment where more specialized rattlesnakes prosper.

Reproduction

Western diamondbacks mate in the spring, and males find females by following pheromone trails. Several males may follow the same female, leading to contests between them, as in many vipers. Because combat is a means of natural selection, there is pressure on males to become bigger. Male diamondbacks are 10 to 15 percent larger than the females.

Females give birth at the end of the summer, four to six months after mating. The young are born without a rattle and remain with their mother for up to16 days until they shed their skin for the first time. Some herpetologists think that the interactions of the young shortly after birth, when they constantly flicker their tongues over each other and over their mother, may help them recognize each other in later life—a skill that could be used when following adults to hibernation dens in the fall and in identifying potential mates.

The Rattle

Early Spanish and Portuguese explorers were the first to remark on the snakes' rattle. One explorer likened it to the sound of little bells (he can never have actually heard a rattler sounding off in anger). Thomas Morton, writing in 1637, says: "There is one creeping beast...that hath a rattle in his tayle, that doth discover his age; for so many years hath he lived, so many joynts are in that rattle, which soundeth (when it is in motion) like pease in a bladder, and this beast is called a rattlesnake." A more modern description of the sound made by one of the larger rattlesnakes is like a pair of castanets knocked together very rapidly, and Morton was slightly wide of the mark about telling the age of the snake from the number of segments

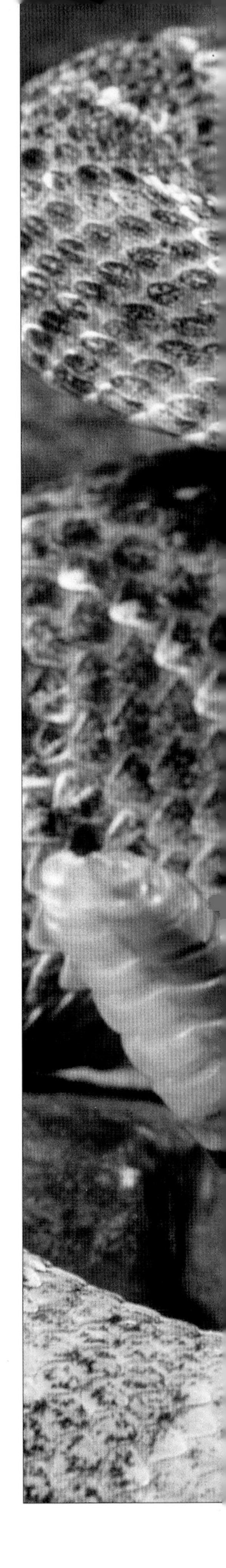